RECENT ADVANCES IN MOBILE ROBOTICS

Edited by **Andon Venelinov Topalov**

Recent Advances in Mobile Robotics

Edited by Andon Venelinov Topalov

Published by InTech

Janeza Trdine 9, 51000 Rijeka, Croatia

Publishing Process Manager Bojana Zelenika

Technical Editor Teodora Smiljanic

Cover Designer InTech Design Team

Additional hard copies can be obtained from orders@intechopen.com

Recent Advances in Mobile Robotics, Edited by Andon Venelinov Topalov
p. cm.
ISBN 978-953-307-909-7

For EU product safety concerns:
IN TECH d.o.o., Prolaz Marije Krucifikse Kozulić 3, 51000 Rijeka, Croatia,
info@intechopen.com or visit our website at intechopen.com.

Meet the editor

Dr. Andon V. Topalov received the MSc degree in Control Engineering from the Faculty of Information Systems, Technologies, and Automation at Moscow State University of Civil Engineering (MGGU) in 1979. He then received his PhD degree in Control Engineering from the Department of Automation and Remote Control at Moscow State Mining University (MGSU), Moscow, in 1984. From 1985 to 1986, he was a Research Fellow in the Research Institute for Electronic Equipment, ZZU AD, Plovdiv, Bulgaria. In 1986, he joined the Department of Control Systems, Technical University of Sofia at the Plovdiv campus, where he is presently a Full Professor. He has held long-term visiting Professor/Scholar positions at various institutions in South Korea, Turkey, Mexico, Greece, Belgium, UK, and Germany. And he has coauthored one book and authored or coauthored more than 80 research papers in conference proceedings and journals. His current research interests are in the fields of intelligent control and robotics.

Contents

Preface

The term *robotics* was coined by the science fiction writer Isaac Asimov in 1942, in his short story *Runaround* (*Astounding Science Fiction*) to denote the technological area addressing the design, construction, operation, and application of robots. The word *robot* itself was first introduced to the public by the Czech play writer Karel Čapek in his play *R.U.R.* (*Rossum's Universal Robots*), published in 1920, and comes from the Czech word *robota,* which means "work". Karel Čapek named his brother - the painter and writer Josef Čapek - as its actual originator.

There is no single correct definition of *robot* which seems to satisfy everyone. In fact, many people have their own definition. According to the *Collins Cobuild Dictionary* "A robot is a machine which is programmed to move and perform certain tasks automatically". *Oxford English Dictionary* defines *robot* as "A machine capable of carrying out a complex series of actions automatically, especially one programmable by a computer" or (especially in science fiction) "a machine resembling a human being and able to replicate certain human movements and functions automatically".

Various techniques have emerged to develop the science of robotics and robots. Mobile robots are the focus of a great deal of current research. Mobile robotics is a young multidisciplinary field involving knowledge from many areas - electrical, electronics and mechanical engineering, computer, cognitive, and social sciences. Being engaged in the design of automated systems, it lies at the intersection of artificial intelligence, computational vision, and robotics. Mobile robots have the capability to move around in their environment and are not fixed to one physical location. Contemporary mobile robots are living up to their name more and more. They are no longer regarded only as mechanisms attached to controls, but represent a combination of manipulative, perceptive, communicative, and cognitive abilities. Nowadays, mobile robots are characterized by an increased autonomy and must negotiate the reality of moving, sensing, and reasoning out their environment. Thanks to the numerous researchers sharing their goals, visions, and results with the community, mobile robotics is becoming a very rich and stimulating field.

The book *Recent Advances in Mobile Robotics* addresses the topic by integrating contributions from many researchers around the globe. It emphasizes the computational methods of programming mobile robots, rather than the methods of constructing the hardware. Its content reflects different complementary aspects of theory and practice, which have recently taken place. The book consists of 21 chapters which are organized into four major parts:

- Visual Perception, Mapping, Robot Localization, and Obstacle Avoidance
- Path Planning and Motion Planning
- Mobile Robots Navigation
- Methods for Control

The book editor is grateful to the authors for their excellent work and interesting contributions. Thanks are also due to the prestigious publisher for their editorial assistance and excellent technical arrangement of the book. We believe that the content of this book will serve as a valuable handbook to those who work in the research and development of mobile robots.

October, 2011

Andon V. Topalov
Control Systems Department
Technical University – Sofia, campus in Plovdiv, Plovdiv,
Bulgaria

Part 1

Visual Perception, Mapping, Robot Localization and Obstacle Avoidance

1

3D Visual Information for Dynamic Objects Detection and Tracking During Mobile Robot Navigation

D.-L. Almanza-Ojeda* and M.-A. Ibarra-Manzano
Digital Signal Processing Laboratory, Electronics Department; DICIS, University of Guanajuato, Salamanca, Guanajuato
Mexico

1. Introduction

An autonomous mobile robot that navigates in outdoor environments requires functional and decisional routines enabling it to supervise the estimation and the performance of all its movements for carrying out an envisaged trajectory. At this end, a robot is usually equipped with several high-performance sensors. However, we are often interested in less complex and low-cost sensors that could provide enough information to detect in real-time when the trajectory is free of dynamic obstacles. In this context, our strategy was focused on visual sensors, particulary on stereo vision since this provides the depth coordinate for allowing a better perception of the environment. Visual perception for robot mobile navigation is a complex function that requires the presence of "salience" or "evident" patrons to identify something that "breaks" the continuous tendency of data. Usually, interesting points or segments are used for evaluating patrons in position, velocity, appearance or other characteristics that allows us forming groups (Lookingbill et al., 2007), (Talukder & Matthies, 2004). Whereas complete feature vectors are more expressive for explaining objects, here we use 3D feature points for proposing a strategy computationally less demanding conserving the main objective of the work: detect and track moving objects in real time.

This chapter presents a strategy for detecting and tracking dynamic objects using a stereo-vision system mounted on a mobile robot. First, a set of interesting points are extracted from the left image. A disparity map, provided by a real-time stereo vision algorithm implemented on FPGA, gives the 3D position of each point. In addition, velocity magnitude and orientation are obtained to characterize the set of points on the space R^6. Groups of dynamic 2D points are formed using the *a contrario* clustering technique in the 4D space and then evaluated on their depth value yielding groups of dynamic 3D-points. Each one of these groups is initialized by a convex contour with the velocity and orientation of the points given a first estimation of the dynamic object position and velocity. Then an active contour defines a more detailed silhouette of the object based on the intensity and depth value inside of the contour. It is well known that active contour techniques require a highly dense computations. Therefore, in order to reduce the time of processing a fixed number of iterations is used at each frame, so the convergence of the object real limits will be incrementally achieved along

*Part of this work was developed when authors were with LAAS-CNRS, Toulouse, France

several frames. A simple and predefined knowledge about the most usual dynamic objects found in urban environments are used to label the growing regions as a rigid or non-rigid object, essentially cars and people. Experiments on detection and tracking of vehicles and people, as well as during occlusion situations with a mobile robot in real world scenarios are presented and discussed.

2. Related works

The issue of moving object detection has been largely studied by the robotic and computer vision community. Proposed strategies use mainly a combination of active and passive sensors mounted on the mobile robot like laser (Vu & Aycard, 2009) with cameras (Katz et al., 2008), or infrared cameras (Matthies et al., 1998), just to name a few. However, a multi-sensor system requires to solve the problem of fusing the data from different sources which often requires more complex estimation cases. Indeed, more information could be acquired using several sensors but these systems are expensive and complex. To overcome those constraints, the proposed solution consist in using one or more cameras as the only source of information in the system (Talukder & Matthies, 2004), (Williamson, 1998). Essentially vision sensors provide enough information for localization and mapping (Sola et al., 2007) or for describing static and moving objects on the environment (Klappstein et al., 2008).
The stereo-vision is one of the most used techniques for reconstructing the 3D (depth) information of a scene from two images, called left and right. This information is acquired from two cameras separated by a previously established distance. The disparity map is a representation that contains the depth information of the scene. It is well known that dense stereo-vision delivers more complete information than sparse stereo-vision but this is a high-processing cost technique which enables to perform in real time using and ordinary computer system. We use a stereo-vision technique in order to detect moving objects but implemented on a re-configurable architecture that maximizes the efficiency of the system.
In the last decade, several works have proposed the development of high-performance architectures to solve the stereo-vision problem i.e. digital signal processing (DSP), field programmable gate arrays (FPGA) or application-specific integrated circuits (ASIC). The ASIC devices are one of the most complicated and expensive solutions, however they afford the best conditions for developing a final commercial system (Woodfill et al., 2006). On the other hand, FPGA have allowed the creation of hardware designs in standard, high-volume parts, thereby amortizing the cost of mask sets and significantly reducing time-to-market for hardware solutions. However, engineering cost and design time for FPGA-based solutions still remain significantly higher than software-based solutions. Designers must frequently iterate the design process in order to achieve system performance requirements and simultaneously minimize the required size of the FPGA. Each iteration of this process takes hours or days to be completed (Schmit et al., 2000). Even if designing with FPGAs is faster than designing ASICs, it has a finite resource capacity which demands clever strategies for adapting versatile real-time systems (Masrani & MacLean, 2006).

2.1 Overall strategy for 3D dynamic object detection from a mobile robot

A seminal work of this strategy was presented in (Almanza-Ojeda et al., 2010) and (Almanza-Ojeda et al., 2011). Whereas the former proposes a monocamera strategy (Almanza-Ojeda et al., 2010), and the latter the fusion with the information provided by inertial (IMU) (Almanza-Ojeda et al., 2011), here we propose an extension of this strategy to a stereo vision images provided by a bank stereo mounted on the robot. The stereo images are

processed by a stereo vision algorithm designed on FPGA that calculates the disparity map which provides the depth at each point of the input image. Thus, we use a combination of the 3D and 2D representation of feature points for grouping them according to similar distribution function in position and velocity. Moreover, these groups of points permit us to initialize an active contour for obtaining the object boundary. Therefore, the contour initialization of the detected points is fundamental in order to properly perform the shape recovering.
A block diagram of the proposed generalized algorithm is depicted in figure 1. A stereo vision module for FPGA based on Census Transform provides the disparity map using left and right images. At the same time, a selection-tracking process of feature points is carried out on a loop. However, rather than considering both stereo vision images during interest point selection, we will consider only left image for obtaining these 2D feature points. Further, each feature point will be associated with a depth value given by the disparity map. At this point it is necessary to point out that clustering task is performed until a small number of images have been processed. According to this, clustering module will receive a spatial-temporal set of points since each feature location and velocity have been accumulated through the time. The clustering method that we use is the *a contrario* method proposed by Veit et. al. (Veit et al., 2007). This clustering method consist in grouping the interesting points which have a "coherent" movement along a short number of consecutive images. Here the term coherent refers to movements that follow a similar and constant magnitude and direction described by the probability density function of the points under evaluation.

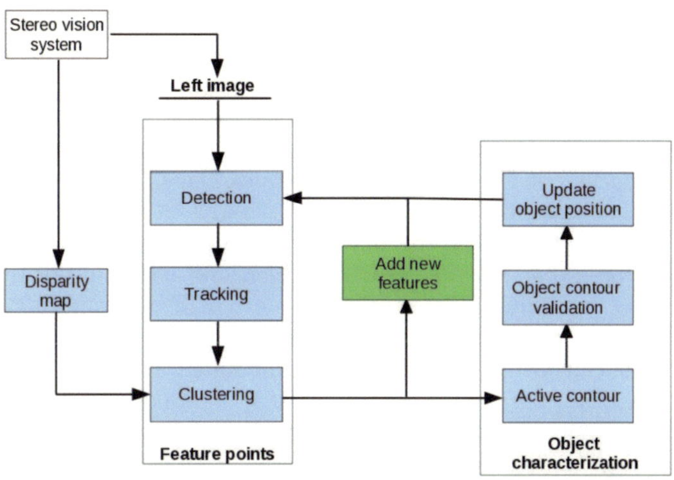

Fig. 1. Global strategy of object detection and identification using a stereo vision system mounted on a mobile robot

Once dynamic group of points have been detected by the clustering process, from this group of points it is possible to delimit a region and an irregular contour in the image with the aim of characterizing a probable dynamic object. Thus, by doing an analysis of intensity inside of this contour, we propose to match the provided region along the image sequence in order to follow the non-rigid objects that have more complicated movement, like pedestrians.
This work provides two distinct contributions: (1) the addition of depth information for performing the clustering of 3D set of dynamic points instead of only 2D as in the *a contrario* clustering technique (Veit et al., 2007), (2) the use of dynamic groups for initializing

an irregular contour active that temporally recovers actual object boundary. The second contribution was preliminary discussed in (Almanza-Ojeda et al., 2011) where the same active contour was initialized for representing object region, however that only uses a snake based on contour information, so nothing is doing with features inside of the region.
The structure of the chapter is as follows. In section 3, we describe the stereo vision module and its performance. Then section 4.1 details interest 2D point selection and how the 3D points are obtained using the disparity map. The grouping technique of the interest points based on the *a contrario* clustering is explained in section 4.2. We present in section 5 object shape recovering using active contours. Section 6 contains our experimental results from indoor an outdoor environments. Finally, we end up with conclusions in section 7.

3. Stereo vision module

We use a disparity map calculated using the Census Transform algorithm (Zabih & Woodfill, 1994) implemented on a programmable device, in this case a FPGA (Field Programmable Gate Array). The architecture of the Census transform algorithm was developed by (Ibarra-Manzano, Devy, Boizard, Lacroix & Fourniols, 2009) in which left and right images acquired from the stereo vision bank are processed for generating up to 325 dense disparity maps of 640 × 480 pixels per second. It is important to point out that most of the vision-based systems do not require high video-frame rates because usually they are implemented on computers or embedded platforms which are not FPGA-based. As this is also our case, we have adapted the disparity map generation to the real-time application required by our system by tuning some configuration parameters in the architecture. A block diagram of the stereo vision algorithm is shown in figure 2. In the following, we will describe in general the calculation of the disparity map based on the Census transform. However, the architectural implementation on the FPGA is a problem that has not dealt with in this work, all these details will be found in (Ibarra-Manzano, Devy, Boizard, Lacroix & Fourniols, 2009), (Ibarra-Manzano & Almanza-Ojeda, 2011).
In the stereo vision algorithm each of the images (right and left) are processed independently in parallel. The process begins with the rectification and correction of the distortion for each image in order to decrease the size of the search of points to a single dimension during disparity calculation. This strategy is known as epipolar restriction in which, once the main axes of the cameras have been aligned in parallel, founding the displacement of the position between the two pixels (one per camera) is reduced to search in each aligned line. That is, if any pair of pixels is visible in both cameras and assuming they are the projection of a single point in the scene, then both pixels must be aligned on the same epipolar line (Ibarra-Manzano, Almanza-Ojeda, Devy, Boizard & Fourniols, 2009). Therefore under this condition, an object location in the scene is reduced to a horizontal translation. Furthermore, the use of the epipolar restriction allows to reduce the complexity and the size of the final architecture.
Next, rectified and corrected images are filtered using an arithmetic mean filter. Once input images have been filtered, they are used to calculate the Census Transform as depicted in the figure 2. This transform is a non-parametric measure used during the matching process for measuring similarities and obtaining the correspondence between the points into the left and right images. A neighborhood of pixels is used for establishing the relationships among them. From the Census Transform, two images are obtained referred to as I_{Cl} and I_{Cr} which represent the left and right Census images. Two pixels extracted from the Census images (one for each image) are compared using the Hamming distance. This comparison

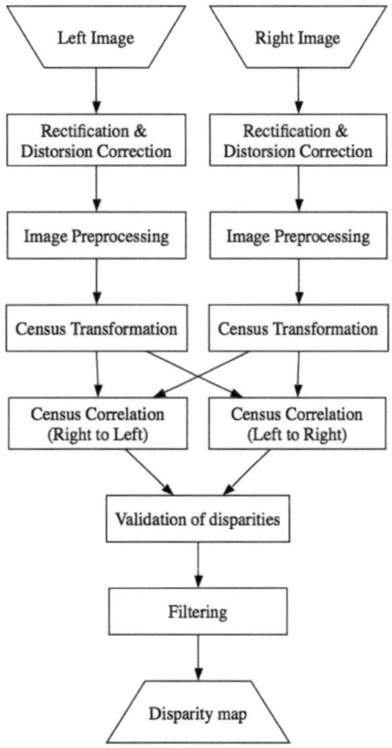

Fig. 2. Block diagram of the stereo vision algorithm (Ibarra-Manzano & Almanza-Ojeda, 2011)

which is called the correlation process allows us to obtain a disparity measure. The disparity measure comes from the similarity maximization function in the same epipolar line for the two Census images. The similarity evaluation is based on the binary comparison between two bit chains calculated by the Census Transform. The correlation process is carried out two times, (left to right then right to left) with the aim of reducing the disparity error and thus complementing the process. Once both disparity measures have been obtained, the disparity measure validation (right to left and left to right) consists of comparing both disparity values and obtaining the absolute difference between them. Before delivering a final disparity image, a novel filtering process is needed for improving its quality. In this final stage, the use of a median spatial filter was found more convenient.

3.1 Disparity map acquired in real time

The final architecture for executing the stereo vision algorithm based on the Census Transform was developed using the level design flow RTL (Ibarra-Manzano, Devy, Boizard, Lacroix & Fourniols, 2009). The architecture was codified in VHDL language using Quartus II workspace and ModelSim. Finally, it was synthesized for an EP2C35F672C6 device contained in the Cyclone IV family of Altera. Table 1 lays out some of the configuration parameters used during the disparity map computation exploited during the dynamic object detection and tracking approach. We would like to highlight that disparity image is computed through the time with a high performance of 30 image per second although detection and tracking approach does not reach this performance. More technical details about the implementation are discussed in section 6.

Parameter	Value
Image size	640×480
Window size	7×7
Disparity max	64
Performance	40
Latency (μs)	206
Area	6,977
Memory size	109 Kb

Table 1. Configuration parameters of the Stereo vision architecture.

The architecture was tested for different navigational scenes using a stereo vision bank, first mounted in a mobile robot and then in a vehicle. In the following, we will describe the obtained results for two operational environments. Figure 3 shows the left and right images acquired from the bank stereo and the associated disparity image delivered by the stereo vision algorithm. Dense disparity image depicts the disparity value in gray color levels in figure 3 (c). By examining this last image, we can determine that if the object is close to the stereo vision bank that means a big disparity value, so it corresponds to a light gray level. Otherwise, if the object is far from the stereo vision bank, the disparity value is low, which corresponds to a dark gray level. In this way, we observe that the gray color which represents the road in the resulting images gradually changes from the light to dark gray level. We point out the right side of the image, where we can see the different tones of gray level corresponding to the building. Since large of the building is located at different depths with respect to the stereo vision bank, in the disparity map corresponding gray color value is assigned from lighter to darker gray tones.

(a) Left image (b) Right image (c) Result from FPGA implementation

Fig. 3. Stereo images acquired from a mobile robot during outdoor navigation: a) left image b) right image and c) the disparity map.

In the second test (see figure 4), the stereo vision bank is mounted on a vehicle that is driven on a highway. This experimental test results in a difficult situation because the vehicle is driven at high-speed during the test. Furthermore, this urban condition requires a big distance between both cameras in the stereo vision bank with the aim of augmenting the field of view. A consequence of having a big distance between the two cameras is represented in figure 4 (a) and (b) in which, while the left image show a car that overtakes our vehicle, this car is out of sight in the right camera. Therefore, the dense disparity map shown in figure 4 (c) does not display a cohesive depth value of the vehicle due to mismatches in the information between both images. Nevertheless, we highlight all the different depths represented by the gray color value in the highway that gradually turns darker until the black color which represents an infinity depth.

(a) Left image (b) Right image (c) Result from FPGA implementation

Fig. 4. Stereo images acquired from a vehicle in the highway: a) left image b) right image and c) the disparity map.

4. Moving features perception and grouping

4.1 Active 3D point selection

When dealing with vision-based approaches, the problem of processing a large quantity of information requires that the system resources be sophisticated and expensive if we want to get a real time performance. As we work with left and right images from the stereo vision bank, processing both images yields to more information about the scene but has also more computation requirements. To overcome this problem, we consider to deal with a sparse set of features resulted by the analysis of the left image that represents the most significant and salience feature points in all the image. We use this sparse proposition because it is important to distribute time of processing among some others essential tasks of the strategy, therefore feature selection process must expend minimal time. For the feature point selection, we use the Shi-Tomasi approach (Shi & Tomasi, 1994). This approach consists in providing the most representative points based on image gradient analysis, i.e. corners, borders, and all the regions with high contrast change. Figure 5 depicts both image gradients in horizontal and vertical directions of an input image in gray color level, from which N best features are selected for describing image content. In particular, this input image has high content of information, therefore, a larger number of points have to be selected in accordance with image-density information. Furthermore, the number N is restricted by the time of processing required for reaching real-time performance. According to this, the number of points does not have to exceed $N = 180$. Once 2D's interesting points have been selected on the left image and the disparity map for the stereo-images computed, obtaining the 3D characterization of the points is straightforward. Each 2D point is associated with a corresponding depth value provided by the disparity image to conform the 3D point representation.

Until now, we have obtain a set of 2D features on the left image at time t and their corresponding 3D characterization. For each feature, displacement vectors are computed through the time by using the Kanade-Lucas-Tomasi tracker, referred to as the KLT

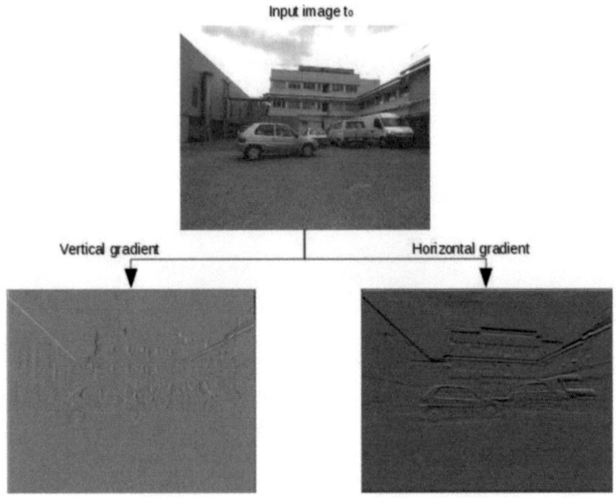

(a) Image gradients

(b) N best features selected

Fig. 5. Image processing for the best interest point selection. a) Gradients in vertical and horizontal direction of the above input image. b) Green points represent the *N* best interesting features resulted by the gradient image analysis.

technique (Shi & Tomasi, 1994), (Lucas & Kanade, 1981). These displacement vectors are used to calculate feature velocities. We are interested in the accumulation of previous position and velocity of the points in order to establish a trail of motion. An example of the accumulation point positions for *N* initial features detected appears in figure 6. Note that, while most of the points are distributed in all the image, a small set of points horizontally aligned can be remarked on the left side of the figure c). These points represent the front part of the vehicle that enters into the field of view of our robot. Once a short number of images have been processed, the accumulated vector displacements and positions of the feature points are evaluated in order to find significant patterns of motion that possibly represent dynamic objects in the scene. A new feature selection task is carried out, as indicated in the green block of figure 1. Further, in the case that any dynamic group of points is found, this information will initialize in the second stage of our strategy, that is the object characterization (right rectangular box in figure 1).

In the following, we explain the a *contrario* method used for clustering position and velocity of the points that possibly describe a mobile object.

(a) First image of the tracking

(b) Last image of the tracking

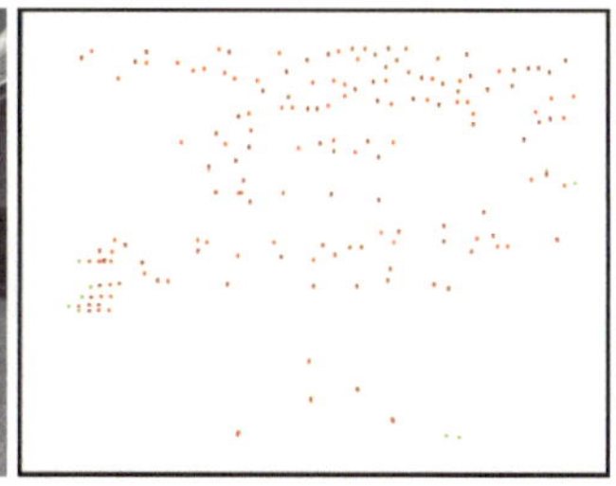

(c) Accumulate position through 4 images

Fig. 6. N feature points initially detected at image a) are tracked through 4 images. Image b) displays the last image during the tracking task. Image c) depicts all the accumulate positions for the N initial points detected calculated by the tracking process.

4.2 Clustering of 3D points cloud

At this point, we have a distributed set of 2D feature points characterized by their position and velocity that have been tracked through a short number of consecutive images. Alternatively, we associated each feature with their corresponding depth position which allows us to manage with a 3D-data set. Following with the global diagram in figure 1, clustering of feature points is the next task to carry out. In order to do that, we use the *a contrario* clustering method proposed in (Veit et al., 2007). This algorithm is based on the Gestalt theory that establishes which groups could be formed based on one or several common characteristics of their elements. In accord to this statement, the *a contrario* clustering technique identifies one group as meaningful if all their elements show a different distribution than an established background random model. Contrary to most clustering techniques, neither initial number of clusters is required nor parameter have to be tuned. These characteristics result very favorable in an unknown environment context where the number of resulted clusters have not been predefined.

In this section we summarize some important concepts of the *a contrario* clustering method, used to group feature points. A detailed description of the method derivation is available in (Desolneux et al., 2008), (Desolneux et al., 2003), (Cao et al., 2007).

4.3 *A contrario* algorithm description

As we mentioned above, a distribution model for the background has to be defined for comparing with the associated distribution of the set of points, here referred to as $V(x,y,v,\theta)$ in R^4. In this work, we use the background model proposed in (Veit et al., 2007) which establishes a random organization of the observations. Therefore, background model elements are independent identically distributed (*iid*) and follow a distribution p. The *iid* nature of random model components proposes an organization with non coherent motion present.

Next, given the input vector $V(x,y,v,\theta)$ from the KLT process (section 4.1), the first objective is to evaluate which elements in V shows a particular distribution contrary to the established distribution p of the background model (that explains "*a contrario*" name). To overcome

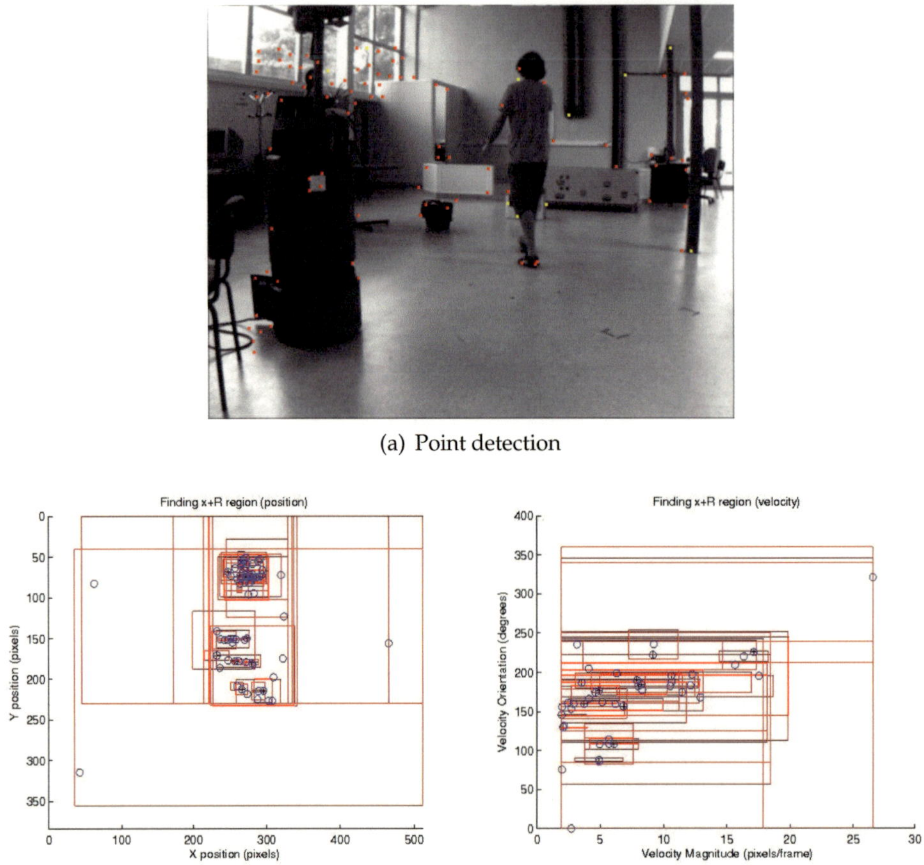

(a) Point detection

(b) Position graph

(c) Polar velocity graph

Fig. 7. Best fit region search. An initial set of points is selected and tracked in an indoor environment displayed in image a). The accumulated locations and velocities of the points are analyzed in order to find the region of points with $NFA\,(G) \leq 1$.

the problem of point by point evaluation, $V(x, y, v, \theta)$ is divided in testing groups with different size of elements using a single linkage method. This method constructs a binary tree where each node represents a candidate group G. Once a different group of points have been established, these will be evaluate using a set of given regions represented by $\mathcal{H}$. This set of regions is formed by a different size of hyper-rectangles that will be used to test the distribution of each data group in G. An example of groups distribution evaluation is depicted in figure 7. Each region $H \in \mathcal{H}$ is centered at each element $X \in G$ to find the region H_X that contains all the elements in G and at the same time this region has to minimize the probability of the background model distribution. This procedure requires that sizes of hyper-rectangles be in function of data range, in our experiments we use 20 different sizes by dimension. The measure of meaningfulness (called Number of False Alarms NFA in referenced work) is given by eq 1.

$$NFA\,(G) = N^2 \cdot |\mathcal{H}| \min_{\substack{X \in G,\\ H \in \mathcal{H},\\ G \subset H_X}} B\,(N-1, n-1, p\,(H_X)) \tag{1}$$

In this equation N represents the number of elements in vector V, $|\mathcal{H}|$ is the cardinality of regions and n is the elements in group test G. The term which appears in the minimum function is the accumulated binomial law, this represents the probability that at least n points including X are inside the region test centered in X (H_X). Distribution p consist of four independent distributions, one for each dimension data. Point positions and velocity orientation follow a uniform distribution because object moving position and direction is arbitrary. On the other hand, velocity magnitude distribution is obtained directly of the empirically histogram of the observed data. So that, joint distribution p will be the product of these four distributions. A group G is said to be meaningful if $NFA\,(G) \leq 1$.
Furthermore two sibling meaningful groups in the binary tree could be belong to the same moving object, then a second evaluation for all the meaningful groups is calculated by Eq. 2.To obtain this new measure, we reuse region group information (dimensions and probability) and just a new region that contains both test groups G_1 and G_2 is calculated. New terms are $N' = N - 2$, number of elements in G_1 and G_2, respectively $n'_2 = n_1 - 1$ and $n'_2 = n_2 - 1$, and term $\mathcal{T}$ which represents the accumulated trinomial law.

$$NFA_G\,(G_1, G_2) = N^4 \cdot |\mathcal{H}|^2\, \mathcal{T}\left(N', n'_1, n'_2, p_1, p_2\right) \tag{2}$$

Both mesures 1 and 2 represent the significance of groups in binary tree. Final clusters are found by exploring all the binary tree and comparing to see if it is more significant to have two moving objects G_1 and G_2 or to fusion it in a group G. Mathematically, $NFA(G) < NFA_G(G_1, G_2)$ where $G_1 \cup G_2 \subset G$. A descriptive result is provided in figure 8. Here blue points correspond to those that could be dynamic but without a well defined motion, so they are associated with the background. On the other hand, green points represent the group in G which shows a different distribution than the random background model. Notice that graph c) displays polar velocity considering magnitude in X-axis and orientation in Y-axis, from there, green point positions are not together because they have orientations among 5° and 355° Further, these points correspond to the vehicle entering on the right side of the image b).

4.4 Depth validation

As previously mentioned, the 3D characterization of points is achieved using the depth value described in the disparity map for each two dimensional feature point. From previous section, a set of dynamic points have been obtained using the clustering process, however this process was performed considering uniquely 2D points. Thanks to the disparity map, we know the depth of each feature in dynamic group, so it is possible to perform a second evaluation looking for similarity in their depths. This depth evaluation is computed in a formally analogous way to that of the *a contrario* clustering. However in this case, the number of regions for testing is considerably reduced since the group of points is already detected in a defined region, so it is only need the evaluation of similar depths of the points around that region avoiding the best region search. Additionally, there is not an associated velocity in Z-axis direction which reduces from hyper-rectangle regions to 3D boxes. Namely x and y denote the point location in the image and z denotes the disparity value.

(a) Initial image (b) Final image

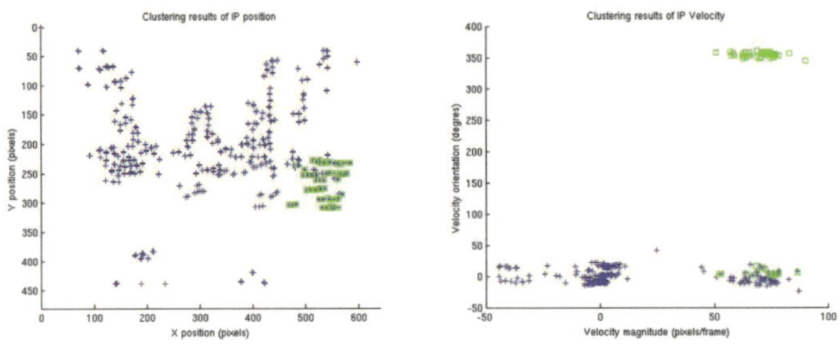

(c) Position points (d) polar-velocity points

Fig. 8. Clustering results after processing 5 consecutive images. a)First image used for feature selection, b) Last image used for obtaining the trails of the points. A unique dynamic group of points is detected (green points) for whose graph c) depicts their position and graph d) their velocity and orientation.

In the next section, we explain how an irregular contour is initialized using the groups of points calculated by the *a contrario* clustering, therefore used for passing from a dynamic set of points to actual object boundary.

5. Object characterization by an active contour

Active models have been widely used in image processing applications, in particular for recovering shapes and the tracking of moving objects (Li et al., 2010), (Paragios & Deriche, 2005). An active contour or snake is a curve which minimizes energy from restrictive external and internal forces in the image, typically calculated from edges, gradient, among others. Essentially, a snake is not thought of to solve the problem of automatic search for prominent contours of the image, but rather for recovering the contour of a form, from an initial position proposed by other mechanisms. That is to say, if the initial contour is relatively close to the solution (for example a contour defined manually by an operator or obtained through any other method), the contour evolves up until minimizing the function of energy defined from the internal and external forces.

One of the main objectives of this work is to track all mobile objects detected without any prior knowledge about object type that one follows. In a real and dynamic context, we expect to find rigid and non rigid mobile objects ie. vehicles, persons. We have shown in section 4.2 the good performance of the *a contrario* method for finding initial sets of dynamic points which correspond to the mobile objects in the scene. However, whereas clustering process deals with feature points, here we describe a strategy for recovering the deformable shape of the objects through the time by considering others features like intensity image gradient inside of contour. Therefore, through this section we will describe fundamental details of the active contours theory, and the object tracking procedure by means of the active contours.

5.1 Active contour initialization

The results obtained in section 4.2 allow the definition of an initial irregular contour that contains totally or partially our interest object. To this end, we take an outer location of points in the detected group for initializing an active contour which delimits the object on the image. An example is displayed in figure 9 [1]: in image (a) a set of points dynamic (in blue) is detected on which correspond to the person in motion on this sequence. From this set of points we have selected those illustrated in color magenta to describe the object contour depicted in yellow. Due to the fact these points are the most farthest from the center then they are the most representative and closest to the object frontiers. For the results show in image 9(a), we have 8 points on the curve and we use for each 4 control points. The value of 4 control points is fixed for each point on the frontier in order to introduce the corners or discontinuities in the curve (Marin-Hernandez, 2004). Initial contour will allow us to obtain a deformable model by a bounded potential shown in figure 9(b). The zone occupied by the object (represented in red color) is separated from the background. The most intens tones inside of the object are used to best adapt initial contour to the actual object silhouette. Therefore, a more detailed object shape is achieved by analyzing internal and external energy in a bounding box that contains the initial contour.

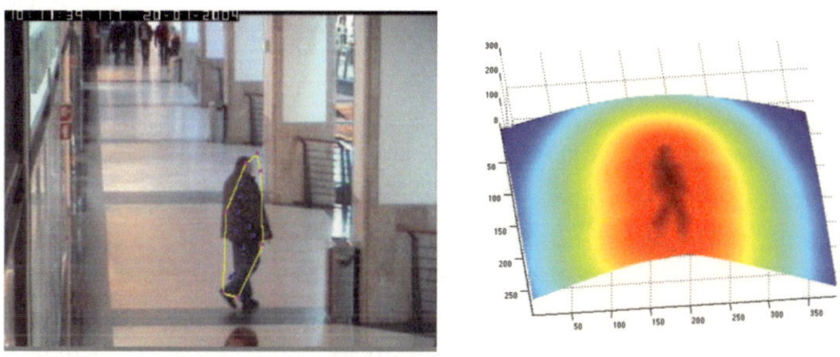

(a) Initial contour of a non-rigid object (b) energy functional

Fig. 9. Test performed with real images acquired by a fixed camera in an indoor environment. a)Initial contour derived from points further away from the center b) Energy functional that concentrates internal and external energy considering the initial contour of a).

5.1.1 Parametric active contours.

At this point, it is necessary to represent the initial contour in 9 by a parametric curve $u(\tau) = (x(\tau), y(\tau))$, $\tau \in [0,1]$, with $u(0) = u(1)$. this contour is deformed through the time domain to minimize the energy expressed by:

[1] This image sequence was downloaded from the web site (Fisher, 2011) provided by EC Funded CAVIAR project/IST 2001 37540

$$E_{snake} = \int_0^1 [E_{int}(u(\tau)) + E_{ext}(u(\tau))] \, d\tau \tag{3}$$

where E_{int} is expressed by two main terms, the first one refers to the elasticity and the second the flexibility, given:

$$E_{int} = \alpha \int_0^1 |u_\tau(\tau)|^2 d\tau + \beta \int_0^1 |u_{\tau\tau}(\tau)|^2 d\tau \tag{4}$$

τ and $\tau\tau$ indexes in the term $u(\tau)$ implies respectively first and second order of derivation. By returning to the equation 3 for defining the term E_{ext} or the energy of the image (Sekhar et al., 2008), as the field of potential P:

$$E_{ext} = \int_0^1 P(u(\tau)) d\tau \tag{5}$$

The potential includes different terms defined from image proprieties like edges, lines, etc. Edges energy is obtained by computing the magnitude of the gradient intensity $|\nabla I|$. Without a good initialization, the energy of the edges will not be enough to locate the objects on noisy or low contrast images. Therefore an additional potential of the regions is added to the edge energy. Generally, the potential of the regions is defined by the mean (μ) and the variance (σ^2) of the pixels intensity in the region. However, other constraints could be added like the object velocities, or other statistics derived from region characteristics (Brox et al., 2010). Because of the real-time constraints, we calculate only some statistics in the region that describe the object, such as the main properties to its implementation in correspondence to the next image. The following section describes the proposed strategy about shape recovering and object tracking using active contours.

5.2 Incremental silhouette definition

We have tested a method based on the work of Chan and Vese (Chan & Vese, 2001) in order to find the silhouette of the object. In this work, the authors give a region of initialization which may contain total or partially the object. The analysis of this initial contour will allow evaluation of the conditions of the energy minimization inside and outside of the contour. An example of a shape recovering of a person in an indoor environment using the Chan and Vese method is shown in figure 10. First row of images display the contour evolution. In the second row the region inside of the contour for each corresponding above image is illustrated. These images are given as input to the process for recovering real object contour. Furthermore, white regions on these images are labeled as occupied locations which avoid the detection of new interesting points inside of it through the time. According to this, new feature selection (detailed in section 4.1) will look for unoccupied locations in the image allowing the detection of incoming objects.

Once a partial contour at image t in figure 10 has been obtained by minimizing eq. 3, we estimate its position at image $t+1$ for starting a new convergence process. The prediction of the region on the next image is always given by the Kalman filter. The vector state of our object is expressed as:

$$\mathbf{x}_0 = \left[\bar{x}, \bar{y}, \bar{v}_x, \bar{v}_y\right]^T \tag{6}$$

Namely, $\bar{x}$, $\bar{y}$ denote the barycenter location and $\bar{v}_x$, $\bar{v}_y$ the means of velocity vector in X and Y direction respectively. Figure 11 illustrates an example of the filter prediction. The barycenter position and the partial converged contour are located in accordance with Kalman

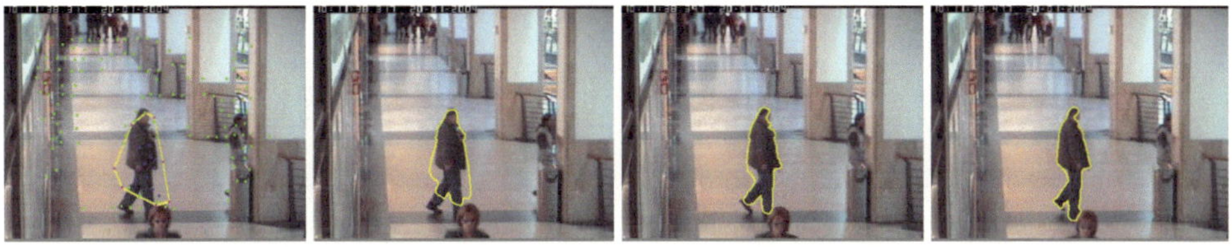

(a) Initial snake at image t_0 (b) 30 iterations at image t_0 (c) Image $t_0 + 2$ (d) Image $t_0 + 4$

(e) Initial mask at image t_0 (f) Mask at image t_0 (g) Mask at image $t_0 + 2$ (h) Mask at image $t_0 + 4$

Fig. 10. Results of the detection and tracking of a non-rigid dynamic object, in this case a pedestrian on the same sequence of figure 9. Here, we have processed 5 consecutive images, executing 30 iterations per image in order to find the actual object shape.

filter prediction. We always consider a velocity constant model and the vector state is obtained from model object displacement. However, in some cases it is first necessary to tune the initial parameters of the filter because there may exist different types of object movements producing an undesired acceleration of the predicted position.

5.3 Object contour validation

The difficulties rising from our proposed strategy points out the sensitivity of the active contours to small discontinuities on the object edges with low-contrast because the energy is configured over the entire image. To overcome these difficulties we evaluate the disparity inside of the region corresponding with the binary mask (that will be presented in the next section figure 15). This evaluation consists in ordering in an ascendant way all the disparity values inside the region designed by the binary mask then we uniquely consider the median of the values. In particular, we consider a valid point of the object region if its disparity value is located up to the 4th percentil in the ordered list (here referred to as $4p$). It follows that our statistical validation for object contour refining can be written as:

$$M = \begin{cases} 1 & depth \geq 4p \\ 0 & otherwise \end{cases} \tag{7}$$

where M represents the binary mask. It is important to remark that the depth value is the inverse value of the disparity therefore in this case the fact of rejecting points located on the first four percentiles represents that these values have a lower disparity value, that is, disparity values of the background points are expecting to be low. This constrain allow us to develop a last validity evaluation for obtaining a most detailed and accurated representation of the object shape. The next section describes our experimental results and presents how the disparity map plays a fundamental role in increasing the efficiency and improve the obtained results.

(a) Object silhouette (b) Estimation for the next image

Fig. 11. Motion estimation using Kalman filter. a) The barycenter location at image t is used for predict its position at the next image. b) The irregular contour is centered at the predicted barycenter position, this will be used as initialization region in incoming images.

6. Experimental results

We evaluate the performance of the proposed strategy for detecting and tracking dynamic objects by carrying out experiments using both secure and mounted cameras on autonomous vehicles. Since, using one secure camera, vibrations, egomotion, among other typical problems of mounted cameras are neglected, first we do an experiment under this controlled condition to verify that our algorithm detects moving objects in real images. Moreover, we have observed that the total convergence of energy equation is computationally demanding. To overcome this constraint, it was necessary to propose that convergence task works uniquely a small number of iterations and re-starting the process at next image from previous results. In our case, we see a significant improvement in the efficiency of computation by using this, even when the most similar contour to actual object shape is found after some iterations.

6.1 Fixed camera case

Figure 12 shows 10 consecutive images in which a person appears in the field of view of one security camera on a Commercial Center. These resulting images refer to the case where the number of iterations for converging the active contour at each image is set to 30. The initial contour obtained from the cluster of dynamic points was displayed in figure 9. The effect of the Kalman filter estimation permit us to achieve a more detailed shape than the initial contour frame by frame. We remark that even if initial contour does not contain all the object, the actual shape is achieved after processing 5 images. It is important to point out that resulting object bounds are almost the same as the real thanks to the high-contrast generated between the person and the background.

6.2 Experiments during mobile robot navigation: rigid objects detection

A second experiment was performed in an outdoor environment where the robot accomplishes a linear trajectory of navigation at low-speed (about 6 m/s). In our experiments, we only use the stereo-vision bank mounted on the mobile robot. The robot cameras provide images with resolution of 640×480 pixels. The egomotion derived from the robot displacements is neglected by verifying similar depth values inside of the region containing the dynamic group of points. Moreover, dynamic groups detected on the road shows different depth values, so they are rejected to define an initial contour. Figure 13 illustrates the detection and tracking of a white vehicle. Note that this vehicle comes from the imaginary epipolar

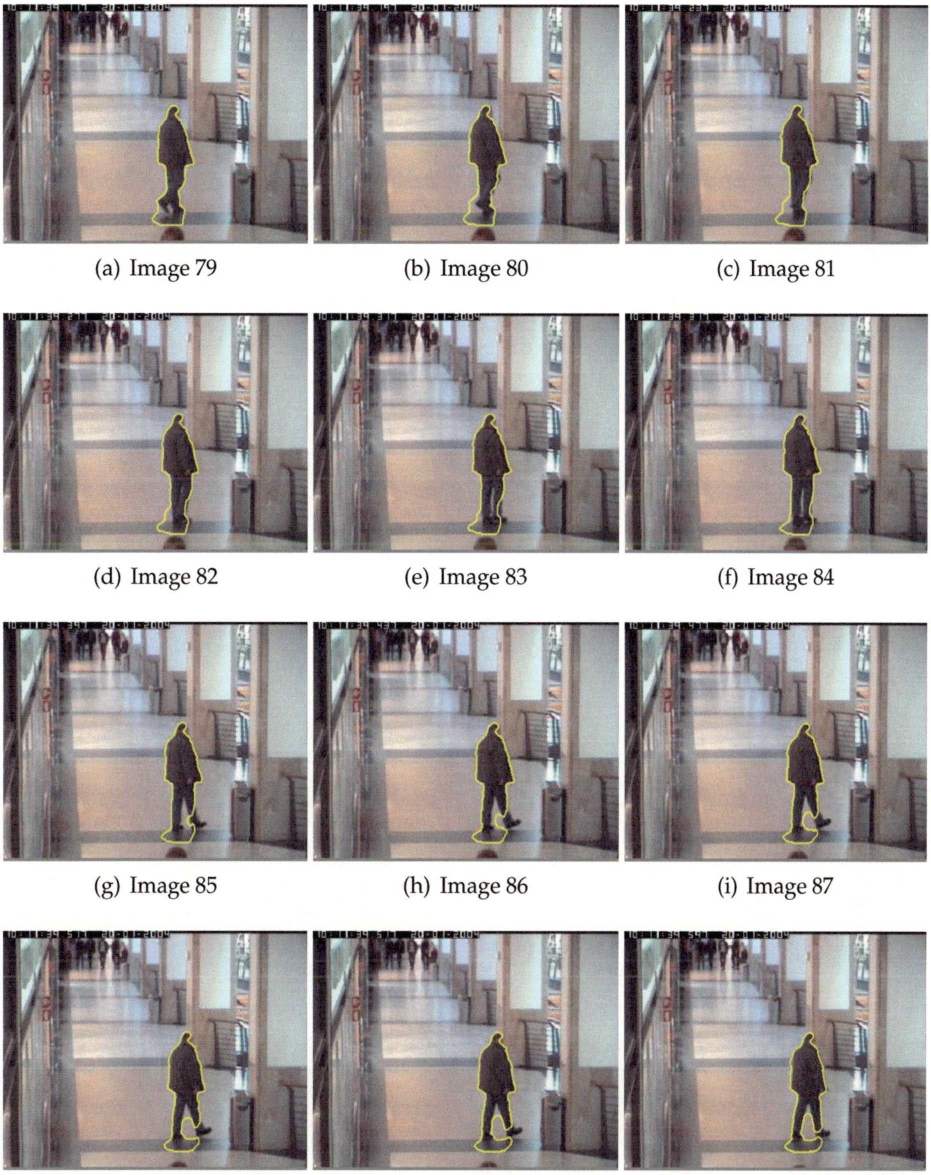

(a) Image 79 (b) Image 80 (c) Image 81

(d) Image 82 (e) Image 83 (f) Image 84

(g) Image 85 (h) Image 86 (i) Image 87

(j) Image 88 (k) Image 89 (l) Image 90

Fig. 12. Detection and tracking of a non-rigid dynamic object along 12 consecutive images. In each image, 30 iterations are used in order to find the object silhouette.

point of image, however it can only be perceived by our strategy at the moment of it is closer to our robot position. Whereas the first experiment was performed under ideal conditions of controlled illumination provided by the indoor environment here this conditions do not hold. This fact notably avoids the computation of a cohesive representation of the energy functional as the one illustrated in figure 9(b). As a consequence, there is only one object at each image of figure 13 but its contour is represented by two separated regions that have similar energy level inside. Furthermore, we analyze the disparity map for obtaining the statistics of disparity values in both regions as mentioned in section 5.3

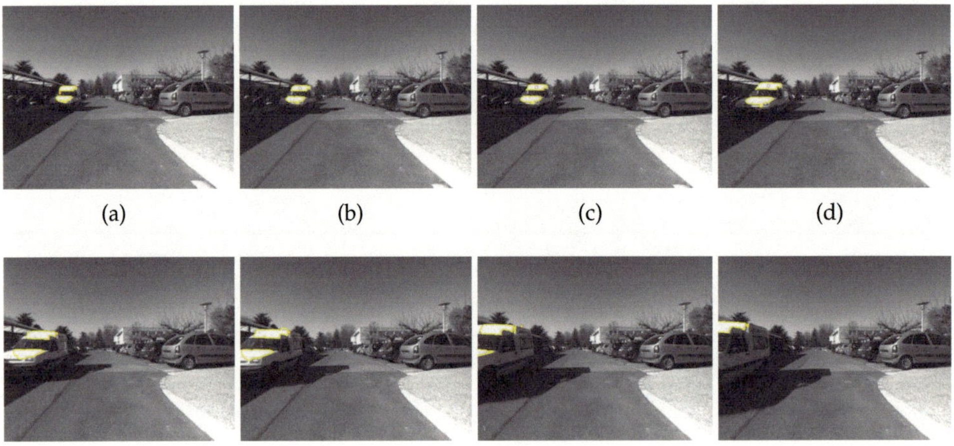

(a) (b) (c) (d)

(e) (f) (g) (h)

Fig. 13. Experimental results in an outdoor environment during robot navigation. Images show the detection and tracking of a rigid object.

6.3 Experiments during mobile robot navigation: non rigid objects detection

In the third experiment, we considered a mobile robot moving in an outdoor environment again but in this case the robot finds a dynamic non-rigid object during its trajectory. The left image of figure 14 displays the frame in which the dynamic object was detected and initialized as a irregular contour, middle image shows the disparity map and right image the respective initial mask. In practice detecting non-rigid objects is more complicated than the previous experiment because a person walking has different motions in his legs than his shoulder or his head. Because of improvements, we use in this experiment 40 iterations per image for converging the active contour. Figure 15 illustrates some resulted images of the tracking performed by our proposed strategy. In this experiment, we found that almost all the person could be covered with the active contour. However by examining the left column

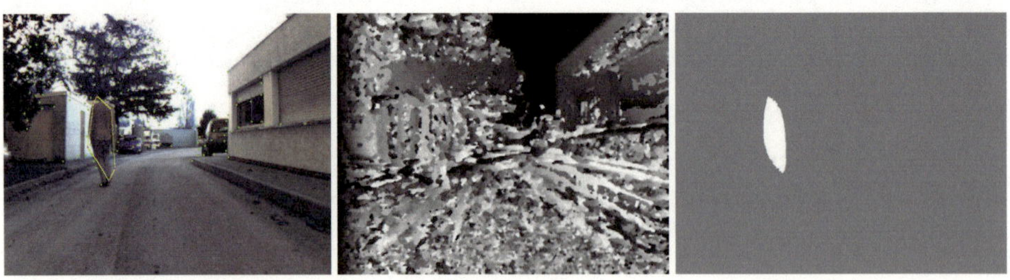

(a) Initial detection (b) disparity map (c) Initial mask

Fig. 14. a) Initial contour derived from the dynamic group of points. b) The corresponding disparity map of the image, c) the initial mask used by the active contour.

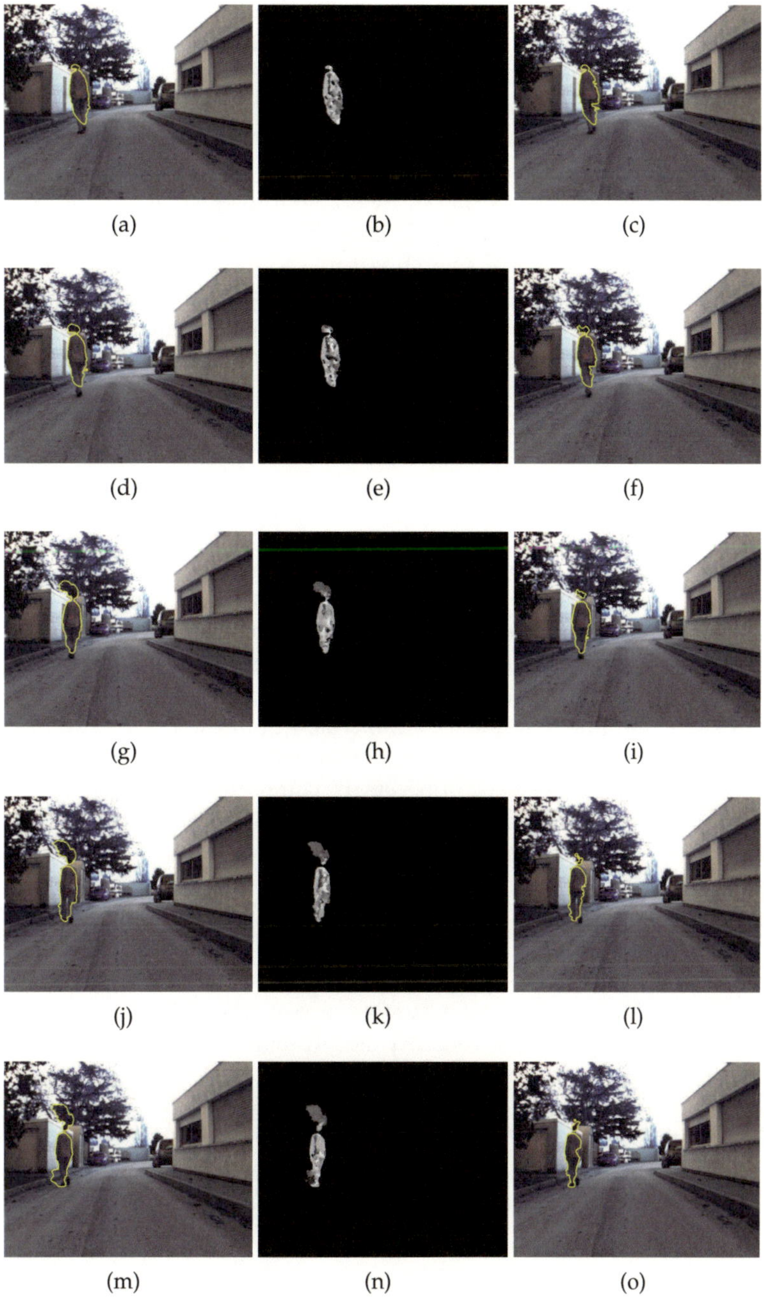

Fig. 15. Images on the left column show the contour obtained by only considering the intensity inside of the region. The middle column isolates the object contour representation and its corresponding disparity values. Images on the right evaluate the depth of the target to boil down extra regions included in the object contour.

of figure 15, we can figure out that part of the background was also included in our object region. To overcome this problem we performed the same statistical calculation of disparity values inside of the region-object as described in section 5.3. By examining the middle column of figure 15, we can figure out that similar depth values are concentrated in actual object contour. We can see that additional zones to this contour that represents the tree and other locations of the background will be rejected using the disparity map since they provide lower values of disparity, that is they are farther away than our interesting moving object.
The proposed algorithm was coded on C/C++ and TCL, however disparity maps are computed for the FPGA by means of an architecture codified in VHDL. The Cyclone IV card calculates 30 disparity maps per second. After several tests, we measure that our algorithm runs around 1 or 1.5 Hz depending of the nature of the environment. From this, note that the disparity is available at higher frequency than our algorithm performance, however we comment that until now the goal of the experiment was to provide an algorithm for detecting and tracking moving objects.

7. Conclusions

In this project, we considered the problem of dynamic object detection from a mobile robot in indoor/outdoor environments of navigation. Specially, the proposed strategy uses only visual-information provided by a stereo-vision bank mounted on the mobile robot. The speed of the robot during navigation is established low to avoid disturbance on the velocity data due to robot ego-motion. Experimental results allow us to realize that the proposed strategy performs a correct and total detection of the rigid and non-rigid objects and it is able to tracking them along the image sequence. Motivated by these results future contributions to this project consist in decreasing the time of computation. Nevertheless, we make some assumptions by avoiding excessive or unnecessary computations ie. the number of selected feature points, number of iterations during the active contour processing, our global algorithm is not able to perform in real time (at least 10 Hz). Significant improvements could be obtained by an emigration of all our algorithm design to embedded architectures like GPU or FPGA devices. Furthermore these kinds of devices provide a high portability towards robotics or autonomous vehicle platforms.
We also comment the difficulties rising from the disparity map constructed by the stereo vision module in which a cohesive and accurate representation of the actual scene have to be improved. To this end, future works consider the addition of a strategy for rejecting "spikes" in the disparity map cased by stereo mismatches.

8. Acknowledgments

This work was partially funded by the CONACyT with the project entitled "Diseno y optimizacion de una arquitectura para la clasificacion de objetos en tiempo real por color y textura basada en FPGA". Authors would like to thank Cyril Roussillon for help provided with the robot.

9. References

Almanza-Ojeda, D., Devy, M. & Herbulot, A. (2010). Visual-based detection and tracking of dynamic obstacles from a mobile robot, *In Proceedings of the 7th International Conference on Informatics in Control, Automation and Robotics (ICINCO 2010)*, Madeira, Portugal.

Almanza-Ojeda, D. L., Devy, M. & Herbulot, A. (2011). Active method for mobile object detection from an embedded camera, based on a contrario clustering, *in* J. A. Cetto, J.-L. Ferrier & J. Filipe (eds), *Informatics in Control, Automation and Robotics*, Vol. 89 of *Lecture Notes in Electrical Engineering*, Springer Berlin Heidelberg, pp. 267–280. 10.1007/978-3-642-19539-6_18.
URL: *http://dx.doi.org/10.1007/978-3-642-19539-6_18*

Brox, T., Rousson, M., Deriche, R. & Weickert, J. (2010). Colour, texture, and motion in level set based segmentation and tracking, *Image Vision Comput.* 28(3): 376–390.

Cao, F., Delon, J., Desolneux, A., Musé, P. & Sur, F. (2007). A unified framework for detecting groups and application to shape recognition, *Journal of Mathematical Imaging and Vision* 27(2): 91–119.

Chan, T. & Vese, L. (2001). Active contours without edges, *Transactions on Image Processing, IEEE* 10(2): 266–277.

Desolneux, A., Moisan, L. & Morel, J.-M. (2003). A grouping principle and four applications, *IEEE Transactions on Pattern Analysis and Machine Intelligence* 25(4): 508–513.

Desolneux, A., Moisan, L. & Morel, J.-M. (2008). *From Gestalt Theory to Image Analysis A Probabilistic Approach*, Vol. 34, Springer Berlin / Heidelberg.

Fisher, R. (2011).
URL: *http://groups.inf.ed.ac.uk/vision/CAVIAR/CAVIARDATA1/*

Ibarra-Manzano, M.-A. & Almanza-Ojeda, D.-L. (2011). *Advances in Stereo Vision*, InTech, chapter High-Speed Architecture Based on FPGA for a Stereo-Vision Algorithm, pp. 71-88.

Ibarra-Manzano, M., Almanza-Ojeda, D.-L., Devy, M., Boizard, J.-L. & Fourniols, J.-Y. (2009). Stereo vision algorithm implementation in fpga using census transform for effective resource optimization, *Digital System Design, Architectures, Methods and Tools, 2009. 12th Euromicro Conference on*, pp. 799 –805.

Ibarra-Manzano, M., Devy, M., Boizard, J.-L., Lacroix, P. & Fourniols, J.-Y. (2009). An efficient reconfigurable architecture to implement dense stereo vision algorithm using high-level synthesis, *2009 International Conference on Field Programmable Logic and Applications*, Prague, Czech Republic, pp. 444–447.

Katz, R., Douillard, B., Nieto, J. & Nebot, E. (2008). A self-supervised architecture for moving obstacles classification, *IEEE/RSJ International Conference on Intelligent Robots and Systems, IROS 2008*, pp. 155–160.

Klappstein, J., Vaudrey, T., Rabe, C., Wedel, A. & Klette, R. (2008). Moving object segmentation using optical flow and depth information, *PSIVT '09: Proceedings of the 3rd Pacific Rim Symposium on Advances in Image and Video Technology*, Springer-Verlag, Berlin, Heidelberg, pp. 611–623.

Li, C., Xu, C., Gui, C. & Fox, M. D. (2010). Distance regularized level set evolution and its application to image segmentation, *IEEE Trans. Image Process.* 19(12): 3243–3254.

Lookingbill, A., Lieb, D. & Thrun, S. (2007). *Autonomous Navigation in Dynamic Environments*, Vol. 35 of *Springer Tracts in Advanced Robotics*, Springer Berlin / Heidelberg, pp. 29–44.

Lucas, B. D. & Kanade, T. (1981). An iterative image registration technique with an application to stereo vision, *Proceedings of DARPA Image Understanding Workshop*, pp. 121–130.

Marin-Hernandez, A. (2004). *Vision dynamique pour la navigation d'un robot mobile.*, PhD thesis, INPT-LAAS-CNRS.

Masrani, D. & MacLean, W. (2006). A Real-Time large disparity range Stereo-System using FPGAs, *Computer Vision Systems, 2006 ICVS '06. IEEE International Conference on*, p. 13.

Matthies, L., Litwin, T., Owens, K., Rankin, A., Murphy, K., Coombs, D., Gilsinn, J., Hong, T., Legowik, S., Nashman, M. & Yoshimi, B. (1998). Performance evaluation of ugv obstacle detection with ccd/flir stereo vision and ladar, *ISIC/CIRA/ISAS Joint Conference* pp. 658–670.

Paragios, N. & Deriche, R. (2005). Geodesic active regions and level set methods for motion estimation and tracking, *Computer Vision and Image Understanding* 97(3): 259 – 282. URL: *http://www.sciencedirect.com/science/article/pii/S1077314204001213*

Schmit, H. H., Cadambi, S., Moe, M. & Goldstein, S. C. (2000). Pipeline reconfigurable fpgas, *Journal of VLSI Signal Processing Systems* 24(2-3): 129–146.

Sekhar, S. C., Aguet, F., Romain, S., Thévenaz, P. & Unser, M. (2008). Parametric b-spline snakes on distance maps—application to segmentation of histology images, *Proceedings of the 16th European Signal Processing Conference, (EUSIPCO2008)* .

Shi, J. & Tomasi, C. (1994). Good features to track, *proceedings of the IEEE Conference on Computer Vision and Pattern Recognition*, pp. 593–600.

Sola, J., Monin, A. & Devy, M. (2007). BiCamSLAM: two times mono is more than stereo, *IEEE International Conference on Robotics Automation (ICRA2007), Rome, Italy*, pp. 4795–4800.

Talukder, A. & Matthies, L. (2004). Real-time detection of moving objects from moving vehicles using dense stereo and optical flow, *proceedings of the International Conference on Intelligent Robots and Systems (IROS2004)*, pp. 3718–3725.

Veit, T., Cao, F. & Bouthemy, P. (2007). Space-time a contrario clustering for detecting coherent motion, *IEEE International Conference on Robotics and Automation, (ICRA07)*, Roma, Italy, pp. 33–39.

Vu, T. & Aycard, O. (2009). Laser-based detection and tracking moving objects using data-driven markov chain monte carlo, *IEEE International Conference on Robotics Automation (ICRA2009), Kobe, Japan.*

Williamson, T. (1998). *A High-Performance Stereo Vision System for Obstacle Detection*, PhD thesis, Robotics Institute, Carnegie Mellon University, Pittsburgh, PA.

Woodfill, J., Gordon, G., Jurasek, D., Brown, T. & Buck, R. (2006). The tyzx DeepSea g2 vision system, ATaskable, embedded stereo camera, *Computer Vision and Pattern Recognition Workshop, 2006. CVPRW '06. Conference on*, p. 126.

Zabih, R. & Woodfill, J. (1994). Non-parametric local transforms for computing visual correspondence, *ECCV '94: Proceedings of the Third European Conference on Computer Vision*, Vol. II, Springer-Verlag New York, Inc., Secaucus, NJ, USA, pp. 151–158.

2

Development of an Autonomous Visual Perception System for Robots Using Object-Based Visual Attention

Yuanlong Yu, George K. I. Mann and Raymond G. Gosine
Faculty of Engineering and Applied Science, Memorial University of Newfoundland
St. John's, NL,
Canada

1. Introduction

Unlike the traditional robotic systems in which the perceptual behaviors are manually designed by programmers for a given task and environment, autonomous perception of the world is one of the challenging issues in the cognitive robotics. It is known that the *selective attention* mechanism serves to link the processes of perception, action and learning (Grossberg, 2007; Tipper et al., 1998). It endows humans with the *cognitive capability* that allows them to *learn* and *think* about how to perceive the environment autonomously. This visual attention based autonomous perception mechanism involves two aspects: *conscious aspect* that directs perception based on the current task and learned knowledge, and *unconscious aspect* that directs perception in the case of facing an unexpected or unusual situation. The *top-down attention* mechanism (Wolfe, 1994) is responsible for the conscious aspect whereas the *bottom-up attention* mechanism (Treisman & Gelade, 1980) corresponds to the unconscious aspect. This paper therefore discusses about how to build an artificial system of autonomous visual perception.

Three fundamental problems are addressed in this paper. The first problem is about pre-attentive segmentation for object-based attention. It is known that attentional selection is either *space-based* or *object-based* (Scholl, 2001). The space-based theory holds that attention is allocated to a spatial location (Posner et al., 1980). The object-based theory, however, posits that some pre-attentive processes serve to segment the field into discrete objects, followed by the attention that deals with one object at a time (Duncan, 1984). This paper proposes that object-based attention has the following three advantages in terms of computations: 1) Object-based attention is more robust than space-based attention since the attentional activation at the object level is estimated by accumulating contributions of all components within that object, 2) attending to an exact object can provide more useful information (e.g., shape and size) to produce the appropriate actions than attending to a spatial location, and 3) the discrete objects obtained by pre-attentive segmentation are required in the case that a global feature (e.g., shape) is selected to guide the top-down attention. Thus this paper adopts the object-based visual attention theory (Duncan, 1984; Scholl, 2001).

Although a few object-based visual attention models have been proposed, such as (Sun, 2008; Sun & Fisher, 2003), developing a pre-attentive segmentation algorithm is still a challenging issue as it is a unsupervised process. This issue includes three types of challenges: 1) The

ability to automatically determine the number of segments (termed as *self-determination*), 2) the computational efficiency, and 3) the robustness to noise. Although K-labeling methods (e.g., normalized cut (Shi & Malik, 2000)) can provide the accuracy and robustness, they are ineffective and inefficient when the number of segments is unknown. In contrast, recent split-and-merge methods (e.g., irregular pyramid based segmentation (Sharon et al., 2006)) are capable of determining the number of segments and computationally efficient, whereas they are not robust to noise. This paper proposes a new pre-attentive segmentation algorithm based on the irregular pyramid technique in order to achieve the self-determination and robustness as well as keep the balance between the accuracy and efficiency.
The second problem is about how to model the attentional selection, i.e., model the cognitive capability of *thinking* about what should be perceived. Compared with the well-developed bottom-up attention models (Itti & Baldi, 2009; Itti et al., 1998), modeling the top-down attention is far from being well-studied. The top-down attention consists of two components: 1) Deduction of task-relevant object given the task and 2) top-down biasing that guides the focus of attention (FOA) to the task-relevant object. Although some top-down methods have been proposed, such as (Navalpakkam & Itti, 2005), several challenging issues require further concerns. Since the first component is greatly dependent on the knowledge representation, it will be discussed in the next paragraph. Regarding the second component, the first issue is about the effectiveness of top-down biasing. The main factor that decays the effectiveness is that the task-relevant object shares some features with the distracters. It indicates that the top-down biasing method should include a mechanism to make sure that the task-relevant object can be discriminated from distracters. The second issue is about the computational efficiency based on the fact that the attention is a fast process to select an object of interest from the image input. Thus it is reasonable to use some low-level features rather than high-level features (e.g., the iconic representation (Rao & Ballard, 1995a)) for top-down biasing. The third one is the adaptivity to automatically determine which feature(s) is used for top-down biasing such that the requirement of manually re-selecting the features for different tasks and environment is eliminated. This paper attempts to address the above issues by using the integrated competition (IC) hypothesis (Duncan et al., 1997) since it not only summarizes a theory of the top-down attention, which can lead to a computational model with effectiveness, efficiency and adaptivity, but also integrates the object-based attention theory. Furthermore, it is known that bottom-up attention and top-down attention work together to decide the attentional selection, but how to combine them is another challenging issue due to the multi-modality of bottom-up saliency and top-down biases. A promising approach to this issue is setting up a unified scale at which they can be combined.
The third problem is about the cognitive capability of autonomously *learning* the knowledge that is used to guide the conscious perceptual behavior. According to the psychological concept, the memory used to store this type of knowledge is called long-term memory (LTM). Regarding this problem, the following four issues are addressed in this paper. The first issue is about the unit of knowledge representations. Object-based vision theory (Duncan, 1984; Scholl, 2001) indicates that a general way of organizing the visual scene is to parcel it into discrete objects, on which perception, action and learning perform. In other words, the internal attentional representations are in the form of objects. Therefore objects are used as the units of the learned knowledge. The second issue is what types of knowledge should be modeled for guiding the conscious perceptual behavior. According to the requirements of the attention mechanism, this paper proposes that the knowledge mainly includes LTM task representations and LTM object representations. The *LTM task representation* embodies the

association between the attended object at the last time and predicted task-relevant object at the current time. In other words, it tells the robot what should be perceived at each time. Thus its objective is to deduce the task-relevant object given the task in the attentional selection stage. The *LTM object representation* embodies the properties of an object. It has two objectives: 1) Directing the top-down biasing given the task-relevant object and 2) directing the post-attentive perception and action selection. The third issue is about how to build their structure in order to realize the objectives of these two representations. This paper employs the *connectionist approach* to model both representations as the self-organization can be more effectively achieved by using the cluster-based structure, although some symbolic approaches (Navalpakkam & Itti, 2005) have been proposed for task representations. The last issue is about how to learn both representations through the duration from an infant robot to a mature one. It indicates that a dynamic, constructive learning algorithm is required to achieve the self-organization, such as generation of new patterns and re-organization of existing patterns. Since this paper focuses on the perception process, only the learning of LTM object representations is presented.

The remainder of this paper is organized as follows. Some related work of modeling visual attention and its applications in robotic perception are reviewed in section 2. The framework of the proposed autonomous visual perception system is given in section 3. Three stages of this proposed system are presented in section 4, section 5 and section 6 respectively. Experimental results are finally given in section 7.

2. Related work

There are mainly four psychological theories of visual attention, which are the basis of computational modeling. Feature integration theory (FIT) (Treisman & Gelade, 1980) is widely used for explaining the space-based bottom-up attention. The FIT asserts that the visual scene is initially coded along a variety of feature dimensions, then attention competition performs in a location-based serial fashion by combining all features spatially, and focal attention finally provides a way to integrate the initially separated features into a whole object. Guided search model (GSM) (Wolfe, 1994) was further proposed to model the space-based top-down attention mechanism in conjunction with bottom-up attention. The GSM posits that the top-down request for a given feature will activate the locations that might contain the given feature. Unlike FIT and GSM, the biased competition (BC) hypothesis (Desimone & Duncan, 1995) asserts that attentional selection, regardless of being space-based or object-based, is a biased competition process. Competition is biased in part by the bottom-up mechanism that favors a local inhomogeneity in the spatial and temporal context and in part by the top-down mechanism that favors items relative to the current task. By extending the BC hypothesis, the IC hypothesis (Duncan, 1998; Duncan et al., 1997) was further presented to explain the object-based attention mechanism. The IC hypothesis holds that any property of an object can be used as a task-relevant feature to guide the top-down attention and the whole object can be attended once the task-relevant feature successfully captures the attention.

A variety of computational models of space-based attention for computer vision have been proposed. A space-based bottom-up attention model was first built in (Itti et al., 1998). The surprise mechanism (Itti & Baldi, 2009; Maier & Steinbach, 2010) was further proposed to model the bottom-up attention in terms of both spatial and temporal context. Itti's model was further extended in (Navalpakkam & Itti, 2005) by modeling the top-down attention mechanism. One contribution of Navalpakkam's model is the symbolic knowledge representations that are used to deduce the task-relevant entities for top-down attention. The other contribution is the multi-scale object representations that are used to bias attentional selection. However, this top-down biasing method might be ineffective in the case that environment contains distracters which share one or some features with the target. Another model that selectively tunes the visual processing networks by a top-down hierarchy of winner-take-all processes was also proposed in (Tsotsos et al., 1995). Some template matching methods such as (Rao et al., 2002), and neural networks based methods, such as (Baluja & Pomerleau, 1997; Hoya, 2004), were also presented for modeling top-down biasing. Recently an interesting computational method that models attention as a Bayesian inference process was reported in (Chikkerur et al., 2010). Some space-based attention model for robots was further proposed in (Belardinelli & Pirri, 2006; Belardinelli et al., 2006; Frintrop, 2005) by integrating both bottom-up and top-down attention.

Above computational models direct attention to a spatial location rather than a perceptual object. An alternative, which draws attention to an object, has been proposed by (Sun & Fisher, 2003). It presents a computational method for grouping-based saliency and a hierarchical framework for attentional selection at different perceptual levels (e.g. a point, a region or an object). Since the pre-attentive segmentation is manually achieved in the original work, Sun's model was further improved in (Sun, 2008) by integrating an automatic segmentation algorithm. Some object-based visual attention models (Aziz et al., 2006; Orabona et al., 2005) have also been presented. However, the top-down attention is not fully achieved in these existing object-based models, e.g., how to get the task-relevant feature is not realized.

Visual attention has been applied in several robotic tasks, such as object recognition (Walther et al., 2004), object tracking (Frintrop & Kessel, 2009), simultaneous localization and mapping (SLAM) (Frintrop & Jensfelt, 2008) and exploration of unknown environment (Carbone et al., 2008). A few general visual perception models (Backer et al., 2001; Breazeal et al., 2001) are also presented by using visual attention. Furthermore, some research (Grossberg, 2005; 2007) has proposed that the adaptive resonance theory (ART) (Carpenter & Grossberg, 2003) can predict the functional link between attention and processes of consciousness, learning, expectation, resonance and synchrony.

3. Framework of the proposed system

The proposed autonomous visual perception system involves three successive stages: pre-attentive processing, attentional selection and post-attentive perception. Fig. 1 illustrates the framework of this proposed system.

Stage 1: The *pre-attentive processing stage* includes two successive steps. The first step is the extraction of pre-attentive features at multiple scales (e.g., nine scales for a 640×480 image). The second step is the pre-attentive segmentation that divides the scene into proto-objects in an unsupervised manner. The *proto-objects* can be defined as uniform regions such that the pixels in the same region are similar. The obtained proto-objects are the fundamental units of attentional selection.

Stage 2: The *attentional selection stage* involves four modules: bottom-up attention, top-down attention, a combination of bottom-up saliency and top-down biases, as well as estimation of proto-object based attentional activation. The bottom-up attention module aims to model the unconscious aspect of the autonomous perception. This module generates a probabilistic location-based bottom-up saliency map. This map shows the conspicuousness of a location compared with others in terms of pre-attentive features. The top-down attention module aims

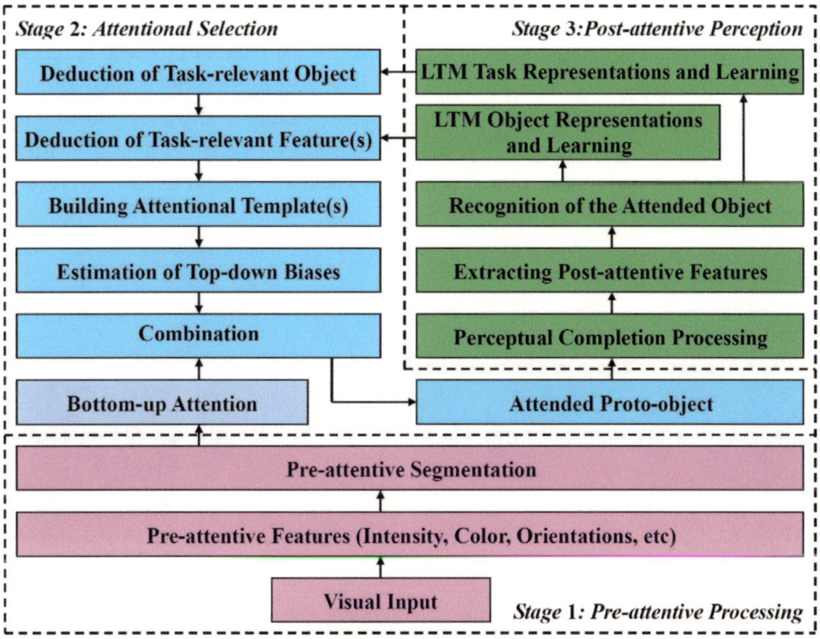

Fig. 1. The framework of the proposed autonomous visual perception system for robots.

to model the conscious aspect of the autonomous perception. This module is modeled based on the IC hypothesis and consists of four steps. *Step 1* is the deduction of the task-relevant object from the corresponding LTM task representation given the task. *Step 2* is the deduction of the task-relevant feature dimension(s) from the corresponding LTM object representation given the task-relevant object. *Step 3* is to build the attentional template(s) in working memory (WM) by recalling the task-relevant feature(s) from LTM. *Step 4* is to estimate a probabilistic location-based top-down bias map by comparing attentional template(s) with corresponding pre-attentive feature(s). The obtained top-down biases and bottom-up saliency are combined in a probabilistic manner to yield a location-based attentional activation map. By combining location-based attentional activation within each proto-object, a proto-object based attentional activation map is finally achieved, based on which the most active proto-object is selected for attention.

Stage 3: The main objective of the *post-attentive perception stage* is to interpret the attended object in more detail. The detailed interpretation aims to produce the appropriate action and learn the corresponding LTM object representation at the current time as well as to guide the top-down attention at the next time. This paper introduces four modules in this stage. The first module is perceptual completion processing. Since an object is always composed of several parts, this module is required to perceive the complete region of the attended object post-attentively. In the following text, the term *attended object* is used to represent one or all of the proto-objects in the complete region being attended. The second module is the extraction of post-attentive features that are a type of representation of the attended object in WM and used for the following two modules. The third module is object recognition. It functions as a decision unit that determines to which LTM object representation and/or to which instance of that representation the attended object belongs. The fourth module is learning of LTM object representations. This module aims to develop the corresponding LTM representation of the attended object. The probabilistic neural network (PNN) is used to build the LTM object representation. Meanwhile, a constructive learning algorithm is also proposed. Note that the LTM task representation is another important module in the post-attentive perception stage. Its learning requires the perception-action training pairs, but this paper focuses on the perception process rather than the action selection process. So this module will be discussed in the future work.

4. Pre-attentive processing

4.1 Extraction of pre-attentive features

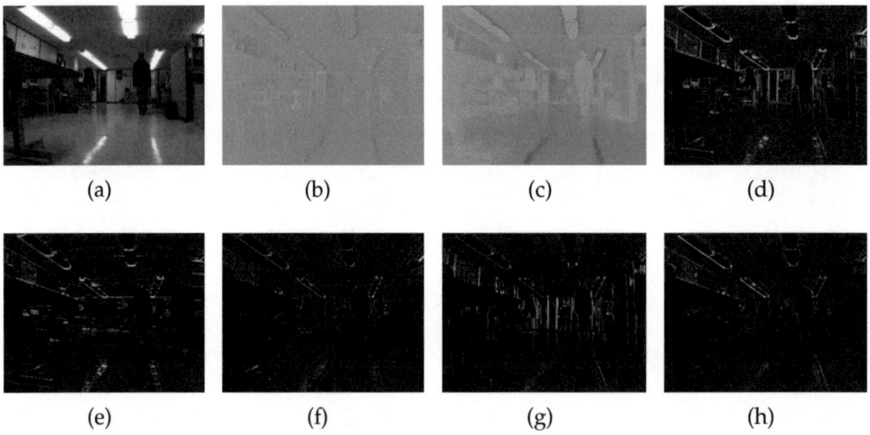

Fig. 2. Pre-attentive features at the original scale. (a) Intensity. (b) Red-green. (c) Blue-yellow. (d) Contour. (e) - (h) Orientation energy in direction $0°, 45°, 90°$ and $135°$ respectively. Brightness represents the energy value.

By using the method in Itti's model (Itti et al., 1998), pre-attentive features are extracted at multiple scales in the following dimensions: intensity $\mathbf{F}_{int}$, red-green $\mathbf{F}_{rg}$, blue-yellow $\mathbf{F}_{by}$, orientation energy $\mathbf{F}_{o_\theta}$ with $\theta \in \{0°, 45°, 90°, 135°\}$, and contour $\mathbf{F}_{ct}$. Symbol $\mathbf{F}$ is used to denote pre-attentive features.
Given 8-bit RGB color components $\mathbf{r}$, $\mathbf{g}$ and $\mathbf{b}$ of the input image, intensity and color pairs at the original scale are extracted as : $\mathbf{F}_{int} = (\mathbf{r}+\mathbf{g}+\mathbf{b})/3$, $\mathbf{F}_{rg} = \mathbf{R}-\mathbf{G}$, $\mathbf{F}_{by} = \mathbf{B}-\mathbf{Y}$, where $\mathbf{R} = \mathbf{r}-(\mathbf{g}+\mathbf{b})/2$, $\mathbf{G} = \mathbf{g}-(\mathbf{r}+\mathbf{b})/2$, $\mathbf{B} = \mathbf{b}-(\mathbf{r}+\mathbf{g})/2$, and $\mathbf{Y} = (\mathbf{r}+\mathbf{g})/2 - |\mathbf{r}-\mathbf{g}|/2 - \mathbf{b}$. Gaussian pyramid (Burt & Adelson, 1983) is used to create the multi-scale intensity and color pairs. The multi-scale orientation energy is extracted using the Gabor pyramid (Greenspan et al., 1994). The contour feature $\mathbf{F}_{ct}(s)$ is approximately estimated by applying a pixel-wise maximum operator over four orientations of orientation energy: $\mathbf{F}_{ct}(s) = \max_{\theta\in\{0°,45°,90°,135°\}} \mathbf{F}_{o_\theta}(s)$, where s denotes the spatial scale. Examples of the extracted pre-attentive features have been shown in Fig. 2.

4.2 Pre-attentive segmentation

This paper proposes a pre-attentive segmentation algorithm by extending the irregular pyramid techniques (Montanvert et al., 1991; Sharon et al., 2000; 2006). As shown in Fig. 3, the pre-attentive segmentation is modeled as a hierarchical accumulation procedure, in which each level of the irregular pyramid is built by accumulating similar local nodes at the level below. The final proto-objects emerge during this hierarchical accumulation process as they are represented by single nodes at some levels. This accumulation process consists of four procedures.

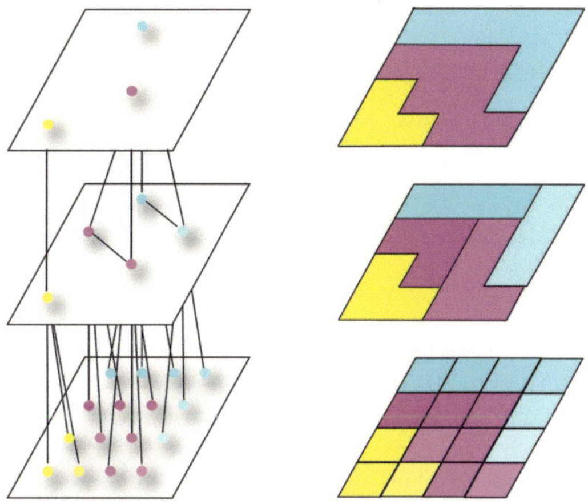

Fig. 3. An illustration of the hierarchical accumulation process in the pre-attentive segmentation. This process is shown from bottom to top. In the left figure, this process is represented by vertices and each circle represents a vertex. In the right figure, this process is represented by image pixels and each block represents an image pixel. The color of each vertex and block represents the feature value. It can be seen that the image is partitioned into three irregular regions once the accumulation process is finished.

Procedure 1 is decimation. A set of surviving nodes (i.e., parent nodes) is selected from the son level to build the parent level. This procedure is constrained by the following two rules (Meer, 1989): 1) Any two neighbor son nodes cannot both survive to the parent level and 2) any son node must have at least one parent node. Instead of the *random values* used in the stochastic pyramid decimation (SPD) algorithm (Jolion, 2003; Meer, 1989), this paper proposes a new recursive similarity-driven algorithm (i.e., the first extension), in which a son node will survive if it has the maximum *similarity* among its neighbors with the constraints of the aforementioned rules. The advantage is the improved segmentation performance since the nodes that can greatly represent their neighbors deterministically survive. As the second extension, *Bhattacharyya distance* (Bhattacharyya, 1943) is used to estimate the similarity between nodes at the same level (i.e., the strength of intra-level edges). One advantage is that the similarity measure is approximately scale-invariant during the accumulation process since Bhattacharyya distance takes into account the correlations of the data. The other advantage is that the probabilistic measure can improve the robustness to noise.

In *procedure 2*, the strength of inter-level edges is estimated. Each son node and its parent nodes are linked by inter-level edges. The strength of these edges is estimated in proportion to the corresponding intra-level strength at the son level by using the method in (Sharon et al., 2000).

Procedure 3 aims to estimate the aggregate features and covariances of each parent node based on the strength of inter-level edges by using the method in (Sharon et al., 2000).

The purpose of *procedure 4* is to search for neighbors of each parent node and simultaneously estimate the strength of intra-level edges at the parent level. As the third extension, a new

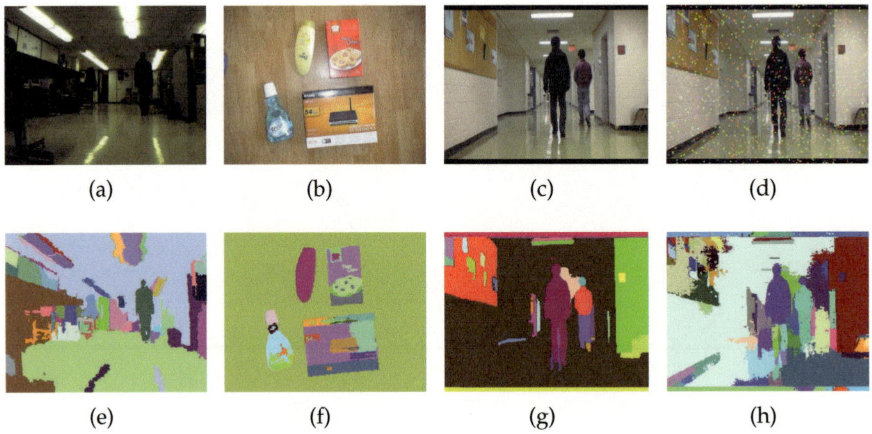

Fig. 4. Results of pre-attentive segmentation. (a)-(d) Original images, where (d) includes salt and pepper noise (noise density:0.1, patch size: 5×5 pixels). (e)-(h) Segmentation results. Each color represents one proto-object in these results.

neighbor search method is proposed by considering not only the graphic constraints but also the similarity constraints. A candidate node is selected as a neighbor of a center node if the similarity between them is above a predefined threshold. Since the similarity measure is scale-invariant, a fixed value of the threshold can be used for most pyramidal levels. The advantage of this method is the improved segmentation performance since the connections between nodes that are located at places with great transition are deterministically cut.

In the case that no neighbors are found for a node, it is labeled as a new proto-object. The construction of the full pyramid is finished once all nodes at a level have no neighbors. The membership of each node at the base level to each proto-object is iteratively calculated from the top pyramidal level to the base level. According to the membership, each node at the base level is finally labeled. The results of the pre-attentive segmentation are shown in Fig. 4.

5. Attentional selection

5.1 Bottom-up attention

The proposed bottom-up attention module is developed by extending Itti's model (Itti et al., 1998). Center-surround differences in terms of pre-attentive features are first calculated to simulate the competition in the spatial context:

$$\mathbf{F}'_f(s_c, s_s) = |\mathbf{F}_f(s_c) \ominus \mathbf{F}_f(s_s)| \tag{1}$$

where $\ominus$ denotes across-scale subtraction, consisting of interpolation of each feature at the surround scale to the center scale and point-by-point difference, $s_c = \{2, 3, 4\}$ and $s_s = s_c + \delta$ with $\delta = \{3, 4\}$ represent the center scales and surround scales respectively, $f \in \{int, rg, by, o_\theta, ct\}$ with $\theta \in \{0°, 45°, 90°, 135°\}$, and $F'_f(s_c, s_s)$ denotes a center-surround difference map.

These center-surround differences in terms of the same feature dimension are then normalized and combined at scale 2, termed as *working scale* and denoted as s_{wk}, using across-scale addition to yield a location-based conspicuity map of that feature dimension:

$$\mathbf{F}_f^s = \mathcal{N}\left(\frac{1}{6}\bigoplus_{s_c=2}^{4}\bigoplus_{s_s=s_c+3}^{s_c+4}\mathcal{N}(\mathbf{F}'_f(s_c,s_s))\right) \tag{2}$$

where $\mathcal{N}$ is the normalization operator, $\bigoplus$ is across-scale addition, consisting of interpolation of each normalized center-surround difference to the working scale and point-by-point addition, $f \in \{int, rg, by, o_\theta, ct\}$, and $\mathbf{F}_f^s$ denotes a location-based conspicuity map.

All conspicuity maps are point-by-point added together to yield a location-based bottom-up saliency map $\mathbf{S}_{bu}$:

$$\mathbf{S}_{bu} = \mathcal{N}\left(\mathbf{F}_{ct}^s + \mathbf{F}_{int}^s + \frac{1}{2}(\mathbf{F}_{rg}^s + \mathbf{F}_{by}^s) + \frac{1}{4}\sum_\theta \mathbf{F}_{o_\theta}^s\right) \tag{3}$$

Given the following two assumptions: 1) the selection process guided by the space-based bottom-up attention is a random event, and 2) the sample space of this random event is composed of all spatial locations in the image, the salience of a spatial location can be used to represent the degree of belief that bottom-up attention selects that location. Therefore, the probability of a spatial location $\mathbf{r}_i$ being attended by the bottom-up attention mechanism can be estimated as:

$$p_{bu}(\mathbf{r}_i) = \frac{S_{bu}(\mathbf{r}_i)}{\sum_{\mathbf{r}_{i'}} S_{bu}(\mathbf{r}_{i'})}, \tag{4}$$

where $p_{bu}(\mathbf{r}_i)$ denotes the probability of a spatial location $\mathbf{r}_i$ being attended by the bottom-up attention, and the denominator $\sum_{\mathbf{r}_{i'}} S_{bu}(\mathbf{r}_{i'})$ is the normalizing constant.

5.2 Top-down attention

5.2.1 LTM task representations and task-relevant objects

The *task-relevant object* can be defined as an object whose occurrence is expected by the task. Consistent with the autonomous mental development (AMD) paradigm (Weng et al., 2001), this paper proposes that actions include external actions that operate effectors and internal actions that predict the next possible attentional state (i.e., attentional prediction). Since the proposed perception system is object-based, the attentional prediction can be seen as the task-relevant object. Thus this paper models the *LTM task representation* as the association between attentional states and attentional prediction and uses it to deduce the task-relevant object.

It can be further proposed that the LTM task representation can be modeled by using a first-order discrete Markov process (FDMP). The FDMP can be expressed as $p(a_{t+1}|a_t)$, where a_t denotes the attentional state at time t and a_{t+1} denotes the attentional prediction for time $t+1$. This definition means that the probability of each attentional prediction for the next time can be estimated given the attentional state at the current time. The discrete attentional states is composed of LTM object representations.

5.2.2 Task-relevant feature

According to the IC hypothesis, it is required to deduce the task-relevant feature from the task-relevant object. This paper defines the *task-relevant feature* as a property that can discriminate the object from others. Although several autonomous factors (e.g., rewards obtained from learning) could be used, this paper uses the *conspicuity* quantity since it is one of the important intrinsic and innate properties of an object for measuring the discriminability. Through a training process that statistically encapsulates the conspicuity quantities obtained under different viewing conditions, a *salience descriptor* is achieved in the LTM object representation (See details in section 6.2 and section 6.3).

Therefore the salience descriptor is used to deduce the task-relevant feature by finding the feature dimension that has the greatest conspicuity. This deduction can be expressed as:

$$(f_{rel}, j_{rel}) = \arg \max_{f \in \{ct,int,rg,by,o_\theta\}} \max_{j \in \{1,2,...,N_j\}} \frac{\overline{\mu}_f^{s,j}}{1+\overline{\sigma}_f^{s,j}}, \quad (5)$$

where N_j denotes the number of parts when $f \in \{int, rg, by, o_\theta\}$ and $N_j = 1$ when $f = ct$, $\overline{\mu}_f^{s,j}$ and $\overline{\sigma}_f^{s,j}$ respectively denote the mean and STD of salience descriptors in terms of a feature f in the LTM representation of the task-relevant object, f_{rel} denotes the *task-relevant feature dimension*, and j_{rel} denotes the index of the *task-relevant part*. The LTM object representation can be seen in section 6.3.

In the proposed system, the most task-relevant feature is first selected for guiding top-down attention. If the post-attentive recognition shows that the attended object is not the target, then the next task-relevant feature is joined. This process does not stop until the attended object is verified or all features are used.

5.2.3 Attentional template

Given the task-relevant feature dimension, its *appearance descriptor* in the LTM representation of the task-relevant object is used to build an attentional template in WM so as to estimate top-down biases. The attentional template is denoted as $\mathbf{F}_f^t$, where $f \in \{ct, int, rg, by, o_\theta\}$. The appearance descriptor will be presented in section 6.3.

5.2.4 Estimation of top-down biases

Bayesian inference is used to estimate the location-based top-down bias, which represents the probability of a spatial location being an instance of the task-relevant object. It can be generally expressed as:

$$p_{td}(\mathbf{r}_i|\mathbf{F}_f^t) = \frac{p_{td}(\mathbf{F}_f^t|\mathbf{r}_i) \times p_{td}(\mathbf{r}_i)}{\sum_{\mathbf{r}_{i'}} p_{td}(\mathbf{F}_f^t|\mathbf{r}_{i'}) \times p_{td}(\mathbf{r}_{i'})}, \quad (6)$$

where $p_{td}(\mathbf{r}_i)$ denotes the prior probability of a location $\mathbf{r}_i$ being attended by the top-down attention, $p_{td}(\mathbf{F}_f^t|\mathbf{r}_i)$ denotes the observation likelihood, $p_{td}(\mathbf{r}_i|\mathbf{F}_f^t)$ is the posterior probability of the location $\mathbf{r}_i$ being attended by the top-down attention given the attentional template $\mathbf{F}_f^t$. Assuming that the prior probability $p_{td}(\mathbf{r}_i)$ is a uniform distribution, Eq. (6) can be simplified into estimating the observation likelihood $p_{td}(\mathbf{F}_f^t|\mathbf{r}_i)$. The detailed estimation of $p_{td}(\mathbf{F}_f^t|\mathbf{r}_i)$ for each feature dimension, including contour, intensity, red-green, blue-yellow and orientations can be seen in our previous object-based visual attention (OVA) model (Yu et al., 2010).

5.2.5 Discussion

Compared with existing top-down attention methods, e.g., (Navalpakkam & Itti, 2005; Rao & Ballard, 1995a), the proposed method has four advantages. The first advantage is effectiveness due to the use of both salience and appearance descriptors. These two descriptors reciprocally aid each other: The salience descriptor guarantees that the task-relevant object can be effectively discriminated from distracters in terms of appearance, while the appearance descriptor can deal with the case that the task-relevant object and distracters have similar task-relevance values but different appearance values. The second advantage is efficiency. The computational complexity of (Rao & Ballard, 1995a) and our method can be approximated as $\mathcal{O}(d_h)$ and $\mathcal{O}(d_f^{few} d_l)$ respectively, where d_h denotes the dimension number of a high-level object representation, e.g., iconic representation (Rao & Ballard, 1995b) used in (Rao & Ballard, 1995a), d_l denotes the dimension number of a pre-attentive feature and d_f^{few} denotes the number of one or a few pre-attentive features used in our method. Since $d_h \gg d_f^{few} d_l$, the computation of our method is much cheaper. The third advantage is adaptability. As shown in (5), the task-relevant feature(s) can be autonomously deduced from the learned LTM representation such that the requirement of redesigning the representation of the task-relevant object for different tasks is eliminated. The fourth advantage is robustness. As shown in (6), the proposed method gives a bias toward the task-relevant object by using Bayes' rule, such that it is robust to work with noise, occlusion and a variety of viewpoints and illuminative effects.

5.3 Combination of bottom-up saliency and top-down biases

Assuming that bottom-up attention and top-down attention are two random events that are independent, the probability of an item being attended can be modeled as the probability of occurrence of either of these two events on that item. Thus, the probabilistic location-based attentional activation, denoted as $p_{attn}(\mathbf{r}_i)$, can be obtained by combining bottom-up saliency and top-down biases:

$$\begin{cases} p_{attn}(\mathbf{r}_i) = p_{bu}(\mathbf{r}_i) + p_{td}(\mathbf{r}_i|\{\mathbf{F}_f^t\}) - p_{bu}(\mathbf{r}_i) \times p_{td}(\mathbf{r}_i|\{\mathbf{F}_f^t\}) & \text{if } w_{bu} = 1 \text{ and } w_{td} = 1 \\ p_{attn}(\mathbf{r}_i) = p_{bu}(\mathbf{r}_i) & \text{if } w_{bu} = 1 \text{ and } w_{td} = 0 \\ p_{attn}(\mathbf{r}_i) = p_{td}(\mathbf{r}_i|\{\mathbf{F}_f^t\}) & \text{if } w_{bu} = 0 \text{ and } w_{td} = 1 \end{cases}, \quad (7)$$

where w_{bu} and w_{td} are two logic variables used as the conscious gating for bottom-up attention and top-down attention respectively and these two variables are set according to the task.

5.4 Proto-object based attentional activation

According to the IC hypothesis, it can be seen that a competitive advantage over an object is produced by directing attention to a spatial location in that object. Thus the probability of a proto-object being attended can be calculated using the *logic or* operator on the location-based probabilities. Furthermore, it can be assumed that two locations being attended are mutually exclusive according to the space-based attention theory (Posner et al., 1980). As a result, the probability of a proto-object $\mathbf{R}_g$ being attended, denoted as $p_{attn}(\mathbf{R}_g)$, can be calculated as:

$$p_{attn}(\mathbf{R}_g) = \frac{1}{N_g} \sum_{\mathbf{r}_i \in \mathbf{R}_g} p_{td}(\mathbf{r}_i|\mathbf{F}_f^t), \quad (8)$$

where $\mathbf{R}_g$ denotes a proto-object, N_g denotes the number of pixels in $\mathbf{R}_g$. The inclusion of $1/N_g$ is to eliminate the influence of the proto-object's size. The FOA is directed to the proto-object with maximal attentional activation.

6. Post-attentive perception

The flow chart of the post-attentive perception can be illustrated in Fig. 5. Four modules, as presented in section 3, are **interactive** during this stage.

6.1 Perceptual completion processing

This module works around the attended proto-object, denoted as $\mathbf{R}^1_{attn}$, to achieve the complete object region. It consists of two steps. The first step is recognition of the attended proto-object. This step explores LTM object representations in order to determine to which LTM object representation the attended proto-object belongs by using the post-attentive features. The extraction of post-attentive features and the recognition algorithm will be presented in section 6.2 and section 6.4 respectively. The matched LTM object representation, denoted as $\mathbf{O}_{attn}$, is then recalled from LTM.

The second step is completion processing:

1. If the local coding of $\mathbf{O}_{attn}$ includes multiple parts, several candidate proto-objects, which are spatially close to $\mathbf{R}^1_{attn}$, are selected from the current scene. They are termed as *neighbors* and denoted as a set $\{\mathbf{R}_n\}$.
2. The local post-attentive features are extracted in each $\mathbf{R}_n$.
3. Each $\mathbf{R}_n$ is recognized using the local post-attentive features and the matched LTM object representation $\mathbf{O}_{attn}$. If it is recognized as a part of $\mathbf{O}_{attn}$, it will be labeled as a part of the attended object. Otherwise, it will be eliminated.
4. Continue *item 2* and *item 3* iteratively until all neighbors have been checked.

These labeled proto-objects constitute the complete region of the attended object, which is denoted as a set $\{\mathbf{R}_{attn}\}$.

6.2 Extraction of post-attentive features

Post-attentive features $\tilde{\mathbf{F}}$ are estimated by using the statistics within the attended object. They consist of *global post-attentive features* $\tilde{\mathbf{F}}_{gb}$ and *local post-attentive features* $\tilde{\mathbf{F}}_{lc}$. Each $\tilde{\mathbf{F}}$ consists of *appearance component* $\tilde{\mathbf{F}}^a$ and *salience component* $\tilde{\mathbf{F}}^s$.

6.2.1 Local post-attentive features

Each proto-object, denoted as $\mathbf{R}^m_{attn}$, in the complete region being attended (i.e., $\mathbf{R}^m_{attn} \in \{\mathbf{R}_{attn}\}$) is the unit for estimating local post-attentive features. They can be estimated as a set that can be expressed as: $\{\tilde{\mathbf{F}}_{lc}\} = \{ (\tilde{\mathbf{F}}^a_{lc}(\mathbf{R}^m_{attn}), \tilde{\mathbf{F}}^s_{lc}(\mathbf{R}^m_{attn}))^T \}_{\forall \mathbf{R}^m_{attn} \in \{\mathbf{R}_{attn}\}}$.

The appearance components in an entry $\tilde{\mathbf{F}}_{lc}$, denoted as $\tilde{\mathbf{F}}^a_{lc} = \{\tilde{\mathbf{F}}^a_f\}$ with $f \in \{int, rg, by, o_\theta\}$, are estimated by using the mean $\tilde{\mu}^{a,m}_f$ of $\mathbf{R}^m_{attn}$ in terms of f, i.e., $\tilde{\mathbf{F}}^a_f(\mathbf{R}^m_{attn}) = \tilde{\mu}^{a,m}_f$.

The salience components, denoted as $\tilde{\mathbf{F}}^s_{lc} = \{\tilde{\mathbf{F}}^s_f\}$ with $f \in \{int, rg, by, o_\theta\}$, can be estimated using the mean of conspicuity $\tilde{\mu}^{s,m}_f$ of a $\mathbf{R}^m_{attn}$ in terms of f, i.e., $\tilde{F}^s_f(\mathbf{R}^m_{attn}) = \tilde{\mu}^{s,m}_f$. The conspicuity quantity F^s_f in terms of f is calculated using (2).

6.2.2 Global post-attentive features

The global post-attentive feature $\tilde{\mathbf{F}}_{gb}$ is estimated after the complete region of the attended object, i.e., $\{\mathbf{R}_{attn}\}$, is obtained. Since the active contour technique (Blake & Isard, 1998; MacCormick, 2000) is used to represent a contour in this paper, the estimation of $\tilde{\mathbf{F}}_{gb}$ includes

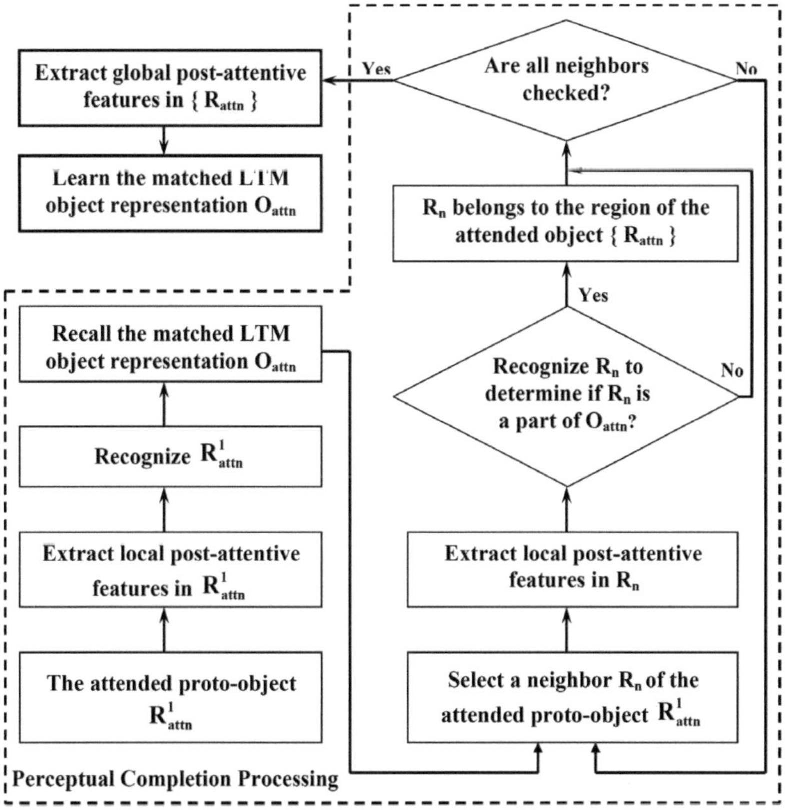

Fig. 5. The flowchart of the post-attentive perception stage.

two steps. The first step is to extract control points, denoted as a set $\{\mathbf{r}_{cp}\}$, of the attended object's contour by using the method in our previous work (Yu et al., 2010). That is, each control point is an entry in the set $\{\tilde{\mathbf{F}}_{gb}\}$. The second step is to estimate the appearance and salience components at these control points, i.e., $\{\tilde{\mathbf{F}}_{gb}\} = \{(\tilde{\mathbf{F}}^a_{gb}(\mathbf{r}_{cp}), \tilde{\mathbf{F}}^s_{gb}(\mathbf{r}_{cp}))^T\}_{\forall \mathbf{r}_{cp}}$. The appearance component of an entry consists of spatial coordinates in the reference frame at a control point, i.e., $\tilde{\mathbf{F}}^a_{gb}(\mathbf{r}_{cp}) = \left(x_{\mathbf{r}_{cp}} \; y_{\mathbf{r}_{cp}} \right)^T$. The salience component of an entry is built by using the conspicuity value $F^s_{ct}(\mathbf{r}_{cp})$ in terms of pre-attentive contour feature at a control point, i.e., $\tilde{\mathbf{F}}^s_{gb}(\mathbf{r}_{cp}) = F^s_{ct}(\mathbf{r}_{cp})$.

6.3 Development of LTM object representations

The LTM object representation also consists of the local coding (denoted as $\mathbf{O}_{lc}$) and global coding (denoted as $\mathbf{O}_{gb}$). Each coding also consists of *appearance descriptors* (denoted as $\mathbf{O}^a$) and *salience descriptors* (denoted as $\mathbf{O}^s$). The PNN (Specht, 1990) is used to build them.

6.3.1 PNN of local coding

The PNN of a local coding $\mathbf{O}_{lc}$ (termed as a *local PNN*) includes three layers. The input layer receives the local post-attentive feature vector $\tilde{\mathbf{F}}_{lc}$. Each radial basis function (RBF) at the

hidden layer represents a part of the learned object and thereby this layer is called a *part layer*. The output layer is a probabilistic mixture of all parts belonging to the object and thereby this layer is called an *object layer*.
The probability distribution of a RBF at the part layer of the local PNN can be expressed as:

$$\begin{aligned} p_j^k(\tilde{\mathbf{F}}_{lc}) &= \mathcal{G}(\tilde{\mathbf{F}}_{lc}; \boldsymbol{\mu}_j^k, \boldsymbol{\Sigma}_j^k) \\ &= \frac{1}{(2\pi)^{\frac{d}{2}} |\boldsymbol{\Sigma}_j^k|^{\frac{1}{2}}} \exp\{-\frac{1}{2}(\tilde{\mathbf{F}}_{lc} - \boldsymbol{\mu}_j^k)^T (\boldsymbol{\Sigma}_j^k)^{-1} (\tilde{\mathbf{F}}_{lc} - \boldsymbol{\mu}_j^k)\} \end{aligned} \tag{9}$$

where $\mathcal{G}$ denotes the Gaussian distribution, $\boldsymbol{\mu}_j^k$ and $\boldsymbol{\Sigma}_j^k$ denote the mean vector and covariance matrix of a RBF, j is the index of a part, k is the index of an object in LTM, and d is the dimension number of a local post-attentive feature $\tilde{\mathbf{F}}_{lc}$. Since all feature dimensions are assumed to be independent, $\boldsymbol{\Sigma}_j^k$ is a diagonal matrix and standard deviation (STD) values of all feature dimensions of a RBF can constitute an STD vector $\boldsymbol{\sigma}_j^k$.
The probabilistic mixture estimation $r^k(\tilde{\mathbf{F}}_{lc})$ at the object layer can be expressed as:

$$r^k(\tilde{\mathbf{F}}_{lc}) = \sum_i \pi_j^k p_j^k(\tilde{\mathbf{F}}_{lc}), \tag{10}$$

where π_j^k denotes the contribution of part j to object k, which holds $\sum_j \pi_j^k = 1$.

6.3.2 PNN of global coding

The PNN for a global coding $\mathbf{O}_{gb}$ (termed as a *global PNN*) also includes three layers. The input layer receives the global post-attentive feature vector $\tilde{\mathbf{F}}_{gb}$. Each node of the hidden layer is a control point along the contour and thereby this layer is called a *control point layer*. The output layer is a probabilistic combination of all control points belonging to the object and thereby this layer is called an *object layer*. The mathematical expression of the global PNN is similar to the local PNN.

6.3.3 Learning of LTM object representations

Since the number of nodes (i.e., the numbers of parts and control points) is unknown and might be dynamically changed during the training course, this paper proposes a dynamical learning algorithm by using both the maximum likelihood estimation (MLE) and a Bayes' classifier to update the local and global PNNs at each time. This proposed dynamical learning algorithm can be summarized as follows. The Bayes' classifier is used to classify the training pattern to an existing LTM pattern. If the training pattern can be classified to an existing LTM pattern at the part level in a local PNN or at the control point level in a global PNN, both appearance and salience descriptors of this existing LTM pattern are updated using MLE. Otherwise, a new LTM pattern is created. Two thresholds τ_1 and τ_2 are introduced to determine the minimum correct classification probability to an existing part and an existing control point respectively. Algorithm 1 shows the learning routine of global and local codings. In the algorithm, a_j^k denotes the occurrence number of an existing pattern indexed by j of object k and it is initialized by 0, N_k denotes the number of parts in the local PNN or control points in the global PNN of object k, $.^2$ denotes the element-by-element square operator, and σ_{init} is a predefined STD value when a new pattern is created.

Algorithm 1 Learning Routine of Local and Global Codings

Given a local or global training pattern $(\tilde{\mathbf{F}}_{lc}, k)$ or $(\tilde{\mathbf{F}}_{gb}, k)$:
Set $\tilde{\mathbf{F}} = \tilde{\mathbf{F}}_{lc}$ or $\tilde{\mathbf{F}} = \tilde{\mathbf{F}}_{gb}$;
Recognize $\tilde{\mathbf{F}}$ to obtain a recognition probability $p_i^k(\tilde{\mathbf{F}})$;
if $p_j^k(\tilde{\mathbf{F}}) \geq \tau_1$ or $\geq \tau_2$ **then**
    // Update part j of object k
    $\sigma_{temp} = \left[a_j^k(\sigma_j^k).^2 + a_j^k(\mu_j^k).^2 + (\tilde{\mathbf{F}}).^2\right] / (a_j^k + 1);$ *// Prepare for updating the STD*
    $\boldsymbol{\mu}_j^k = (a_j^k \boldsymbol{\mu}_j^k + \tilde{\mathbf{F}}) / (a_j^k + 1);$ *// Update the mean vector*
    $\sigma_j^k = [\sigma_{temp}^d - (\mu_j^k).^2].^{-\frac{1}{2}};$ *// Update the STD*
    $a_j^k = a_j^k + 1;$ *// Increment the occurrence number*
else
    // Create a new part i of object k
    Set $N_k = N_k + 1; \quad i = N_k;$
    $\boldsymbol{\mu}_j^k = \tilde{\mathbf{F}}; \quad \boldsymbol{\sigma}_j^k = \boldsymbol{\sigma}_{init}; \quad a_j^k = 1;$ *// Set the initial mean, STD and occurrence number*
end if
$\forall j$: $\pi_j^k = a_j^k / \sum_{j'} a_{j'}^k.$ *// Normalize weights π*

6.4 Object recognition

Due to the page limitation, the object recognition module can be summarized as follows. It can be modeled at two levels. The first one is the object level. The purpose of this level is to recognize to which LTM object an attended pattern belongs. The second one is the part level or control point level. Recognition at this level is performed given an LTM object to which the attended pattern belongs. Thus, the purpose of this level is to recognize to which part in a local PNN or to which control point in a global PNN an attended pattern belongs. At each level, object recognition can generally be modeled as a decision unit by using Bayes' theorem. Assuming that the prior probability is equal for all LTM patterns at each level, the observation likelihood can be seen as the posterior probability.

7. Experiments

This proposed autonomous visual perception system is tested in the task of object detection. The unconscious perception path (i.e., the bottom-up attention module) can be used to detect a salient object, such as a landmark, whereas the conscious perception path (i.e., the top-down attention module) can be used to detect the task-relevant object, i.e., the expected target. Thus the unconscious and conscious aspects are tested in two robotics tasks respectively: One is detecting a salient object and the other is detecting a task-relevant object.

7.1 Detecting a salient object

The salient object is an unusual or unexpected object and the current task has no prediction about its occurrence. There are three objectives in this task. The first objective is to illustrate the unconscious capability of the proposed perception system. The second objective is to show the advantages of using object-based visual attention for perception by comparing it with the space-based visual attention methods. The third objective is to show the advantage of integrating the contour feature into the bottom-up competition module. The result is that an object that has a conspicuous shape compared with its neighbors can be detected. Two experiments are shown in this section, including the detection of an object that is conspicuous in colors and in contour respectively.

7.1.1 Experimental setup

Artificial images are used in the experiments. The frame size of all images is 640×480 pixels. In order to show the robustness of the proposed perception system, these images are obtained using different settings, including noise, spatial transformation and changes of lighting. The noisy images are manually obtained by adding salt and pepper noise patches (noise density: 0.1 $\sim$ 0.15, patch size: 10×10 pixels $\sim$ 15×15 pixels) into original r, g and b color channels respectively. The experimental results are compared with the results of Itti's model (i.e., space-based bottom-up attention) (Itti et al., 1998) and Sun's model (i.e., object-based bottom-up attention) (Sun & Fisher, 2003).

7.1.2 An object conspicuous in colors

The first experiment is detecting an object that is conspicuous to its neighbors in terms of colors and all other features are approximately the same between the object and its neighbors. The experimental results are shown in Fig. 6. The salient object is the red ball in this experiment. Results of the proposed perception system are shown in Fig. 6(d), which indicate that this proposed perception system can detect the object that is conspicuous to its neighbors in terms of colors in different settings. Results of Itti's model and Sun's model are shown in Fig. 6(e) and Fig. 6(f) respectively. It can be seen that Itti's model fails to detect the salient object when noise is added to the image, as shown in column 2 in Fig. 6(e). This indicates that the proposed object-based visual perception system is more robust to noise than the space-based visual perception methods.

7.1.3 An object conspicuous in contour

The second experiment is detecting an object that is conspicuous to its neighbors in terms of contour and all other features are approximately the same between the object and its neighbors. The experimental results are shown in Fig. 7. In this experiment, the salient object is the triangle. Detection results of the proposed perception system are shown in Fig. 7(d), which indicate that the proposed perception system can detect the object that is conspicuous to its neighbors in terms of contour in different settings. Detection results of Itti's model and Sun's model are shown in Fig. 7(e) and Fig. 7(f) respectively. It can be seen that both Itti's model and Sun's model fail to detect the salient object when noise is added to the image, as shown in column 2 in Fig. 7(e) and Fig. 7(f) respectively. This experiment indicates that the proposed object-based visual perception system is capable of detecting the object conspicuous in terms of contour in different settings due to the inclusion of contour conspicuity in the proposed bottom-up attention module.

7.2 Detecting a task-relevant object

It is an important ability for robots to accurately detect a task-relevant object (i.e., target) in the cluttered environment. According to the proposed perception system, the detection procedure consists of two phases: a learning phase and a detection phase. The objective of the learning phase is to develop the LTM representation of the target. The objective of the detection phase is to detect the target by using the learned LTM representation of the target. The detection phase can be implemented as a two-stage process. The first stage is attentional selection: The

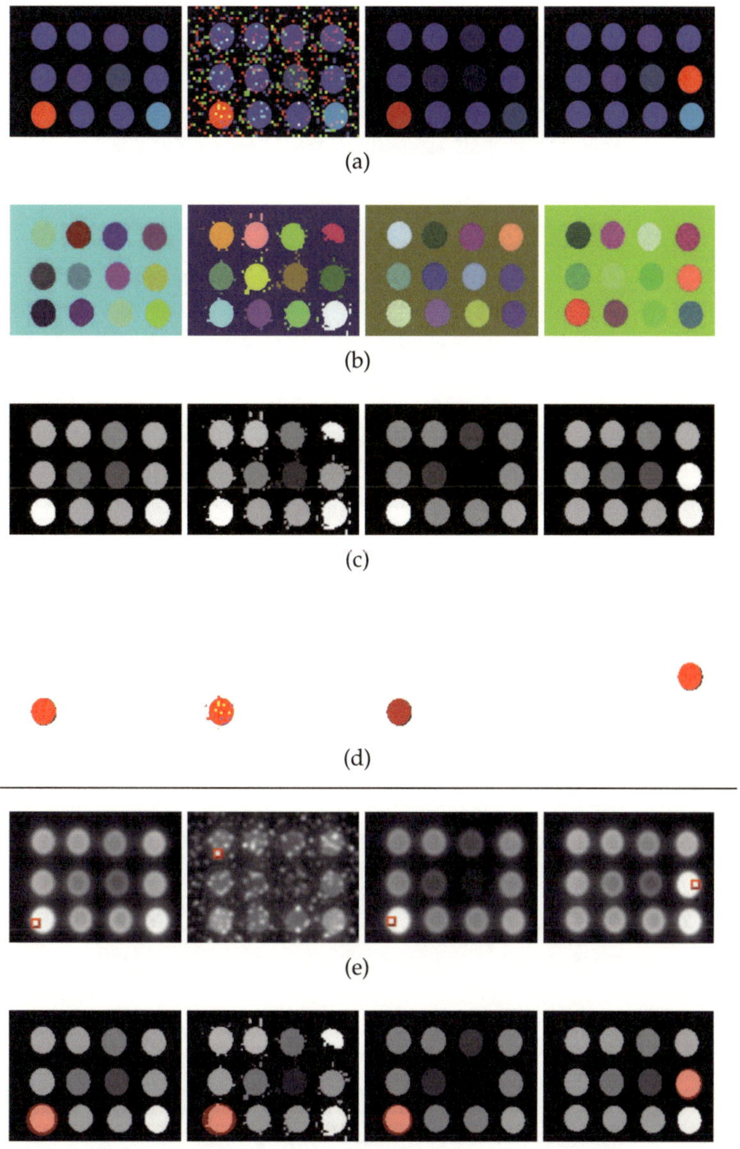

Fig. 6. Detection of a salient object, which is conspicuous to its neighbors in terms of colors. Each column represents a type of experimental setting. Column 1 is a typical setting. Column 2 is a noise setting of column 1. Column 3 is a different lighting setting with respect to column 1. Column 4 is a spatial transformation setting with respect to column 1. Row (a): Original input images. Row (b): Pre-attentive segmentation. Each color represents one proto-object. Row (c): Proto-object based attentional activation map. Row (d): The complete region being attended. Row (e): Detection results using Itti's model. The red rectangles highlight the attended location. Row (f): Detection results using Sun's model. The red circles highlight the attended object.

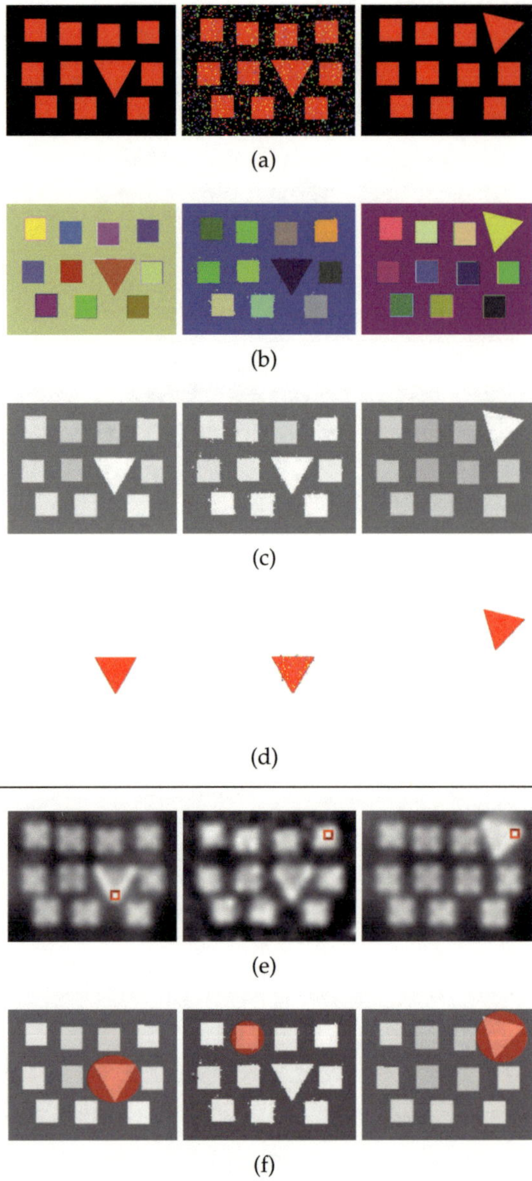

Fig. 7. Detection of a salient object, which is conspicuous to its neighbors in terms of contour. Each column represents a type of experimental setting. Column 1 is a typical setting. Column 2 is a noise setting of column 1. Column 3 is a spatial transformation setting with respect to column 1. Row (a): Original input images. Row (b): Pre-attentive segmentation. Each color represents one proto-object. Row (c): Proto-object based attentional activation map. Row (d): The complete region being attended. Row (e): Detection results using Itti's model. The red rectangles highlight the attended location. Row (f): Detection results using Sun's model. The red circles highlight the attended object.

task-relevant feature(s) of the target is used to guide attentional selection through top-down biasing to obtain an attended object. The second stage is post-attentive recognition: The attended object is recognized using the target's LTM representation to check if it is the target. If not, another procedure of attentional selection is performed by using more task-relevant features.

7.2.1 Experimental setup

Two objects are used to test the proposed method of detecting a task-relevant object: a book and a human. Images and videos are obtained under different settings, including noise, transformation, lighting changes and occlusion. For training for the book, 20 images are used. For testing for the book, 50 images are used. The size of each image is 640×480 pixels. For detecting the human, three videos are obtained by a moving robot. Two different office environments have been used. Video 1 and video 2 are obtained in office scene 1 with low and high lighting conditions respectively. Video 3 is obtained in office scene 2. All three videos contain a total of 650 image frames, in which 20 image frames are selected from video 1 and video 2 for training and the rest of the 630 image frames are used for testing. The size of each frame in these videos is 1024×768 pixels. It is important to note that each test image includes not only a target but also various distracters. The noisy images are manually obtained by adding salt and pepper noise patches (noise density: 0.1, patch size: 5×5 pixels) into original r, g and b color channels respectively.
The results of the proposed method are compared with the results of Itti's model (Itti et al., 1998) (i.e., a space-based bottom-up attention model) and Navalpakkam's model (Navalpakkam & Itti, 2005) (i.e., a space-based top-down attention model) respectively.

7.2.2 Task 1

The first task is to detect the book that has multiple parts. The learned LTM representation of the book is shown in Table 1, which has shown that the book has two parts and the blue-yellow feature in the first part can be deduced as the task-relevant feature dimension since the value $\mu^s/(1+\sigma^s)$ of this feature is maximal. Detection results of the proposed perception system are shown in Fig. 8(d). It can be seen that the book is successfully detected. Results of Itti's model and Navalpakkam's model, as shown in Fig. 8(e) and Fig. 8(f) respectively, show that these models fail to detect the target in some cases.

7.2.3 Task 2

The second task is to detect a human. Table 2 has shown that the human has two parts (including face and body) and the contour feature can be deduced as the task-relevant feature dimension since the value $\mu^s/(1+\sigma^s)$ of this feature is maximal. Detection results of the proposed perception system are shown in Fig. 9(d). It can be seen that the human is successfully detected. Results of Itti's model and Navalpakkam's model, as shown in Fig. 9(e) and Fig. 9(f) respectively, show that these models fail to detect the target in most cases.

7.2.4 Performance evaluation

Performance of detecting task-relevant objects is evaluated using true positive rate (TPR) and false positive rate (FPR), which are calculated as:

$$TPR = TP/nP, \tag{11}$$

f	j	μ^a	σ^a	μ^s	σ^s	$\mu^s/(1+\sigma^s)$
ct	1	-		75.0	19.7	3.6
int	1	106.6	5.8	27.9	14.5	1.8
rg	1	22.1	8.7	199.6	18.2	10.4
by	1	-108.0	9.1	215.6	8.7	**22.2**
o_{0°	1	N/A	N/A	41.8	9.8	3.9
o_{45°	1	N/A	N/A	41.4	12.8	3.0
o_{90°	1	N/A	N/A	34.7	16.3	2.0
o_{135°	1	N/A	N/A	46.5	15.7	2.8
int	2	60.5	8.2	80.0	5.7	11.9
rg	2	0.4	4.3	18.3	6.4	2.5
by	2	120.8	6.7	194.7	8.1	21.4
o_{0°	2	N/A	N/A	48.5	11.1	4.0
o_{45°	2	N/A	N/A	53.8	9.9	4.9
o_{90°	2	N/A	N/A	38.4	14.6	2.5
o_{135°	2	N/A	N/A	59.4	20.3	2.8

Table 1. Learned LTM object representation of the book. f denotes a pre-attentive feature dimension. j denotes the index of a part. The definitions of μ^a, σ^a, μ^s and σ^s can be seen in section 5.2.2.

f	j	μ^a	σ^a	μ^s	σ^s	$\mu^s/(1+\sigma^s)$
ct	1	-		68.3	6.9	**8.6**
int	1	28.4	21.7	18.8	13.9	1.3
rg	1	-7.0	7.1	28.6	10.8	2.4
by	1	10.9	5.4	48.4	10.9	4.1
o_{0°	1	N/A	N/A	33.4	6.7	4.3
o_{45°	1	N/A	N/A	39.8	11.4	3.2
o_{90°	1	N/A	N/A	37.4	6.1	5.3
o_{135°	1	N/A	N/A	37.5	13.5	2.6
int	2	52.0	12.5	25.6	15.6	1.5
rg	2	-2.3	17.4	49.5	18.8	2.5
by	2	-29.3	6.9	60.4	22.3	2.6
o_{0°	2	N/A	N/A	12.1	6.6	1.6
o_{45°	2	N/A	N/A	16.5	8.3	1.8
o_{90°	2	N/A	N/A	15.0	7.9	1.7
o_{135°	2	N/A	N/A	17.2	8.1	1.9

Table 2. Learned LTM object representation of the human. f denotes a pre-attentive feature dimension. j denotes the index of a part. The definitions of μ^a, σ^a, μ^s and σ^s can be seen in section 5.2.2.

$$FPR = FP/nN, \tag{12}$$

where nP and nN are numbers of positive and negative objects respectively in the testing image set, TP and FP are numbers of true positives and false positives. The positive object is the target to be detected and the negative objects are distracters in the scene.

Detection performance of the proposed perception system and other visual attention based methods is shown in Table 3. Note that "Naval's" represents Navalpakkam's method.

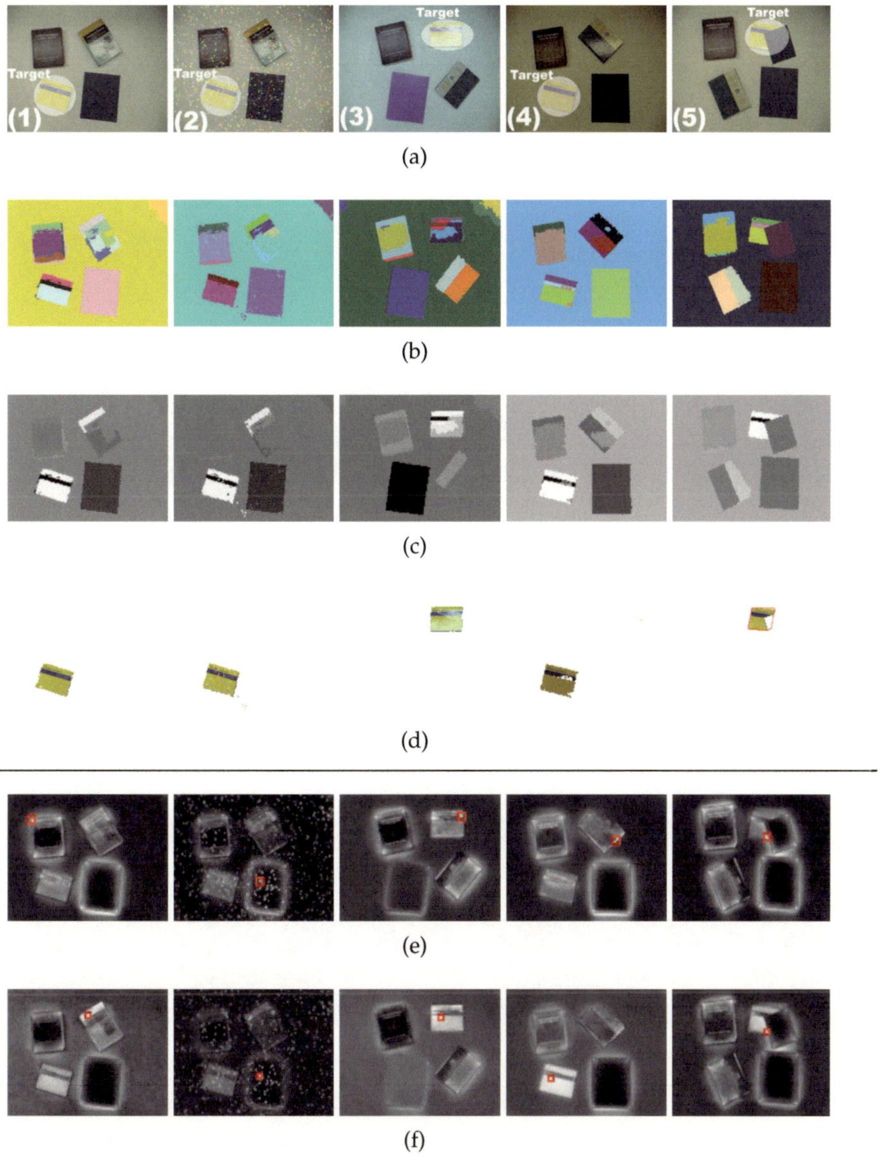

Fig. 8. Detection of the book. Each column represents a type of experimental setting. Column 1 is a typical setting. Column 2 is a noise setting of column 1. Column 3 is a spatial transformation (including translation and rotation) setting with respect to column 1. Column 4 is a different lighting setting with respect to column 1. Column 5 is an occlusion setting. Row (a): Original input images. Row (b): Pre-attentive segmentation. Each color represents one proto-object. Row (c): Proto-object based attentional activation map. Brightness represents the attentional activation value. Row (d): The complete region of the target. The red contour in the occlusion case represents the illusory contour (Lee & Nguyen, 2001), which shows the post-attentive perceptual completion effect. Row (e): Detection results using Itti's model. The red rectangle highlights the most salient location. Row (f): Detection results using Navalpakkam's model. The red rectangle highlights the most salient location.

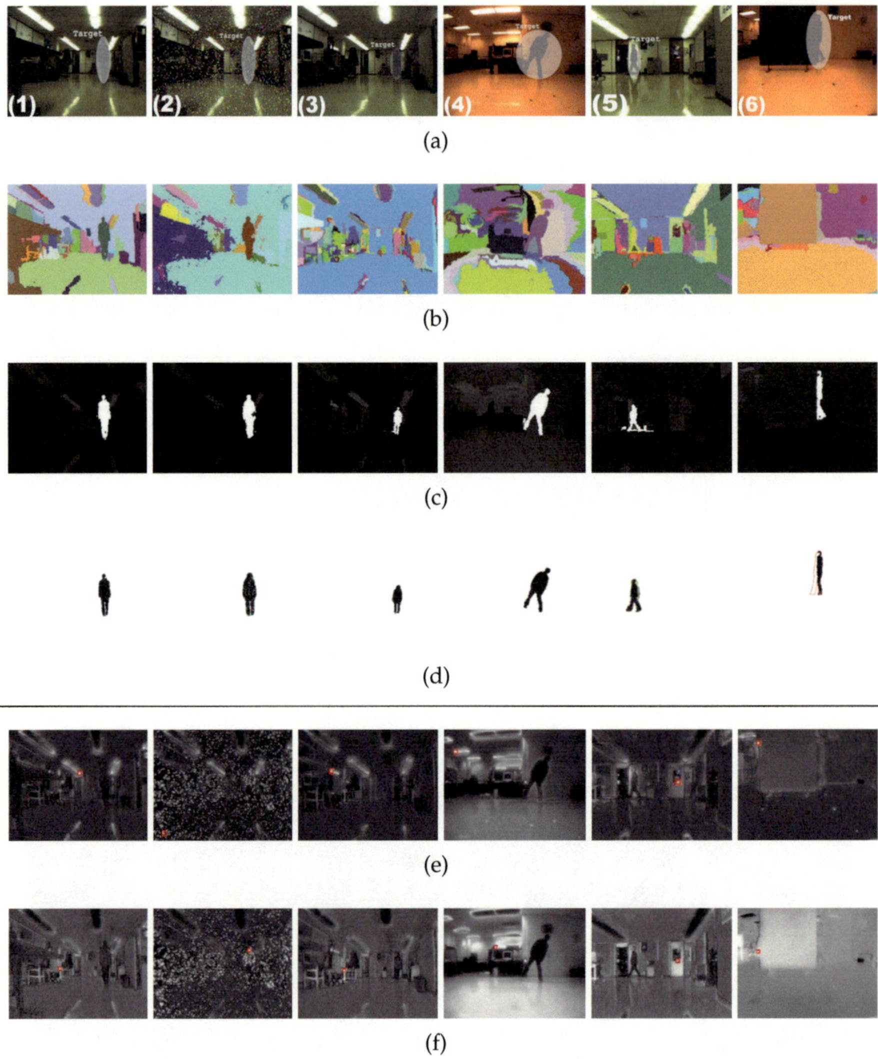

Fig. 9. Detection of the human in the cluttered environment. Each column represents a type of experimental setting. Column 1 is a typical setting (from video 1). Column 2 is a noise setting of column 1. Column 3 is a scaling setting with respect to column 1 (from video 1). Column 4 is a rotation setting with respect to column 1 (from video 3). Column 5 is a different lighting setting with respect to column 1 (from video 2). Column 6 is an occlusion setting (from video 3). Row (a): Original input images. Row (b): Pre-attentive segmentation. Each color represents one proto-object. Row (c): Proto-object based attentional activation map. Brightness represents the attentional activation value. Row (d): The complete region of the target. The red contour in the occlusion case represents the illusory contour (Lee & Nguyen, 2001), which shows the post-attentive perceptual completion effect. Row (e): Detection results using Itti's model. The red rectangle highlights the most salient location. Row (f): Detection results using Navalpakkam's model. The red rectangle highlights the most salient location.

Task	Method	*TP*	*FP*	*nP*	*nN*	*TPR* (%)	*FPR* (%)
1	Proposed	47	3	50	244	94.00	1.23
	Itti's	16	34	50	244	32.00	13.93
	Naval's	41	9	50	244	82.00	3.69
2	Proposed	581	49	630	30949	92.22	0.16
	Itti's	5	625	630	30949	0.79	2.02
	Naval's	36	594	630	30949	5.71	1.92

Table 3. Performance of detecting task-relevant objects.

8. Conclusion

This paper has presented an autonomous visual perception system for robots using the object-based visual attention mechanism. This perception system provides the following four contributions. The first contribution is that the attentional selection stage supplies robots with the cognitive capability of knowing how to perceive the environment according to the current task and situation, such that this perception system is adaptive and general to any task and environment. The second contribution is the top-down attention method using the IC hypothesis. Since the task-relevant feature(s) are conspicuous, low-level and statistical, this top-down biasing method is more effective, efficient and robust than other methods. The third contribution is the PNN based LTM object representation. This LTM object representation can probabilistically embody various instances of that object, such that it is robust and discriminative for top-down attention and object recognition. The fourth contribution is the pre-attentive segmentation algorithm. This algorithm extends the irregular pyramid techniques by integrating a scale-invariant probabilistic similarity measure, a similarity-driven decimation method and a similarity-driven neighbor search method. It provides rapid and satisfactory results of pre-attentive segmentation for object-based visual attention. Based on these contributions, this perception system has been successfully tested in the robotic task of object detection under different experimental settings.

The future work includes the integration of the bottom-up attention in the temporal context and experiments of the combination of bottom-up and top-down attention.

9. References

Aziz, M. Z., Mertsching, B., Shafik, M. S. E.-N. & Stemmer, R. (2006). Evaluation of visual attention models for robots, *Proceedings of the 4th IEEE Conference on Computer Vision Systems*, p. 20.

Backer, G., Mertsching, B. & Bollmann, M. (2001). Data- and model-driven gaze control for an active-vision system, *IEEE Transactions on Pattern Analysis and Machine Intelligence* 23(12): 1415–1429.

Baluja, S. & Pomerleau, D. (1997). Dynamic relevance: Vision-based focus of attention using artificial neural networks, *Artificial Intelligence* 97: 381–395.

Belardinelli, A. & Pirri, F. (2006). A biologically plausible robot attention model, based on space and time, *Cognitive Processing* 7(Supplement 5): 11–14.

Belardinelli, A., Pirri, F. & Carbone, A. (2006). Robot task-driven attention, *Proceedings of the International Symposium on Practical Cognitive Agents and Robots*, pp. 117–128.

Bhattacharyya, A. (1943). On a measure of divergence between two statistical populations defined by their probability distributions, *Bulletin of the Calcutta Mathematical Society* 35: 99–109.

Blake, A. & Isard, M. (1998). *Active Contours: The Application of Techniques from Graphics, Vision, Control Theory and Statistics to Visual Tracking of Shapes in Motion*, Springer-Verlag New York, Secaucus, NJ.

Breazeal, C., Edsinger, A., Fitzpatrick, P. & Scassellati, B. (2001). Active vision for sociable robots, *IEEE Transactions on Systems, Man, and Cybernetics - Part A: Systems and Humans* 31(5): 443–453.

Burt, P. J. & Adelson, E. H. (1983). The laplacian pyramid as a compact image code, *IEEE Transactions on Communications* 31(4): 532–540.

Carbone, A., Finzi, A. & Orlandini, A. (2008). Model-based control architecture for attentive robots in rescue scenarios, *Autonomous Robots* 24(1): 87–120.

Carpenter, G. A. & Grossberg, S. (2003). Adaptive resonance theory, *in* M. A. Arbib (ed.), *The Handbook of Brain Theory and Neural Networks, Second Edition*, MIT Press, Cambridge, MA, pp. 87–90.

Chikkerur, S., Serre, T., Tan, C. & Poggio, T. (2010). What and where: A bayesian inference theory of attention, *Visual Research* 50(22): 2233–2247.

Desimone, R. & Duncan, J. (1995). Neural mechanisms of selective visual attention, *Annual Reviews of Neuroscience* 18: 193–222.

Duncan, J. (1984). Selective attention and the organization of visual information, *Journal of Experimental Psychology: General* 113(4): 501–517.

Duncan, J. (1998). Converging levels of analysis in the cognitive neuroscience of visual attention, *Philosophical Transactions of The Royal Society Lond B: Biological Sciences* 353(1373): 1307–1317.

Duncan, J., Humphreys, G. & Ward, R. (1997). Competitive brain activity in visual attention, *Current Opinion in Neurobiology* 7(2): 255–261.

Frintrop, S. (2005). *VOCUS: A visual attention system for object detection and goal-directed search*, PhD thesis, University of Bonn, Germany.

Frintrop, S. & Jensfelt, P. (2008). Attentional landmarks and active gaze control for visual slam, *IEEE Transactions on Robotics* 24(5): 1054–1065.

Frintrop, S. & Kessel, M. (2009). Most salient region tracking, *Proceedings of the IEEE International Conference on Robotics and Automation (ICRA)*, pp. 1869–1874.

Greenspan, A. G., Belongie, S., Goodman, R., Perona, P., Rakshit, S. & Anderson, C. H. (1994). Overcomplete steerable pyramid filters and rotation invariance, *Proceedings of IEEE Computer Vision and Pattern Recognition*, pp. 222–228.

Grossberg, S. (2005). Linking attention to learning, expectation, competition, and consciousness, *in* L. Itti, G. Rees & J. Tsotsos (eds), *Neurobiology of Attention*, Elsevier, San Diego, CA, pp. 652–662.

Grossberg, S. (2007). Consciousness clears the mind, *Neural Networks* 20: 1040–1053.

Hoya, T. (2004). Notions of intuition and attention modeled by a hierarchically arranged generalized regression neural network, *IEEE Transactions on Systems, Man, and Cybernetics - Part B: Cybernetics* 34(1): 200–209.

Itti, L. & Baldi, P. (2009). Bayesian surprise attracts human attention, *Visual Research* 49(10): 1295–1306.

Itti, L., Koch, C. & Niebur, E. (1998). A model of saliency-based visual attention for rapid scene analysis, *IEEE Transactions on Pattern Analysis and Machine Intelligence* 20(11): 1254–1259.

Jolion, J. M. (2003). Stochastic pyramid revisited, *Pattern Recognition Letters* 24(8): 1035–1042.

Lee, T. S. & Nguyen, M. (2001). Dynamics of subjective contour formation in the early visual cortex, *Proceedings of the Natual Academy of Sciences of the United States of America* 98(4): 1907–1911.

MacCormick, J. (2000). *Probabilistic modelling and stochastic algorithms for visual localisation and tracking*, PhD thesis, Department of Engineering Science, University of Oxford.

Maier, W. & Steinbach, E. (2010). A probabilistic appearance representation and its application to surprise detection in cognitive robots, *IEEE Transactions on Autonomous Mental Development* 2(4): 267–281.

Meer, P. (1989). Stochastic image pyramids, *Computer Vision, Graphics, and Image Processing* 45(3): 269–294.

Montanvert, A., Meer, P. & Rosenfeld, A. (1991). Hierarchical image analysis using irregular tessellations, *IEEE Transactions on Pattern Analysis and Machine Intelligence* 13(4): 307–316.

Navalpakkam, V. & Itti, L. (2005). Modeling the influence of task on attention, *Vision Research* 45(2): 205–231.

Orabona, F., Metta, G. & Sandini, G. (2005). Object-based visual attention: a model for a behaving robot, *Proceedings of the IEEE Computer Society Conference on Computer Vision and Patter Recognition (CVPR)*, p. 89.

Posner, M. I., Snyder, C. R. R. & Davidson, B. J. (1980). Attention and the detection of signals, *Journal of Experimental Psychology: General* 14(2): 160–174.

Rao, R. P. N. & Ballard, D. H. (1995a). An active vision architecture based on iconic representations, *Artificial Intelligence* 78: 461–505.

Rao, R. P. N. & Ballard, D. H. (1995b). Object indexing using an iconic sparse distributed memory, *Proc. the 5th Intl. Conf. Computer Vision*, pp. 24–31.

Rao, R. P. N., Zelinsky, G. J., Hayhoe, M. M. & Ballard, D. H. (2002). Eye movements in iconic visual search, *Vision Research* 42: 1447–1463.

Scholl, B. J. (2001). Objects and attention: the state of the art, *Cognition* 80(1-2): 1–46.

Sharon, E., Brandt, A. & Basri, R. (2000). Fast multiscale image segmentation, *Proceedings of IEEE Conference on Computer Vision and Pattern Recognition (CVPR)*, pp. 70–77.

Sharon, E., Galun, M., Sharon, D., Basri, R. & Brandt, A. (2006). Hierarchy and adaptivity in segmenting visual scenes, *Nature* 442: 810–813.

Shi, J. & Malik, J. (2000). Normalized cuts and image segmentation, *IEEE Transactions on Pattern Analysis and Machine Intelligence* 22(8): 888–905.

Specht, D. F. (1990). Probabilistic neural networks, *Neural Networks* 3(1): 109–118.

Sun, Y. (2008). A computer vision model for visual-object-based attention and eye movements, *Computer Vision and Image Understanding* 112(2): 126–142.

Sun, Y. & Fisher, R. (2003). Object-based visual attention for computer vision, *Artificial Intelligence* 146(1): 77–123.

Tipper, S. P., Howard, L. A. & Houghton, G. (1998). Action-based mechanisms of attention, *Philosophical Transactions: Biological Sciences* 353(1373): 1385–1393.

Treisman, A. M. & Gelade, G. (1980). A feature integration theory of attention, *Cognition Psychology* 12(1–2): 507–545.

Tsotsos, J. K., Culhane, S. M., Wai, W. Y. K., Lai, Y., Davis, N. & Nuflo, F. (1995). Modelling visual attention via selective tuning, *Artificial Intelligence* 78: 282–299.

Walther, D., Rutishauser, U., Koch, C. & Perona, P. (2004). On the usefulness of attention for object recognition, *Workshop on Attention and Performance in Computational Vision at ECCV*, pp. 96–103.

Weng, J., McClelland, J., Pentland, A., Sporns, O., Stockman, I. & Thelen, E. (2001). Autonomous mental development by robots and animals, *Science* 291(5504): 599–600.

Wolfe, J. M. (1994). Guided search 2.0: A revised model of visual search, *Psychonomic Bulletin and Review* 1(2): 202–238.

Yu, Y., Mann, G. K. I. & Gosine, R. G. (2010). An object-based visual attention model for robotic applications, *IEEE Transactions on Systems, Man and Cybernetics, Part B: Cybernetics* 40(5): 1398–1412.

3

Three-Dimensional Environment Modeling Based on Structure from Motion with Point and Line Features by Using Omnidirectional Camera

Ryosuke Kawanishi, Atsushi Yamashita and Toru Kaneko
Shizuoka University
Japan

1. Introduction

Three-dimensional map is available for autonomous robot navigation (path planning, self-localization and object recognition). In unknown environment, robots should measure environments and construct their maps by themselves.
Three-dimensional measurement using image data makes it possible to construct an environment map (Davison, 2003). However, many environmental images are needed if we use a conventional camera having a limited field of view (Ishiguro et al., 1992). Then, an omnidirectional camera is available for wide-ranging measurement, because it has a panoramic field of view (Fig. 1). Many researchers showed that an omnidirectional camera is effective in measurement and recognition in environment (Bunschoten & Krose, 2003; Geyer & Daniilidis, 2003; Gluckman & Nayar, 1998).

Fig. 1. Omnidirectional camera equipped with a hyperboloid mirror. The left figure shows an acquired image.

Our proposed method is based on structure from motion. Previous methods based on structure from motion often use feature points to estimate camera movement and measure environment (Rachmielowski et al., 2006; Kawanishi et al., 2009). However, many non-textured objects may exist in surrounding environments of mobile robots. It is hard to extract enough number of feature points from non-textured objects. Therefore, in an environment having non-textured objects, it is difficult to construct its map by using feature points only.

Then, line features should be utilized for environment measurement, because non-texture objects often have straight-lines. As examples of previous works using lines, a method for precise camera movement estimation by using stereo camera (Chandraker et al., 2009), a method for buildings reconstruction by using orthogonal lines (Schindler, 2006) and so on (Bartoli & Sturm, 2005; Smith et al., 2006; Mariottini & Prattichizzo, 2007) have been proposed.
However, there is a prerequisite on previous line detections of them. A method must obtain a vanishing point (Schindler, 2006) or a pair of end points of the straight-line (Smith et al., 2006). Some of previous line detection is only for a normal camera (Chandraker et al., 2009). Alternatively, some previous methods obtain line correspondences by hand (Bartoli & Sturm, 2005; Mariottini & Prattichizzo, 2007).
We propose a method for straight-line extraction and tracking on distorted omnidirectional images. The method does not require a vanishing point and end points of straight-lines. These straight-lines are regarded as infinite lines in the measurement process (Spacek, 1986). Therefore, the proposed method can measure straight-lines even if a part of the line is covered by obstacles during its tracking.
Our proposed method measures feature points together with straight-lines. If only straight-lines are used for camera movement estimation, a non-linear problem must be solved. However, camera movement can be estimated easily by a linear solution with point correspondences. Moreover, although few numbers of straight-lines may be extracted from textured objects, many feature points will be extracted from them. Therefore, we can measure the environment densely by using both feature points and straight-lines.
The process of our proposed method is mentioned below (Fig. 2). First, feature points and straight-lines are extracted and tracked along an acquired omnidirectional image sequence. Camera movement is estimated by point-based Structure from Motion. The estimated camera movement is used for an initial value for line-based measurement.

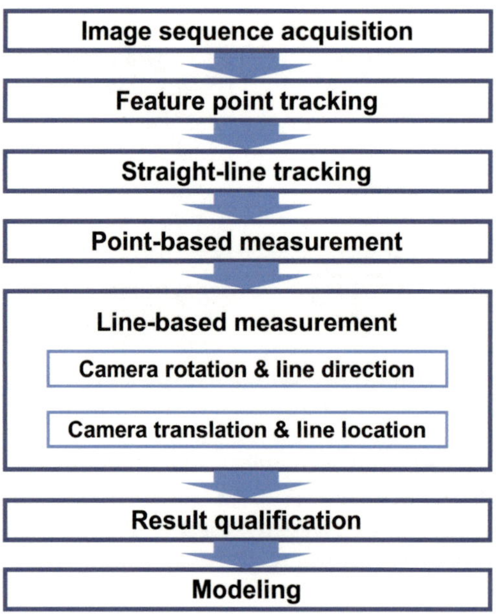

Fig. 2. Procedure of our proposed method.

The proposed line-based measurement is divided into two phases. At the first phase, camera rotation and line directions are optimized. Line correspondence makes it possible to estimate camera rotation independently of camera translation (Spacek, 1986). Camera rotation can be estimated by using 3-D line directions. At the second phase, camera translation and 3-D line location are optimized. The optimization is based on Bundle adjustment (Triggs et al., 1999). Some of measurement results have low accuracy. The proposed method rejects such results.
Measurement results of feature points and straight-lines are integrated. Triangular meshes are generated from the integrated measurement data. By texture-mapping to these meshes, a three-dimensional environment model is constructed.

2. Coordinate system of omnidirectional camera

The coordinate system of the omnidirectional camera is shown in Fig. 3. A ray heading to image coordinates (u,v) from the camera lens is reflected on a hyperboloid mirror. In this paper, the reflected vector is called a ray vector. The extension lines of all ray vectors intersect at the focal point of the hyperboloid mirror. The ray vector $\mathbf{r}$ is calculated by the following equations.

$$\mathbf{r} = \begin{bmatrix} \lambda(u - c_x)p_x \\ \lambda(v - c_y)p_y \\ \lambda f - 2\gamma \end{bmatrix} \tag{1}$$

$$\lambda = \frac{\alpha^2\left(f\gamma + \beta\sqrt{u^2 + v^2 + f^2}\right)}{\alpha^2 f^2 - \beta^2\left(u^2 + v^2\right)} \tag{2}$$

where c_x and c_y are the center coordinates of the omnidirectional image, p_x and p_y are pixel size, f is the focal length, α, β and γ are hyperboloid parameters. These parameters are calibrated in advance.

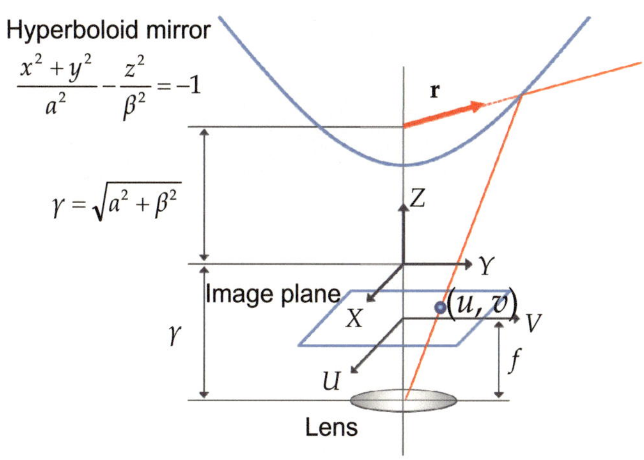

Fig. 3. The coordinate system of the omnidirectional camera. Ray vector $\mathbf{r}$ is defined as a unit vector which starts from the focal point of a hyperboloid mirror.

3. Feature tracking

3.1 Point tracking

Feature points are tracked along an omnidirectional image sequence by KLT tracker (Shi & Tomasi, 1994). These points are used for initial estimation of camera movement and measurement for textured objects. An example of feature point tracking is shown in Fig. 4.

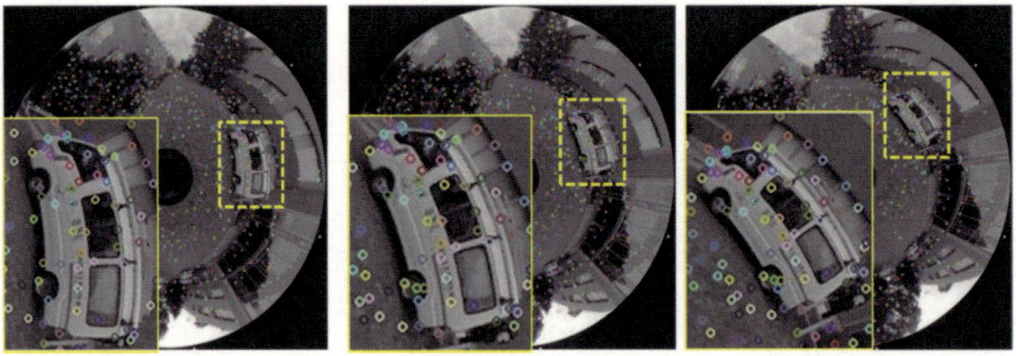

(a) 0 frame. (b) 10 frame. (c) 20 frame.

Fig. 4. An example of feature point tracking. Points which have the same color show corresponding points.

3.2 Straight-line tracking

Straight-lines are extracted from a distorted omnidirectional image. The proposed method obtains edge points by Canny edge detector (Canny, 1986). An example of edge point detection is shown in Fig. 5 (a) and (b).

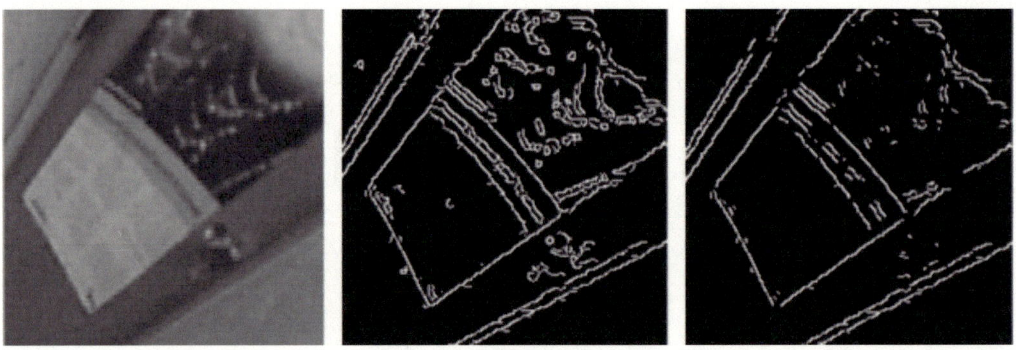

(a) (b) (c)

Fig. 5. Edge segment extraction. (a) Input image. (b) Detected canny edge points. (c) Edge segments are separated by rejecting corner points.

To separate each straight-line, corner points are rejected as shown in Fig. 5 (c). Corner points are detected by using two eigenvalues of the Hessian of the image. Hessian matrix is calculated by the following equation.

$$\mathbf{H} = \begin{bmatrix} {I_x}^2 & I_x I_y \\ I_x I_y & {I_y}^2 \end{bmatrix} \tag{3}$$

where I_x and I_y are derivatives of image I. An edge point which has large value of the ratio of eigenvalues is regarded as a point locating on a line. In the proposed method, if the ratio is smaller than 10, the edge point is rejected as a corner point. This process provides us separated edge segments.
A least square plane is calculated from ray vectors of edge points which belong to an edge segment. If the edge segment constitutes a straight-line, these ray vectors are located on a plane (Fig. 6). Therefore, an edge segment which has a small least square error is regarded as a straight-line. The proposed method is able to extract straight-lines, even if an edge segment looks like a curved line in a distorted omnidirectional image.

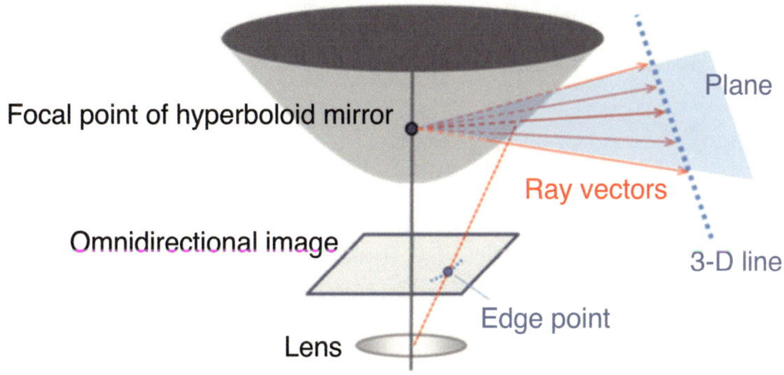

Fig. 6. The relationship between a straight-line and a ray vector.

The maximum number of edge points which satisfy equation (4) is calculated by using RANSAC (Fischler & Bolles, 1981).

$$\left(\mathbf{r}_{i,j}{}^{\mathrm{T}}\mathbf{n}_i\right)^2 < l_{\mathrm{th}} \tag{4}$$

where l_{th} is a threshold. $\mathbf{r}_{i,j}$ is a ray vector heading to an edge point j included in an edge segment i. $\mathbf{n}_i$ is the normal vector of the least square plane calculated from the edge segment i. If over half the edge points of the edge segment i satisfy equation (4), the edge segment is determined as a straight-line. The threshold l_{th} is calculated by the following equation.

$$l_{\mathrm{th}} = \cos^2\left(\frac{2\pi}{2r_{\mathrm{m}}\pi}\right) = \cos^2\left(\frac{1}{r_{\mathrm{m}}}\right) \tag{5}$$

where r_{m} is the radius of projected mirror circumference in an omnidirectional image. A threshold l_{th} allows angle error within $1/r_{\mathrm{m}}$ [rad]. It means that an angle error between a ray vector and a straight-line is within 1 pixel.
Straight-lines are tracked along an omnidirectional image sequence. The proposed method extracts points at constant intervals on a straight-line detected in the current frame (Fig. 7 (a) and (b)). These points are tracked to the next frame by KLT tracker (Fig. 7 (d)). Edge segments are detected in the next frame (Fig. 7 (c)). The edge point closest to the tracked point is selected as a corresponding edge point (Fig. 7 (e)). The edge segment which has themaximum number of corresponding edge points is regarded as a corresponding edge segment (Fig. 7 (f)). If an edge segment corresponds to several lines, a line which has larger number of corresponding edge points is selected.

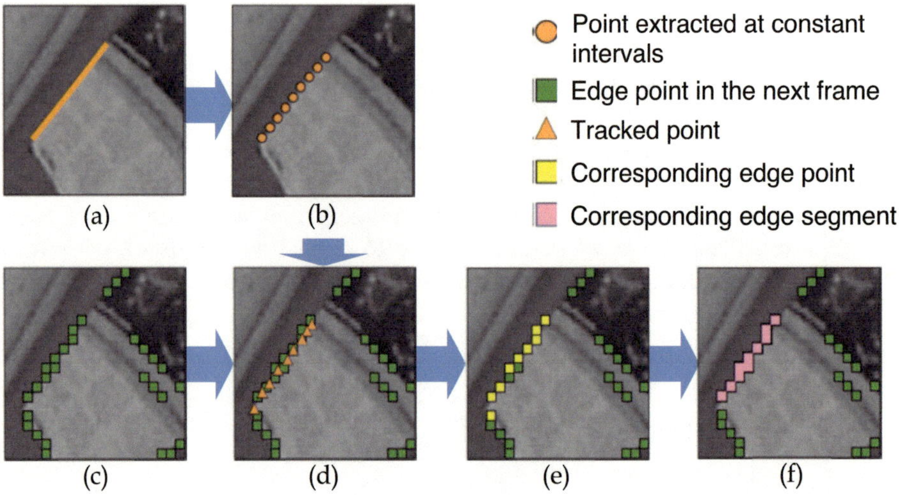

Fig. 7. Searching for a corresponding edge segment in the next frame. (a) Straight-line extracted in the current frame. (b) Points extracted at constant intervals on the line. (b) Edge segments in the next frame. (c) Points (b) are tracked between the current frame and the next frame. (d) Corresponding edge points. (e) Corresponding edge segment.

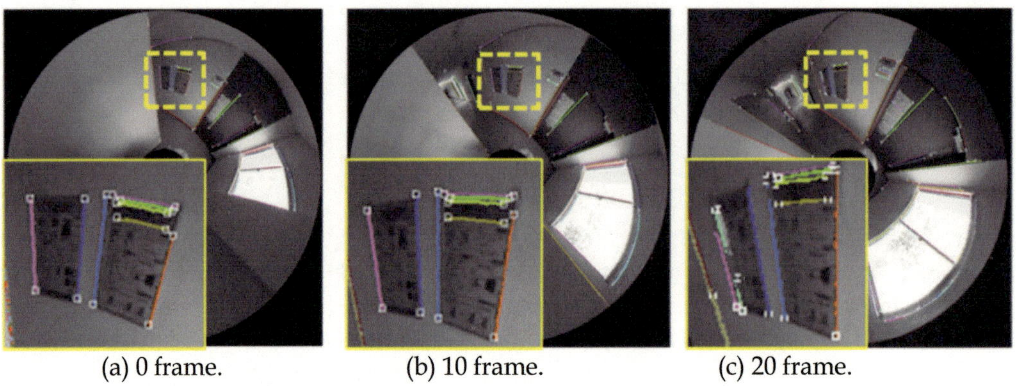

Fig. 8. An example of straight-line tracking. Lines which have a same color show corresponding lines. Although an end point of a line is shown as a white square, it is not used for straight-line detection.

Matching point search on a line has the aperture problem (Nakayama, 1988). However, it is not difficult for the proposed method to obtain corresponding lines, because it does not require point-to-point matching. By continuing the above processes, straight-lines are tracked along the omnidirectional image sequence. An example of line tracking is shown in Fig. 8.

4. Environment measurement

4.1 Point-based measurement

Camera movement is estimated by a point-based method (Kawanishi et al., 2009). The method is based on eight-point algorithm (Hartley, 1997).

An essential matrix $\mathbf{E}$ is calculated from ray vectors of corresponding feature points. An essential matrix $\mathbf{E}$ and ray vectors satisfy the following equation.

$$\mathbf{r}_i \mathbf{E} \mathbf{r}'_i = 0 \tag{6}$$

where ray vectors $\mathbf{r}_i = [x_i, y_i, z_i]^{\mathrm{T}}$ and $\mathbf{r}'_i = [x'_i, y'_i, z'_i]^{\mathrm{T}}$ are those of the corresponding point in two images. Camera rotation matrix $\mathbf{R}$ and translation vector $\mathbf{t}$ are calculated from essential matrix $\mathbf{E}$ by singular value decomposition. Equation (6) is transformed as follows.

$$\mathbf{u}^{\mathrm{T}} \mathbf{e} = 0 \tag{7}$$

where

$$\mathbf{u} = [x_i x'_i, y_i x'_i, z_i x'_i, x_i y'_i, y_i y'_i, z_i y'_i, x_i z'_i, y_i z'_i, z_i z'_i]^{\mathrm{T}},$$

$$\mathbf{e} = [e_{11}, e_{12}, e_{13}, e_{21}, e_{22}, e_{23}, e_{31}, e_{32}, e_{33}]^{\mathrm{T}}.$$

e_{ab} is the row a and column b element of Essential matrix $\mathbf{E}$. The matrix $\mathbf{E}$ is obtained by solving simultaneous equations for more than eight pairs of corresponding ray vectors.

$$J = \|\mathbf{U}\mathbf{e}\|^2 \rightarrow \min \tag{8}$$

where $\mathbf{U} = [\mathbf{u}_1, \mathbf{u}_2, \cdots, \mathbf{u}_n]^{\mathrm{T}}$. n is the number of feature points. $\mathbf{e}$ is calculated as the eigenvector of the smallest eigenvalues of $\mathbf{U}^{\mathrm{T}}\mathbf{U}$. Estimated camera movement in this process is used as an initial value for line-based measurement. However, not all feature points tracked in the image sequence correspond satisfactorily due to image noise, etc. Mistracked feature points should be rejected. The proposed method rejects these points as outliers by using RANSAC algorithm (Fischler & Bolles, 1981).

4.2 Line-based measurement

Estimated camera movement is optimized by using straight-lines. A straight-line is represented as infinite lines by using its direction vector $\mathbf{d}^w$ and location vector $\mathbf{l}^w$ ($\mathbf{l}^w + k\mathbf{d}^w$, k is a factor). The superscript w means that the vector is in world coordinate system. As a prerequisite for line-based measurement, at least, more than 3 images and 3 pairs of corresponding lines (at least one line is not parallel to others) are needed. In the first step, camera rotation and line directions are estimated. The step is independent of camera translation and line locations estimation. In the next step, camera translation and line locations are optimized by a method based on Bundle adjustment (Triggs et al., 1999). In these phases, initial value of 3-D line direction and location are required. These initial values are calculated from line correspondences and initial camera movements.

4.2.1 Camera rotation and 3-D line direction optimization

Our proposed method calculates a normal vector $\mathbf{n}_i^c$ of a least square plane calculated from an edge segment i in Section 3.2. The superscript c means that the vector is in a camera coordinate system at camera position c. Camera rotation depends on 3-D line direction vector $\mathbf{d}_i^w$ and normal vector $\mathbf{n}_i^c$. By using initial values of camera rotation and normal vectors $\mathbf{n}_i^c$, a sum of errors $E_\mathbf{R}$ between camera rotation matrix $\mathbf{R}_c^w$ and 3-D line direction $\mathbf{d}_i^w$ are calculated as shown in the following equation.

$$E_\mathbf{R} = \sum_c \sum_i \left(\left(\mathbf{R}_c^{w\mathrm{T}} \mathbf{n}_i^c \right)^\mathrm{T} \mathbf{d}_i^w \right)^2 \tag{9}$$

where, $\mathbf{R}_c^w$ is a rotation matrix from the world coordinate system to camera coordinate system c. Here, $\mathbf{d}_i^w$ and $\mathbf{n}_i^c$ are unit vectors. The relationship between a direction vector and a normal vector is shown in Fig. 9. Camera rotations and line directions are optimized by minimizing $E_\mathbf{R}$. Levenburg-Marquardt method is used for the minimization.

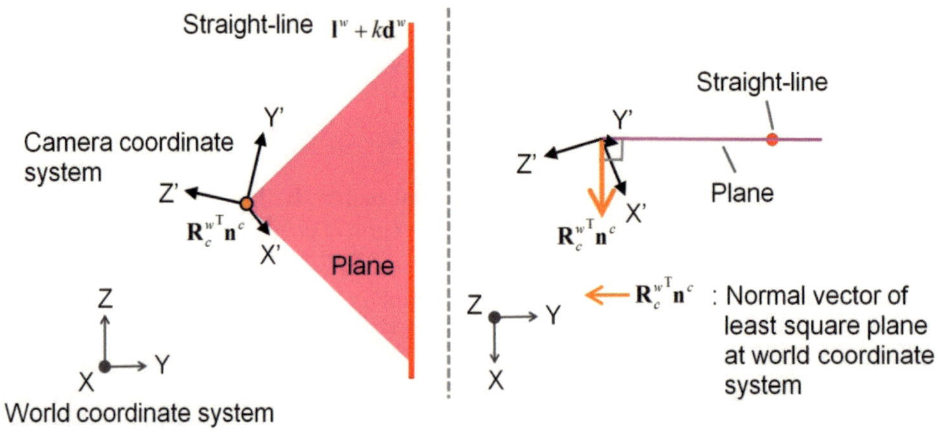

Fig. 9. Relationship between a direction vector of straight-line and a normal vector of a least square plane.

4.2.2 Camera translation and 3-D line location optimization

Camera translation vector $\mathbf{t}_c^w$ and 3-D line location $\mathbf{l}_i^w$ are optimized by Bundle adjustment (Triggs et al., 1999). The method estimates camera movements by minimizing reprojection errors. The projection error of the straight-line is calculated as an angle error between two vectors on a plane which is orthogonal to the line direction. The sum of reprojection errors of straight-lines $E_\mathbf{t}$ is calculated by the following equation.

$$E_\mathbf{t} = \sum_c \sum_i \left(1 - \hat{\mathbf{l}}_i^{c\mathrm{T}} \mathbf{g}_i^c \right)^2 \tag{10}$$

where $\mathbf{g}_i^c$ is a vector located on the plane $\mathbf{n}_i^c$, and it crosses the 3-D line at a right angle. Thus, $\mathbf{g}_i^c$ satisfies $\mathbf{R}_c^{w\mathrm{T}}\mathbf{g}_i^c \perp \mathbf{d}_i^w$ and $\mathbf{g}_i^c \perp \mathbf{n}_i^c$. $\hat{\mathbf{l}}_i^c$ is a vector which connects the camera position c and the 3-D line location with the shortest distance. $\hat{\mathbf{l}}_i^c$ is calculated by the following equation.

$$\hat{\mathbf{l}}_i^c = \mathbf{R}_c^w \left(B_{i,c}\mathbf{d}_i^w + \mathbf{l}_i^w - \mathbf{t}_c^w \right) \Big/ \left\| B_{i,c}\mathbf{d}_i^w + \mathbf{l}_i^w - \mathbf{t}_c^w \right\| \tag{11}$$

where $B_{i,c}$ is a factor which shows a location on the 3-D line. $B_{i,c}$ satisfies the following equation.

$$\left\| \left(A_{i,c}\mathbf{R}_c^{w\mathrm{T}}\mathbf{g}_i^c + \mathbf{t}_c^w \right) - \left(B_{i,c}\mathbf{d}_i^w + \mathbf{l}_i^w \right) \right\| \to \min \tag{12}$$

The relationship between these vectors is shown in Fig. 10.

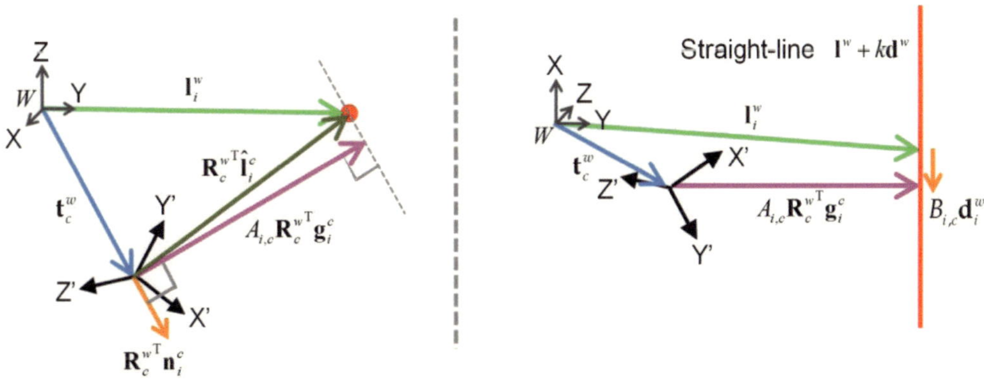

Fig. 10. Relationship between camera translation vector and 3-D line location vector.

The sum of reprojection errors of straight-lines E_{t} is minimized by a convergent calculation based on Levenburg-Marquardt method. In these two optimization steps, lines which have large error are rejected as outliers by RANSAC algorithm.

In the proposed method, 3-D lines are represented as uniformly-spaced points $\mathbf{e}_{i,n}^w$. 3-D coordinates of these points are calculated by the following equation.

$$\mathbf{e}_{i,n}^w = hn\mathbf{d}_i^w + \mathbf{l}_i^w \tag{13}$$

where h is a uniform distance and n is an integer number, respectively. 3-D coordinates of $\mathbf{e}_{i,n}^w$ is reprojected to the image sequence. When the 2-D coordinates of the reprojection point are close to the corresponding edge segment enough, the point is added into measurement data.

By using estimated camera movement, 3-D coordinates of feature points which have the minimal reprojection error are calculated and integrated with straight-line measurement data.

4.3 Result qualification

Measurement data which have low accuracy should be rejected before 3-D model construction. Measurement accuracy of the feature point is evaluated by following equations.

$$a_{i,m} = \left\| \frac{\partial \mathbf{p}_{i,m}}{\partial u_{c1,i}} \right\| + \left\| \frac{\partial \mathbf{p}_{i,m}}{\partial v_{c1,i}} \right\| + \left\| \frac{\partial \mathbf{p}_{i,m}}{\partial u_{c2,i}} \right\| + \left\| \frac{\partial \mathbf{p}_{i,m}}{\partial v_{c2,i}} \right\| \tag{14}$$

where $\mathbf{p}_{i,m}$ is 3-D coordinates calculated from corresponding feature points between camera position c_1 and c_2. $\left(u_{c1,i}, v_{c1,i}\right)$ and $\left(u_{c2,i}, v_{c2,i}\right)$ are image coordinates of feature points. The method calculates $a_{i,m}$ of all camera position combination. If the smallest value $a_{\min,i}$ is larger than a given threshold a_{th}, the feature point is rejected as a measurement result which has low accuracy.

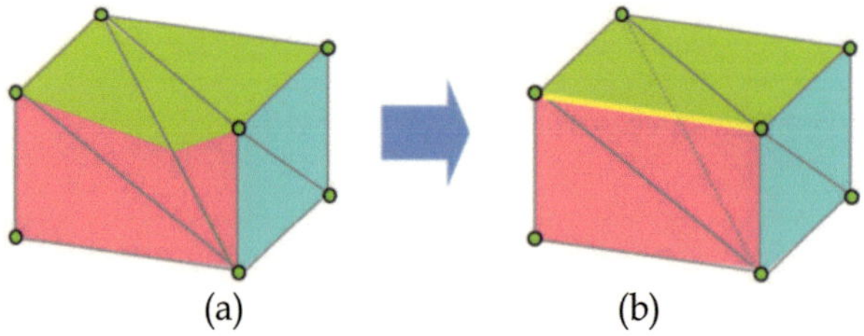

Fig. 11. Triangular mesh generation and its optimization. (a) Triangular meshes generated by Delaunay triangulation. (b) Optimized triangular meshes.

Measurement accuracy of straight-line is evaluated by equation (14), too. In this evaluation, $\mathbf{p}_{i,m}$ is the middle point of the line connecting two vectors $\mathbf{g}_i^{c1}$ and $\mathbf{g}_i^{c2}$ at the shortest distance. Image coordinates $(u_{c1,i}, v_{c1,i})$ and $(u_{c2,i}, v_{c2,i})$ are reprojection points of these vectors to images acquired at camera position c_1 and c_2.

5. Model construction

Triangular meshes are generated from integrated measurement data by using the 3-D Delaunay triangulation. However, Delaunay triangulation generates a triangular mesh which contradicts a physical shape because the triangular mesh does not consider the shape of the measurement object. Therefore, we apply the triangular optimization method (Nakatsuji et al., 2005) to the triangular mesh (Fig. 11). The method adapts the triangular mesh to the physical shape by detecting a texture distortion. By texture mapping to these meshes, a 3D environment model is constructed.

6. Experiments

First, accuracy of line-based measurement is evaluated. Measurement objects are lengthwise-lines on a flat wall shown in Fig. 12. The reason for including crosswise-lines is that the proposed method needs lines having different direction. The moving distance of the camera was about 2m. The number of input images is 72. An input image size is 2496×1664 pixels.

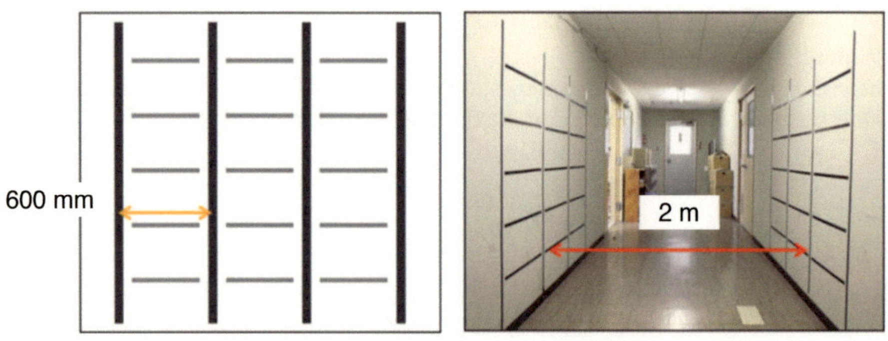

(a) Object. (b) Environment.

Fig. 12. Measurement objects place on flat walls. Vertical lines are measured for accuracy evaluation. Level lines are set for camera movement estimation.

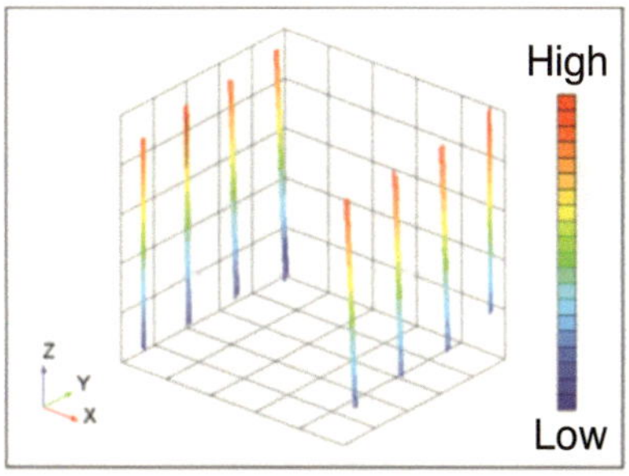

(a) Bird's-eye.

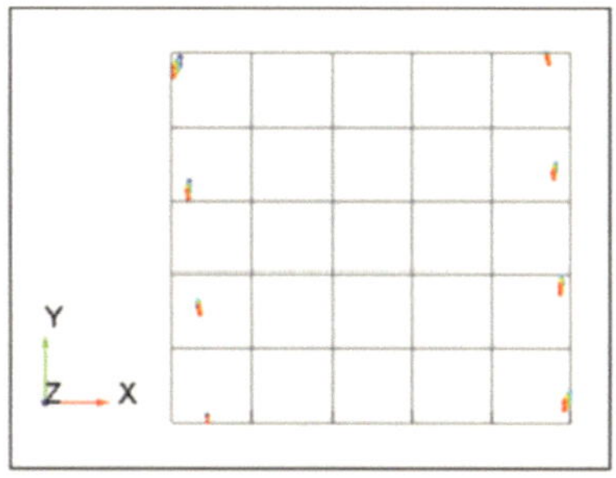

(b) Top.

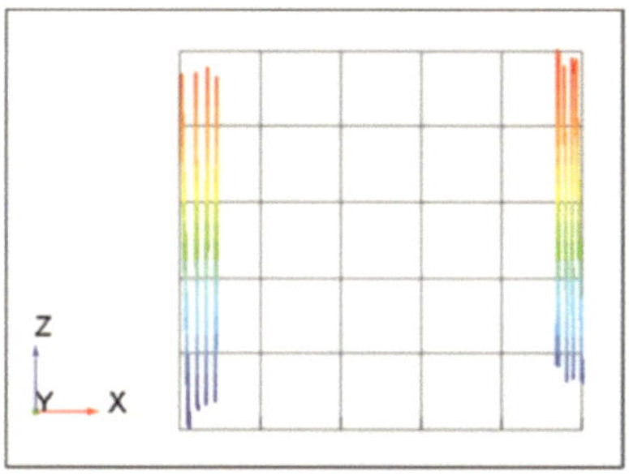

(c) Front.

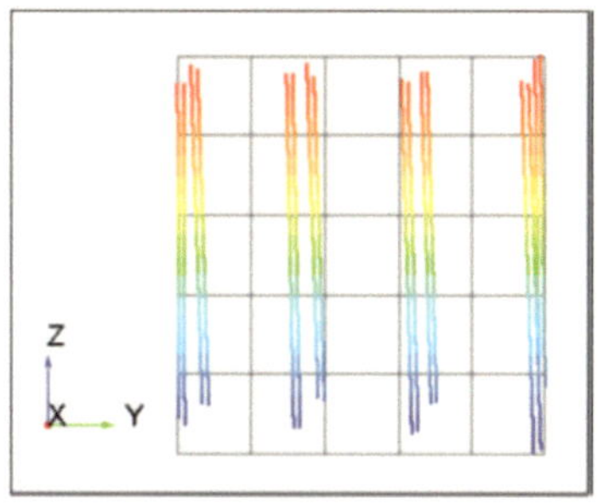

(d) Side.

Fig. 13. Measurement result for accuracy evaluation. Vertical lines are measured precisely. Measurement results of level lines are removed as low measurement accuracy data.

	Angle (direction) [deg]	Angle (plane) [deg]	Depth [mm]
Standard deviation	1.2	-	2.3
Maximum	1.9	1.5	7.5

Table 1. Evaluation result of vertical line measurements.

A measurement result is shown in Fig. 13. A measurement result which has larger Z-coordinate value is displayed in red, and smaller one is displayed in blue. Angles and depth errors were calculated for evaluation of measurement accuracy in Table 1. An angle error of calculated line directions is 1.2 degree standard deviation. Its maximum error is 1.9 degree. An angle error between two flat walls estimated from measurement data is within 1.5 degrees. A depth error between an estimated flat wall and reconstructed lines has 2.3 mm standard deviation. Its maximum error is 7.5 mm. This experiment shows that our proposed method has sufficient accuracy to accomplish static obstacle avoidance, self-localization.

Next, we experimented in an environment including non-textured objects as shown in Fig. 14. We used 84 omnidirectional images. Measurement results of feature points and straight-lines are shown in Fig. 15. The blue marks in the figure show the camera trajectory. Although feature point measurement results are sparse, straight-lines can be measured densely. This experimental result shows that our proposed method is effective for a non-textured environment.

(a) Non-textured environment.

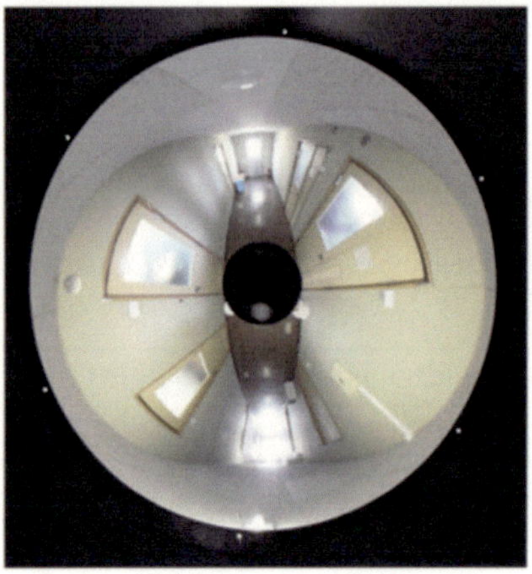

(b) Input image.

Fig. 14. Non-textured environment. We cannot get enough feature points in the environment because there are few features.

Modeling result is shown in Fig. 16. Images having a view-point which is different from camera observation points can be acquired. A model constructed from feature point measurement data is only a small part of this environment (Fig. 16(a) and (c)). Meanwhile, edge measurement data makes it possible to construct a non-textured environment model (Fig. 16}(b) and (d)).

We also experimented in an outdoor environment including textured objects (trees and so on) and non-textured objects (buildings and so on) as shown in Fig. 17. We used 240 omnidirectional images. An integrated measurement result is shown in Fig. 18. As one of textured objects, the shape of ground surface is measured by point-based measurement. As non-textured objects, the shape of the building is measured by line-based measurement.

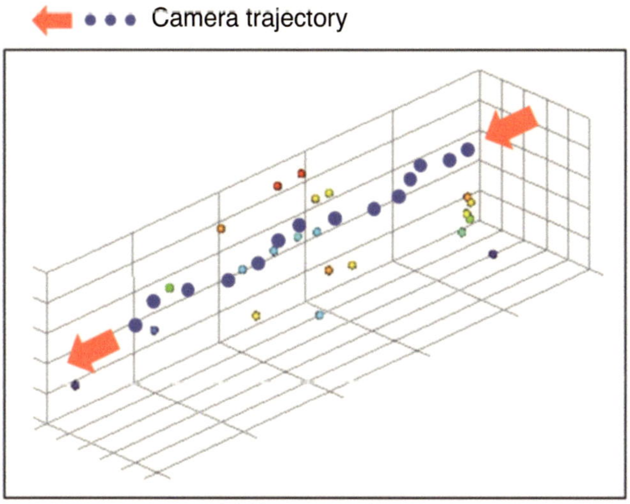

(a) Feature points.

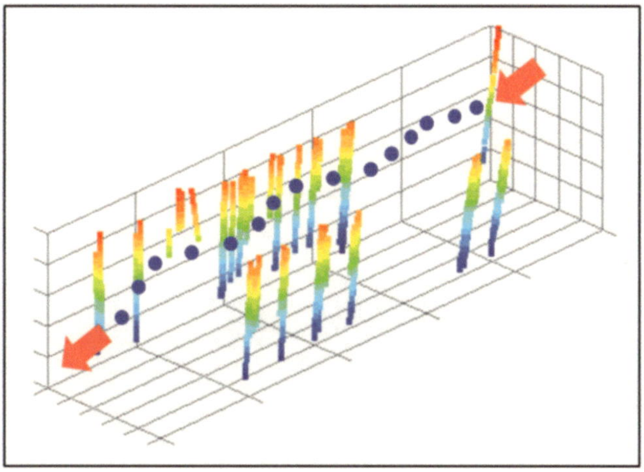

(b) Straight-lines.

Fig. 15. Measurement results of non-textured environment. (a) Although camera movement estimation is possible, we get sparse measurement results. (b) Straight-lines make it possible to measure non-textured environments densely.

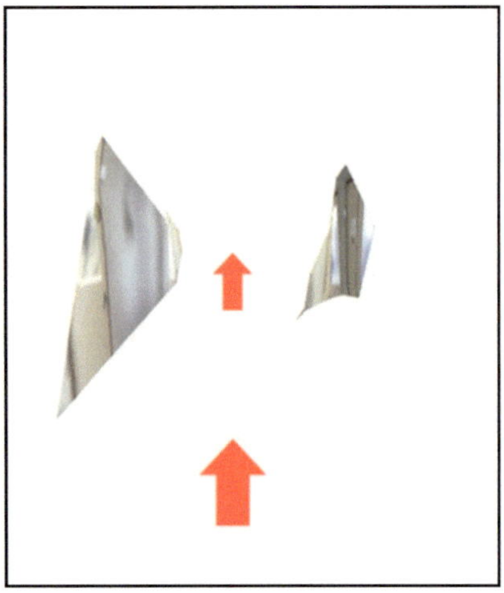

(a) Front (feature point only).

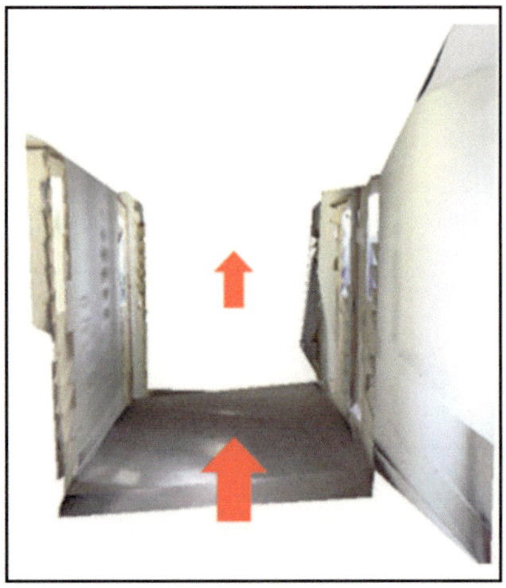

(b) Front (with straight-line).

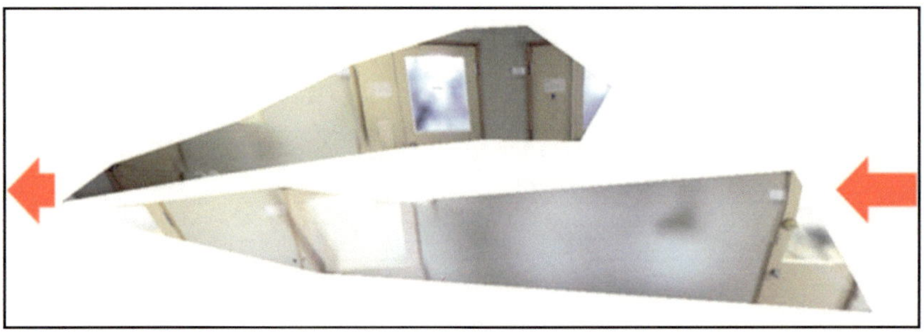

(c) Bird's-eye (feature point only).

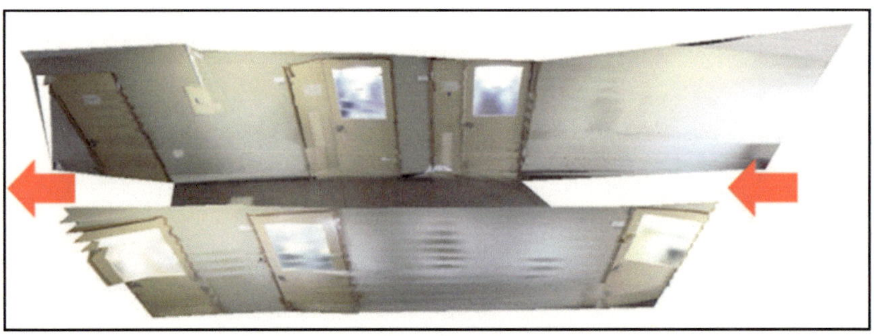

(d) Bird's-eye (with straight-line)

Fig. 16. Modeling results of non-textured environment. We can construction a model including many non-textured objects by the method with straight-lines.

Modeling result is shown in Fig. 19. By the combination of point-based measurement and line-based measurement, our proposed method can construct a model of 3-D environment including textured and non-textured objects. Experimental results showed the effectiveness of our proposed method.

(a) Environment. (b) Input image.

Fig. 17. Outdoor environment. There are textured objects (trees, tiles and so on) and non-textured objects (walls etc.)

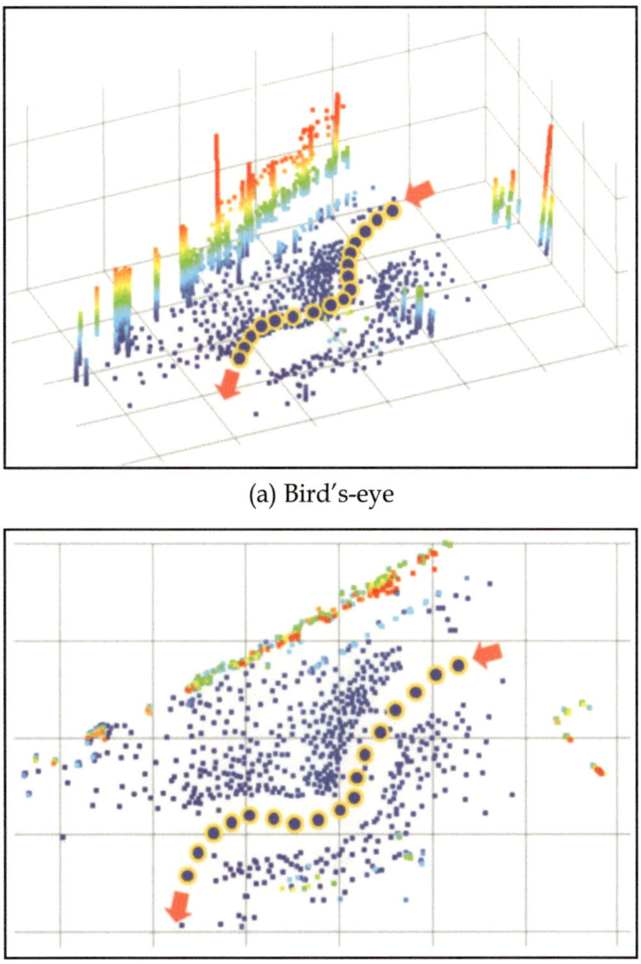

(a) Bird's-eye

(b) Top

Fig. 18. Measurement result of outdoor environment.

(a) Bird's-eye.

(b) Right.

(c) Front.

Fig. 19. Modeling results of outdoor environment.

7. Conclusions

We proposed an environment modeling method based on structure from motion using both feature points and straight-lines by using an omnidirectional camera. Experimental results showed that our proposed method is effective in environment including both textured and non-textured objects.

As future works, the precision improvement of edge tracking is necessary. Moreover, we should evaluate difference of camera movement estimation accuracy between point-based measurement and edge-based measurement. Further, edge position correlation should be used for increasing measurement stability.

8. References

Davison, A. J. (2003). Real-Time Simultaneous Localisation and Mapping with a Single Camera, *Proceedings of the 9th IEEE International Conference on Computer Vision*, Vol. 2, pp. 1403-1410, 2003.

Ishiguro, H. & Yamamoto, M. & Tsuji, S. (1992). Omni-Directional Stereo, *IEEE Transactions on Pattern Analysis and Machine Intelligence*, Vol. 14, No. 2, pp. 257-262, 1992.

Gluckman, J. & Nayar, S. K. (1998). Ego-motion and Omnidirectional Cameras, *Proceedings of the 6th International Conference on Computer Vision*, pp. 999-1005, 1998.

Bunschoten, R. & Krose, B. (2003). Robust Scene Reconstruction from an Omnidirectional Vision System, *IEEE Transactions on Robotics and Automation*, Vol. 19, No. 2, pp. 351-357, 2003.

Geyer, C. & Daniilidis, K. (2003). Omnidirectional Video, *The Visual Computer*, Vol. 19, No. 6, pp. 405-416, 2003.

Rachmielowski, A. & Cobzas, D. & Jagersand, M. (2006). Robust SSD Tracking with Incremental 3D Structure Estimation, *Proceedings of the 3rd Canadian Conference on Computer and Robot Vision*, pp. 1-8, 2006.

Kawanishi, R.; Yamashita, A. & Kaneko, T. (2009). Three-Dimensional Environment Model Construction from an Omnidirectional Image Sequence, *Journal of Robotics and Mechatronics*, Vol. 21, No. 5, pp. 574-582, 2009.

Chandraker, M. & Lim, J. & Kriegman, D. (2009). Moving in Stereo: Efficient Structure and Motion Using Lines, *Proceedings of the 12th IEEE International Conference on Computer Vision*, pp. 1741-1748, 2009.

Schindler, G.; Krishnamurthy, P. & Dellaert, F. (2006). Line-Based Structure from Motion for Urban Environments, *Proceedings of the 3rd International Symposium on 3D Data Processing, Visualization, and Transmission*, pp. 846-853, 2006.

Bartoli, A. & Sturm, P. (2005). Structure-from-motion using lines: representation, triangulation, and bundle adjustment, *Computer Vision and Image Understanding*, Vol. 100, Issue 3, pp. 416-441, 2005.

Smith, P. & Reid, I. & Davison, A. (2006). Real Time Monocular SLAM with Straight lines, *Proceedings of the 17th British Machine Vision Conference*, pp. 17-26, 2006.

Mariottini, G. L. & Prattichizzo, D. (2007). Uncalibrated video compass for mobile robots from paracatadioptric line images, *Proceedings of the 2007 IEEE/RSJ International Conference on Intelligent Robots and Systems*, pp. 226-231, 2007.

Spacek, L. A. (1986). Edge Detection and Motion Detection, *Image and Vision Computing*, Vol. 4, Issue 1, pp. 43-56, 1986.

Triggs, B. & McLauchlan, P. & Hartley, R. & Fitzgibbon, A. (1999). Bundle Adjustment -A Modern Synthesis, *Proceedings of the International Workshop on Vision Algorithms: Theory and Practice*, Springer-Verlag LNCS 1883, pp. 298-372, 1999.

Shi, J. & Tomasi, C. (1994). Good Features to Track, *Proceedings of the 1994 IEEE Computer Society Conference on Computer Vision and Pattern Recognition*, pp. 593-600, 1994.

Canny, J. F. (1986). A Computational Approach to Edge Detection, *IEEE Transactions on Pattern Analysis and Machine Intelligence*, Vol. PAMI-8, No. 6, pp. 679-698, 1986.

Fischler, M. A. & Bolles, R. C. (1981). Random Sample Consensus: A Paradigm for Model Fitting with Applications to Image Analysis and Automated Cartography, *Communications of the ACM*, Vol. 24, No. 6, pp. 381-395, 1981.

Nakayama, K. & Silverman, G. (1988). The Aperture Problem -- II. Spatial Integration of Velocity Information Along Contours, *Vision Research*, Vol. 28, No. 6, pp. 747-753, 1988.

Hartley, R. (1997). In defence of the eight-point algorithm, *IEEE Transactions on Pattern Analysis and Machine Intelligence*, Vol. 19, No. 6, pp. 580-593, 1997.

Nakatsuji, A. & Sugaya, Y. & Kanatani, K. (2005), Optimizing a Triangular Mesh for Shape Reconstruction from Images, *IEICE Transactions on Information and Systems*, Vol. E88-D, No. 10, pp. 2269-2276, 2005.

4

Mobile Robot Position Determination

Farouk Azizi and Nasser Houshangi
Purdue University Calumet
USA

1. Introduction

One of the most important reasons for the popularity of mobile robots in industrial manufacturing is their capability to move and operate freely. In order for the robots to perform to the expectations in manufacturing, their position and orientation must be determined accurately. In addition, there is a strong tendency to grant more autonomy to robots when they operate in hazardous or unknown environments which also requires accurate position determination. Mobile robots are usually divided into two categories of legged and wheeled robots. In this chapter, we focus on wheeled mobile robots.
Techniques used for position determination of wheeled mobile robots (or simply, mobile robots) are classified into two main groups: relative positioning and absolute positioning (Borenstein, 1996, 1997). In relative positioning, robot's position and orientation will be determined using relative sensors such as encoders attached to the wheels or navigation systems integrated with the robots. Absolute positioning techniques are referred to the methods utilizing a reference for position determination. The Global Positioning Systems, magnetic compass, active beacons are examples of absolute positioning systems.
Calculating position from wheel rotations using the encoders attached to the robot's wheels is called Odometry. Although odometry is the first and most fundamental approach for position determination, due to inherent errors, it is not an accurate method. As a solution to this problem, usually odometry errors are modeled using two different methods of benchmarks and multiple sensors. In this chapter, we will discuss odometry and two different methods to model and estimate odometry errors. At the end, an example for mobile robot position determination using multiple sensors will be presented.

2. Odometry

There is a consensus among researcher that odometry is the vital technique for mobile robot position determinations. The governing equations of odometry are based on converting rotational motion of robot wheels to a translational motion (Borenstein, 1996, 1997). Although odometry is an inexpensive method for position determination, it has several inherent issues. One issue is that errors accumulate over time and consequently make odometry unreliable over time. The odometry errors are mainly classified as systematic and nonsystematic (Borenstein, 1996). The source of systematic errors usually caused by:

- Average of both wheel diameters differs from nominal diameter
- Misalignment of wheels

- Uncertainty about the effective wheelbase (due to non-point wheel contact with the floor)
- Limited encoder resolution
- Limited encoder sampling rate

Nonsystematic errors are caused by the following conditions:

- Travel over uneven floors
- Travel over unexpected objects on the floor
- Wheel-slippage

Since these errors drastically affect the accuracy of odometry over short and long distances, it is empirical to accurately estimate the errors. The techniques used to overcome problems with odometry can be categorized to benchmark techniques and utilizing multiple sensors.

2.1 Correcting odometry using benchmark techniques

Benchmark techniques are based on testing robots over some predefined paths. In each experiment, the actual position and orientation of the robot is compared with the theoretical final position and orientation of the robot. Using the robot's kinematic equation a series of correction factors are derived to be incorporated in future position determination.

Borenstein and Feng proposed a test bench to model and estimate systematic errors (Borenstein, 1995). The method is called UMBmark and is based on running a robot with differential wheel drives on a square clockwise and counter clockwise paths for several times and compare the recording with the actual final position and orientation of the robot to generate the odometry errors. The errors are then classified as type A and type B. Type A error is the error that increases (or decreases) amount of rotation of the robot during the square path experiment in both clockwise and counter clockwise direction while type B error is the kind of error that increases (decreases) amount of rotation in one direction (clockwise or counter clockwise) and decreases (increases) amount of rotation in other direction (counter clockwise or clockwise). The robot's geometric relations are used to calculate the wheelbase and distance errors in terms of type A and B errors. Finally, the results are used to find two correction factors for right and left wheels of the robot to be applied to odometry calculation to improve the positioning system. The advantage of this method lies in the fact that mechanical inaccuracies such as wheel diameters can be modeled and compensated for. However, the method is only suitable to overcome systematic errors and does not estimate nonsystematic errors.

Chong and Kleeman modified UMBmark to model odometry positioning system more effectively by offering a mechanism to include odometry's nonsystematic errors (Chong, 1997). This approach uses robot's geometric and dynamical equation in a more fundamental way to generate an error covariance matrix for estimating the odometry errors. The technique also utilizes an unloaded wheel installed independently on linear bearings of each wheel to minimize wheel slippage and motion distortion. As a result of these modifications, odometry errors are minimized to a level that could be considered a zero-mean white noise signal. The other important characteristic of this method is that the errors calculated at each given time are uncorrelated to the next or from previous errors. The systematic errors are calibrated using UMBmark test. To model non-systematic errors, it is assumed that the robot motion can be divided to small segments of translational and rotational motions. For each segment the associated covariance matrix is calculated based on the physical model of the motion. The covariance matrix is updated using the previous calculation of the matrix. This method does not use any signal processing filter to estimate the errors. The main drawback of the method is the use of unloaded wheels along with drive wheels which limits the robot motion significantly.

Antonelli and coworkers proposed a method to calibrate odometry in mobile robots with differential drive (Antonelli, 2005). This method is based on estimating some of odometry parameters using linear identification problems for the parameters. These parameters come from kinematic equations of the odometry which are the sources of odometry's systematic errors. The estimation model is derived from the least square technique which allows a numerical analysis of the recorded data. Unlike UMBmark, this method is based on a more flexible platform and does not require a predefined path for the robot. Since the precision of the robot position and orientation at any given time is very important in estimating the robot position at the next moment, a vision-based measurement system is used to record robot's position and orientation. The experimental results shows this technique provide less error in estimating the robot position and orientation when compared to UMBMark technique.

Larsen and co-workers reported a method to estimate odometry systematic errors using Augmented Kalman Filter (Larsen, 1998). This technique estimates the wheel diameter and distance traveled with a sub-percent precision by augmenting the Kalman filter with a correction factor. The correction factor changes the noise covariance matrix to rely on the measurements more than the model by placing more confidence on the new readings. The augmented Kalman filter uses encoder readings as inputs and vision measurements as the observations. The only condition to achieve such precision is to use more sensitive measurement system to measure the robot position and orientation.

Martinelli and Siegwart proposed a method to estimate both systematic and nonsystematic errors odometry errors during the navigation (Martinelli, 2003). The systematic errors are estimated using augmented Kalman filter technique proposed in (Larsen, 1998) and the nonsystematic errors are estimated using another Kalman Filter. The second Kalman filter observations is called Observable Filter (OF) come from the estimates of the robot configuration parameters of the segmented Kalman filter. The main idea of Observable filters is to estimate the mean and variance of the observable variables of the robot parameters which are characterizing the odometry error.

2.2 Correcting odometry using multiple sensors

The other method to estimate the odometry errors is to integrate odometry with information from another sensor. The information from another sensor eventually is used to reset odometry errors especially during long runs. In many studies Global Positioning System (GPS), Inertial Navigation System (INS), compass, vision and sonar have been used in conjunction with odometry for position determination. In most cases Kalman filter or a derivation of Kalman filter, such as Indirect Kalman filter (IKF), Extended Kalman filter (EKF) and Unscented Kalman filter (UKF) has been used to integrate the information. In the following, we discuss few works have been done in this direction.

Park and coworkers employed a dead reckoning navigation system using differential encoders installed on the robot wheels and a gyroscope which is attached to robot (Park, 1997). The approach is based on estimation and compensation of the errors from the differential encoders and the gyroscope angular rate output. An Indirect Kalman filter (IKF) is used to integrate heading information from gyroscope and odometry. The output of IKF is used to compensate the errors associated with heading information in odometry as well as the error in gyroscope readings. The improved heading has been used in some formalism to give more accurate position and heading of the mobile robot. The work is followed by introducing a system of linear differential equations for each one of position errors on x- and y- directions, heading rate error, left and right wheel encoder errors, gyro scale factor error, and gyro bias drift error and wheel base error. The differential matrix derived from that linear differential equation is the equation that is used as the input for Indirect Kalman filter. The advantage of this method is that by including both the encoder and the gyroscope errors as the input of Kalman filter, both odometry and gyroscope errors can be estimated.

Cui and Ge in their work utilized Differential Global Positioning System (DGPS) as the basic positioning system for autonomous vehicle (Cui, 2001). GPS is currently used in many applications of land vehicle navigation and it is based on using information about the location and direction of motion of a moving vehicle which received from different satellites orbiting the earth. The expected accuracy of this technique is less than 0.5 meters. DGPS is similar to GPS but it uses two or more different satellite signals to localize a moving vehicle with more accuracy. One of the problems with both GPS and DGPS is when these methods are used to track a moving vehicle in an urban canyon. In such an environment, tall buildings prevent GPS to receive signals from the satellites. Therefore, it's critical to decrease the number of required satellite signals when the vehicle is moving in urban canyons. To achieve this goal, a constrained method was used that approximately modeled the path of the moving vehicle in an urban canyons environment as pieces of lines which is the cross section of two different surfaces. Each surface can be covered by two satellites at each moment. This will decrease the number of different satellite signals needed for vehicle localization to two. However, the method can switch back to use more than two satellite signals once the vehicle is not in moving in an urban canyon. In this method a pseudorange measurement is proposed to calculate distance from each satellite. Since the GPS receiver is modeled so that the measurement equation is nonlinear, the Extended Kalman Filter is used to the augmented matrix generated from the measurements to estimate the vehicle's position. While this work has not used odometry as the basic method for positioning of the vehicle, it decreases the number of required satellite signals to estimate the vehicle position when it travels in urban canyons. Recently Chae and coworkers have reported a method that uses EKF to efficiently integrate data from DGPS and INS sensors (Chae, 2010). The proposed method is designed to cover the task of mobile robot position determination during the times the robot navigates in urban areas with very tall buildings in which GPS signals are inaccessible.

Borenstein and Feng in another work have introduced a method called gyrodometry (Bronstein, 1996). In this method, the odometry is corrected for systematic errors using UMBmark. The approach intends also to get around non-systematic errors such as the errors generated when the robot traverses on a bump. The nonsystematic errors impact the robot for short period and is not detected by odometry.Once the robot faced a bump or another source of nonsystematic errors, odometry is replaced by gyro-based positioning system which doesn't fail in that situation and can provide a reliable information as the robot passes the bump. In practice both sets of information are available all the time and it is assumed that two sets of data, as long as there is no bump or object ahead of the robot, return almost the same information. In fact the procedure starts when odometry data differs substantially from gyro data. At the end of the time period robot traversing on a bump, the odometry is set by gyro positioning system. Afterwards, the odometry is the main positioning system and uses the correct new starting point. The gyro output drifts over time and this was compensated using the method proposed by Barshan and White (Barshan, 1993, 1994, 1995). Using this method both types of odometry errors are compensated and as result more reliable position and orientation are calculated. This approach is simple and doesn't need any sensor integration. While the method is reliable for the conditions it was tested for, it does not include the situation where odometry fails due to other nonsystematic errors such as moving on an uneven surface or in presence of wheel slippage.

Using ultrasonic beacons is another option to improve odometry which is proposed by Kleeman (Kleeman, 1992). In this approach active ultrasonic beacons are used to locate the robot at any given time. The information from beacons was used as a reliable position to reset the odometry. It is well known that ultrasonic measurements have random errors which don't accumulate over time but in contrast are not smooth and require to be compensated. In this method, an ultrasonic localizer is used which has six ultrasonic beacons and a transmitter controller which sequences the firing of the beacons in a cyclic manner. Beacons have 150 milliseconds intervals to have the previous pulse reverberation settled. Beacon 1 is distinguished by transmitting two bursts 3 milliseconds apart. The robot has an onboard receiver which is composed of eight ultrasonic receiver arranged at 45 degrees intervals. In modeling the positioning system, in addition to position of robot in both x- and y-directions, velocity of the robot, beacon cycle time, speed of sound and beacon firing time are introduced as the states of the system. An Iterative Extended Kalman Filter (IEKF) is utilized to integrate two sets of data from the beacons and from odometry. This method provides reliable information about the robot position in a constructed environment. However, it has several important drawbacks which make it unsuitable in many situations. For example, there is a problem with having delayed time arrival of an ultrasonic beacon due to an indirect path incorporating reflection off obstacles, walls, ceiling or floor. These echoed arrival time should be identified and must not be taken into account in estimation otherwise IEKF result in erroneous state estimation. In addition, one could easily point out that this method is appropriate only for indoor applications when ultrasonic beacons can be installed.

Barshan and Whyte have developed an Inertial Navigation system to navigate outdoor mobile robots (Barshan, 1993, 1994, 1995). In this system, three solid states gyroscope, a triaxial accelerometer and two Electrolevel tilt sensor are used. One of the important results of this work was the development of a method to model drift error of inertial sensors as an exponential function with time. Modeling the drift error for each sensor was done by leaving the sensor on a surface motionless. The sensor's readings were recorded until it was stabled. The sensor's readings are then modeled as an exponential function. The proposed Navigation system has been tested on radar-equipped land vehicles. The orientation from inertial sensors is reliable for 10 minutes. However, in the presence of electromagnetic fields information regarding the heading is valid only for 10 seconds. Although, using the inertial sensors reliably (after the drift errors were modeled) for 10 minutes is an improvement for vehicle position determination, it is not good enough when compared with systems which integrate two or more sensors. It is assumed that in all cases the errors associated with inertial sensor have been modeled and were taken in to account before the position and orientation of the vehicle was calculated.

Gyro's scale factor is provided by manufacturer to convert gyro output (digital or analog signal) to degrees per second. The scale factor is not a fixed and varies slightly with gyro's temperature, direction of rotation and rate of rotation. Therefore, modeling the variation of scale factor is critical to accurate calculation of gyro's angular information. Borenstein has used odometry and a gyroscope which was corrected for its scale factor variation along with its drift error (Borenstein, 1998). In addition, an Indirect Kalman Filter is used to model all errors corresponding with this method.

Roumeliotis and Bekey have utilized multiple sensors to navigate a mobile robot with two drive wheel and one cast (Roumeliotis, 2000). In this work one sensor (a potentiometer) was installed on rear wheel, to measure robot's steering angle. Additional sensors were: 1) two encoders on each wheel, 2) one single axis gyroscope, and 3) one sun sensor. A sun sensor is capable of measuring of the robot's absolute orientation based on sun position. Such sensor could be an effective alternative in applications such as Mars explorations where there is no access to GPS signals or strong magnetic field. It is indicated that sun sensor data should be used as often as one fifth of other sensors to achieve better estimation. Also this research shows that, by excluding robot's angular velocity and acceleration and translational acceleration into the estimation system state (in simulations), the system is very sensitive to changes in orientation caused by external sources such as slippage. Therefore to reduce this effect, robot angular velocity and angular acceleration as well as the robot translational acceleration are included into the estimation system's state.

Amarasinghe and co-workers have reported a method to integrate information from a laser range finder and a laser camera to navigate a mobile robot (Amarasinghe, 2010). In this technique, a camera is used to detect landmarks and the position of the landmark is measured by the laser range finder using laser camera. The information from two sensors is integrated in a Simultaneous Localization and Mapping (SLAM) platform supported by an Extended Kalman Filter (EKF). While the main advantage of this work is using appropriate sensors for detecting the landmarks and calculating the robot's position, it provides unreliable information in an indoor setting with no landmarks or in an unfavorable lighting condition such as a smoke-filled environment.

2.2.1 An example of using multiple sensors for position determination

In this section an example of position determination using sensor integration is discussed. The mobile considered is a wheeled robot with two driving wheels in the front and one dummy wheel in the back. The mobile robot position and orientation is represented by vector $\underline{P}(kT_s) = [\underline{p}(kT_s), \psi(kT_s)]^T$ at time step kT_s, where k is a positive integer and T_s is the sampling period. For simplicity, T_s will be dropped from notation after this point and the mobile robot is assumed to be traveling on a flat surface. The vector $\underline{P}(k) = [\underline{p}(k), \psi(k)]^T$ specifies the Cartesian coordinates of the robot, and the $\psi(k)$ defines the orientation (heading) of the robot. The position and orientation vector $\underline{P}(k)$ is updated in each sampling period during the robot's motion. Two coordinate frames are assigned as shown in figure 1. Body coordinate frame which is attached to the robot and moves with the robot. World coordinate frame, which sometimes is called navigation coordinate frame, is fixed at a reference point and the robot position is measured with respect to this frame. The superscript 'O' is used to denote the information obtained from odometry. The robot position ${}^{o}\underline{p}(k)$ is determined incrementally by

$$ {}^{o}\underline{p}(k+1) = {}^{o}\underline{p}(k) + R_{\psi o}(k)\underline{v}^{b}(k)T_s \tag{1} $$

where $\underline{v}^b(k) = [v_x^b(k) \;\; v_y^b(k)]^T$, is the robot's velocity in the body coordinate frame, and $R_{\psi_o}(k)$ denotes the rotational transformation matrix from the body coordinate frame to the world coordinate frame defined by

$$R_{\psi_o}(k) = \begin{bmatrix} \cos\psi_o(k) & -\sin\psi_o(k) \\ \sin\psi_o(k) & \cos\psi_o(k) \end{bmatrix} \tag{2}$$

where $\psi_o(k)$ is heading of mobile robot based on odometry. The velocity component $v_y^b(k)$ is assumed to be zero because of forward motion of the mobile robot and $v_x^b(k)$ is calculated by

$$v_x^b(k) = \frac{1}{2}(v_{er}(k) + v_{el}(k)) \tag{3}$$

where $v_{er}(k)$ is the measured translational velocity of robot's right wheel and $v_{el}(k)$ is the left wheel's translational velocity. Figure 1 shows the geometric relations between robot's position and orientation in three consequent samples.

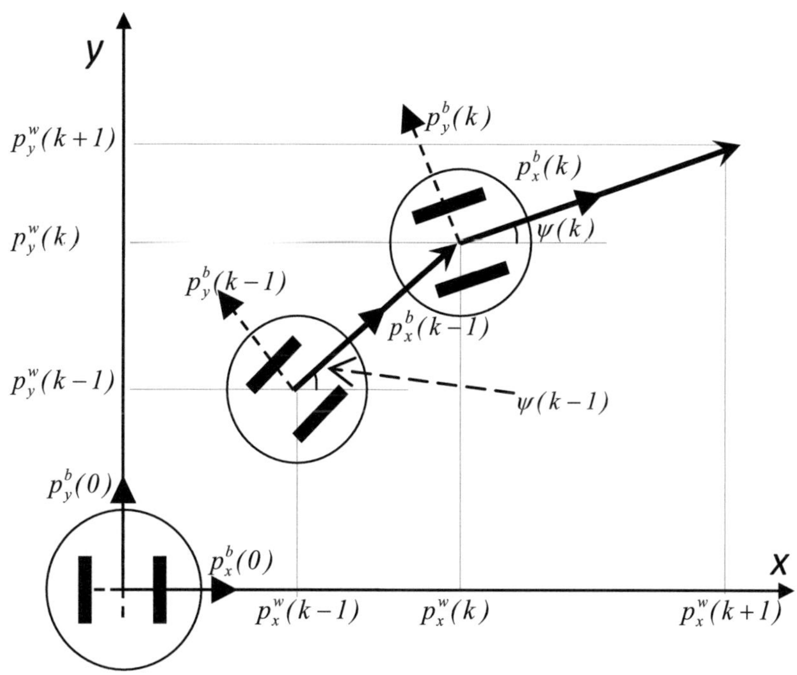

Fig. 1. Position determination based on dead reckoning.

The heading angle $\psi_o(k)$ is calculated by

$$\underline{v}^w(k) = R_{\psi_o}(k)\underline{v}^b(k) \tag{4}$$

$$\psi_o(k) = \tan^{-1}(v_y^w(k) / v_x^w(k)) \tag{5}$$

where $\underline{v}^w(k) = [v_x^w(k) \quad v_y^w(k)]^T$ is the translation velocity of mobile robot in world coordinate frame. The orientation based on odometry is updated by

$$\psi_o(k+1) = \psi_o(k) + \dot{\psi}_o(k)T_s \tag{6}$$

where the rate of change of orientation $\dot{\psi}_o(k)$ is calculated by

$$\dot{\psi}_o(k) = \frac{v_{er}(k) - v_{el}(k)}{b} \tag{7}$$

where b is the robot's wheelbase.

To alleviate the errors from odometry, the information from second sensor is integrated with odometry. Since based on the equations (1) and (4), the robot's heading plays an important role in position determination, a single axis gyroscope is used to determine the heading of the robot. The gyro's drift error is modeled using the technique introduced by Barshan (Barshan, 1993-1995). Figure 2 shows the gyro's output when the gyro examined on a flat and stationary surface for 14 hours. The result suggests that the gyro output increases exponentially when it was stationary during the test. Gyro's output consists of gyro accurate angle rate, the modeled error and a white noise:

$$\dot{\psi}_G(k) = \dot{\psi}_G^a(k) + e_m(k) + \eta_G(k) \tag{8}$$

where $\dot{\psi}_G^a(k)$ is actual heading change rate of mobile robot based on gyroscope reading, $\dot{\psi}_G(k)$ is heading change rate of mobile robot based on gyroscope reading, $e_m(k)$ is the gyroscope bias error, and $\eta_G(k)$ is the associate white noise.

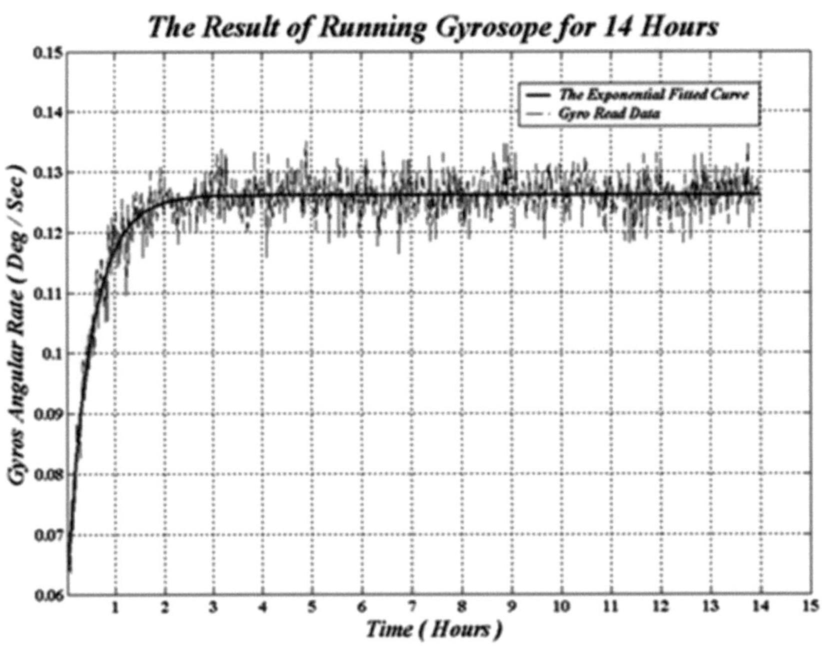

Fig. 2. The gyroscope's output for 14 hours.

A nonlinear parametric model for bias error was fitted to the data from gyroscope using least square fit method:

$$e_m(t) = C_1(1 - e^{-t/T}) + C_2 \tag{9}$$

$$\dot{e}_m(t) = \frac{C_1 + C_2}{T} + \frac{1}{T} e_m(t) \tag{10}$$

with initial conditions $e_m(0) = 0$ and $e_m(0) = C_1 / T$. After discretizing (10) becomes:

$$e_m(k+1) = \frac{T}{T+T_s} e_m(k) + \frac{T}{T+T_s}(C_1 + C_2) \tag{11}$$

The best fitting parameter value to experimental data obtained from gyroscope with zero input for 14 hours sampled every second are C_1=*0.06441, T=30.65 s,* and C_2=*0.06332*.
If we assume that the original position is known, the next position of the robot in the world frame can be determined using equation (1) by replacing the robot's heading with the information calculated from the gyro.
The mobile robot position ${}^{G}\underline{p}(k)$ in world coordinate frame based on gyro's readings is estimated by

$${}^{G}\underline{p}^{w}(k+1) = \begin{bmatrix} {}^{G}p_x^{w}(k+1) \\ {}^{G}p_y^{w}(k+1) \end{bmatrix} = \begin{bmatrix} {}^{G}p_x^{w}(k) \\ {}^{G}p_y^{w}(k) \end{bmatrix} + \begin{bmatrix} \cos\psi_G^a(k) & -\sin\psi_G^a(k) \\ \sin\psi_G^a(k) & \cos\psi_G^a(k) \end{bmatrix} \begin{bmatrix} v_x^b(k) \\ v_y^b(k) \end{bmatrix} T_s \tag{12}$$

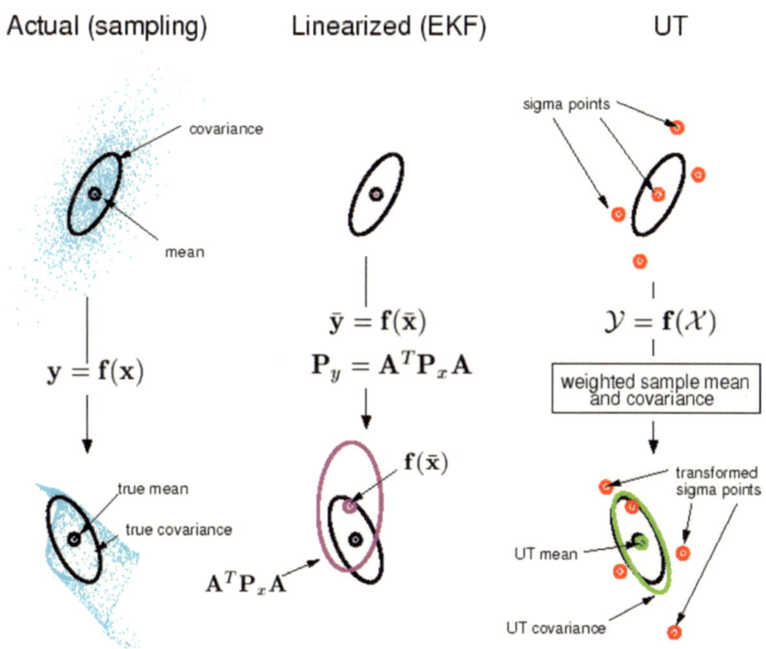

Fig. 3. Unscented Transformation (Wan, 2000).

The next step is to effectively integrate the information from odometry and gyroscope so that the odometry errors can be accurately estimated. As shown, the Kalman filter gives optimal state estimation to a linear system (Kalman, 1982). The EKF is an extension of the Kalman filter for nonlinear functions. The EKF is based on linearizing the nonlinear functions and applying Kalman filter for the estimation. Due to nonlinear nature of errors associated with odometry and other sensors, many studies have used Extended Kalman filter to integrate two or more sets of data for more accurate position determination. However, since the level of nonlinearity for odometry's nonsystematic error is high, the EKF may not be the best choice. Unscented Kalman filter (UKF) is another extension for Kalman filter which can provide accurate estimation for systems with high nonlinearity such as our system.

An important aspect of the UKF is that it can be applied to nonlinear functions without a need for linearizing (Julier, 1995, 1996). It is much simpler than EKF to implement because it doesn't require the calculation of Jacobian matrix which may provide some difficulties in the implementation.

The UKF is based on Unscented Transform (UT) that transforms the domain of the nonlinear function to another space of sigma points that has the same statistics as the original points (Julier, 2000). The UT is capable of estimating the statistics of a random variable defined by a nonlinear function (Julier, 2000). Figure 3 shows the general idea of the UT (Wan, 2000). In this transformation, a set of points called sigma point, are chosen such that it has mean of $\overline{x}$ and covariance of P_x .The first two moments of x can be calculated by choosing *2n+1* sample points (also called sigma points) as follow

$$\chi_0(k|k) = \overline{x} \tag{13}$$

$$\chi_i(k|k) = \overline{x} + (\sqrt{(n+\kappa)P_x})_i \quad i = 1,..,n \tag{14}$$

$$\chi_i(k|k) = \overline{x} - (\sqrt{(n+\kappa)P_x})_i \quad i = n+1,..,2n \tag{15}$$

$$W_0 = \kappa/(n+\kappa) \tag{16}$$

$$W_i = 1/\{2(n+\kappa)\} \quad i = 1,..,2n \tag{17}$$

where κ is scaling factor, $(\sqrt{(n+\kappa)P_x})_i$ is the i-th row or column of matrix square root of $(n+\kappa)P_x$ and W_i s are the weight associated with *i*–th sigma point. It is known that

$$\sum_{i=0}^{2n} W_i = 1 \tag{18}$$

The sigma point are propagated through the nonlinear function of $f(x)$ as

$$Y_i = f(\chi_i(k|k)) \quad i = 0,..,2n \tag{19}$$

Mean of output of function $f(x)$ can be determined as

$$\overline{y} = f(Y_i) \tag{20}$$

Therefore, it is possible to estimate the statistics of a nonlinear function without taking derivatives from the function as it is needed in the case of EKF. This makes implementation much easier than EKF which needs linearizing of the nonlinear function around the operating points and calculation of the Jacobian matrices. The parameter κ provides another degree of freedom to eliminate the estimation error for higher order statistics. It can be positive or negative but choosing a negative number returns a non-positive semi-definite estimation for P_{yy} (Wan, 2000).

Fig. 4. The mobile robot with gyroscope.

Julier and Uhlmann (Julier, 1995, 1996) were first to introduce the UT as a technique for the estimation of nonlinear systems. Their work was based on the intuition that, with a fixed number of parameters it should be easier to approximate a Gaussian distribution than it is to approximate an arbitrary nonlinear function or transformation. This could be easily drawn from the way the sigma points of the UT are being calculated. There are few versions for the UKF which they differ from each other based on whether to include the random noise into the calculations.

Houshangi and Azizi have used the UKF to successfully integrate information from odometry and a single axis gyroscope (Houshangi, 2005, 2006; Azizi, 2004). The mobile robot used in their work is a Pioneer 2-Dxe from ActivMedia Robotics Corporation. Figure 4 illustrates the robot and the single axis gyroscope used as the second sensor for position determination. The robot is a two-wheel drive and incremental encoders are installed on each wheel. These wheels have pneumatic rubber tires which give better mobility but potentially are additional source of error for positioning based on odometry such as inequality of wheel diameters which inescapable. One rear caster is used to stabilize the robot's motion and standing.

The difference between robot position and orientation comes from odometry and robot position and orientation are calculated by

$$e_x(k) = {}^{G}P_x^w(k) - {}^{O}P_x^w(k) \tag{21-1}$$

$$e_y(k) = {}^{G}P_y^w(k) - {}^{O}P_y^w(k) \tag{21-2}$$

$$e_\psi(k) = \psi_G^a(k) - \psi_o(k) \tag{21-3}$$

Each one of these errors has a modeled part and an associate noise as shown in equations (20):

$$e_x(k) = e_{mx}(k) + \eta_x(k) \tag{22-1}$$

$$e_y(k) = e_{my}(k) + \eta_y(k) \tag{22-2}$$

$$e_\psi(k) = e_{m\psi}(k) + \eta_\psi(k) \tag{22-3}$$

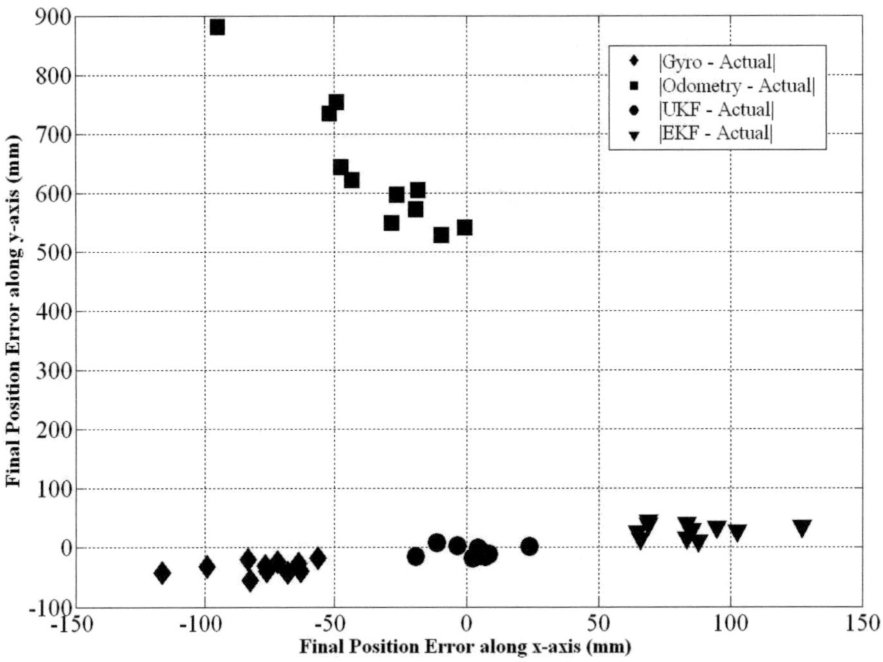

Fig. 5. Robot's position along the x-axis.

Three independent UKF described in equations (11-18) are applied to each one of the errors defined in equation (22). The estimated errors are included when an accurate position and orientation for the robot were calculated in the next sampling interval. The results were compared to the results of applying EKF to equation (22).

To evaluate the approach an experiment was designed so that the robot was run over a 3-meter square for eleven runs. The final position of the robot was calculated with respect to the actual final position. As shown in Figure 5, the UKF returns more accurate results compared to the EKF and better than using odometry or gyroscope alone. This is partly due the nature of odometry errors which are nonlinear and can be estimated more effectively using UKF rather than EKF.

3. Conclusions

Mobile robot position determination is vital for the effectiveness of robot's operation regardless of the task. Odometry is the fundamental technique for position determination but with many issues described in this chapter. The odometry's errors can be estimated using benchmark techniques and utilizing multiple sensors. Kalman Filtering is the most popular technique to integrate information from multiple sensors. However, most of physical and engineering phenomena are nonlinear and it is necessary to modify Kalman filter to estimate nonlinear systems. The EKF is able to deal with nonlinear systems and is used for mobile robot position determination. As was discussed, the odometry errors have high nonlinearity in nature. The newer extension of Kalman Filter using a transform called Unscented Transform provides another alternative for sensor integration. An example was provided using UKF to integrate gyro data and odometry. The implementation results indicated utilizing multiple sensors using UKF provided more accurate position even as compared to EKF.

4. References

Borenstein, J. & Feng. L. (1996) Measurement and Correction of Systematic Odometry Errors in Mobile Robots. *IEEE Journal of Robotics and Automation*, pp. 869-880, Vol 12, No 6, (December 1996).

Borenstein, J., Everett, H.R., Feng, L., & Wehe, D. (1997) Mobile Robot Positioning: Sensors and Techniques. *Invited paper for the* Journal of Robotic Systems, Special Issue on Mobile Robots. pp. 231-249, Vol. 14, No. 4, (April 1997).

Borenstein, J. & Feng. L. (1995) UMBmark: A Benchmark Test for Measuring Odometry Errors in Mobile Robots. *Proceedings of SPIE Conference on Mobile Robots*, Philadelphia, (October, 1995).

Chong, K.S. & Kleeman, L. (1997) Accurate Odometry and Error Modeling for a Mobile Robot. *Proceeding of the, IEEE Intl. Conference on Robotics and Automation*, Albuquerque, New Mexico, pp. 2783-2788, (April 1997)

Antonelli, G, Chiaverini, S. & Fusco, G. (2005) A calibration method for odometry of mobile robots based on the least-squares technique: theory and experimental validation. *IEEE Transactions on Robotics*, vol. 21, pp. 994–1004, (Oct. 2005).

Larsen, T. D., Bak, M., Andersen, N. A. & Ravn, O. (1998) Location estimation for an autonomously guided vehicle using an augmented Kalman filter to autocalibrate the odometry. *First International Conference on Multisource-Multisensor Information Fusion (FUSION'98)*, (1998).

Martinelli, R. & Siegwart, R. (2003) Estimating the Odometry Error of a Mobile Robot During Navigation *European Conference on Mobile Robots*, Poland (2003).

Park, K. C., Chung, H., Choi, J., and Lee, J.G. (1997) Dead Reckoning Navigation for an Autonomous Mobile Robot Using a Differential Encoder and Gyroscope", Proc. of the 8th Intl. Conference on Advanced Robotics, Monterey, CA, July 1997, 441-446

Cui, Y.J. & Ge, S.S. (2001) Autonomous vehicle positioning with GPS in urban canyon environments. *Proc. of IEEE International Conference on Robotics and Automation*, pp. 1105-1110, Seoul, Korea (Feb. 2001)

Chae, H., Christiand, Choi, S., Yu W & Cho, J. (2010) Autonomous Navigation of Mobile Robot based on DGPS/INS Sensor Fusion by EKF in Semi-outdoor Structured Environment. *Proceedings of the IEEE/RSJ International Conference on Intelligent Robots and Systems (IROS)*, pp. 1222-1227 (Oct. 2010)

Bronstein, J. & Feng, L. (1996) Gyrodometry: A New Method for Combining Data from Gyros and Odometry in Mobile robots. *Proceeding of the 1996 IEEE International*

Conference on Robotics and Automation, pp. 423–428, Minneapolis, Minnesota, (April 1996)

Kleeman, L. (1992) Optimal estimation of position and heading for mobile robots using ultrasonic beacons and dead-reckoning. *IEEE International Conference on Robotics and Automation*, Nice, France, pp 2582-2587, (May 1992)

Barshan, B., and Durrant–Whyte, H. F. (1993) An Inertial Navigation System for a for Mobile Robot. *Proceeding of 1993 IEEE/RSJ International Conference on Intelligent Robots and Systems*, pp. 2243 - 2247, Yokohama, Japan, (July 1993)

Barshan, B., and Durrant–Whyte, H. F. (1994) Evaluation of a Solid - State Gyroscope for Mobile Robots Application *IEEE Transactions on Instrumentation and Measurement*, pp. 61 - 67, Vol. 44, No. 1, (February 1994)

Barshan, B. & Durrant-Whyte, H. F. (1995)Inertial Navigation Systems for Mobile Robots. *IEEE Transactions on Robotics and Automation*, pp. 328 - 342, Vol. 11, No. 3, (June 1995)

Borenstein, J. (1998) Experimental Evaluation of the Fiber Optics Gyroscope for Improving Dead-reckoning Accuracy in Mobile Robots *Proceeding Of the IEEE International Conference on Robotics and Automation*, 3456 - 3461, Leuven, Belgium, (May 1998)

Roumeliotis, S.I. & Bekey, G.A. (2000) Bayesian Estimation and Kalman Filtering: A unified Framework for Mobile Robot Localization. *Proceedings of IEEE International Conference on Robotics and Automation*, pp. 2985-2992, San Fransisco, (April 2000).

Amarasinghe, D., Mann, G.K.I. & Gosine, R.G. (2010) Landmark detection and localization for mobile robot applications: a multisensor approach. *Robotica,* pp. 663–673 (2010) volume 28.

Kalman, R. E. (1982) A New Approach to Linear Filtering and Prediction Problems. *Transaction of the ASME – Journal of Basic Engineering, (Series D)*, pp. 35 – 45.

Julier, S. J. & Uhlmann, J. K. &Durrant (1995) Whyte A New Approach for Filtering Nonlinear Systems. *Proceeding of the American Control Conference*, pp. 1628 - 1632, Seattle, Washington, 1995.

Julier, S. J. & Uhlmann, J. K. (1996) A New Method for Approximating Nonlinear Transformation of Probability Distributions. *Tech. Report, Robotic Research Group, Department of Engineering Science*, University of Oxford, 1996

Julier, S. J. & Uhlmann, J. K. (1997) A New Extension of the Kalman Filter to Nonlinear Systems. Proceeding of Aerospace: *The 11th International Symposium on Aerospace/ Defense Sensing, Simulation and Controls*, pp. 182 – 193, Orlando, Florida, (2000)

Julier, S. J. (2002) The scaled unscented transformation. *Proceedings of the American Control Conference*, Vol. 6, pp. 4555–4559, (2002)

Wan, E.A. & van der Merwe, R. (2000) The Unscented Kalman Filter for Nonlinear Estimation. *Proceedings of Symposium on Adaptive Systems for Signal Processing, Communication and Control*, Lake Louise, Alberta, Canada, (Oct. 2000)

Azizi, F. & Houshangi, N. (2004) Mobile robot position determination using data from gyro and odometry. *Canadian Conference on Electrical Engineering*, Vol. 2, pp.719 - 722, (May 2004).

Houshangi, N. & Azizi, F. (2005) Accurate mobile robot position determination using unscented Kalman filter. *Canadian Conference on Electrical Engineering*, pp. 846 – 851, (May 2005)

Houshangi, N. & Azizi, F. (2006) Mobile Robot Position Determination Using Data Integration of Odometry and Gyroscope. *World Automation Congress*. (2006), pp. 1–8

5

Vision Based Obstacle Avoidance Techniques

Mehmet Serdar Guzel and Robert Bicker
Newcastle University,
United Kingdom

1. Introduction

Vision is one of the most powerful and popular sensing method used for autonomous navigation. Compared with other on-board sensing techniques, vision based approaches to navigation continue to demand a lot of attention from the mobile robot research community. This is largely due to its ability to provide detailed information about the environment, which may not be available using combinations of other types of sensors. One of the key research problems in mobile robot navigation is the focus on obstacle avoidance methods. In order to cope this problem, most autonomous navigation systems rely on range data for obstacle detection. Ultrasonic sensors, laser rangefinders and stereo vision techniques are widely used for estimating the range data. However all of these have drawbacks. Ultrasonic sensors suffer from poor angular resolution. Laser range finders and stereo vision systems are quite expensive, and computational complexity of the stereo vision systems is another key challenge (Saitoh et al., 2009). In addition to their individual shortcomings, Range sensors are also unable to distinguish between different types of ground surfaces, such as they are not capable of differentiating between the sidewalk pavement and adjacent flat grassy areas. The computational complexity of the avoidance algorithms and the cost of the sensors are the most critical aspects for real time applications. Monocular vision based systems avoid these problems and are able to provide appropriate solution to the obstacle avoidance problem. There are two fundamental groups of vision based obstacle avoidance techniques; those that compute the apparent motion, and those that rely on the appearance of individual pixels for monocular vision based obstacle avoidance systems. First group is called as Optical flow based techniques, and the main idea behind this technique is to control the robot using optical flow, from which heading of the observer and time-to-contact values are obtained (Guzel & Bicker, 2010). One way of the control using these values is by acting to achieve a certain type of flow. For instance, to maintain ambient orientation, the type of Optic flow required is no flow at all. If some flow is detected, then the robot should change the forces produced by its effectors so as to minimize this flow, based on Law of Control (Contreras, 2007).

A second group is called Appearance Based methods rely on basic image processing techniques, and consist of detecting pixels different in appearance than that of the ground and classifying them as obstacles. The algorithm performs in real-time, provides a high-resolution obstacle image, and operates in a variety of environments (DeSouza & Kak, 2002). The main advantages of these two conventional methods are their ease of implementation and high availability for real time applications. However optical flow *based* methods suffer from two major problems, which are the illumination problem that varies with time and the problem of motion discontinuities induced by objects moving with respect to other objects or the background (Contreras, 2007). Various integrated methods for solving these problems have been proposed; nevertheless it is still a key challenge to employ optical flow for mobile robot navigation. Furthermore, appearance based methods also suffer from illumination problems and, are highly sensitive to floor stains, as well as to the physical structure of the terrain.

Consequently, while having significant performance advantages, there are certain drawbacks which restrict the applicability of these methods. In order to solve those challenges, a novel obstacle avoidance method is introduced in this chapter. The method is principally designed to fuse a Scale invariant features transform (SIFT) algorithm (Lowe, 1999), and template matching with a convolution mask technique, using a *Fuzzy Logic approach.* As opposed to the *Appearance based methods,* previously mentioned, an occupancy map of the environment is generated with respect to the local features and a template. The experimental results reveal that the proposed obstacle avoidance technique allows the robot to move efficiently within its environment and to successfully attain its local goals.

This chapter is organized as follows. In Section 2, the background knowledge to the conventional methods is briefly introduced. In Section 3, new technique is introduced. Section 4 provides the implementation of the behaviour-based robot and the experiment results from both the real and simulation experiments. Section 5 provides a summary of the work.

2. Background

In this section, optical flow based navigation techniques will first be outlined, followed by a brief introduction of the appearance based methods.

2.1 Optical flow

Optical flow, illustrated in Fig. 1, is an approximation to the motion field, summarizing the temporal change in an image sequence. The main idea behind the technique assumes that for a given scene point, the corresponding image point intensity I remain constant over time, which is referred as *conservation of image intensity* (Atcheson et al., 2009). Therefore, if two consecutive images have been obtained at the following time intervals, the basic idea is to detect the motion using image differencing. If any scene point projects onto image point (x, y) at time t and onto image point $(x + \delta x, y + \delta y)$ at time $(t + \delta t)$, the following equation is inferred based on the conservative of image intensity assumption.

$$I(x, y, t) = I(x + \delta x, y + \delta y, t + \delta t) \tag{1}$$

Expanding the right-hand side of the Eq. 1 using a Taylor series about (x, y, t), and ignoring the higher order terms then by rearrangement gives the following expression.

$$\delta x \frac{\partial I}{\partial x} + \delta y \frac{\partial I}{\partial y} + \delta t \frac{\partial I}{\partial t} = 0 \tag{2}$$

A simpler expression, is obtained by dividing by δt throughout and movement along the horizontal $(\frac{\delta x}{\delta t})$, and vertical $(\frac{\delta y}{\delta t})$ directions are u and v respectively. Having these rearrangements and denoting partial derivatives of I by I_x, I_y and I_t gives the differential flow equation shown in following expressions:

$$I_x u + I_y v + I_t = 0 \tag{3}$$

where, I_x, I_y and I_t are the partial derivatives of image brightness with respect to x, y and t, respectively. Having one equation in two unknowns δx, δy for each pixel is an aperture problem of the optical flow algorithms. To find the optical flow another set of equations is needed, using some additional constraint. All optical flow methods introduce additional conditions for estimating the actual flow. There are several methods employed to determine optical flow, namely: Block-based methods, differential methods, Phase Correlation and General variational methods (Atcheson et al., 2009). Differential methods are widely used for navigation tasks, and are mainly based on partial derivatives of the image signal and/or the sought flow field and higher-order partial derivatives. One of those methods is used to estimate flow vectors to steer the robots.

2.1.1 The optical flow method proposed by Horn and Schunk

Horn and Schunk proposed one of the most important optical flow methods using an gradient based approach (Horn & Schunck, 1981). According to their methodology, a regularizing term associated with smoothness is added to the general flow equation, as illustrated in Equation 3, in which neighbouring pixels have the same velocity as moving objects, so the brightness pattern of an image changes regularly. This constraint is demonstrated by minimizing the squares of gradient magnitudes. Smoothness of an optical flow area can also be calculated by determining the Laplacian of optical flow vectors speed both horizontal and vertical directions denoted by u and w respectively, illustrated in following expressions:

$$\nabla^2 u = \frac{\partial^2 u}{\partial x^2} + \frac{\partial^2 u}{\partial y^2}$$

$$\nabla^2 v = \frac{\partial^2 u}{\partial x^2} + \frac{\partial^2 u}{\partial y^2} \tag{4}$$

Where $E_i = \frac{\partial I}{\partial x}\delta x + \frac{\partial I}{\partial y}\delta y + \frac{\partial I}{\partial t}$ and $E_s = \nabla^2 u + \nabla^2 v$. The aim is to minimize the total error given by the following expressions that includes σ as the regularization parameter, controlling the association between the detail and the smoothness. High values of σ makes the smoothness constraint dominate and leads to a smoother flow

$$\iint \left(E_i^{\ 2} + \sigma^2 E_s^{\ 2}\right) dxdy \tag{5}$$

Horn and Schunk, can be used as the main reference to understand and solve the given error function, from which a pair of equations for each point can be obtained. Direct solution of these equations such as Gaus-Jordan Elimination (Bogacki, 2005) would be very costly. Instead, an iterative Gauss Seidel, approach is used to reduce the cost and obtain the flow vectors, as follows (Horn & Schunck, 1981):

$$u^{n+1} = \bar{u}^n - \left(\frac{I_x \bar{u}^n + I_y \bar{u}^n + I_t}{\sigma^2 + I_x^2 + I_y^2}\right); \; v^{n+1} = \bar{v}^n - \left(\frac{I_x \bar{v}^n + I_y \bar{v}^n + I_t}{\sigma^2 + I_x^2 + I_y^2}\right) \tag{6}$$

where I_x I_y and I_t are the partial derivatives with respect to x,y and t respectively, and the superscript *n+1* denotes the next iteration, which is to be calculated and *n* is the last calculated result (Horn & Schunck, 1981).

Fig. 1. Optical Flow vectors (Guzel & Bicker, 2010).

2.2 Optical flow for mobile robot navigation

Flow vectors are utilized to navigate autonomous systems based on the Balance Strategy (Souhila & Karim, 2007), shown in the following equation, and the depth information which is extracted from the image sequence using Focus of Expansion(FOE) and Time To Contact values(TTC) (Souhila & Karim, 2007). The fundamental idea behind the Balance strategy is that of motion parallax, when the agent is translating, closer objects give rise to faster motion across the retina than farther objects. It also takes advantage of perspective in that closer objects also take up more of the field of view, biasing the average towards their associated flow (Contreras, 2007). The agent turns away from the side of greater flow. This control law is formulated by:

$$\Delta(F_l - F_r) = \left(\frac{\sum|w_L| - \sum|w_R|}{\sum|w_L| + \sum|w_R|}\right) \tag{7}$$

where $\sum|w_L|$ and $\sum|w_R|$ are the sum of the magnitudes of optical flow in the visual hemi fields on both sides of the robot's body. The following expression gives the new heading angle

$$\theta_{new} = (\Delta(F_l - F_r) \times k) \tag{8}$$

where k, is a constant, and used to convert the obtained result to an appropriate control parameter to steer the robot.

2.3 Appearance-based methods

Appearance based methods that identify locations on the basis of sensory similarities are a promising possible solution to mobile robot navigation. The main idea behind the strategy is to head the robot towards the obstacle-free position using similarities between the template and the active images (F. Vassallo et al., 2000). The similarity between the image patterns can be obtained by using feature detectors, involving corner based detectors, region based detectors and distribution based descriptors (Alper et al., 2006). However, most of these techniques consume a lot of process on time which is not appropriate for real time systems. In order to handle this problem in mobile robot applications, algorithms are designed based on the appearance of individual pixels. The classification of the obstacles is carried out by using the pixel difference between the template and active image patterns.In principally; any pixel that differs in appearance from the ground is classified as an obstacle. However, the method requires three assumptions that are reasonable for a variety of indoor and outdoor environments which are (Saitoh et al., 2009):

a. Obstacles must be different in appearance from the ground.
b. The ground must be flat.
c. There must be no overhanging obstacles.

The first assumption is to distinguish obstacles from the ground, while the second and third assumptions are required to estimate the distances between detected obstacles and the robot. There are several models for representing colour. The main model is the RGB(Red, Green, Blue) model which is used in monitor screens and most image file formats however, colour information for RGB model is very noisy at low Intensity. The RGB format is mostly converted to a HSV (Hue, Saturation, and Value). In HSV, Hue is what humans perceive as colour, S is saturation and Value is related to brightness, (or HIS(Hue, Intensity, Saturation) model) and in HIS, H and S represents the same as parameters in HSV colour models but I is an intensity value with a range between [0,1] where 0 is black and white is 1. These colour spaces are assumed to be less sensitive to noise and lighting conditions. The flow chart of the appearance based obstacle detection systems is illustrated in Figure 2. The input image is first convolved with a smoothing filter to reduce the noise effects, and then smoothed image is converted to *HIS, HSV* or any related *colour space* with respect to the developed algorithm (Fazl-Ersi & Tsotsos, 2009). A reference area is obtained from this image which might be any shape of geometry such as trapezoidal, triangle or square, and histogram values of this reference area are generated (Saitoh et al., 2009). Finally, a comparison between the reference image and the current image is made using some predefined threshold values. For instance, assume that the bin value, Hist($H(x, y)$), of the generated histogram and the threshold value,T_H, are compared, where $H(x, y)$ is the H value at pixel (x, y). If Hist($H(x, y)$) > T_H then the pixel (x, y) is classified into the safe region, or else it is classified into the obstacle region. In order to simply the problem, the results are represented in a binary image in which the safe path is represented with white but the obstacles are represented with black, as illustrated in Figure 3.However, identifying places purely on the basis of sensory similarity is too simplistic; different places may look very similar, even with a rich sensing methodology due to lighting conditions, shadows on illumination Furthermore, for dynamic environments there might be unexpected stains on the ground which may be the detected as an obstacle and leads the robot to an unsafe path. An example with respect to this case is illustrated in Figure 4.

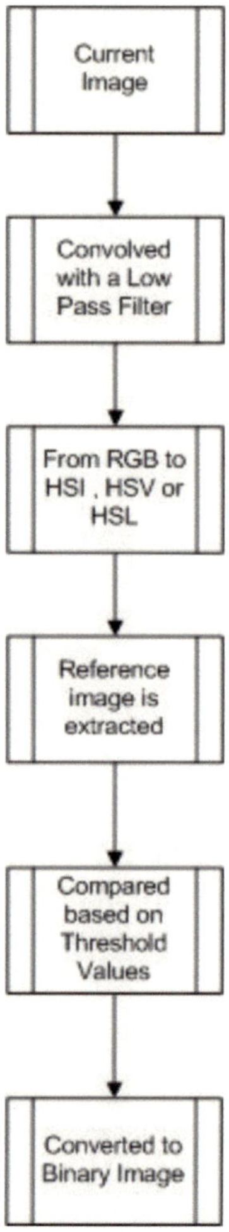

Fig. 2. Flow chart of the Appearance Based obstacle detection algorithm.

Fig. 3. Appearance based obstacle detection method.

Fig. 4. Effects of lighting conditions and unexpected stains on the floor.

3. SIFT and template matching based obstacle avoidance strategy

In order to cope with the drawbacks of the conventional appearance based methods, a novel feature matching based technique, comprising a Scale Invariant Feature Transform and Template matching with a convolution mask, will be discussed in this section. The detail of the control algorithms with respect to these techniques is illustrated in Figure 5. Before introducing the proposed control algorithm and fusion technique, essential background knowledge regarding the SIFT and Template matching will be presented.

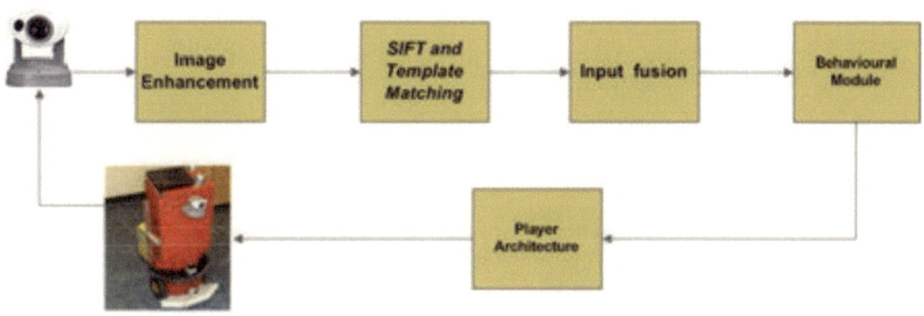

Fig. 5. Control architecture of the obstacle avoidance system.

3.1 Scale invariant feature transform (SIFT)

The Scale Invariant Feature Transform formerly abbreviated as SIFT is an algorithm in computer vision to detect and describe local features in images. The algorithm was published by Lowe (Lowe, 1999a, 2004b), and since then has been accepted as one of the most powerful local feature detection technique. The most notable improvements provided by SIFT are invariance to scale and rotation, and accuracy in feature point localization and matching. The evaluations carried out proposes that SIFT-based descriptors, which are region-based, are the most robust and distinctive, and are therefore best suited for feature matching. A summary of the SIFT methodology is illustrated in Fig. 6. (Lowe, 1999a, 2004b). The initial state of this algorithm is Scale space extreme detection where the interest points, which are called key-points in the SIFT framework, are detected. For this, the image is convolved using Gaussian filters, proved the only possible scale-space kernel, at different

Problem	Technique	Advantage
key localization / scale / rotation	DoG / scale - space pyramid / orientation assignment	accuracy, stability, scale & rotational invariance
geometric distortion	blurring / resampling of local image orientation planes	affine invariance
indexing and matching	nearest neighbour / Best Bin First search	Efficiency / speed
Cluster identification	Hough Transform voting	reliable pose models
Model verification / outlier detection	Linear least squares	better error tolerance with fewer matches
Hypothesis acceptance	Bayesian Probability analysis	reliability

Fig. 6. SIFT Methodology (Lowe, 1999a, 2004b).

scales, and then the difference of successive Gaussian-blurred images are obtained, illustrated in Figure 7. The convolved images are grouped by octave which corresponds to doubling the value of standard deviation of the Gaussian distribution (σ).The Convolution of the image at scale $k\sigma$ with a Gaussian filter is expressed as follows:

$$L(x, y, k\sigma) = G(x, y, k\sigma) * I(x, y) \tag{9}$$

where,

$$G(x, y, \sigma) = \frac{1}{(2\pi\sigma^2)} exp^{-(x^2+y^2)/2\sigma^2} \tag{10}$$

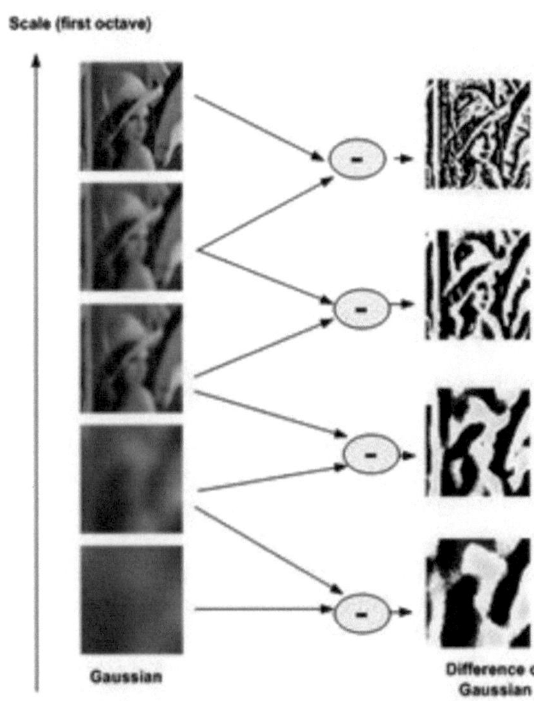

Fig. 7. Gaussian images are subtracted to produce the DoG.

When DoG images have been obtained, key- points are identified as local minima/maxima of the DoG images across scales. This is done by comparing each pixel in the DoG images to its eight neighbors at the same scale and nine corresponding neighboring pixels in each of the neighboring scales. If the pixel value is the maximum or minimum among all compared pixels, it is selected as a candidate keypoint', as shown in Figure 8.
Following steps are Key-point localization and Orientation assignment (Lowe, 1999a, 2004b). After key-point orientation has been completed, each key specifies stable 2D coordinates, comprising x, y, scale and orientation. Finally, a signature, local descriptor, is computed as a set of orientation histograms on 4x4 pixel neighbourhoods. Histograms have 8 bins each, and each descriptor contains an array of 4 histograms around the key-point. This leads to a SIFT feature vector with 8x4x4 = 128 elements, illustrated in Figure 9. This vector is normalized to enhance invariance to changes in illumination.

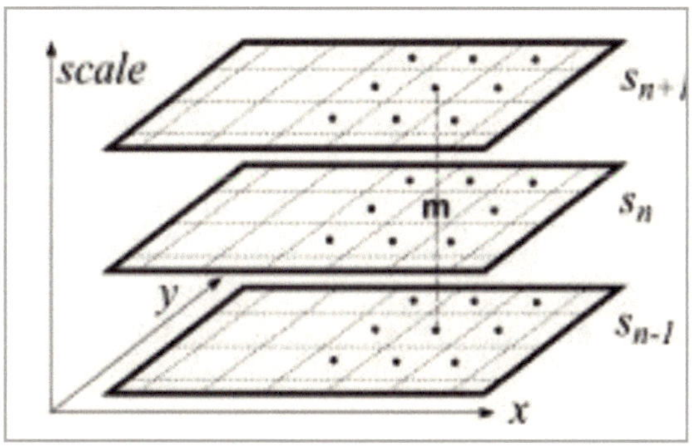

Fig. 8. A key-point is defined as any value in the DoG.

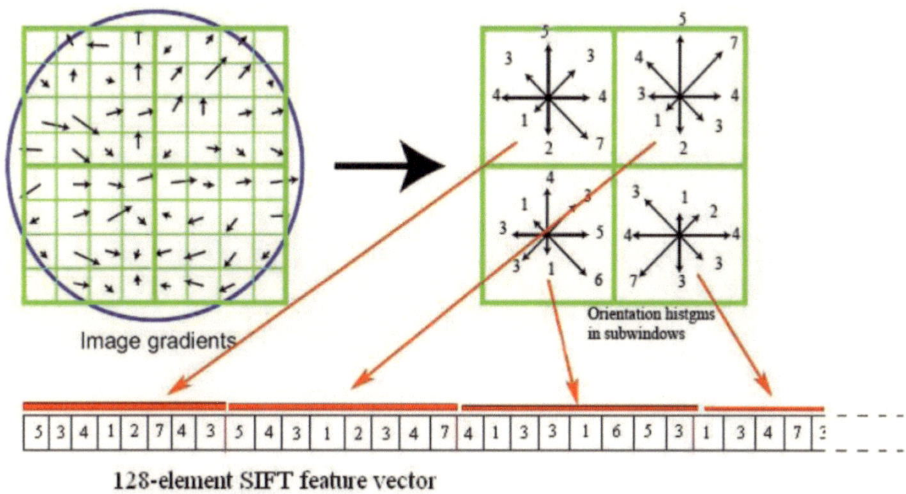

Fig. 9. SIFT Feature Descriptor.

3.1.1 SIFT matching

Feature vectors which extracted from the SIFT algorithm to solve common computer vision problems, comprising object detection, 3D scene modeling, recognition and tracking, robot localization and mapping. This procedure requires an appropriate and fast matching algorithm. An example with respect to the SIFT matching is illustrated in Figure 10. The main matching algorithm is able to find each key-point by identifying its nearest neighbor in the database of key-points from training images. The nearest neighbor is defined as the key-point with minimum Euclidean distance for the invariant descriptor vector, as previously discussed.However, many features may not be matched correctly due to background clutter in natural images or may not have correct match in the training database, and hence, mismatches should be discarded in order to obtain accurate results. Global thresholding on distance to the closest feature does not perform well, as some descriptors are much more discriminative than others. Alternatively, a more effective measure is obtained by comparing the distance of the closest neighbor to that of the second-closest neighbor, which performs well. An appropriate threshold value regarding this comparison, called distance ratio, is employed to reject false matches while increasing the correct matches. This value varies from 0.1 to 0.9, depending on the application type. In this case, 0.7 is employed with respect to the experimental results, which eliminates %90 of the false matches while discarding almost %10 correct matches. In addition, to reject rest of all false matches, an essential statistical method is applied to the matching space, fundamentally using the rate between scale and orientation parameters of the feature vectors. According to this method, even though the matching is validated between any two feature vectors, if the scale or the orientation parameter rate between them is more than a threshold value, matching is discarded; this procedure performs robustly to decrease false matches over all data sets (Lowe, 1999a,2004b).

Fig. 10. SIFT matching.

3.2 Template matching

Template matching is a simple and popular technique in computer vision and image processing to find small parts of an image which match a template image. It can be used in mobile robot navigation or as a way to detect edges or objects in images; an example with respect to this technique is illustrated in Figure 11. A basic method of template matching uses a convolution mask which can be easily performed on grey images. The convolution output will be the highest at places where the image structure matches the mask structure, i.e. where large image values get multiplied by large mask values. This method is normally implemented by first picking out a part of the search image to use as a template.

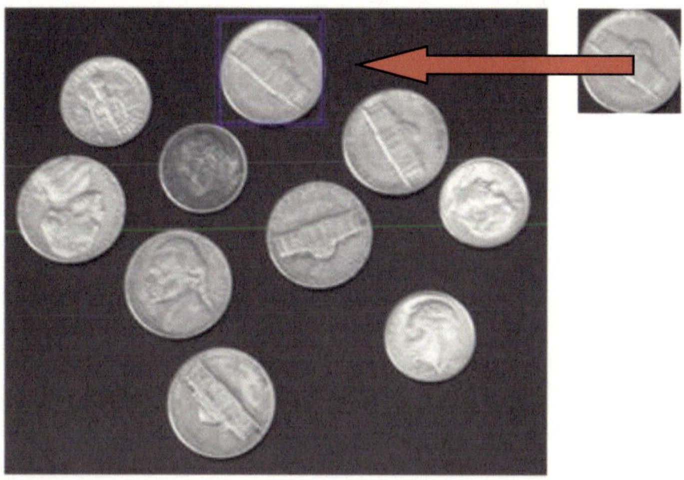

Fig. 11. Template Matching.

For instance, the input and output images are called I(x, y) and O(x, y) respectively, where (x, y) represent the coordinates of each pixel in the images and the template is called $T(x_t, y_t)$, where (x_t, y_t) represent the coordinates of each pixel in the template. The technique simply moves the centre of the template $T(x_t, y_t)$ over each (x, y) point in the search image and calculates the sum of products between the coefficients in I(x, y) and $T(x_t, y_t)$ over the whole area spanned by the template. As all possible positions of the template with respect to the Input image are considered, the position with the highest score is the best position, and which is represented in the output image. There are several techniques to handle translation problem; these include using SSD (Sum of squared differences), CC (Cross Correlation) and SAD (Sum of absolute differences) (Wen-Chia & Chin-Hsing, 2009). One of the most powerful and accurate of those is CC, which basically measures the similarity of two variables and defined as follows (9):

$$Cor = \frac{\sum_{i=0}^{N-1}(x_i-\dot{x})\times(y_i-\dot{y})}{\sqrt{\sum_{i=0}^{N-1}(x_i-\dot{x})^2\times\sum_{i=0}^{N-1}(y_i-\dot{y})^2}} \qquad (11)$$

Where N is the template image size; $\dot{x}$ and $\dot{y}$ represents average gray level in the template and source image respectively. The goal is to find the corresponding (correlated) pixel within a certain disparity range that minimizes the associated error and maximizes the similarity. This matching process involves computation of the similarity measure for each disparity value, followed by an aggregation and optimization step (ZitovÃ¡ & Flusser, 2003). An example related to correlation based technique is illustrated in Figure 12.

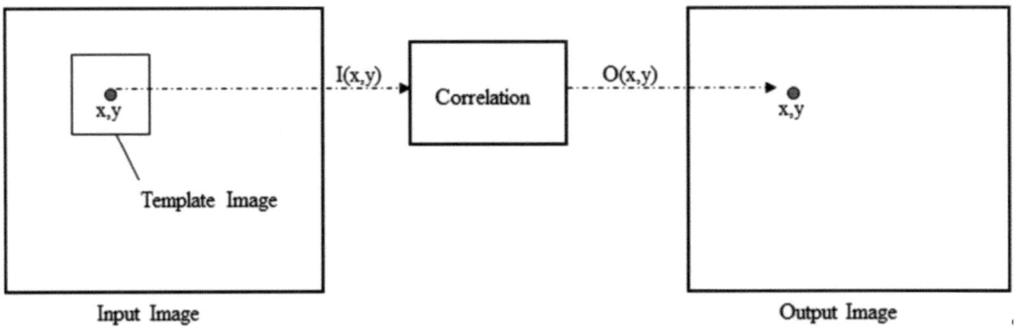

Fig. 12. Correlation technique based Template Matching.

3.3 Obstacle avoidance using SIFT and appearance based

Appearance based methods have significant processing and performance advantages which make them a good alternative for vision based obstacle avoidance problems. However as mentioned previously, there are certain drawbacks, which *restricts the applicability of these methods*. In order to handle these drawbacks, the results of the conventional method is improved by using the results of *SIFT* based feature matching approach. The flowchart diagram of the proposed algorithm is illustrated in Figure 13. First the acquired image is smoothed using a Gaussian filter to eliminate the noise in the image. Then a copy of the original image is converted to PGM image format which is required to carry out the *SIFT* matching process. Both images are divided into 16 sub-images composed of 44×36 pixels. For each sub images, template matching using cross correlation and *SIFT* matching are performed against reference images, illustrating the safe route, simultaneously. The results for each segment are fused using fuzzy logic to build up a sufficiently accurate occupation map of the environment.

The robot, which employs a Subsumption Architecture (Brooks, 1986), is successfully directed along a collision-free path using this map. To evaluate the performance of the proposed algorithm, it is applied to a test case, as illustrated in Figure 14. The matching results, shown in Figure 15, indicate that both techniques generate similar results under ideal conditions, involving low illumination changes and flat ground. However, template matching may fail to provide accurate matching results against illumination problem, and non-flat surfaces. In order to handle these cases, *SIFT* matching provides a reliable matching strategy, which is able to match the extracted features, invariant to scale, orientation, affine distortion, and partially invariant to illumination changes, with high accuracy. An example illustrating this condition can be seen in Figure 16, comprising 9th and 10th sub-image sequences. Despite the illumination problem and non-flat ground, the SIFT matching performs well for these cases illustrated in Figure 17. Consequently; the results indicate that instead of using each method separately, fusion of them generates more reliable results. A fuzzy logic based approach with respect to this fusion procedure will be discussed in the following section.

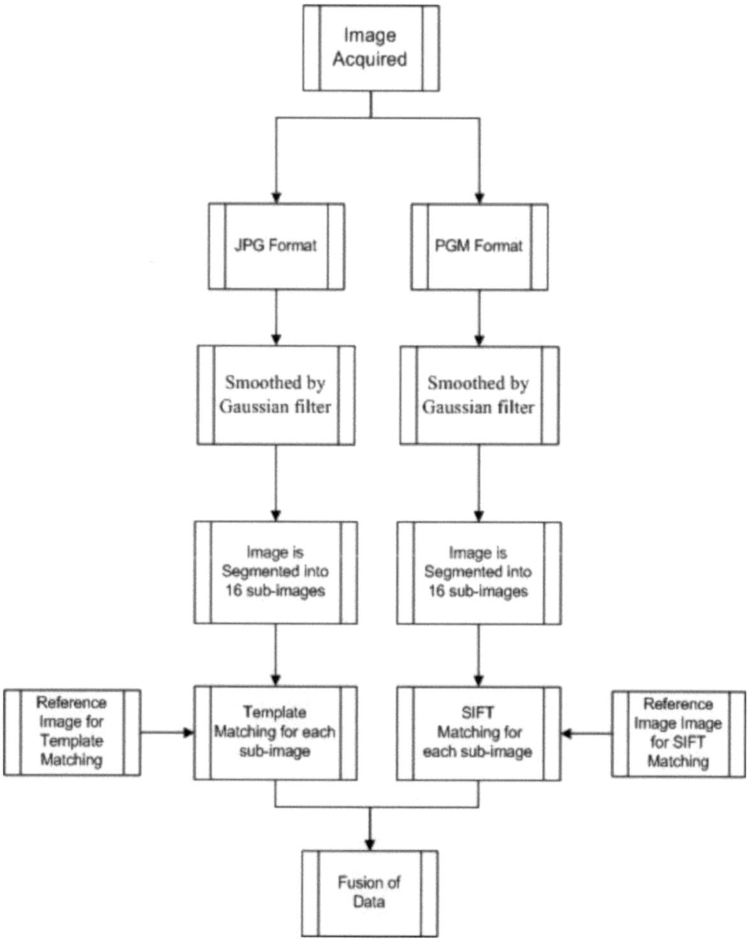

Fig. 13. An example from the testing environment.

Fig. 14. An example from the testing environment.

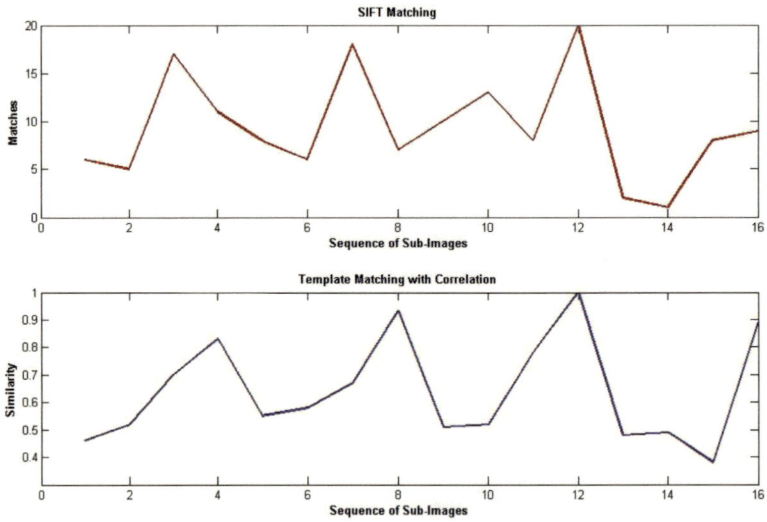

Fig. 15. Comparison of two algorithms.

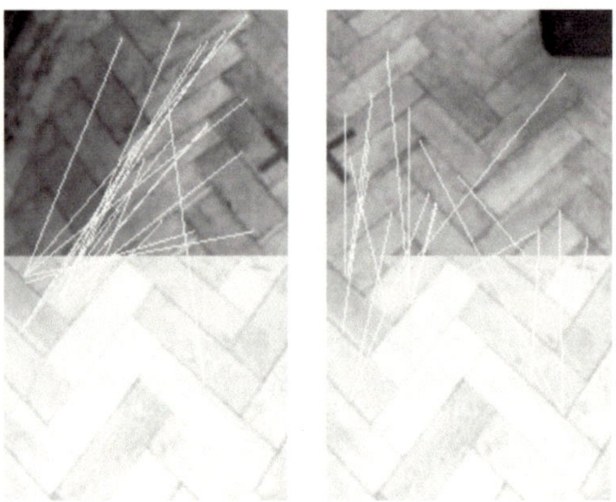

(a) 9th sub image (b) 10th sub image

Fig. 16. Matching results of two algorithms.

Fig. 17. False matching generated by SIFT matching.

3.4 Fuzzy logic

Fuzzy logic (FL) is a form of many-valued logic derived from fuzzy set theory to deal with reasoning that is robust and approximate rather than brittle and exact. In contrast with two-valued Boolean logic, FL deals with degrees of membership and degrees of truth. FL uses the continuum of logical values between 0 (completely false) and 1 (completely true). FL has been utilized as a problem-solving control system by several researchers for different problems. Since, FL lends itself to implementation in systems ranging from simple, small, embedded micro-controllers to large, networked, multi-channel PC or workstation-based data acquisition and control systems.

FL is fundamentally easy to implement and provide faster and more consistent results than conventional control methods. In this study, a FL based control system based on the Mamdani method is designed to fuse given algorithms. The basic configuration of a fuzzy-logic system is composed of three parts: *Fuzzification*, *Inference Mechanism* and *Deffuzification (Driankov, 1987)*. These will be presented and associated with the fusion problem in the following parts.

3.4.1 Fuzzification

Fuzzfication comprises a scale of transformation of input data of a current process into a normalised domain. This process requires the identification of two parts: the first part defines the fuzzy variables that correspond to the system input variables. The second part is to define the fuzzy sets of the input variables and their representative membership functions including the ranges of the data. Membership function, may cross the boundary of another fuzzy membership function. Each membership function may be triangular, a trapezoidal or bell shaped, as illustrated in Figure 18. The choice of the fuzzy sets is based on expert opinion using natural language terms that describe the fuzzy values. In this study triangle and trapezoid models are utilized to design membership functions of input and output values.

Fuzzy logic uses intersection, union, and complement operations to represent the standard common operators of AND, OR, and NOT, respectively. The most common method used to calculate intersection and union operations are the Minimum and Maximum functions. For the fuzzy sets M and N which are subsets of the universe X, the following definitions are proposed to represent the AND, OR, and NOT operators, respectively (Ross & Hoboken, 2004) (see Figure 19).

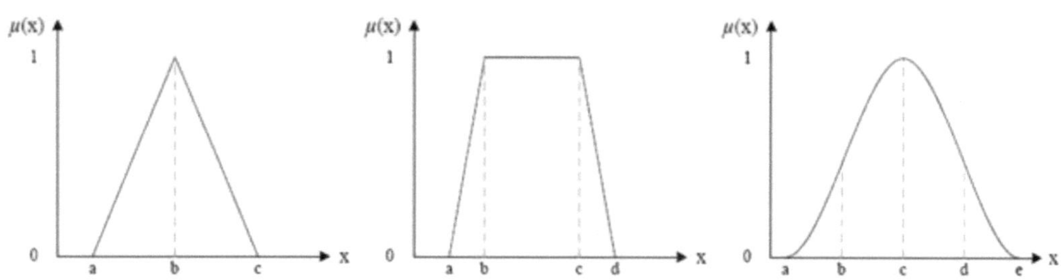

(a) (b) (c)

Fig. 18. Membership function shapes, (a) Triangular, (b) Trapezoidal, (c) Gaussian.

The triangular function has three parameters which can be defined as follows:

$$\mu(x) = \begin{cases} 0, & x < a \\ \frac{x-a}{b-a}, & a \leq x < b \\ \frac{c-x}{c-b}, & b \leq x \leq c \\ 0, & x > c \end{cases} \tag{12}$$

The trapezoidal function incorporates four parameters can be represented as:

$$\mu(x) = \begin{cases} 0, & x < a \\ \frac{x-a}{b-a}, & a \leq x < b \\ 1, & b \leq x \leq c \\ \frac{d-x}{d-c}, & c < x \leq d \\ 0, & x > d \end{cases} \tag{13}$$

3.4.2 Inference mechanism

The generation of the fuzzy rules is a first step which depends on the knowledge and experience of the human operators, the fuzzy model of the plant concerned, and an analysis of the system. The rule-base is composed of two parts namely, the IF-part and the THEN-part. The IF-part is the antecedent part where rules are defined to describe the system state in terms of a combination of fuzzy propositions while the THEN-part is the consequent part which forms the desired conclusion of the output variable. Afterwards, fuzzy inference provides the conclusion of the rule-base and forms the intermediate stage between the fuzzification and defuzzification of the fuzzy system. There are two methods used to find the rules conclusion namely Max-Min inference and Max-Product inference. Max-Min utilizes the Minimum operator to combine the antecedent of the IF-THEN rules which produces modified fuzzy sets for the outputs. These modified sets are then combined using the Maximum operator and Max-Product inference and utilizes the standard Product operator to combine the antecedent of the IF-THEN rules. Then the Maximum operator is used to combine these modified sets (Ross & Hoboken, 2004).

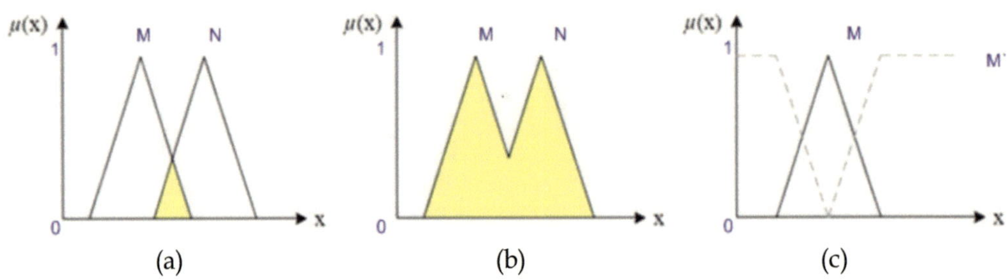

Fig. 19. Fuzzy Operators, (a) AND, (b) OR, (c) NOT

3.4.3 Defuzzification

Defuzzification is the process of mapping from a space of inferred fuzzy control action to a space of non-fuzzy control actions where the calculated crisp value is that which best represents the inferred control action. Several methods can be used to calculate this crisp value such as the Centre-of-Area, Centre-of-Largest-Area, Centre-of-Sums, and Mean-of-Maximum. These methods are based on two basic mechanisms: Centroid and Maximum. The centroid methods are based on finding a balance point while the Maximum methods search for the highest peak of weight (area) of each fuzzy set (Ross & Hoboken, 2004). Centre-of-Sum is used in this study which is faster than many defuzzification methods, and is not restricted to symmetric membership functions. This process performs the algebraic sum of individual output fuzzy sets instead of their union, illustrated in the following equation:

$$\mu^* = \frac{\int y \sum_{n=1}^{k} \mu_n(y) d_y}{\int \sum_{n=1}^{k} \mu_n(y) d_y} \quad (14)$$

3.4.4 Fusion of algorithms with fuzzy logic

The fundamental architecture of the proposed system is illustrated in Figure 20. The FL controller has two inputs, namely SIFT and Correlation which involves matching strength and similarity rate respectively. The output of the controller generates an appropriate turning rate (w) to avoid obstacle. In order to adapt the results of vision based algorithms efficiently as well as to operate the robot smoothly, the image is divided into n clusters (sub-images), based on the resolution of the image, its parts, and for each cluster a fuzzy fusion algorithm is applied. SIFT matching value of each cluster is rescaled by multiplying with n which normalizes the input for Fuzzy Inference System (FIS). Final turning rate is calculated to sum up all clusters, considering the sign of each part which is defined as follows:

$$w = \sum_{i=1}^{n} w_i \,, \; n \; is \; even, left \; clusters \; are \; positive \; and \; right \; ones \; are \; negative \quad (15)$$

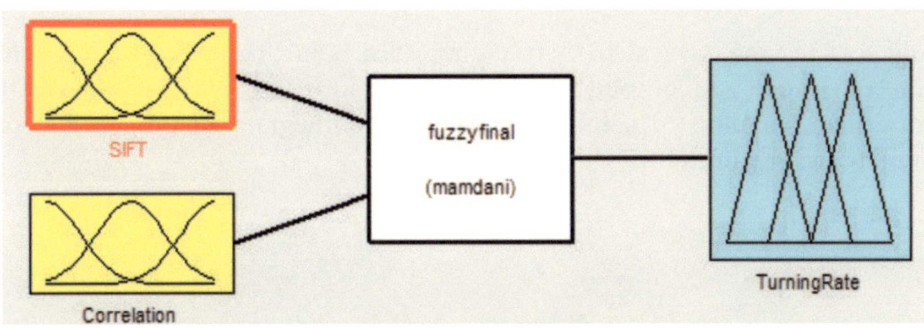

Fig. 20. Architecture of Fuzzy control system for fusing operation.

To generate a value of turning rate (w), a Max-Min type FIS and a Centre-of-Sums defuzification method were used, as previously discussed. The first step is to design membership functions for fuzzification and defuzzification processes. While, several researchers utilize different membership function shapes regarding to the problem in various applications. The trapezoidal and triangular shapes have been selected in this work to simplify the computation. However, it should be noted that there are no precise methods to adjust the membership functions. Table 1 provides an example of the selection of the fuzzy terms that describe linguistic variables used in this study. For this study three membership functions are defined *namely: SIFT* (s_m), *Correlation* (c_s) *and Turning Rate* (w). *SIFT* function, representing the matching strength between the reference image and the current image, is illustrated in Figure 21(a). Whereas the correlation function represents the similarity between the reference and current images, as illustrated in Figure 21(b). Turning Rate is the output function which represents the angular velocity value to steer the robot whilst avoiding obstacles (see Figure 22). The next step is to design appropriate fuzzy rules depending on the detail of each Fuzzy Inference System (FIS). A set of experiments were carried out until the outputs are judged to satisfy each different situations. Table 2 displays the fuzzy rules for the given problem.

Linguistic Variables	Linguistic Terms
SIFT	Weak, Medium, Strong
Correlation	Weak, Medium, Strong
Turning Rate	Straight, Less, Medium, Sharp

Table 1. Linguistics variables and their linguistics terms.

Inputs	**Correlation (c_s)**		
SIFT (s_m)	Weak	Medium	Strong
Weak	Sharp	Medium	Less
Medium	Sharp	Medium	Straight
Strong	Medium	Less	Straight

Table 2. Fuzzy rule-base for Turning rate (w).

Test results lead to tune the system by changing rules, adjusting the membership functions shapes of both input and outputs. Once the procedure has been run several times, a consistent system is attained. The following section will integrate the proposed algorithm to a behavioral based architecture.

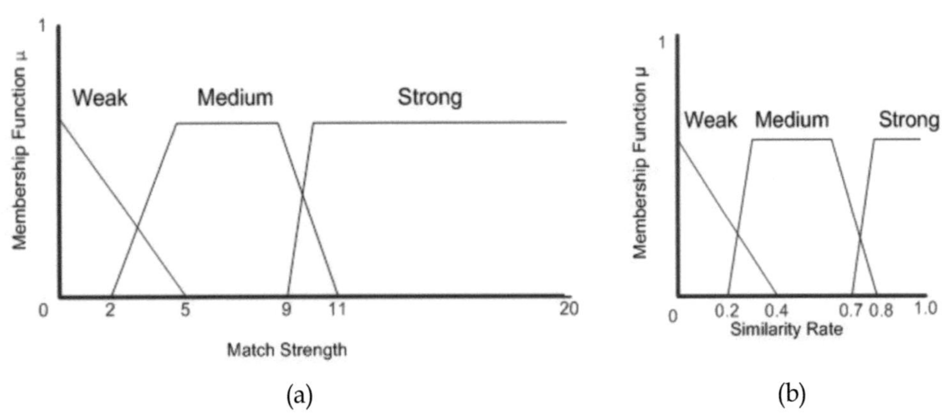

Fig. 21. Membership function, (a) 'input' SIFT matching, (b) 'input' Correlation.

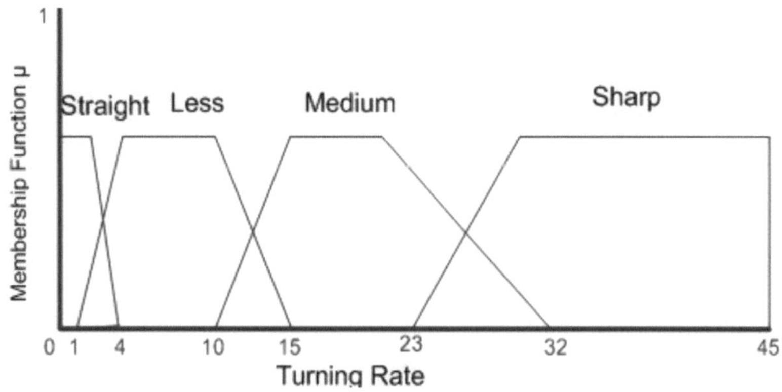

Fig. 22. Membership function, 'output' Turning rate.

4. Evaluation and implementation of the proposed system

In order to evaluate the performance of the algorithm, it is integrated using a behavioral based architecture. There are several approaches to designing a behavioural-based architecture depending on the required task. In this study, the architecture has been designed based on the subsumption architecture in which each layer or behaviour implements a particular goal of the robot and higher layers are increasingly abstract. Each layer's goal subsumes that of the underlying layer, and their interaction with each other will be illustrated by using finite state machines (FSM) which defines several states (behaviours) that represents a current situation for the robot. Certain events from the outside of the world can change the state. For instance, the robot could have a *Goto* state whereby it is moving about the environment trying to get closer to its goal. When any obstacle is detected nearby, the state may change from *Goto* to *Obstacle Avoidance*, and the avoidance algorithm will move the robot away from the obstacle. When the obstacle has been avoided, the robot state will change back to the *Goto*. The architecture, designed for this study, comprises three behaviors, namely: *Goto, Avoid_ Obstacle* and *Finish*. *Goto* behavior steers the robot to a specific goal position, *Avoid_Obstacle behaviour* utilizes the proposed vision based intelligent algorithm to avoid obstacles, and *Finish* behavior is merely enabled after the goal is found, and the robot is stopped. FSM diagram of the system is illustrated in Figure 23.

The system was developed using a Pioneer 3-DX mobile robot, with an on-board IntelPentium 1.8 GHz (Mobile) processor, and includes 256 Mbytes of RAM memory, as shown in Figure 24 (a). The mobile robot used in this study has been developed as a part of the Intelligent Robot Swarm for Attendance, Recognition, Cleaning and Delivery (IWARD) project. An Axis-213 camera, 25 frame rate, was integrated into this system. The software architecture of the proposed system is supported by CIMG Library and Player Architecture, which are open-source software projects. All experiments were conducted in an area of the Robotics and Automation Research Laboratory of Newcastle University, which has physical dimensions of 15.60m x 17.55m, as illustrated in Figure 24 (b). The camera tilted down 30 degrees to detect the floor precisely, and a reference image was taken from this environment. To evaluate the performance of the system, several different scenarios were performed and four of them will be discussed in this section.

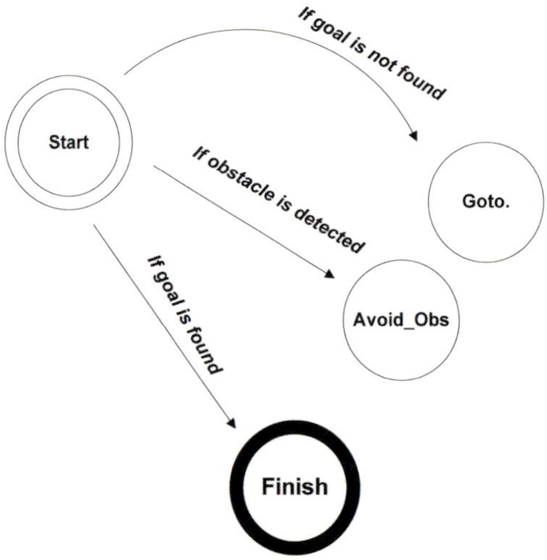

Fig. 23. FSM diagram of the behavioral architecture.

(a) (b)

Fig. 24. Test environment, (a) Pioneer 3-DX robot with Axis-213, (b) Robotics and Automation Laboratory

Scenario 1: The mobile robot is required to navigate from 'Start Position' (-6, 0) along a forward direction in a partially cluttered envioremen and avoids one obstacle located along its path as shown in Figure 25 (a). Performance of the image processing algorithms during the task is given Figure 25 (b). The robot's GoTo behaviour steers the robot forward direction.

Scenario 2: The mobile robot navigates from 'Start Position' (-6, 0) along a forward direction in a partially cluttered envioremen and has to avoid two obstacles located along its path as shown in Figure 26 (a) and evaluation of the image processing algorithms illustrated in Figure 26 (b). The robot's GoTo behaviour steers the robot forward direction.

Scenario 3: The mobile robot navigates from 'Start Position' (-7, 0) along a forward direction, two obstacles located on both sides of its path as shown in Figure 27 (a). Performance of the image processing algorithms are illustrated in Figure 27 (b). The robot's GoTo behaviour steers the robot forward direction.
Scenario 4: The mobile robot is required to navigate from 'Start Position' (-6, 0) to the 'Goal Position' (3, 0.5) in a partially cluttered enviorement as shown in Figure 28(a). Performance of the image processing algorithms during the task is given Figure 28 (b). The robot's GoTo behaviour steers the robot to a specific position which evaluates the performance of the obstacle avoidance algortihm with a Wavefront path planning algorithm (Barraquand & Latombe, 1991).
Figure 25 (a) presents the navigation results for scenario 1, in which the robot steers forward until it percives the obstacle. The robot avoids the obstacle succesfully and then keeps going forward until it detects the wall, after which it avoids the wall and continues heading forward until it encounters the door. The robot succesfully avoids the door and continues moving.

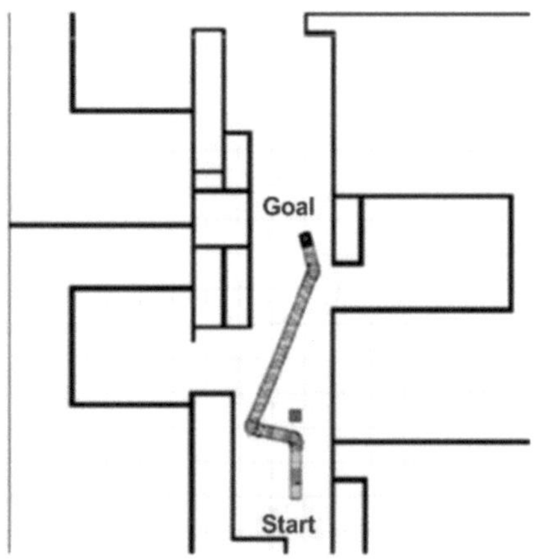

(a)

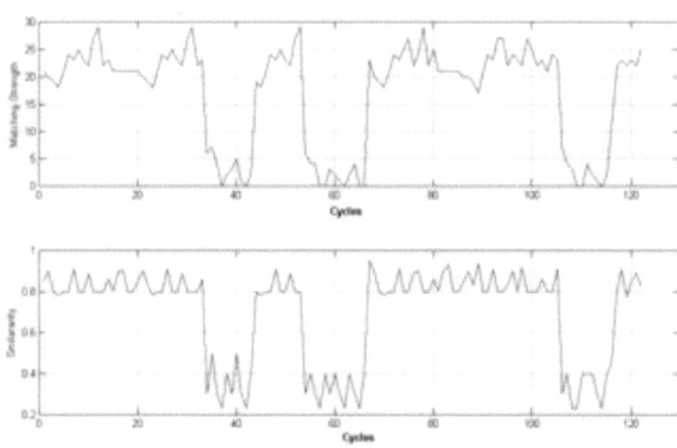

(b)

Fig. 25. Scenario 1, (a) Estimated Trajectory, (b) Performance of the image processing algorithms.

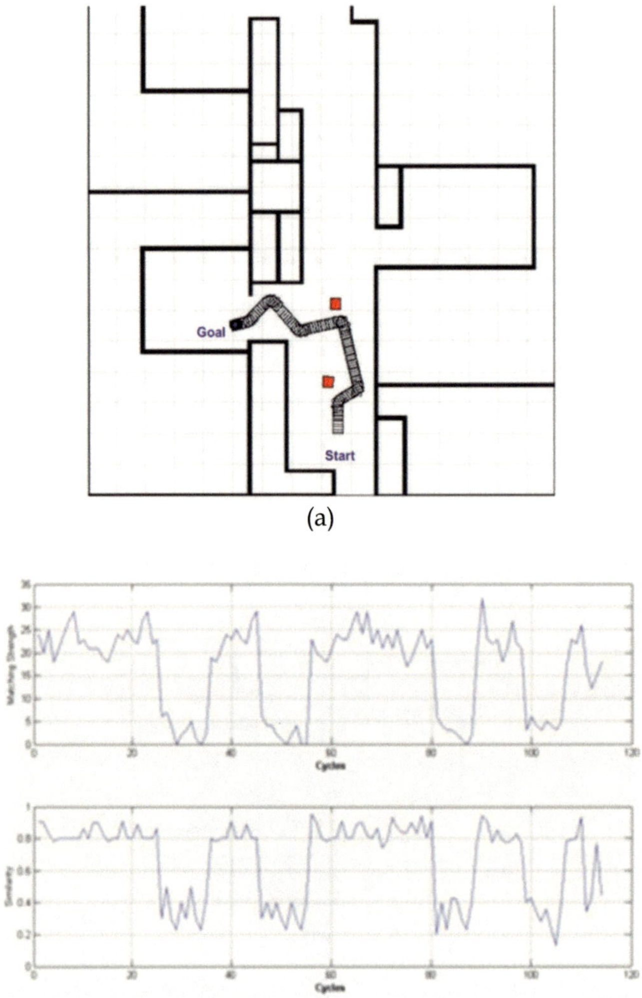

Fig. 26. Scenario 2, (a) Estimated trajectory, (b) Performance of the image processing algorithms.

Figure 26(a) displays the test results for scenario 2. Two obstacles are located along the robot's path in the forward direction. The robot navigates until it detects the first obstacle, then it avoids the obstacle and steers forward. Having detected the wall, the robot ceases forward motion and avoids the wall, after which, it moves forward until it detects the second obstacle, which it avoids succesfully. Subsequently, the robot avoids the door and the wall respectively.It finally moves forward until it is ceased. The third simulation is given Figure 27(a) where the robot attempts to move forward while passing the gap between two obstacles. It keeps moving forward until the obstacles are detected, initially turns left, followed by a right maneuver. After the robot succesfully avoids the objects it then resumes its path and continues moving forward.

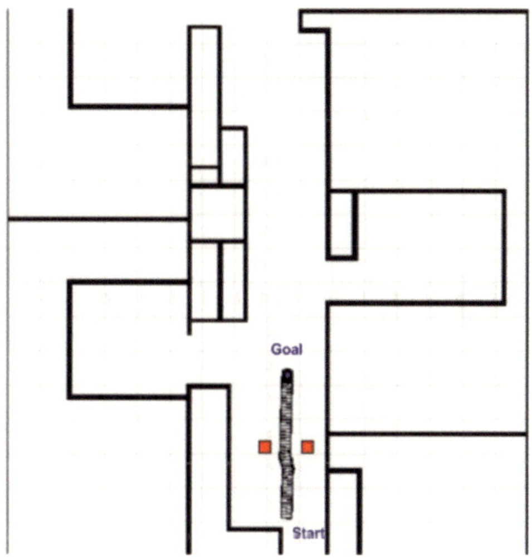

(a)

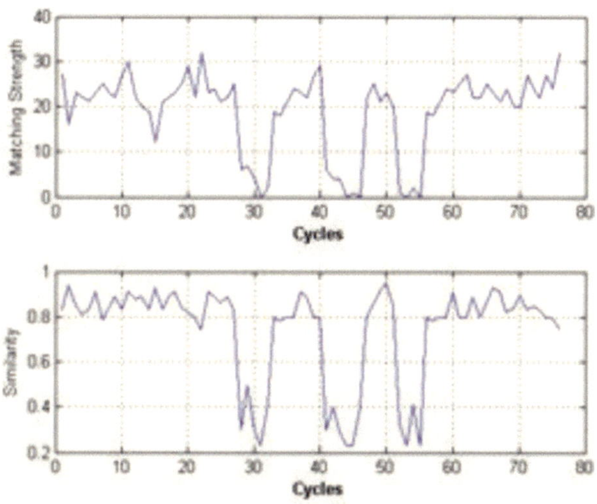

(b)

Fig. 27. Scenario 3, (a) Estimated trajectory, (b) Performance of the image processing algorithms.

Figure 28(a) displays the test results for scenario 4. The aim of this scenario is to test the avoidance strategy with a Wavefront path planning algortihm (Barraquand & Latombe, 1991) provided by the Player Architecutre. The robot navigates towards the 'first waypoint' until it perceives the obstacle. The robot avoids the obstacle and resumes its desired path. As there are no obstacles located along the rest of its path to the goal, the robot, therefore navigates directly to its goal.

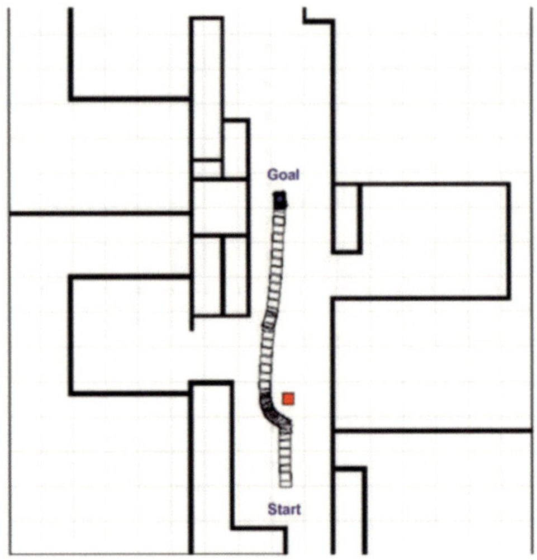

(a)

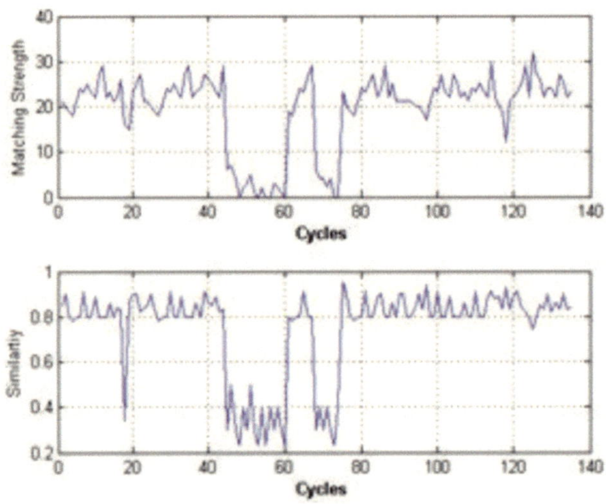

(b)

Fig. 28. Scenario 4, (a) Estimated trajectory, (b) Performance of the image processing algorithms.

5. Conclusion

The aim of this research was that it should be possible to develop a robust intelligent vision based obstacle avoidance system that can be adapted to realistic navigation scenarios. Most of the previously proposed techniques suffer from many problems. One of the most popular of those is *Apperance-based* methods which basically consist of detecting pixels different in appearance than the ground and classifying them as obstacles. Conventional *Appearance-*

based methods utilize simple template matching techniques which are fast but highly sensitive to lighting conditions and structure of the terrains. In addition, *Optical flow-based* methodologies,, the pattern of apparent motion of objects in a visual scene caused by the relative motion relying on apparent motion, are also highly sensitive to lighting conditions and suffer from nose.

To overcome those problems, a new obstacle avoidance method has been proposed which is inspired from feature matching principal. *SIFT-based* descriptors outperform other local descriptors on both textured and structured scenes, with the difference in performance larger on the textured scene. Accordingly, conventional *SIFT* algorithm, which is able to perform with high accuracy and reasonable processing time, is employed to match images, and it is finally adapted to an obstacle avoidance technique. According to the techniques, reference image, presenting the free path, is matched with the current image during the navigation to estimate the steering direction. However, preliminary experimental results reveals that despite the *SIFT* algorithm performs better than conventional methods with stained environment and compensates lighting failures, they may fail or produce several mismatched features due to low resolution or vibration caused by navigation over rough terrains. Therefore instead of using each method individually, conventional template matching method and *SIFT-based* descriptor are fused by using an appropriate *Fuzzy Fusion* algorithm which increases the accuracy and compensates the errors caused by lighting conditions and stains.

In order to verify the performance of the proposed obstacle avoidance algorithm, Real experiments to guide a *Pioneer 3-DX mobile robot* in a partially cluttered environment are presented, Results validate that proposed method provides an alternative and robust solution for mobile robots using a single low-cost camera as the only sensor to avoid obstacles.

6. References

Alper, Y., Omar, J. & Mubarak, S. (2006) Object tracking: A survey. *ACM Comput. Surv.*, 38, 13.

Atcheson, B., Heidrich, W. & Ihrke, I. (2009) An evaluation of optical flow algorithms for background oriented schlieren imaging. *Experiments in Fluids,* 46, 467-476.

BArraquand, J. & Latombe, J.-C. (1991) Robot motion planning. A distributed representation approach. *International Journal of Robotics Research,* 10, 628-649.

Bogacki, P. (2005) HINGES - An illustration of Gauss-Jordan reduction. *Journal of Online Mathematics and its Applications.*

Brooks, R. A. (1986) ROBUST LAYERED CONTROL SYSTEM FOR A MOBILE ROBOT. *IEEE journal of robotics and automation,* RA-2, 14-23.

Contreras, E. B. (2007) A biologically inspired solution for an evolved simulated agent. *Proceedings of GECCO 2007: Genetic and Evolutionary Computation Conference.* London.

Desouza, G. N. & Kak, A. C. (2002) Vision for mobile robot navigation: A survey. *IEEE Transactions on Pattern Analysis and Machine Intelligence,* 24, 237-267.

Driankov, D. (1987) Inference with a single fuzzy conditional proposition. *Fuzzy Sets and Systems,* 24, 51-63.

F. Vassallo, R., Schneebeli, H. J. & Santos-Victor, J. (2000) Visual servoing and appearance for navigation. *Robotics and Autonomous Systems,* 31, 87-97.

Fazl-Ersi, E. & Tsotsos, J. K. (2009) Region classification for robust floor detection in indoor environments. *Lecture Notes in Computer Science (including subseries Lecture Notes in Artificial Intelligence and Lecture Notes in Bioinformatics).* Halifax, NS.

Guzel, M. S. & Bicker, R. (2010) Optical flow based system design for mobile robots. *2010 IEEE Conference on Robotics, Automation and Mechatronics, RAM 2010.* Singapore.

Horn, B. K. P. & Schunck, B. G. (1981) Determining optical flow. *Artificial Intelligence,* 17, 185-203.

Lowe, D. G. (1999) Object recognition from local scale-invariant features. *Proceedings of the IEEE International Conference on Computer Vision.* Kerkyra, Greece, IEEE.

Ross, T. J. & Hoboken, N. J. (2004) *Fuzzy Logic with engineering applications* Hoboken, NJ:Wiley.

Saitoh, T., Tada, N. & Konishi, R. (2009) Indoor mobile robot navigation by central following based on monocular vision. *IEEJ Transactions on Electronics, Information and Systems,* 129, 1576-1584+18.

Souhila, K. & Karim, A. (2007) Optical flow based robot obstacle avoidance. *International Journal of Advanced Robotic Systems,* 4, 13-16.

Wen-Chia, L. & Chin-Hsing, C. (2009) A Fast Template Matching Method for Rotation Invariance Using Two-Stage Process. *Intelligent Information Hiding and Multimedia Signal Processing, 2009. IIH-MSP '09. Fifth International Conference on.*

Zitovã¡, B. & Flusser, J. (2003) Image registration methods: A survey. *Image and Vision Computing,* 21, 977-1000.

6

Non-Rigid Obstacle Avoidance for Mobile Robots

Junghee Park and Jeong S. Choi
Korea Military Academy
Korea

1. Introduction

In mobile robotics, obstacle avoidance problem has been studied as a classical issue. In practical applications, a mobile robot should move to a goal in the environment where obstacles coexist. It is necessary for a mobile robot to safely arrive at its goal without being collided with the obstacles. A variety of obstacle avoidance algorithms have been developed for a long time. These algorithms have tried to resolve the collisions with different kinds of obstacles. This chapter classifies the types of obstacles and presents characteristics of each type. We focus on the new obstacle type that has been less studied before, and apply representative avoidance strategies to solve this new type of obstacle avoidance problem. The results of different strategies would be compared and evaluated in this chapter.

2. Classification of obstacles

This section summarizes previous points of view on obstacles, and presents a new angle on obstacles for advanced robot navigation. A robotic system often suffers from severe damage as result of a physical collision with obstacles. Thus, there is no doubt that robots should have a perfect competency to avoid all kinds of obstacles. In robotics, an obstacle is generally defined as a physical object that is in the way or that makes it hard for robot to move freely in a space, and thereby classified into *static and moving obstacles*: the former are sometimes subdivided into stationary (or fixed) and movable ones of which position can be changed by force from a robot. Also, moving robots are often separated from moving obstacles, since their motion can be changed by themselves.
On the basis of this definition, various strategies have been presented to resolve anticipated physical collisions systematically. Since the majority of strategies focus on generating immediate reaction to environments, obstacles are considered static for an instant although they are moving in most previous studies. Recently, some studies formalized way to explicitly consider current velocity of moving obstacles for coping with real collisions caused by the static assumption. This chapter views obstacles from a different standpoint and introduces a concept of *non-rigid obstacles*, which differs from the conformable obstacle in that the concept covers various physical entities as well as the shape (or boundary) of obstacle. Fig. 1 shows these views on obstacles and details are discussed below.

2.1 Rigid obstacles

Early studies on robot motion assumed that a robot is the only moving object in the workspace and all obstacles are fixed and distributed in a workspace (Latombe, 1991). Under the

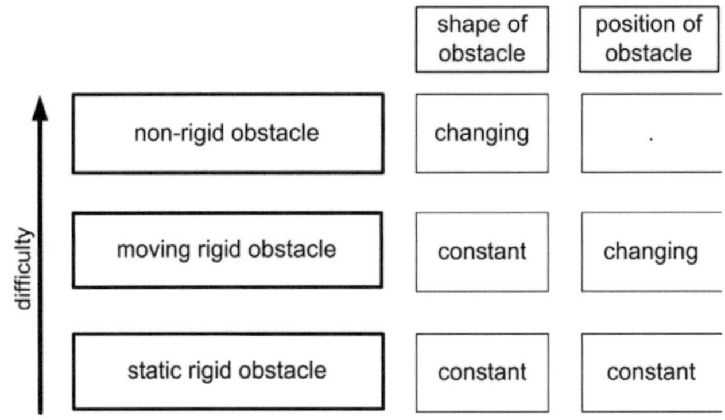

Fig. 1. Classification of obstacles

assumption, obstacle avoidance is automatically solved if we solve the global path planning problem in which the robot is assumed to have complete knowledge of the environment and generates a full path from its start to goal. The famous examples include the following approaches: probabilistic road map (Kavraki et al., 1996), cell decomposition (Boissonnat & Yvinec, 1998), grid cells (Hee & Jr., 2009), and rapidly-exploring randomizing tree (LaValle, 1998). To realize the approaches, several systematic tools for path generation were developed: visibility graph, Voronoi diagram, random sampling method, gradient method, node-edge graph, and mathematical programming (Ge & McCarthy, 1990). The final objective of the approaches is to find the global optimal (or shortest) one in the path candidates given by an optimization algorithm, such as Dijkstra and A* method.

Some studies relaxed the assumption of complete knowledge about environments to take into consideration the case that there is no device to provide the complete knowledge. Accordingly, the studies focused on following a detour (or local path) to avoid a detected static obstacle, instead of finding the shortest path to its goal. As a result, the obstacle avoidance is separated from the global path planning, which implies that it can be performed by on-line planning. The simplest strategy for the static obstacle avoidance with a limited sensing range may be the bug algorithm in which the robot is controlled to simply follow the contour of an obstacle in the robot's way (Lumelsky & Skewis, 1990). As with the bug algorithm, other avoidance strategies involving such obstacles create an immediate reaction to nearby environments for robot navigation. The popular methods include vector field histogram (Borenstein & Koren, 1991), nearness diagram (Minguez & Montano, 2004), fuzzy rules (Lee & Wang, 1994), force (or potential) fields (Wang et al., 2006), and dynamic window (Fox et al., 1997). In fact, these tools have been widely used for even the following moving obstacle avoidance problem by virtue of their simplicity in formulation and low operational speed of robots which permits the assumption that moving obstacles can be seen as static ones for a short time.

Moving obstacle avoidance differs from the static case in that an irreversible dimension - time - should be added into problem formulation. Like the static obstacle avoidance problem, the solution for the moving case can be computed globally or locally according to the assumption of the complete knowledge of environments. The configuration-time (CT) space presented in fig. 2 and 3 is a well-known tool for global problem solving. The CT space is a three-dimensional space in which a time axis is added into a xy-plane where a robot is moving. Static obstacles are represented as a straight pillar as shown in fig. 2 when they are represented with respect to CT space. On the other hands, if a moving obstacle is modeled as a circle, its representation with respect to CT space is an oblique cylinder as shown in fig. 3,

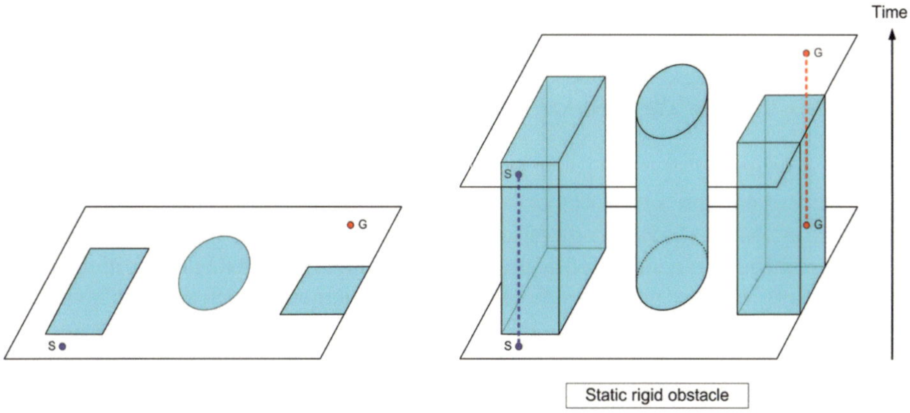

Fig. 2. A configuration-time space of static rigid obstacles

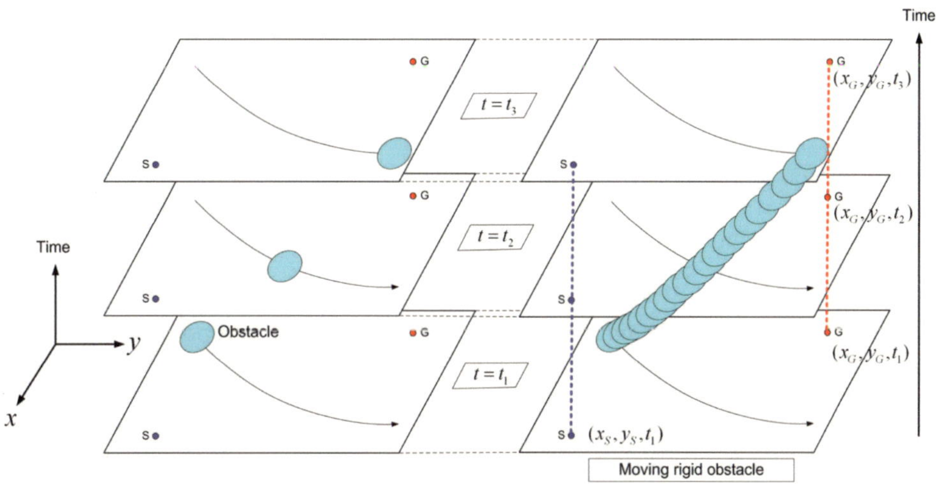

Fig. 3. A configuration-time space of moving rigid obstacles

because the position of the obstacle is varied with time along its path. These representations indicate that the problem of global moving obstacle avoidance is reduced to finding a 3-D geometric path obeying two conditions: safety and time efficiency. Sampling, combinatorial, and velocity-tuning methods have been introduced to solve the geometric problem (Lavalle, 2006).

Local moving obstacle avoidance is split into two groups, according to whether or not obstacles can be seen static for a short time, as mentioned above. Several recent papers argued that the problem can be made more realistic by explicitly considering the robot's velocity and acceleration, and included the constraints to problem formulation. The collision representation tools based on the consideration include the velocity obstacle (VO) (Fiorini & Shiller, 1993), inevitable collision states (ICS) (Martinez-Gomez & Fraichard, 2009) and the triangular collision object (TCO) (Choi et al., 2010). The major difference is that object shapes representing potential collisions are determined by obstacle's velocity, unlike the tools for the static obstacle avoidance problem. In the benchmarking (Martinez-Gomez & Fraichard, 2009), it was shown that the method using the VO is superior to the methods for static obstacle avoidance because it takes into account objects' future behavior, and suggested that the future motion of moving obstacles should explicitly been deal with. So far, we have briefly reviewed conventional rigid obstacle types and avoidance strategies which are popularly used in robotics. Now turn our attention to a new type, non-rigid obstacle in the following section.

2.2 Non-rigid obstacles

In this section, we explain a new obstacle type, non-rigid obstacle. This concept emerges as a consequence of the need for non-physical collisions, such as observation. More specifically, robots only need to have considered a single physical element, obstacle's boundary, for safe navigation to date, but they now have to consider other elements, such as agent's visibility, for smart navigation. Thus, to accommodate such a need, we here define a new term, *non-rigid obstacle*, which is an abstract object of which boundary is changeable and determined by an element involving an obstacle.

We explain this concept, non-rigid obstacle, through a moving observer with omnidirectional view. In a infiltration mission, the robot should covertly trespass on the environment without being detected by sentries. As a reverse example, in a security mission, the most important thing is that robots are not observed (or detected) by the intruder until it reaches the intruder (Park et al., 2010). In this way, the robot can covertly capture or besiege the intruder who would try to run away from the robot when detecting the approaching robot. It causes that robots can closely approach the intruder without being detected, and the intruder cannot recognize himself being besieged until the robots construct a perfect siege. In these cases, the robot should avoid the intruder's field of view during the navigation, rather than its physical boundary, which is called *stealth navigation*. This signifies that the proposed concept of non-rigid obstacle could be exceptionally helpful in planning an advanced robot motion.

The visible area determined by the field is a changeable element, especially in a cluttered indoor environment. So, we can regard the area as an abstract non-rigid obstacle that robots should avoid. This kind of obstacle can be generated by any element the obstacle has, which is a remarkable extension of the conventional concept on obstacles. Fig. 4 illustrates an example of the changes of visible areas with time traveled (or z-axis) in a CT space. As shown in the figure, the shape (or boundary) of visible area is greatly varied with the observer's motion which is a set of position parameterized by time. For this reason, it is needed to develop a new strategy to avoid the highly changeable obstacle, which will be discussed in detail in Section 3. If the robot succeeds in the avoidance with a plausible strategy, it gives a completely new value to a robotic system.

3. Avoidance strategies for non-rigid obstacle avoidance

As we mentioned in section 2, researches on obstacle avoidance problem have been focused on dealing with rigid (static or moving) obstacles. On the other hand, researches on avoidance strategies of non-rigid obstacles scarcely exist except grid-based methods (Marzouqi & Jarvis, 2006; Teng et al., 1993) a few decades ago. These methods were based on the grid map which approximates the environment to spatially sampled space. In this grid map, all possible movements of the robot were taken into consideration to find the optimal(fastest) motion of the robot to the goal. This brought about high burden of computation time. In addition, the arrival time of the planned motion and the computation time highly depended on the grid size.

Since then, any other strategies have not been developed for non-rigid obstacle avoidance. Even though other methods have not been suggested, we can apply rigid obstacle avoidance

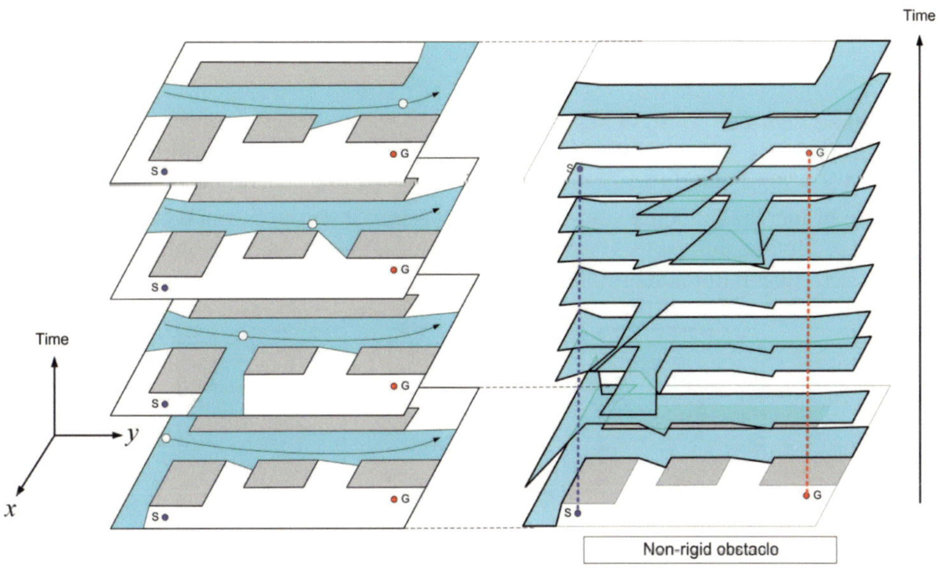

Fig. 4. A configuration-time space of non-rigid obstacles

algorithms to the problem of non-rigid obstacle avoidance. On a big scale, these algorithms can be classified into four categories: reactive method, grid-based method, randomization method, and path-velocity decomposed method. We will explain this classification of algorithms and how they can be applied to non-rigid obstacle avoidance.

3.1 Reactive method

Reactive methods utilize nearby environment information for robot navigation, i.e., the robot has a sensing capability with a limited coverage. In the case where the robot is sent to a unknown region for exploration or reconnaissance, it is reasonable to assume that information of the environment is not known beforehand. In addition, if the robot has no communication connection with other sensors on the environment, then it has to rely on the nearby information acquired from its own sensor. In this case, the robot should find a suitable motion by making full use of given information in order to attain two purposes: avoiding obstacles and moving to its goal. In other words, it can be widely used when the information of the environment is limited and not deterministic, whereas other methods, which will be explained in following sections, are based on the assumption that the information of environment is known in advance.

This reactive method can also be called short-term planning because it plans a few next steps of robot motion. The planner focuses on creation of immediate respondence of the robot in order to avoid obstacles and move to the goal. On a rough view, the reactive method yields instantaneous moving direction by synthesizing attractive attribute to the goal and repulsive attribute from obstacles. In non-rigid obstacle avoidance, the robot is planned to be repulsive from non-rigid obstacles. As mentioned in section 2.1, there have been many researches to find reactive respondence of the robot, such as vector field histogram(VFH) (Borenstein & Koren, 1991), force field (Wang et al., 2006), dynamic window (Fox et al., 1997), and behavior-based robotics (Arkin, 1985). In this chapter, we implemented one of representative reactive methods, vector field histogram, for comparison. It constructs a polar histogram, which contains nearby environment information, and finds suitable detour direction towards its goal.

3.2 Grid-based method: (Teng et al., 1993)

Grid-based methods use spatially sampled environment map as mentioned earlier. This algorithm has a merit that all possible robot motion can be considered in the grid map at the expense of increasing computation time. Therefore it can guarantee optimal solution in the grid map. That is, this method can have the completeness even though it is limited to the given grid map which depends on the shape and the size of grids.

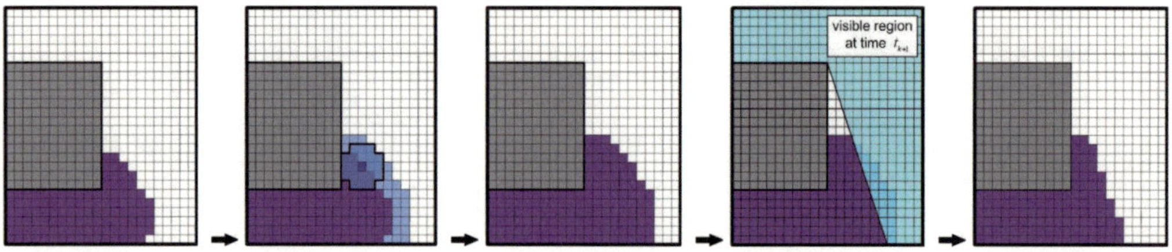

(a) Safely reachable region at time t_k

(b) Each boundary point is dilated in amount of the movable distance of the robot in one time step

(c) Reachable region at time t_{k+1}, which is dilated from the safely reachable region at time t_k

(d) The dilated region should be trimmed in order to avoid visibility from the observer

(e) Safely reachable region at time t_{k+1}

Fig. 5. Construction procedure of safely reachable region(purple) in grid-based method

Differently from other methods, this grid-based algorithm was designated for solving this non-rigid obstacle avoidance and introduced in (Teng et al., 1993). In this algorithm, the concept of the *safely reachable region* was introduced and used. It represents the region that the robot is able to arrive without being detected by the observer at each time step. The planner sequentially constructs safely reachable region of each time step until the last safely reachable region contains the final goal. Fig. 5 represents the procedure for obtaining safely reachable region from a previous one. Let the purple region of fig. 5(a) be the safely reachable region at time t_k. This region is dilated to the reachable region from it one time step later, which is shown in fig. 5(c). This dilated region is trimmed by the visible area of the observer at time t_{k+1}. Then it becomes the region that the robot can arrive without being detected by the observer at time t_{k+1}, which is shown in fig. 5(e). In this way, all possible safely reachable grid points were calculated at each time, and the shortest time for arrival is yielded.

3.3 Randomization method: modified RRT

In order to avoid a high computational burden of brute-force technique, randomization methods have been developed in robot motion planning. In these methods, free configurations are randomly selected and discriminated whether they can be via configurations that the robot is able to safely pass by. These kinds of algorithms have been developed for a long time, such as probabilistic road map(RPM) (Kavraki et al., 1996) and rapidly-exploring randomizing tree(RRT) (LaValle, 1998). These randomization methods could be usefully adopted for solving non-rigid obstacle avoidance problem in order to prevent burgeoning computational burden. In this chapter, we have modified RRT algorithm in order to cope with non-rigid obstacles which change its shape while time passes. The modified RRT algorithm for solving non-rigid obstacle avoidance problem is described in table 1.

```
s_init ← (t_init, q_init);
G.init(s_init);
while
   q_rand ← RAND_FREE_CONF();
   s_near = (t_near, q_near) ← NEAREST_VERTEX(q_rand,G);
   v_rand ← RAND_VEL();
   if CHECK_SAFETY(s_near,q_rand,v_rand) then
      t_new ← NEW_TIME(s_near,q_rand,v_rand);
      s_new ← (t_new, q_rand);
      G.add_vertex(s_new);
      G.add_edge(s_near,s_new);
      v_g_rand ← RAND_VEL();
      if CHECK_SAFETY(s_new,q_goal,v_g_rand) then
         t_arrival ← NEW_TIME(s_new,q_goal,v_g_rand);
         s_goal ← (t_arrival, q_goal);
         G.add_vertex(s_goal);
         G.add_edge(s_new,s_goal);
         break;
      end if;
   end if;
end while;
```

Table 1. Construction procedure of rapidly-exploring randomizing tree G for non-rigid obstacle avoidance

In the algorithm, the tree G is constructed with randomly selected vertexes. Each vertex $\mathbf{s}$ consists of time t and configuration $\mathbf{q}$, i.e., $\mathbf{s} = (t, \mathbf{q})$. In non-rigid obstacle avoidance problem, the time t when the robot passes by the configuration $\mathbf{q}$ should be coupled with $\mathbf{q}$ as a vertex. Thus, the time t also should be selected randomly. In this algorithm, t is calculated as the time at which the robot is able to arrive at $\mathbf{q}$ with the randomly selected velocity. Each randomly selected vertex becomes component of the tree G, and the tree G is continuously extended until new selected vertex is able to be linked with the goal vertex. After construction of tree G is completed, the connection from the start vertex to goal vertex is determined as the robot's movement.

In table 1, RAND_FREE_CONF() returns randomly selected free configuration. NEAREST_VERTEX(q,G) returns the vertex $\mathbf{s}_r = (t_r, \mathbf{q}_r)$ in G, which has the minimum distance from $\mathbf{q}$ to $\mathbf{q}_r$. RAND_VEL() is the function that yield randomly selected velocity v which should satisfy the robot's physical constraints, i.e., $0 \le v \le v_{max}$. In this procedure, the probability distribution function $f_V(v)$ for selecting v was designed as follows because it is beneficial for the robot to move as fast as possible.

$$f_V(v) = \frac{2v}{v_{max}^2}, \quad 0 \le v \le v_{max}$$

CHECK_SAFETY($\mathbf{s}_1$,$\mathbf{q}_2$,v) returns whether it is possible that the robot is able to safely move from the configuration $\mathbf{q}_1$ of vertex $\mathbf{s}_1$ to $\mathbf{q}_2$ with the velocity v or not. NEW_TIME($\mathbf{s}_1$,$\mathbf{q}_2$,v) returns the time t_r taken for the robot to move from $\mathbf{q}_1$ of $\mathbf{s}_1$ to $\mathbf{q}_2$ with the speed v. In this chapter, the configuration $\mathbf{q}$ of the robot is two dimensional point $\mathbf{p}$ because the robot is assumed to be holonomic and has circular shape. In this case, returned time t_r is calculated as follows.

$$t_r = t_1 + \frac{\| \mathbf{p}_2 - \mathbf{p}_1 \|}{v}$$

3.4 Path-velocity decomposed method

Lastly, we can consider the path-velocity decomposed(PVD) method. This concept was firstly introduced in (Lee & Lee, 1987) and has been developed for resolving collisions of multiple robots in (Akella & Hurchinson, 2002; Lavalle & Hurchinson, 1998). This method decomposes the robot motion into the path, which is a geometric specification of a curve in the configuration space, and the velocity profile, which is a series of velocities according to the time following the path. In other words, the PVD method firstly plans the path, and adjusts the velocity of the robot following this path. It reduces computation burden by not guaranteeing the safety condition when planning the path at the cost of optimality. This safety condition is considered only in planning of velocity profile. If we predefine the geometric road map (topological graph) on the environment, we can easily plan the robot's path on the road map and find the velocity profile that guarantees safely movement of the robot following the path. This method used for solving moving rigid obstacle avoidance also can be applied to non-rigid obstacle avoidance problem without difficulty.

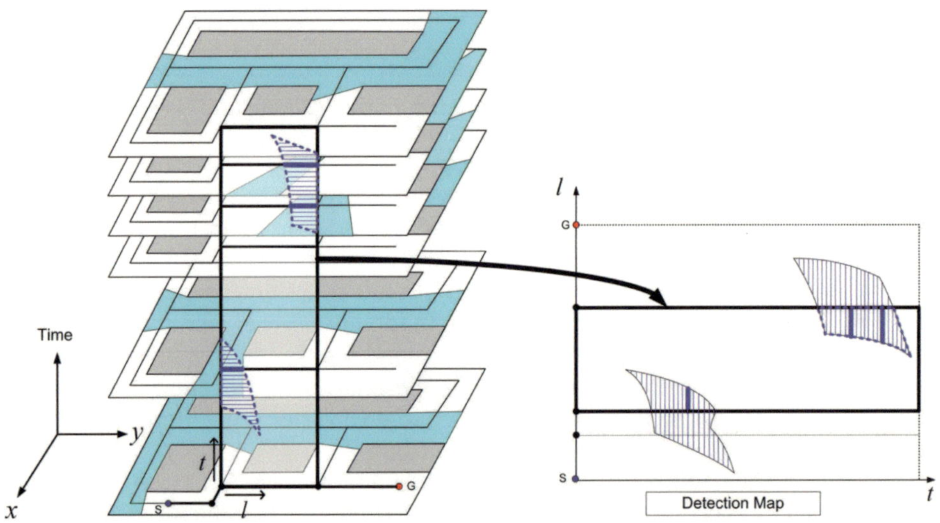

Fig. 6. Construction of the detection map

(Park et al., 2010) has proposed the concept of detection map. The detection map is a two-dimensional(time and point on the path) map that represents whether each time and point on the path is detected by the observer or not. It depends on the path chosen on the road map. Fig. 6 shows the construction of the detection map on the predetermined path. When the detected intervals on the path are accumulated according to the time, the detection map is constructed. In the detection map, the shaped regions represents the time and points on the path that are detected by the observer, which is called detection region. We can draw the robot's trajectory on the detection map without intersecting these detection regions in order to make the robot safely move to its goal. After the detection map is constructed, the safely velocity profile can be found in a shorter time because moving direction is restricted to one dimension.

The planning method of the velocity profile can be selected in a various way. (Park et al., 2010) applied the grid-based method to this velocity profile planning. The concept in (Teng et al., 1993), which constructs safely reachable region by dilation and trimming in order to find the optimal solution on the grid map, is adopted to solve the similar problem with the one-dimensional moving direction. The points on path is sampled with uniform interval and examined whether it is safely reachable by the robot for each discretized time step. This grid-based method also finds the velocity profile that has the shortest arrival time. We note that this optimality is valid only when the predetermined path could not be modified by the planner.

The randomization method could be proposed in velocity profile planning. Without considering all possible safe movement of the robot, we can find the robot's safe trajectory by randomly selecting via points on the detection map. In this chapter, we applied the modified RRT method explained in section 3.3 t velocity profile planning. In this application, the one-dimensional value is randomly selected as via point that the robot would pass. The RRT would be constructed on the detection map without intersecting detection regions. The detection map makes it easy to discriminate whether tree link is safe or not.

The method of selecting path on the road map also could be suitably adopted. (Park et al., 2010) has selected the path that has the shortest moving distance. If the safe velocity profile did not exist on the selected path, the next shortest path was examined in consecutive order. However, it often led to a situation that not a few path had to be examined because this method completely ignored the safety condition on the path planning step. We proposed the method to select the path highly possible to have safe trajectory of the robot. We designed the cost of each link on the road map as follows. Let $E_{i,j}$ be the link between node i and node j on the road map.

$$cost(E_{i,j}) = length(E_{i,j}) * (detection\_ratio(E_{i,j}))$$

$$\texttt{where } detection\_ratio(E_{i,j}) = \frac{area\ of\ detection\ region\ on\ the\ detection\ map\ of\ E_{i,j}}{area\ of\ whole\ detection\ map\ of\ E_{i,j}}$$

Compared to the method which considered only path distance, this cost function makes the path planner to consider not only distance but also safely condition. If we select the path that has the least cost on the road map, this path would be relatively less detected by the observer. If a safe velocity profile is not found on the selected path, then the other path which has next least cost could be found and examined again.

4. Simulation results

We have conducted a computer simulation on the environment in fig. 7. In this simulation, the visible area of the observer is considered as a example of the non-rigid obstacle. The robot is planned to move stealthy from the observer. The observer is patrolling on the environment following a specific path with the speed of 2.0m/s. It is assumed that the robot has a maximum speed of 4.0m/s and it can change its speed and its moving direction instantaneously. We assumed that the information of the environment and the non-rigid obstacle is known to the

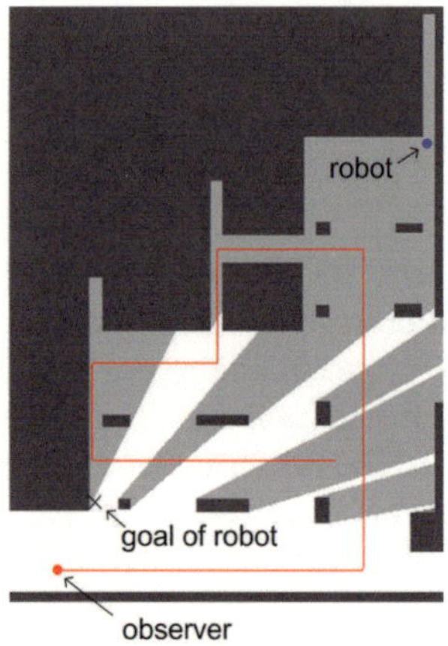

Fig. 7. Simulation environment (27.5m x 36.5m) and the patrol path of the observer

robot. This computer simulation is conducted on a commercial computer with 3.34GHz CPU and 3.00GB RAM.

Table 2 shows the simulation results according to algorithms from reactive method to path-velocity decomposed method. In table 2, the *exposure rate* means a ratio of the time interval that the robot is detected by the observer to the whole arrival time. In addition, we define *safety* as the distance between the robot and the point which is contained in the visible area of the observer and is the closest to the robot. When the robot is contained in the visible area of the observer, i.e., the robot is detected by the observer, the safety value is set to zero. In table 2, the average value of safety during the operation time is revealed. Also, the computation times of each method are shown in table in order to analyze practical implementation feasibility of each method.

	Reactive (VFH)	Grid-based (Teng et al., 1993)	Randomization (Modified RRT)	Path-velocity decomposed	
				Grid-based (Park et al., 2010)	Randomization (Modified RRT)
Arrival time	10.8s	10.4s	20.1s	16.2s	16.6s
Exposure rate	3.0%	0.0%	0.0%	0.0%	0.0%
Average safety	2.97m	2.88m	3.14m	3.32m	3.40m
Computation time	0.01s	33.67s	17.15s	path planning: 3.70s	
				0.01s	0.01s

Table 2. Simulation results

In table 2, the exposure rate had non-zero value only when the reactive method was applied. This is because the reactive method does not utilize whole information of visible areas of the observer. It assumes that the robot is able to sense only nearby information of environment and visible area. In this simulation, the robot's sensing coverage is assumed to be omnidirectional with 3.0m radius. Therefore, it is possible that the robot is placed in the situation that it is too late for avoiding the visible area that emerges suddenly near by. At the cost of this dangerous situation, the reactive method does require a negligible amount of the computation time. More detailed results are shown in fig. 8 and fig. 9. As we can see in fig. 8(b), and 8(c), the robot's speed and direction were changed extremely for a whole operation time. In fig. 9, the robot detects the non-rigid obstacle in the rear of itself at time 5.7(s), nevertheless, it could not avoid the obstacle due to the lack of speed, and it collided with the obstacle at time 6.0(s). We note that this chapter assumes that non-rigid obstacle is also detected by the sensor of the robot in order to apply the reactive method for rigid obstacle avoidance to non-rigid obstacle avoidance problem. In real application, some non-rigid obstacle such as observer's field of view could not be detected by the sensor. As this result shows, the robot has a difficulty in stealth navigation when it has a limited sensing capability of the environment.

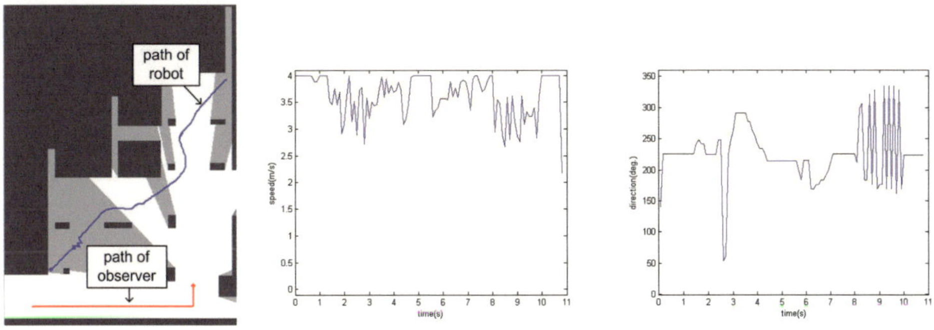

(a) path (b) speed (c) direction

Fig. 8. The reactive method (VFH) results

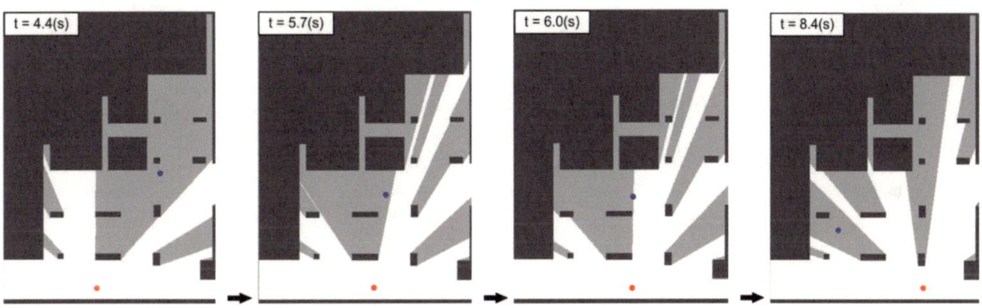

(a) (b) (c) (d)

Fig. 9. The movement of observer(red) and robot(blue) in the reactive method result

From the results in table 2, we can find that the arrival time was the least when the grid-based method was applied. This is because the grid-based method considers all possible motions of the robot and finds the fastest motion among them, i.e., it guarantees the optimality and completeness. However, these are valid only on given grid map. That is, if the grid map is set in a different way, the robot's fastest motion and the arrival time would become different. Since it examines all grid cells whether they are possibly reached safely by the robot at each time, it needs high computational burden. In table 2, the computation time of the grid-based method is higher than that of any other methods. More detailed results are shown in fig. 10. The speed and direction of the robot were not changed extremely. Between time 6.8(s) and 9.1(s), the robot remained stationary in the shaded region until the obstacle was away from itself. The path in fig. 10(a) shows that the robot was able to almost directly traverse the environment to the goal without being detected by the observer. The performance in path length and arrival time is the best in the grid-based method. Fig. 11 shows safely reachable regions changing in accordance with the passed time.

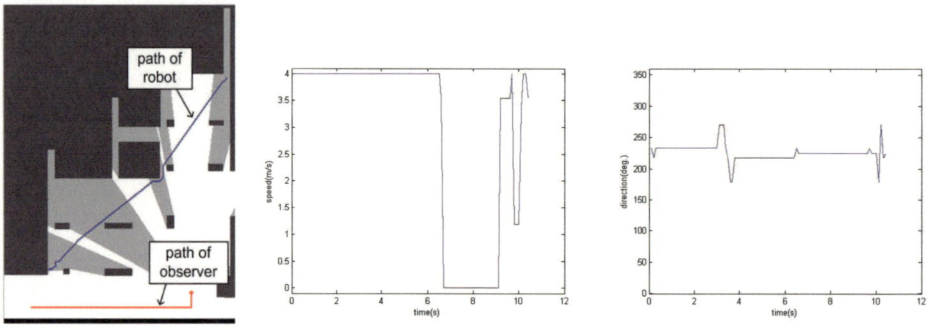

(a) path (b) speed (c) direction

Fig. 10. The grid-based method results(grid size is $64cm^2$/cell)

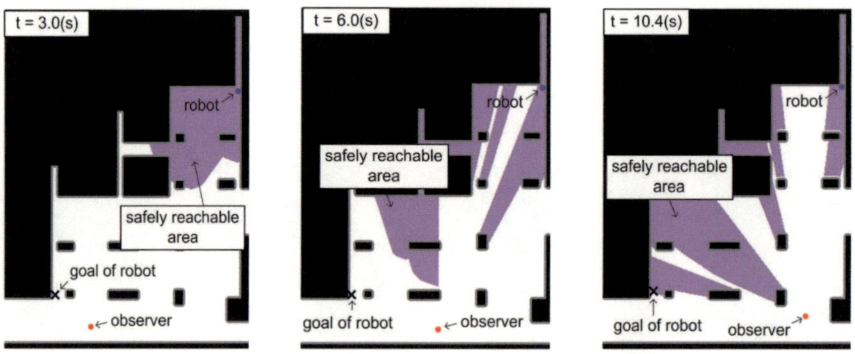

(a) safely reachable region at time t=3.0(s) (b) safely reachable region at time t=6.0(s) (c) safely reachable region at time t=10.4(s)

Fig. 11. Safely reachable regions of the grid-based method in fig. 10

The result of randomization method in table 2 has the lowest performance in the arrival time. This result is not deterministic and can be changed in each trial. There is possibility that relatively better performance can be yielded, but it could not be guaranteed. This method does not purpose on greatly reducing the arrival time, but it concerns a practical computation time. In table 2, modified RRT algorithm needed not a low computation time. This is because it needs relatively much time to check that the tree link candidate is safe for the movement of the robot. In order to reduce the computation time in the randomized method, it is necessary to develop a specialized randomization method to solve the non-rigid obstacle avoidance problem. Fig. 12 shows the result of modified RRT method. In fig. 12(a) and 12(c), the path of the robot was planned as zigzagged shape because via points are selected randomly. As shown in fig. 12(b), the segments between via points have respectively uniform speeds which are also selected randomly.

Lastly, we implemented the PVD method in the same environment. The road map of the environment was predefined, and the path with the least cost on the road map was selected as shown in fig. 13(a). We can see that the path with the least cost is selected as series of

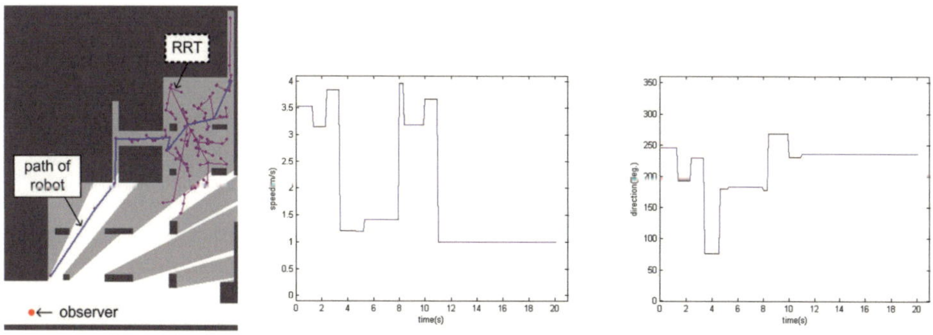

(a) path (b) speed (c) direction

Fig. 12. The randomized method (modified RRT) results

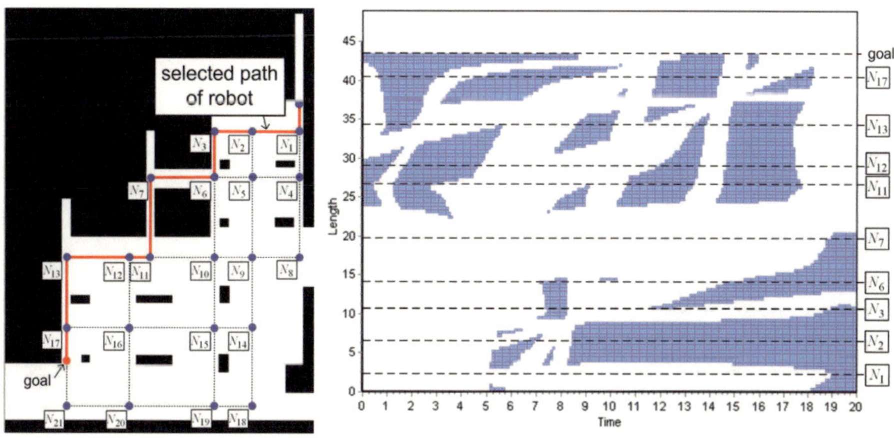

(a) road map on the selected environment and selected path (b) detection map on the selected path

Fig. 13. The road map and the detection map

edges that are in the vicinity of walls, because they have relatively small detection ratios. The detection map corresponding to the selected path was constructed as shown in fig. 13(b). In the detection map, the colored regions, which are detection regions, are the set of time and point that is detected by the observer on the determined path. The trajectory of the robot following this path could be planned on the detection map without traversing detection regions.

As shown in table 2, the arrival time of the grid-based result in PVD method was larger than that of the grid-based method. This is because the PVD method restricts the path to be selected on the predefined roadmap, not among whole possible movement. Therefore, the PVD method has the lower performance than the grid-based method even though it also finds the optimal solution on the determined path. However, its computation time is much smaller than the grid-based method because it reduces the dimension of the grid map. This planned result is shown in fig. 14(a). The movement of the observer and the robot in this result is shown in fig. 15. In fig. 15(b) and 15(c), we can see that the robot remains stationary because

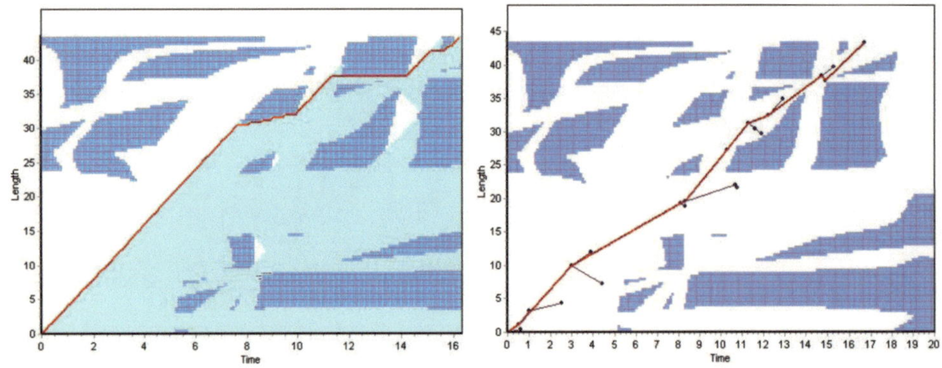

(a) planned trajectory of the grid-based method (b) planned trajectory of the randomization method

Fig. 14. The trajectory results of two detail velocity profile planning algorithm in the path-velocity decomposed(PVD) method

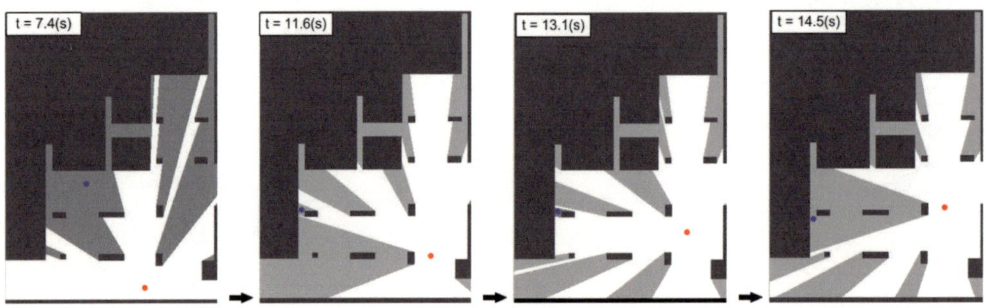

(a) (b) (c) (d)

Fig. 15. The movement of observer(red) and robot(blue) in the grid-based result of path-velocity decomposed method

the observer's field of view obstructs the front and the rear of the robot. In fig. 15(d), the robot started to move after the observer's field of view is away from itself.
In the case of the randomization result in PVD method in table 2, it not only needs much lower computation time but also yields the better performance than the modified RRT methods. The randomization methods cannot guarantee the optimality, and its performance is on a case by case basis. Therefore, it is possible that the case restricting path in PVD method has the shorter arrival time than the case having whole opportunity of path selection. This planned result is shown in fig. 14(b).

5. Conclusion

In this chapter, we classified the type of obstacles into three categories: static rigid obstacles, moving rigid obstacles, and non-rigid obstacles. However, most researches of obstacle avoidance have tried to solve the rigid obstacles. In reality, the non-rigid obstacle avoidance problem is often emerged and needed to be solved, such as stealth navigation. We adopted the representative algorithms of classic rigid obstacle avoidance, and applied to non-rigid obstacle avoidance problem. We examined how the inherent characteristics of each algorithms are revealed in non-rigid obstacle avoidance and evaluated the performance. The result in the reactive method informed us that limited information of the environment and obstacle greatly hinders the robot from avoiding non-rigid obstacles. The grid-based method could find the optimal safe motion of the robot at the cost of high computation burden. The randomization method gave priority to reducing computational burden over improving the performance. Finally, the path-velocity decomposed(PVD) method greatly reduced computation burden by abandoning the diversity of path selection. This method does not cause great loss in arrival time. This method makes the robot avoid the obstacle by adjusting its speed. In velocity profile planning, several detail methods, such as the grid-based method and the randomization method, can be applied with practical computation time. When complete information of the environment and the obstacle is given, this PVD method could be efficiently used.

6. References

Akella, S. & Hurchinson, S. (2002). Coordinating the motions of multiple robots with specified trajectories, *Proceedings of IEEE International Conference of Robotics and Automation*, New Orleans, LA, pp. 3337–3344.

Arkin, R. C. (1985). *Behavior-based Robotics*, MIT Press.

Boissonnat, J. D. & Yvinec, M. (1998). *Algorithm Geometry*, Cambridge University Press.

Borenstein, J. & Koren, Y. (1991). The vector field histogram - fast obstacle avoidance for mobile robot, *Journal of Robotics and Automation* Vol. 7(No. 3): 278–288.

Choi, J. S., Eoh, G., Kim, J., Yoon, Y., Park, J. & Lee, B. H. (2010). Analytic collision anticipation technology considering agents' future behavior, *Proceedings of IEEE International Conference on Robotics and Systems*, pp. 1656–1661.

Fiorini, P. & Shiller, Z. (1993). Motion planning in dynamic environments using the relative velocity paradigm, *Proceedings of IEEE International Conference on Robotics and Automation*, Vol. Vol. 1, pp. 560–566.

Fox, D., Burgard, W. & Thrun, S. (1997). The dynamic window approach to collision avoidance, *IEEE Robotics and Automation Magazine* Vol. 4(No. 1): 23–33.

Ge, Q. J. & McCarthy, J. M. (1990). An algebraic formulation of configuration-space obstacles for special robots, *Proceedings of International Conference on Robotics and Automation*, pp. 1542–1547.

Hee, L. & Jr., M. H. A. (2009). Mobile robots navigation, mapping, and localization part i, *Encyclopedia of Artificial Intelligence* pp. 1072–1079.

Kavraki, L. E., Svestka, P., Latombe, J. C. & Overmars, M. H. (1996). Probabilistic roadmaps for path planning in high-dimensional configuration space, *IEEE Transactions on Robotics and Automation* Vol. 12(No. 4): 566–580.

Latombe, J. C. (1991). *Robot Motion Planning*, Kluwer academic publishers.

LaValle, S. M. (1998). Rapidly-exploring random trees: A new tool for path planning, *Technical report 98-11, Computer Science Department, Iowa State University* .

Lavalle, S. M. (2006). *Planning Algorithms*, Cambridge University Press.

Lavalle, S. M. & Hurchinson, S. A. (1998). Optimal motion planning for multiple robots having independent goals, *IEEE Transactions on Robotics and Automation* Vol. 14(No. 6): 912–925.

Lee, B. H. & Lee, C. S. G. (1987). Collision-free motion planning of two robots, *IEEE Transactions on Systems Man and Cybernetics* Vol. 17(No. 1): 21–31.

Lee, P. S. & Wang, L. L. (1994). Collision avoidance by fuzzy logic for agv navigation, *Journal of Robotic Systems* Vol. 11(No. 8): 743–760.

Lumelsky, V. & Skewis, T. (1990). Incorporating range sensing in the robot navigation function, *IEEE Transactions on Systems Man and Cybernetics* Vol. 20: 1058–1068.

Martinez-Gomez, L. & Fraichard, T. (2009). Collision avoidance in dynamic environments: an ics-based solution and its comparative evaluation, *Proceedings of IEEE International Conference on Robotics and Automation*, Kobe, pp. 100–105.

Marzouqi, M. S. & Jarvis, R. A. (2006). New visibility-based path-planning approach for covert robotics navigation, *IEEE Transactions on Systems Man and Cybernetics* Vol. 24(No. 6): 759–773.

Minguez, J. & Montano, L. (2004). Nearness diagram navigation: collision avoidance in troublesome scenarios, *IEEE Transactions on Robots and Automation* Vol. 20(No. 1): 45–59.

Park, J., Choi, J. S., Kim, J., Ji, S.-H. & Lee, B. H. (2010). Long-term stealth navigation in a security zone where the movement of the invader is monitored, *International Journal of Control, Automation and Systems* Vol. 8(No. 3): 604–614.

Teng, Y. A., Dementhon, D. & Davis, L. S. (1993). Stealth terrain navigation, *IEEE Transactions on Systems Man and Cybernetics* Vol. 23(No. 1): 96–113.

Wang, D., Liu, D. & Dissanayake, G. (2006). A variable speed force field method for multi-robot collaboration, *Proceedings of IEEE International Conference on Intelligent Robots and Systems*, pp. 2697–2702.

Part 2

Path Planning and Motion Planning

7

Path Planning of Mobile Robot in Relative Velocity Coordinates

Yang Chen[1,2,3], Jianda Han[1] and Liying Yang[1,3]
[1]*State Key Laboratory of Robotics, Shenyang Institute of Automation, Chinese Academy of Sciences, Shenyang*
[2]*Department of Information Science and Engineering, Wuhan University of Science and Technology, Wuhan*
[3]*Graduate School, Chinese Academy of Sciences, Beijing*
China

1. Introduction

Path planning of mobile robot in dynamic environment is one of the most challenging issues. To be more specific, path planning in multi-obstacle avoidance environment is defined as: given a vehicle *A* and a target *G* that are moving, planning a trajectory that will allow the vehicle to catch the target satisfy some specified constrains while avoiding obstacle *O*, and each of the obstacles can be either mobile or immobile in the environment. The corresponding problem is named target-pursuit and obstacles-avoidance (TPOA) and will be researched extensively in this chapter.

The traditional method, such as probability road map, can achieve a successful path in 2D static environments. The planning process using this method generally consists of two phases: a construction and a query phase. In construction stage, the workspace of the robot is sampled randomly for generating candidate waypoints. In the query stage, the waypoints between the start and goal position are connected to be a graph, and the path is obtained by some searching algorithm, such as Dijkstra, A* algorithm and so on. Hraba researched the 3D application of probability road map where A* algorithm is used to find the near-optimal path (Hrabar, 2006). Although probability road map method is provably probabilistically complete (Ladd & Kavraki, 2004), it does not deal with the environment where the information is time-varying. The underlying reason is that this method only focuses on the certain environment. Once some uncertainty appears in the robot workspace, probability road map can not update with the changing environment and plan a valid trajectory for the mobile robot, never an optimal path.

Artificial potential field is another traditional method which is generally used in both 2D and 3D environment. The mechanism that the robot is driven by attractive and repulsive force in a cooperative way is simple and often works efficiently even in dynamic environment. Kitamura et al. construct the path planning model based on the artificial potential field in three-dimensional space which is described by octree (Kitamura et al, 1995). Traditionally, artificial potential field applies in two dimensions extensively. Also some other field concepts are invented. For example, there are harmonic potential functions (Kim & Khosla, 1992; Fahimi et al, 2009; Cocaud et al, 2008; Zhang & Valavanis, 1997), hydrodynamics (Liu et al, 2007), gradient field (Konolige, 2000), and virtual force field (Oh et al., 2007). Unfortunately, path planning approach based on the function family of potential field cannot obtain the optimal objective function which, in turn, cannot guarantee the desired optimal trajectories. Additionally, some of them, for example, potential panel method and harmonic potential function, can plan a path for an autonomous vehicle, but the computing burdens is huge and real time performance hardly satisfies the practical requirement.

Inspired by biological intelligence, many approaches, such as ant colony optimization (ACO) (Chen et al., 2008), particle swarm optimization (PSO) (Jung et al., 2006), genetic algorithm (GA) (Allaire et al., 2009), evolution algorithm (EA) (Zhao & Murthy, 2007), and their combination, are introduced to solve the path planning problem. They mostly rely on the stochastic searching, known as non-deterministic algorithm. These methods will eventually find an optimal solution, but no estimate on the time of convergence can be given. Thus, it may take long even infinite time to find the best solution. Furthermore, all of them have a great number of parameters to tune and that is never an easy job particularly when users are short of prior knowledge. Some comparisons (Allaire et al., 2009; Krenzke, 2006) show that the expensive calculations limit their real application.

Another kind of method is mathematic programming. Based on the relative velocity coordinates (RVCs), linear programming (LP) and mixed integer linear programming (MILP) can be employed for path planning problem. For example, Wang et al. converted the 2D path planning of an unmanned underwater vehicle to constrained optimization or semi-infinite constrained optimization problem (Wang et al., 2000). Zu et al. discussed the path planning in 2D and an LP method was proposed for the problem of dynamic target pursuit and obstacle avoidance (Zu et al., 2006). This method tried to plan the variables of the linear acceleration and the angular acceleration of a ground vehicle. Schouwenaars et al. proposed a MILP formulation with receding horizon strategy where a minimum velocity and a limited turn rate of aircraft are constrained (Schouwenaars et al., 2004). However, these results still focused on the two dimensions.

In this chapter, we consider the same problem of TPOA but in three dimensions. To our problem, the uncertain environment has one target or some of them, denoted by *G*, and many obstacles *O* that are all velocity-changeable with their moving actions. The aim of the vehicle *A* is to find an optimal path for pursuing the target while, at the same time, avoiding collision threaten from obstacles. The position and velocity of movers, including the target and obstacles, are assumed known or estimated at current time. To be more specific, it is assumed that the noise contained in the data can be eliminated with certain filters. However, this paper has no intend to discuss the filtering algorithm for achieving the feasible data of on-board sensors. The trajectory from a start to a destination location typically needs to be computed gradually over time in the fashion of receding horizon (Schouwenaars et al., 2004) in which a new waypoint of the total path is computed at each time step by solving a linear programming problem. The target and obstacles, probably static or dynamic, are modeled by spheres having a certain radius for their impact area, and the vehicle is modeled by a mass point. In order to optimize the vehicle's acceleration online, we construct a linear programming model in Cartesian orthogonal coordinates based on relative velocity space (Fiorini & Shiller, 1998).

This chapter is organized as follows. First, the relative velocity coordinates are introduced in both two dimensions and three dimensions and then the TPOA principles are introduced, including obstacle-avoidance and target-pursuit principles. The formulation of the standard linear programming is mathematically introduced in Section 3.1, and the path planning model is described in detail in Section 3.2. Simulations of path planning in 2D and 3D are

A, G, O	Vehicle, target, obstacle
τ	Planning period
k	Time step
N	Sum of obstacles
$\mathbf{V}_A$	Vehicle velocity
$V_A = \lVert \mathbf{V}_A \rVert$	Magnitude of the vehicle velocity
$\Delta \mathbf{V}_A$	Vehicle acceleration
$\mathbf{V}_O$	Obstacle velocity
$V_O = \lVert \mathbf{V}_O \rVert$	Magnitude of the obstacle velocity
$\mathbf{V}_G$	Target velocity
$V_G = \lVert \mathbf{V}_G \rVert$	Magnitude of the target velocity
$\mathbf{V}_{AO} = \mathbf{V}_A - \mathbf{V}_O$	Vehicle's relative velocity to the obstacle
$V = \lVert \mathbf{V}_{AO} \rVert$	Magnitude of the vehicle velocity relative to the obstacle
$\Delta \mathbf{V}_{AO}$	Vehicle acceleration relative to the obstacle
$\mathbf{V}_{AG} = \mathbf{V}_A - \mathbf{V}_G$	Vehicle velocity relative to the target G
$V_{AG} = \lVert \mathbf{V}_{AG} \rVert$	Magnitude of the vehicle's velocity relative to the target
$\Delta \mathbf{V}_{AG}$	Vehicle acceleration relative to the target
$\mathbf{L}_{AO}$	Vehicle position relative to the obstacle
$L = \lVert \mathbf{L}_{AO} \rVert$	Magnitude of the vehicle position relative to the obstacle
$\mathbf{L}_{AG}$	Vehicle position relative to target
$L_{AG} = \lVert \mathbf{L}_{AG} \rVert$	Magnitude of the vehicle position relative to the target
$P = \mathbf{V}_{AO}^{\mathrm{T}} \mathbf{L}_{AO}$	Temporal variable
$P_{AG} = \mathbf{V}_{AG}^{\mathrm{T}} \mathbf{L}_{AG}$	Temporal variable

Table 1. Nomenclatures used in this chapter (see Fig.1 and Fig. 2).

shown in Section 4.1 and 4.2, respectively. In Section 5, an application about the multiple task assignment (MTA) is used to verify the method proposed in this chapter. Finally, a brief conclusion is drawn in Section 6.

2. Relative velocity coordinates and the TPOA principles

2.1 Relative velocity coordinates

RVCs are constructed on the mass point of the vehicle as shown in Fig.1 and Fig.2. Fig.1 shows the relative coordinates when the obstacle-avoiding is considered while Fig.2 shows the scenario where the target-pursuing is considered. In these figures, the target G and obstacle O, which are 2D or 3D movers, are assumed as circles or spheres having certain radiuses which are denoted by R_Os. As mentioned above, the relative parameters are measurable or estimable for the current planning time, by using certain on-board sensors. For the convenience of description, the relative velocity and relative position between the vehicle A and the target G are denoted by V_{AG} and L_{AG}. Similarly, parameters about the obstacle O are denoted by V_{AO} and L_{AO}. Some other nomenclatures are listed in Table 1.

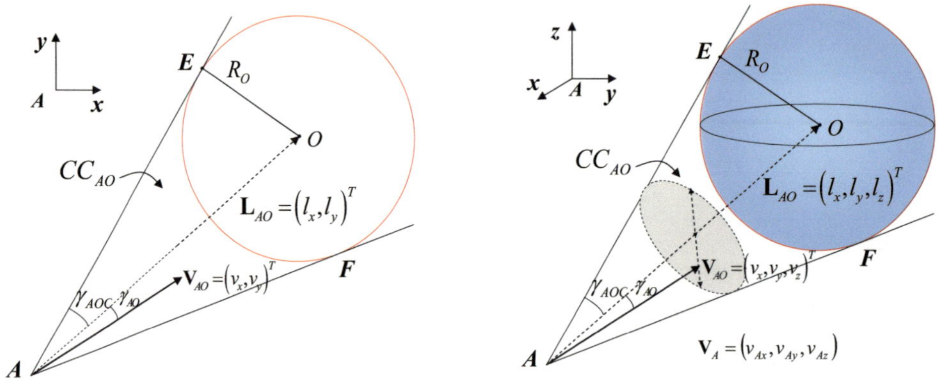

(a) Obstacle avoidance in two dimensions

(b) Obstacle avoidance in three dimensions

Fig. 1. Geometrical representation of the relative parameters in the relative coordinates when the obstacle-avoiding problem is considered. (a) and (b) show the definition of *relative obstacle angle* in 2D scenario and 3D scenario, respectively.

We define the *relative obstacle angle*, γ_{AO}, as the angle between $\mathbf{V}_{AO}$ and $\mathbf{L}_{AO}$. $\gamma_{AO} \in [0, \pi]$. *Collision cone* is defined as an area in which the vehicle will collide with an obstacle by current velocity $\mathbf{V}_{AO}$, and it is denoted by CC_{AO}. Generally, CC_{AO} is in the cone AEF, as shown in Fig.1. The half cone angle of CC_{AO}, γ_{AOC}, is defined as *collision region angle*, i.e.

$$\gamma_{AOC} = \arcsin\left(\frac{R_O}{L_{AO}}\right) \tag{1}$$

Similar to the definition in the obstacle-avoiding scenario in Fig.1, the angle between the relative velocity $\mathbf{V}_{AG}$ and relative distance $\mathbf{L}_{AG}$ is defined as *relative target angle*, denoted by $\gamma_{AG} \in [0,\pi]$, as shown in Fig.2. The *pursuit cone*, as well as the *pursuit region angle*, is defined in the same as obstacle avoidance scenario.

Some assumptions should be declared as the prerequisites of the planning principles. First, in order to guarantee the robot to catch the target successfully, the maximum velocity of the robot is assumed to be larger than that of the target and the obstacles. Due to this precondition, the vehicle A has the ability to catch the target, as well as avoid the collision with obstacles. Second, the target and the obstacles are supposed to keep their speed as constants during the planning period, τ. In fact, this hypothesis is usually accepted because τ is short enough for a real vehicle. Owe to the mechanism of numerical approximation, the path planning problem can be solved and optimized with a receding horizon fashion in

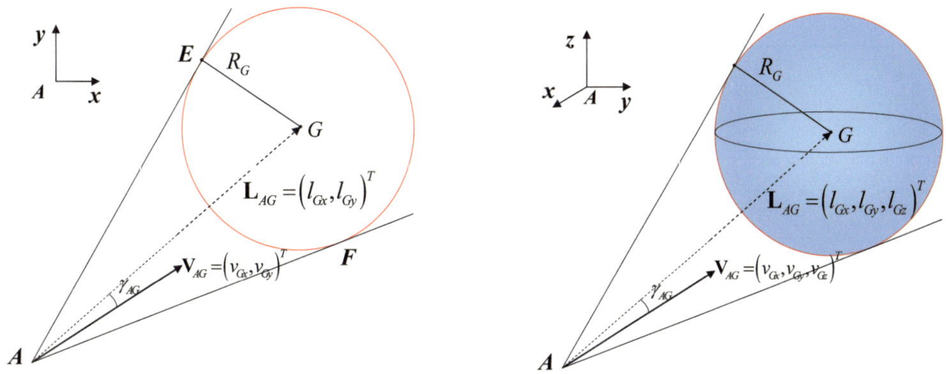

(a) Target pursuit in two dimensions

(b) Target pursuit in three dimensions

Fig. 2. Geometrical representation of the relative parameters in the relative coordinates when the target-pursuing problem is considered. (a) and (b) show the definition of *relative target angle* in 2D scenario and 3D scenario, respectively.

which a new waypoint of the total path is computed gradually over time. Here, we obtain the following equivalence.

$$\Delta\mathbf{V}_{AO} = \Delta\mathbf{V}_{AG} = \Delta\mathbf{V}_{A} \quad (2)$$

2.2 Obstacle-avoidance principle

For each obstacle O, under the assumption that its velocity is constant in τ, it will be avoided if the γ_{AO} is large enough to make the $\mathbf{V}_{AO}$ out of the CC_{AO} over the interval τ when the vehicle moves from time step k to k+1. This fact suggests that the obstacle-avoiding principle hold the following inequality. That is

$$\gamma_{AOC(k)} \le \gamma_{AO(k+1)} \le \pi \quad (3)$$

where $\gamma_{AOC(k)}$ is the *collision region angle* in time step k, which is shown in Fig.1. If there are multi obstacles, Eq. (3) changes to

$$\gamma_{AOCi(k)} \le \gamma_{AOi(k+1)} \le \pi \quad (4)$$

where the subscript i denotes the label of the obstacles. i=1, 2… N. N stands for the number of obstacles.

2.3 Target-pursuit principle

The $\mathbf{V}_{AG}$ can be resolved into a pair of orthogonal components, as shown in Fig. 3. The vehicle is expected to tune its velocity to the optimum. Only when the velocity direction of the vehicle is identical to $\mathbf{L}_{AG}$, it will not lose its target. In the meanwhile, the vehicle should better minimize the magnitude of $\mathbf{V}_T$. Consequently, the policy for the optimal velocity is obvious in that $\mathbf{V}_T$ should be minimized while $\mathbf{V}_C$ maximized. We formulate these principles as two cost functions. They are

$$\min : V_T = \|\mathbf{V}_{AG}\| \sin \gamma_{AG} \tag{5}$$

and

$$\max : V_C = \|\mathbf{V}_{AG}\| \cos \gamma_{AG} \tag{6}$$

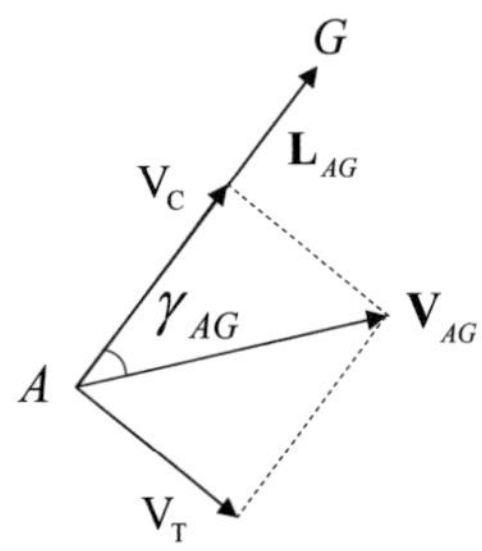

Fig. 3. The resolution of $\mathbf{V}_{AG}$. The velocity of the vehicle relative to the target is resolved into two orthogonal components, $\mathbf{V}_C$ and $\mathbf{V}_T$. One component, $\mathbf{V}_C$, is in the direction of $\mathbf{L}_{AG}$ and pointed to the target. The other, $\mathbf{V}_T$, is orthogonal to $\mathbf{L}_{AG}$.

3. Linear programming model

In this section, we first introduce some knowledge about the mathematical description of the standard linear programming model. Then, the mentioned principles for path planning are decomposed and linearized according to the linear prototype of the standard model.

3.1 Standard linear programming model

The standard Linear Programming is composed of a cost function and some constrains, all of which need to be affine and linear (Boyd & Vandenberghe, 2004). The following is the standard and inequality form.

$$\begin{aligned} &\min_{\mathbf{x}} : \mathbf{c}^T\mathbf{x} \\ &\text{subject to:} \mathbf{D}\mathbf{x} \le \mathbf{h} \end{aligned} \tag{7}$$

where $\mathbf{D} \in R^{m \times n}$. The vector $\mathbf{x} \in R^{n \times 1}$ denotes the variables. In the standard linear programming formulation, the vector $\mathbf{c} \in R^{n \times 1}$, representing the coefficients, the vector $\mathbf{h} \in R^{m \times 1}$, and the matrix $\mathbf{D}$ are all known when the model is going to be solved by some method.
During the decades, several methods have been invented to solve the optimization as shown in (7). From those methods, simplex algorithm and interior-point method are two most famous algorithms. More information about these algorithms can be found in (Boyd & Vandenberghe, 2004). In this chapter, an open library named Qsopt is used where the primal and dual simplex algorithms are imbedded. For the corresponding procedures, the user can specify the particular variant of the algorithm which is applied (David et al., 2011).

3.2 Linear programming model for path planning problem

In this subsection, the problem of obstacle-avoidance is formulated and we obtain some inequality constraints after we introduce two slack variables. Additionally, the formulation of the target-pursuit problem is transferred to some cost functions. The final linear programming model consists of a weighted sum of the cost functions and a set of linear constraints. It is

$$\min: J = \sum_j \omega_j d_j + \omega_{v1} q_1 + \omega_{v2} q_2 \tag{8}$$

satisfying

$$\sum_j \omega_j + \omega_{v1} + \omega_{v2} = 1 \tag{9}$$

where ω_j, ω_{v1}, $\omega_{v2} \geq 0$. The letter j denotes the subscript of the components of $\mathbf{V}_{AG}$. If the path planning is operated in two-dimensional environment, j belongs to a two-element-set, i.e., $j \in \{x, y\}$. And if in three-dimensional environment, j belongs to a three-element-set, i.e., $j \in \{x, y, z\}$. The variables d_j, q_1, q_2 will be explained later.

3.2.1 Linearization of the avoidance constrain

From Fig.1, we get the *relative obstacle angle* function γ_{AO}.

$$\gamma_{AO} = \arccos\left(\frac{\mathbf{V}_{AO} \cdot \mathbf{L}_{AO}}{\|\mathbf{V}_{AO}\| \cdot \|\mathbf{L}_{AO}\|}\right) = \arccos\left(\frac{P}{VL}\right) \tag{10}$$

The definitions of V, L, and P refer to Table 1. Here, Eq. (10) is linearized by Taylor's theorem and we obtain the following equation.

$$\begin{aligned} \gamma_{AO(k+1)} &= \gamma_{AO(k)} + \Delta\gamma_{AO(k)} \\ &= \gamma_{AO(k)} + \tau \left(\frac{d\gamma_{AO}}{d\mathbf{V}_{AO}}\right)_{(k)} \frac{d\mathbf{V}_{AO}}{dt} + o\left(\|\Delta\mathbf{V}_{AO}\|\right) \end{aligned} \tag{11}$$

$$\Delta\gamma_{AO(k)} = -\frac{\tau}{\sqrt{V^2L^2 - P^2}\Big|_{(k)}} \left(\mathbf{L}_{AO} - \frac{P}{V^2}\mathbf{V}_{AO}\right)_{(k)} \Delta\mathbf{V}_{AO} \tag{12}$$

Let $\gamma_{AO(k+1)}$ represent the *relative obstacle angle* in time step k+1 after vehicle's movement. The variables, $\Delta\mathbf{V}_{AO}$, in (11), are that we are trying to plan in step k for step k+1. If there are multiple obstacles, Eq. (12) changes to Eq. (13).

$$\Delta\gamma_{AOi(k)} = -\frac{\tau}{\sqrt{V^2L^2 - P^2}\Big|_{i,(k)}} \left(\mathbf{L}_{AO} - \frac{P}{V^2}\mathbf{V}_{AO}\right)_{i,(k)} \Delta\mathbf{V}_{AO} \tag{13}$$

where

$$\gamma_{AOCi} = \arcsin\left(\frac{R_{Oi}}{L_i}\right) \tag{14}$$

3.2.2 Minimize the relative distance

In this subsection, the relative distance between the robot and the target-objective is minimized. By tuning the velocity of the robot, the relative distance between the robot and target will became smaller through each step time. This distance is resolved into two or three elements by the axis number. That is to minimize the elements of the vector $\mathbf{L}_{AG}$, i.e.,

$$\min : \left\| l_{Gj} - \left(v_{Gj}\tau + \Delta v_{Gj}\tau^2 \right) \right\| \tag{15}$$

where l_{Gj} is the element of $\mathbf{L}_{AG}$ and v_{Gj} is the element of $\mathbf{V}_{AG}$. See Fig.2. Δv_{Gj} denotes the acceleration of v_{Gj}. $j \in \{x, y\}$ is for two-dimensional environment and $j \in \{x, y, z\}$ is for three-dimensional environment. The inherited objective function is derived from Eq. (15).

$$\begin{aligned} &\min : d_j \\ &\text{subject to} : -d_j \le l_{Gj} - \left(v_{Gj}\tau + \Delta v_{Gj}\tau^2 \right) \le d_j \end{aligned} \tag{16}$$

where Δv_{Gj} is the variable that we hope to compute in the linear programming model, as stated in Eq. (2). In a manner of preserving convexity of the problem, $d_j \ge 0$ is the newly introduced slack variable associated with each element (Boyd & Vandenberghe, 2004).

3.2.3 Optimize the relative velocity

The relative velocity between the robot and the target-objective needs to optimize on the consideration of target-pursuit principle. We respectively discuss the optimization of the pair component, $\mathbf{V}_T$ and $\mathbf{V}_C$, as shown in Fig. 3.

(1) Minimize the magnitude of the component $\mathbf{V}_T$. We compute $\mathbf{V}_T$ with

$$V_T = \sqrt{V_{AG}^2 - \frac{P_{AG}^2}{L_{AG}^2}} \tag{17}$$

So the optimization of (5) is equal to

$$\min : g\left(\vec{V}_{AG}\right) = V_T^2 = V_{AG}^2 - \frac{P_{AG}^2}{L_{AG}^2} \tag{18}$$

Using Taylor's theorem and introducing slack variable, q_1, we get the new formulation for optimization.

$$\begin{aligned} &\min : q_1 \\ &\text{subject to:} \quad 0 \le g\left(\mathbf{V}_{AG}\right) + \nabla g \Delta \mathbf{V}_{AG}^T \le q_1 \\ &\qquad\qquad\quad 0 \le q_1 \le V_{AG\max}^2 \end{aligned} \tag{19}$$

where $\Delta\mathbf{V}_{AG}$ is the variable vector that we hope to include in the linear programming model. q_1 is the slack variable. ∇g represents the grades of the function g(.). It is computed with

$$\nabla g = 2\tau \left(\mathbf{V}_{AG} - \frac{P_{AG}}{L_{AG}^2} \mathbf{L}_{AG} \right) \tag{20}$$

$V_{AG\max}$ is the maximum of the relative velocity between the robot and the target. We estimate it by $V_{AG\max} = 2V_{A\max}$. $V_{A\max}$ denotes the upper boundary of V_A.

(2) Maximize the magnitude of the component $\mathbf{V}_C$.
Since we can maximize $\mathbf{V}_C$ by minimizing -$\mathbf{V}_C$, we refer to a minimize problem with affine cost function and constraint functions as a linear programming. Consequently, the problem described by (6) can be rewritten as

$$\min : -V_C = \|\mathbf{V}_{AG}\| \cos \gamma_{AG} \tag{21}$$

Linearized using Taylor's theorem, Eq. (21) changes to a standard linear programming problem. That is

$$\begin{aligned} &\min : q_2 \\ &\text{subject to: } -V_C - \nabla V_C \Delta \mathbf{V}_{AG}^{\mathrm{T}} \le q_2 \\ &\qquad -V_{AG\max} \le q_2 \le V_{AG\max} \end{aligned} \tag{22}$$

where $\Delta\mathbf{V}_{AG}$ is the variable that we are trying to include in our model. q_2 is also a slack variable. Here, the grades is computed with $\nabla\mathbf{V}_C = \tau\, \mathbf{L}_{AG}/\mathrm{L}_{AG}$.

3.3 Other constraints
One of the advantages of the LP method is that various constraints can easily be added in the opening constraint set. Here we provide some demonstrations of constraints that are transformed from the dynamics, sensor data, and searching boundaries.

3.3.1 The constraints from the kinematics and dynamics
The kinematics and dynamics are extremely simplified in the path planning model where only the bound of the velocity and acceleration are added in the feasible set. These bound can be described mathematically as

$$\begin{cases} -V_{A\max} \le v_{Aj} \le V_{A\max} \\ -\Delta_{\max} \le \Delta v_j \le \Delta_{\max} \end{cases} \tag{23}$$

where v_{Aj} denotes the components of $\mathbf{V}_A$, as shown in Fig. 1. $\Delta_{\max}$ denotes the max magnitude of the changing of v_{Aj} in each period τ.

$$-V_{A\max} \le v_{Aj} + \Delta v_j \tau \le V_{A\max} \tag{24}$$

We get the net constraints (25) from the simultaneous equations (23) and (24).

$$\max\left\{-\Delta_{\max}, -\frac{1}{\tau}\left(V_{A\max} + v_{Aj}\right)\right\} \le \Delta v_j \le \min\left\{\Delta_{\max}, \frac{1}{\tau}\left(V_{A\max} - v_{Aj}\right)\right\} \tag{25}$$

3.3.2 The constraints from the limit of the sensor
All sensors can not detect the area that is beyond their capability. In real application, the obstacle that has been avoided by the vehicle will possibly disappear from the sensor's operating region. Similarly, any new obstacle from long distance is possible entering the detecting area. On this point, we evaluate the threat of each obstacle in the environment and propose a threat factor for each of them to estimate the performance. More specially, if the relative distance between the vehicle and the obstacle O_i satisfies $\mathrm{L}_{AOi} \ge \mathrm{L}_{\min i}$, this

obstacle is assigned λ_i=0. L_{mini} is the threshold. On the contrary, O_i is assigned λ_i=1 if L_{AOi}<L_{mini}.
In addition, the *relative obstacle angle* has impact on the vehicle. If the velocity vector of the vehicle satisfies $\gamma_{AOi} \geq \gamma_{AOCi} + \Delta\gamma_{AOimax}$, the vehicle can be looked as moving from the obstacle O_i, λ_i will be assigned zero.

3.3.3 The constraints from the searching region

The domain of the variables is generally computed in a square in 2D or a cube in 3D for the interceptive magnitude limit of the acceleration on each axis. In fact, the acceleration on each axis is coupled. So if the domain is treated as a circle in 2D (Richards & How, 2002) or a sphere in 3D, the algorithm will lower its conservation. See Fig. 4. We denote the domain as *dom*. That is

$$\Delta \mathbf{V}_A \in dom \tag{26}$$

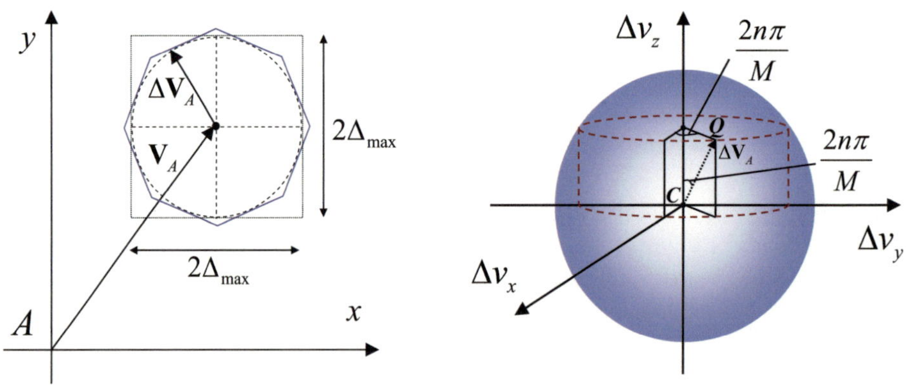

(a) Searching region in 2D; (b) Searching region in 3D

Fig. 4. Searching region for acceleration. (a) The domain of the acceleration will be approximated by multiple lines in 2D. (b) The domain of the acceleration will be approximated by multiple planes in 3D.

We try to approximate the *dom* with multiple lines in 2D, or multiple planes in 3D. That is

$$\sin\frac{2m\pi}{M}\Delta v_x + \cos\frac{2m\pi}{M}\Delta v_y \leq \Delta_{max} \tag{27}$$

where *m*=0,1,2,..., *M*-1. Here, *M* represents the quantity of the line used for approximation. In three-dimensional environment,

$$\sin\frac{2m\pi}{M}\cos\frac{2n\pi}{M}\Delta v_x + \sin\frac{2m\pi}{M}\sin\frac{2n\pi}{M}\Delta v_y + \cos\frac{2m\pi}{M}\Delta v_z \leq \Delta_{max} \tag{28}$$

where *m*, *n*=0,1,2,...,*M*-1. Here, *M* represents the quantity of the plane used for approximation. At this point, the linear programming model is formulized. The objective function is (8). All constrains are recapitulated as the following.

$$\begin{cases} \lambda_i \gamma_{AOCi} \le \lambda_i \left(\gamma_{AOi} + \Delta \gamma_{AOi} \right) \le \pi \\ -d_j \le l_j - \left(v_j \tau + \Delta v_j \tau^2 \right) \le d_j \\ 0 \le g(\mathbf{V}_{AG}) + \nabla g \cdot \Delta \mathbf{V}_{AG}^T \le q_1 \\ -\mathrm{V}_C - \nabla \mathbf{V}_C \Delta \mathbf{V}_{AG}^T \le q_2 \\ \max\left\{ -\Delta_{\max}, -\frac{1}{\tau}\left(\mathrm{V}_{A\max} + v_{Aj} \right) \right\} \le \Delta v_j \le \min\left\{ \Delta_{\max}, \frac{1}{\tau}\left(\mathrm{V}_{A\max} - v_{Aj} \right) \right\} \\ \Delta \mathbf{V}_A \in dom \end{cases} \tag{29}$$

4. Path planning simulation for TPOA simulation

The approach proposed in this chapter is simulated with three obstacles and the results are given scenarios of in 2D and 3D, respectively. See Fig.5 and Fig.6. All the simulations run on the platform of WinXP/Pentium IV 2.53 GHz/2G RAM. A linear programming solver, named QSopt, is called from the library (David et al., 2011). We run the examples with three obstacles and give the results in two-dimensional environment and three-dimensional environment, respectively.

4.1 Path planning in 2D

According to the LP model, as shown with Eq. (29), all initial parameters are listed in Table 2. Assuming that the maximal velocity of the vehicle in 2D is 50cm/s, and the maximal acceleration of the vehicle is 350cm/s^2. The panning period is τ=20ms. The three parameters will be kept the same in the following simulation.

Fig. 5(a)~(d) show the key scenarios when avoiding the three obstacles. Fig. 5(a) shows the situation when the robot avoiding SO. It is obvious that the robot can rapidly adjust its velocity to the most favorable one and go on pursuing the target. The planner calls LP algorithm about 13 times in this stage. Fig. 5(b) shows the case while the robot has to avoid the moving obstacle MO1. At this time, the robot turns left to avoid MO1 because the over speeding is accessible for the robot. Fig. 5(c) shows the different decision that the robot selected to avoid MO2 from its back. Two conclusions can be drawn from this simulation. First, the robot is indeed qualified the avoiding and pursuing ability in uncertain environment. Second, the robot can adjust its velocity autonomously, including the magnitude and the direction, and adapt the optimal decision to avoid the obstacle and to catch the target. The whole pursuing process of this example lasts about 640ms and each period of calculation of our algorithm takes only about 0.469ms which is a low time complexity for real-time application.

	Initial Position (cm)	Initial Velocity (cm/s)	Radius (cm)
Robot (A)	(0,0)	(10,0)	N/A
Target (G)	(1000,1000)	(-12,0)	50
Static Obstacle (SO)	(300,300)	N/A	100
Moving Obstacle (MO1)	(800,600)	(-20,0)	50
Moving Obstacle (MO2)	(300,900)	(6,-3)	50

Table 2. The initial parameters for simulation in 2D.

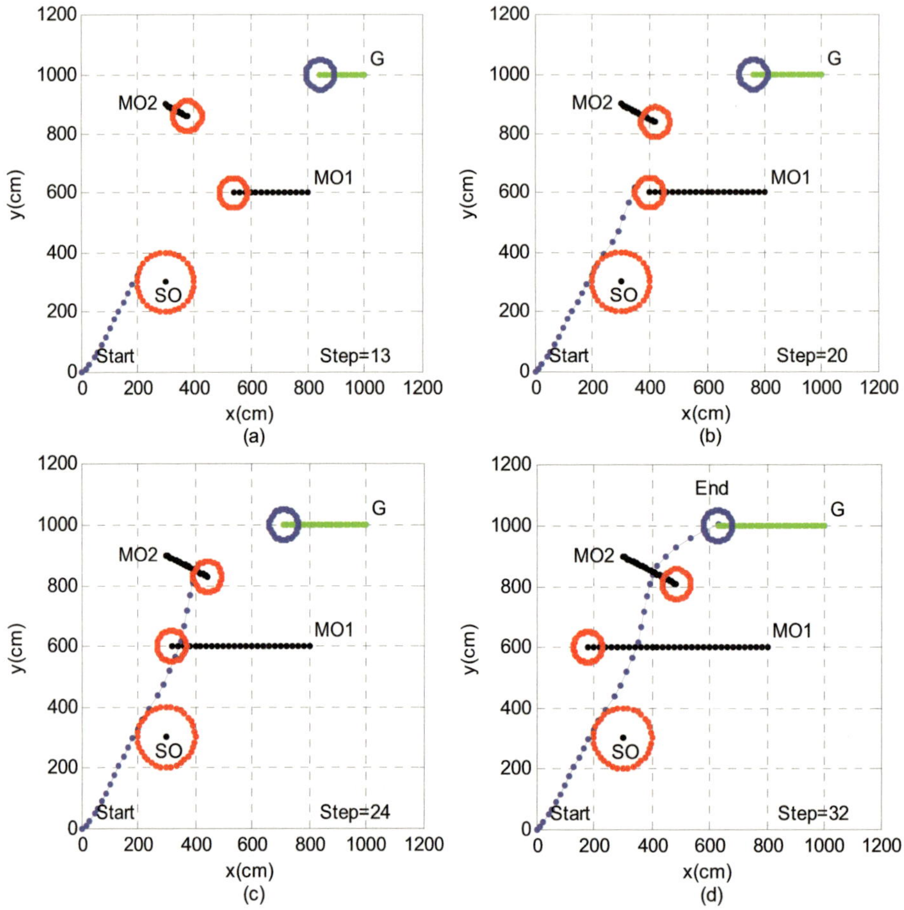

Fig. 5. Simulation in dynamic environment of 2D. The dot-line denotes the trajectories of the robot, the obstacles, and the target. And the circles with dot-line edge are the shape profile of any mover. SO represents static obstacle. MO1 and MO2 represent moving obstacles. *G* represents the moving target.

4.2 Path planning in 3D

For the specific 3D environment, assuming that there is one static obstacle SO1, and two moving obstacles, MO1 and MO2. All the initial parameters are listed in Table 3.

	Initial Position (cm)	Initial Velocity (cm/s)	Radius (cm)
Robot (A)	(0,1000,0)	(15,-15,0)	N/A
Target (G)	(1000,0,1000)	(-5,5,0)	50
Static Obstacle (SO)	(300,700,300)	N/A	100
Moving Obstacle (MO1)	(450,500,900)	(0,0,-10)	80
Moving Obstacle (MO2)	(450,250,850)	(0,10,0)	50

Table 3. The initial parameters for simulation in 3D.

Fig. 6 shows the results. At the beginning, SO is on the line between the robot and the target, and the robot is blocked. In the following time, the robot avoids MO1 and MO2 at step=19 and step=24, respectively. See Fig. 6(b) and (c). It is evident that the velocity-decision of the robot is optimized online. The time complexities of this simulation are 0.556ms in every period averagely.

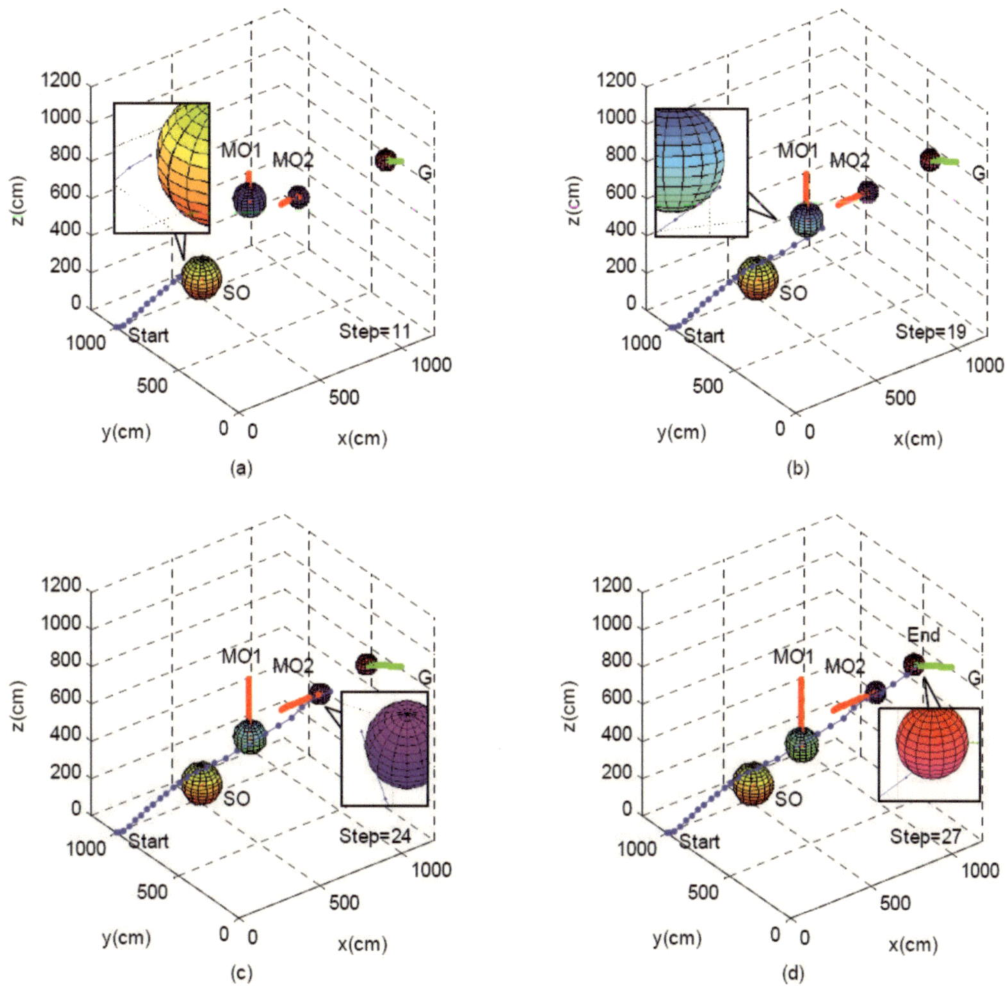

Fig. 6. Simulation in dynamic environment of 3D. The dot-line denotes the trajectories, and the sphere denotes the target and the obstacles.

5. Path planning for multiple tasks planning (MTP): an application

MTP problem is a variant of multiple TPOA. It can be stated that given N targets and N vehicles, let each vehicle can only pursue one target and each target is pursued exactly by one vehicle at any interval. The task is finished until all the targets have been caught. Then, the goal of MTP problem is to assign the tasks to vehicles within each interval online as well as complete the whole mission as fast as possible.

In this section, MTP problem is decomposed into two consecutive models, i.e., the task-assignment model and the path planning model. The path planning model has already been introduced above. With respect to the task-assignment model, a minimax assignment criterion is proposed to direct the task assignment. Under this assignment criterion, the optimal assignment will cost the least time to catch all the targets from the current measurable information.

Some simulations in Fig. 7 are given to show the efficiency of the method. Vehicles pursuing target are assigned according to the minimax assignment. According to the criterion, the current task assignment prefers to finish the whole mission fastest (Yang et al., 2009).

5.1 Assignment model under minimax assignment criterion

In the assignment problem, the payment in each vehicle for a target is assumed to be known. The problem is how to assign the target for each particular vehicle that the total pursuit mission can be completed as fast as possible.

Let x_{ij} be the n^2 0-1 decision variables, where $x_{ij}=1$ represents vehicle i for target j; otherwise, $x_{ij}=0$. Because the "payment" is understood as time, it is important to minimize the maximal time expended by vehicle for its pursuing process. This task assignment problem may be described as the minimax assignment model mathematically.

$$\begin{aligned} &\min : z = \max_{i,j}\{c_{ij}x_{ij}\} \\ &\text{subject to: } \sum_{i=1}^{n} x_{ij} = 1 \quad (j = 1,\dots,n) \\ &\sum_{j=1}^{n} x_{ij} = 1 \quad (i = 1,\dots,n) \\ &x_{ij}=0 \text{ or } 1,\ \ i,j \in \{1,\dots,n\} \end{aligned} \tag{30}$$

where c_{ij} represents the payment in vehicle i for target j and will be given in Section 5.3. The elements c_{ij} give an n^2 binary cost matrix. This is a classic integer programming where the objective function is nonlinear and difficult to solve.

A solution method for the above minimax assignment problem named the operations on matrix is proposed. This solution method finds the solution directly from the cost matrix $C_n = (c_{ij})$ and the objective function (Yang et al., 2008).

5.2 Global cost function

We give the cost function of this MTP problem as the following. Let N_v, N_T, and N_o be the number of vehicles, targets and obstacles, respectively. The payment of the vehicle-i to pursue the target-j in the assignment problem is written as:

$$c_{ij} = \frac{d_{ij} + \xi_1 \sum_{k=1,k\neq i}^{N_v+N_o} d_{kij}}{\Delta v_{ij}} + \frac{\xi_2 \left|\Delta\Phi_{ij}\right|}{\omega_i} \tag{31}$$

where d_{ij}, Δv_{ij} and $\Delta\Phi_{ij}$ are, respectively, the distance, velocity difference and heading difference between vehicle-i (V-i) and target-j (T-j). ω_i is the maximum turning rate of V-i; d_{kij} is the "additional distance" due to the obstacles and the vehicles other than V-i itself (with respect to vehicle-i, not only the obstacles, but also the other vehicles are its obstacles). ξ_1 and ξ_2 are positive constants. the c_{ij} in Eq. (33) indicates the time possibly needed by V-i to pursue T-j, and the "extended distance" consists of the linear distance, the angle distance, as well as the possible obstacle avoidance between V-i and T-j (Yang et al., 2009).

5.3 Simulation of MTP in 2D

The simulations demonstrated here include three robots, three moving target and three moving obstacles in two-dimensional environment (see Fig. 7). All the initial parameters are listed in Table 4.

	Initial Position (cm)	Initial Velocity (cm/s)	Radius (cm)
Robot (R1)	(0,200)	(17,0)	N/A
Robot (R2)	(0,0)	(17,0)	N/A
Robot (R3)	(200,0)	(17,0)	N/A
Target (G1)	(900,600)	(-11,9)	40
Target (G2)	(800,800)	(-10,10)	40
Target (G3)	(600,900)	(-9,11)	40
Moving Obstacle (MO1)	(850,450)	(-11,9)	60
Moving Obstacle (MO2)	(750,650)	(-10,10)	60
Moving Obstacle (MO3)	(550,750)	(-9,11)	60

Table 4. The initial parameters for MTP simulation in 2D.

In order to test the performance of the proposed linear programming model in the presence of different-motion vehicles, the following modifications are made during the pursuit process.
① at k=0, the maximal robot velocities are $v_{1\text{-max}}$=30 cm/s, $v_{2\text{-max}}$=68 cm/s, $v_{3\text{-max}}$=45 cm/s; ② at k=16, the maximal robot velocities are $v_{1\text{-max}}$=30 cm/s, $v_{2\text{-max}}$=68 cm/s, $v_{3\text{-max}}$=22.5 cm/s; ③ at k=31, the maximal robot velocities are $v_{1\text{-max}}$=42.5 cm/s, $v_{2\text{-max}}$=68 cm/s, $v_{3\text{-max}}$=22.5 cm/s.

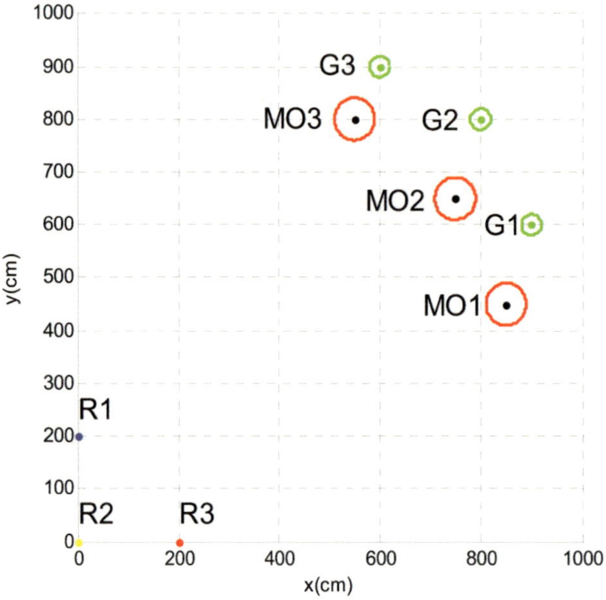

(a)

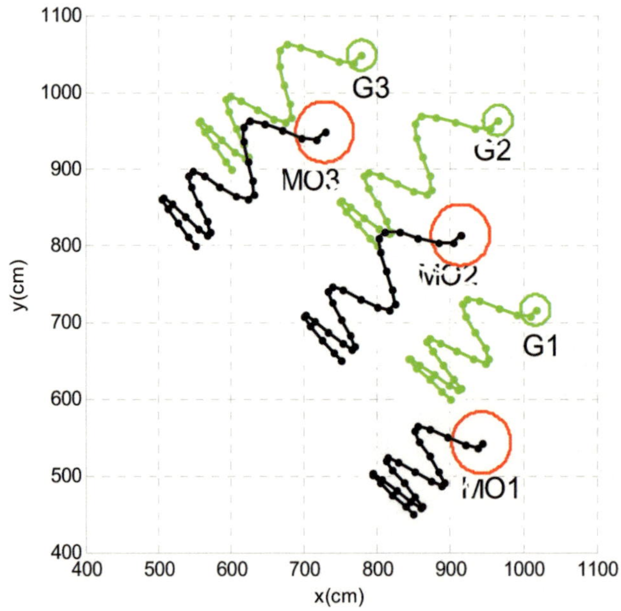

(b)

Fig. 7. Robots, targets and obstacles.
(a) Initial positions at k=0; (b) The target and obstacle trajectories till k=30. The targets and the obstacles are moving in a pso-sine curve.

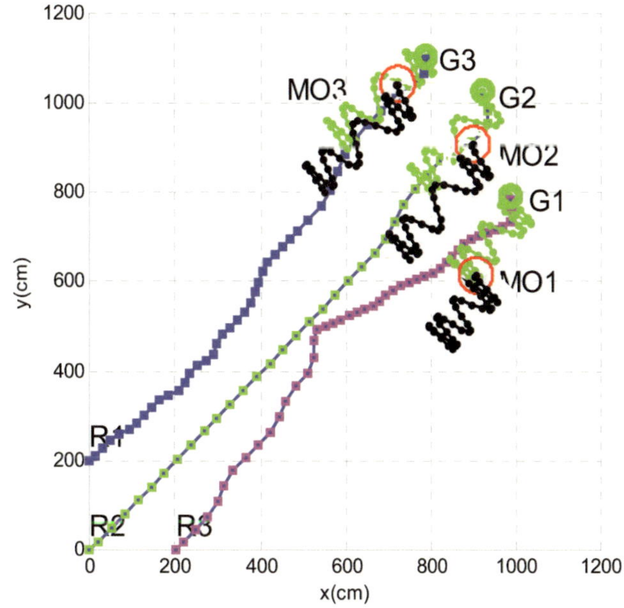

(a)

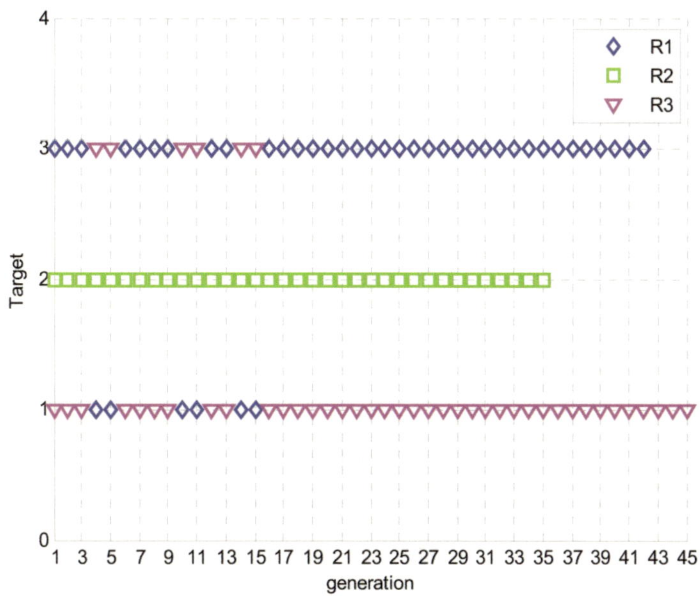

(b)

Fig. 8. Simulation of MTP in 2D. (a) Planning trajectories; (b) robot-target assignment pairs.

As Fig. 8 shows, the pair assignment alters between {R1-G3, R2-G2, R3-G1} and {R1-G1, R2-G2, R3-G3} at initial 15 steps according to the minimax assignment computation. At time step 16, assignment is kept for {R1-G3, R2-G2, R3-G1} due to that R3 is the slowest robot after its maximal velocity reducing 50%, and it can only catch G1. R2 is the fastest robot, so it pursues the fastest target G2. When k=35 and k=42, target G2 and G3 are caught by R2 and R1, respectively. Finally, G1 is caught at k=45 by R3, so the mission is completed successfully. The dot-lines are the planned trajectories of the robots and the colorful dots distinguish every step. From the figures we can see that the robots avoid the obstacles and catch all the moving targets successfully.

6. Conclusion

Path planning is a fundamental problem in robot application. In order to solve the path planning in dynamic environment, this chapter proposes a method based on LP/MILP to plan the acceleration of the robot in relative velocity coordinates. This method has the uniform formulation for 2D and 3D environment and it can give the information of the optimal velocity and acceleration in the view of the special cost function. Multiple constrains, such as the bounds of velocity, acceleration, and sensors, are included in the LP model and the MILP model. We also can add other constrains easily.

A particular application of this method is discussed for the problem of the multi-task planning where several robots are set to pursuit several targets. In the classical cooperation problem, the targets are assumed to be dynamic, similar to the TPOA problem. In the solution of MTP problem, a minimax assignment criterion and a global cost function are proposed to direct the task assignment.

Many simulations about the path planning in 2D/3D and in the multi-task planning requirements are taken to verify this novel method. The results show the low computing load of this method so it is potential to apply in real time manner.

7. Acknowledgment

This work was partially supported by Natural Science Foundation of China (NSFC) under grant #61035005 and #61075087, Hubei Provincial Education Department Foundation of China under grant #Q20111105, and Hubei Provincial Science and Technology Department Foundation of China under grant #2010CDA005.

8. References

Allaire, F. C. J.; Tarbouchi, M.; Labont´e G.; et al. (2009). FPGA implementation of genetic algorithm for UAV real-time path planning. *Journal of Intelligent and Robotic Systems*, Vol.54, No.1-3, pp.495-510, ISSN 09210296

Boyd S. & Vandenberghe L. (2004). Convex Optimization. Cambridge University Press, ISBN: 0521833787, New York

Chen, M.; Wu, Q. X. & Jiang, C. S. (2008). A modified ant optimization algorithm for path planning of UCAV. *Applied Soft Computing Journal*, Vol.8, No.4, pp.1712-1718, ISSN: 15684946

Chen, Y. & Han J. (2010). LP-Based Path Planning for Target Pursuit and Obstacle Avoidance in 3D Relative Coordinates. *Proceeding of American Control Conference*, pp.5394-5399, ISBN-13: 9781424474264, Baltimore, MD, USA, June 30- July 2, 2010

Chen, Yang; Zhao, Xingang; Zhang, Chan; et al. (2010). Relative coordination 3D trajectory generation based on the trimmed ACO. *Proceedings of International Conference on*

Electrical and Control Engineering, pp.1531-1536, ISBN-13: 9780769540313, Wuhan, China, June 26-28, 2010

Cocaud, C.; Jnifene, A. & Kim, B. (2008). Environment mapping using hybrid octree knowledge for UAV trajectory planning. *Canadian Journal of Remote Sensing*, Vol.34, No.4, pp.405-417, ISSN 07038992

David, A.; William, C.; Sanjeeb, D. & Monika, M. (2011). *QSopt Linear Programming Solver* [Online], June 1, 2011, available from: <http://www2.isye.gatech.edu/~wcook/qsopt/index.html>

Fahimi, F.; Nataraj, C & Ashrafiuon, H. (2009). Real-time obstacle avoidance for multiple mobile robots. *Robotica*, Vol.27, No.2, pp.189-198, ISSN 02635747

Fiorini, P. & Shiller, Z. (1998). Motion Planning in Dynamic Environments Using Velocity Obstacles. *International Journal of Robotics Research*, Vol. 17, No.7, pp.760-772. ISSN 02783649

Hrabar, S. E. (2006). Vision-based 3D navigation for an autonomous helicopter, PhD thesis, University of Southern California, United States

Jung, L. F.; Knutzon, J. S.; Oliver, J. H, et al. (2006). Three-dimensional path planning of unmanned aerial vehicles using particle swarm optimization, *11th AIAA/ISSMO Multidisciplinary Analysis and Optimization Conference*, pp.992-1001, ISBN-10: 1563478234, Portsmouth, VA, USA, September 6-8, 2006

Kim, J. O. & Khosla, P. K. (1992). Real-time obstacle avoidance using harmonic potential functions. *IEEE Transactions on Robotics and Automation*, Vol.8, No.3, pp.338-349, ISSN 1042296X

Kitamura, Y.; Tanaka, T.; Kishino, F.; et al. (1995). 3-D path planning in a dynamic environment using an octree and an artificial potential field, *IEEE/RSJ International Conference on Intelligent Robots and Systems*, Vol.2, pp.474-481, Pittsburgh, PA, USA, August 5-9, 1995

Konolige, K. (2000). A gradient method for realtime robot control, *IEEE/RSJ International Conference on Intelligent Robots and Systems*, Vol.1, pp.639-646, Piscataway, NJ, USA, October 31-November 5, 2000

Krenzke, T. (2006). Ant Colony Optimization for agile motion planning. Master thesis. Massachusetts Institute of Technology, United States

Ladd, A. M. & Kavraki, L. E. (2004). Measure theoretic analysis of probabilistic path planning, *IEEE Transactions on Robotics and Automation*, Vol.20, No.2, pp.229-242, ISSN 1042296X

Liu, C.; Wei, Z. & Liu, C. (2007). A new algorithm for mobile robot obstacle avoidance based on hydrodynamics, *IEEE International Conference on Automation and Logistics*, pp.2310-2313, ISBN-10: 1424415314, Jinan, China, August 18-21, 2007

Nie, Y. Y.; Su, L. J. & Li, C. (2003). An Isometric Surface Method for Integer Linear Programming, *International Journal of Computer Mathematics*, Vol.80, No.7, pp.835-844, ISSN 00207160

Oh, T. S.; Shin, Y. S.; Yun, S. Y., et al. (2007). A feature information based VPH for local path planning with obstacle avoidance of the mobile robot, *4th International Conference on Mechatronics and Information Technology*. Vol.6794. Bellingham, WA, USA: SPIE

Richards, A. & How, J. P. (2002). Aircraft trajectory planning with collision avoidance using mixed integer linear programming. *Proceeding of American Control Conference*, pp. 936-1941, ISSN 07431619, Anchorage, AK, USA, May 8-10, 2002

Schouwenaars, T.; How, J. & Feron, E. (2004). Receding Horizon Path Planning with Implicit Safety Guarantees. *Proceeding of the American Control Conference*, pp. 5576-5581, ISSN 07431619, Boston, MA, USA, June 30 - July 2, 2004

Schumacher, C. J.; Chandler, P. R.; Pachter, M.; et al. (2004). Constrained optimization for UAV task assignment, *Proceedings of the AIAA guidance, navigation, and control conference*, AIAA 2004-5352, Providence, RI., 2004

Shima, T.; Rasmussen, S. J.; Sparks, A. G.; et al. (2006). Multiple task assignments for cooperating uninhabited aerial Vehicles using genetic algorithms, *Computers & Operations Research*, Vol.33, No.11, pp.3252-3269, ISSN 03050548

Wang, Y.; Lane, D. M. & Falconer, G. J. (2000). Two novel approaches for unmanned underwater vehicle path planning: constrained optimisation and semi-infinite constrained optimization. *Robotica*, Vol.18, No.2, pp.123-142, ISSN 02635747

Yang, L. Y.; Nie, M. H.; Wu, Z. W.; et al. (2008). Modeling and Solution for Assignment Problem, *International Journal of Mathematical Models and Methods in Applied Sciences*, Vol. 2, No. 2, pp.205-212, ISSN: 1998-0140

Yang, L. Y.; Wu, C. D.; Han, J. D.; et al. (2009). The task assignment solution for multi-target pursuit problem. *International Conference on Computational Intelligence and Software Engineering*, pp.1–6, ISBN-13 9781424445073, Wuhan, China, December 11-13, 2009

Zhang, Y. & Valavanis K P. (1997). A 3-D potential panel method for robot motion planning. *Robotica*, Vol.15, No.4, pp.421-434, ISSN 02635747

Zhao, L. & Murthy, V. R. (2007). Optimal flight path planner for an unmanned helicopter by evolutionary algorithms, *AIAA Guidance, Navigation and Control Conference,* Vol.4, pp.3716-3739, ISBN-10: 1563479044, Hilton Head, SC, USA, August 20-23, 2007

Zu, D.; Han, J. & Tan, D. (2006). Acceleration Space LP for the Path Planning of Dynamic Target Pursuit and Obstacle Avoidance. *Proceedings of the 6th World Congress on Intelligent Control and Automation*, Vol.2, pp.9084-9088, ISBN-10 1424403324, Dalian, China, June 21-23, 2006

8

Reachable Sets for Simple Models of Car Motion

Andrey Fedotov[1], Valerii Patsko[1] and Varvara Turova[2]
[1]*Institute of Mathematics and Mechanics, Ural Branch of the Russian Academy of Sciences*
[2]*Mathematics Centre, Technische Universität München*
[1]*Russia*
[2]*Germany*

1. Introduction

In 1889, Andrey Andreevich Markov published a paper in "Soobscenija Charkovskogo Matematiceskogo Obscestva" where he considered four mathematical problems related to the design of railways. The simplest among these problems (and the first one in course of the presentation) is described as follows. Find a minimum length curve between two points in the plane provided that the curvature radius of the curve should not be less than a given quantity and the tangent to the curve should have a given direction at the initial point.
In 1951, Rufus Philip Isaacs submitted his first Rand Corporation Report on differential game theory where he stated and lined out a solution to the "homicidal chauffeur" problem. In that problem, a "car" with a bounded turning radius and a constant magnitude of the linear velocity tries as soon as possible to approach an avoiding the encounter "pedestrian". At the initial time, the direction of the car velocity is given.
In 1957, in American Journal of Mathematics, Lester Eli Dubins considered a problem in the plane on finding among smooth curves of bounded curvature a minimum length curve connecting two given points provided that the outgoing direction at the first point and incoming direction at the second point are specified.
Obviously, if one takes a particular case of Isaacs' problem with the immovable "pedestrian", then the "car" will minimize the length of the curve with the bounded turning radius. The arising task coincides with the problem considered by A. A. Markov. The difference from the problem by L. E. Dubins is in the absence of a specified direction at the incoming point. The fixation of incoming and outgoing directions presents in the other three problems by A. A. Markov. However, they contain additional conditions inherent to the railway construction.
In such a way the notion of a "car" which moves only forward and has bounded turning radius appeared.
In 1990, in Pacific Journal of Mathematics, James Alexander Reeds and Lawrence Alan Shepp considered an optimization problem where the object with bounded turning radius and constant magnitude of the linear velocity can instantaneously change the direction of motion to the opposite one. In a similar way, carts move around storage rooms. Thus, the model of the car that can move forward and backward has appeared.

Two models (the forward moving car with bounded turning radius; the forward and backward moving car with bounded turning radius) gave rise to numerous literature where various optimization problems related to the transportation are studied. More sophisticated models in which a moving object is considered in a more realistic way have arisen (controlled wheel, bicycle, car with two chassis, car with a trailer). Optimization problems related to such complicated tasks are extremely difficult. The simplest models serve as a "sample" showing where the situation is easy and where it becomes complex.
One of the key notion in the mathematical control theory (Pontryagin et al., 1962), (Lee & Markus, 1967), (Agrachev & Sachkov, 2004) is reachable set. The reachable set is a collection of states which can be attained within the framework of the motion model under consideration. The reachable set at given time describes a collection of states realizable at a specified time instant. The reachable set by given time is a collection of states that can be obtained on the whole time interval from an initial time instant to a specified one.
This paper is devoted to the investigation of reachable sets at given and by given time for simplest models of car motion.

2. Simple models of car motion

The simplest car dynamics are described in dimensionless variables by the following system of equations

$$\dot{x} = \sin\theta, \;\; \dot{y} = \cos\theta, \;\; \dot{\theta} = u; \quad |u| \leq 1. \tag{1}$$

The variables x, y specify the center mass position in the two-dimensional plane; θ is the angle between the direction of the velocity vector and that of the vertical axis y measured clockwise from the latter. The value u acts as a control input. The control $u(t)$ specifies the instantaneous angular velocity of the vector $(\dot{x}(t), \dot{y}(t))$ of linear velocity and is bounded as $|u(t)| \leq 1$.
The motion trajectories in the plane x, y are curves of bounded curvature. The paper (Markov, 1889) considers four optimization problems related to curves of bounded curvature. The first problem (Markov, 1889), p. 250, can be interpreted as a time-optimal control problem for car dynamics (1). Also, the main theorem (Dubins, 1957), p. 515, allows an interpretation in the context of time-optimal problem for such a car. In works on theoretical robotics (Laumond, 1998), an object with dynamics (1) is called Dubins' car. Model (1) is often utilized in differential game problem formulations (Isaacs, 1951; 1965).
Next in complexity is the car model by Reeds and Shepp (Reeds & Shepp, 1990):

$$\dot{x} = w\sin\theta, \;\; \dot{y} = w\cos\theta, \;\; \dot{\theta} = u; \quad |u| \leq 1, \; |w| \leq 1. \tag{2}$$

Control u changes the angular velocity, control w is responsible for instantaneous changes of the linear velocity magnitude. In particular, the car can instantaneously change its motion direction to the opposite one. The angle θ is the angle between the direction of the axis y and the direction of the forward motion of the car.
It is natural to consider control dynamics where the control w is from the interval $[a, 1]$:

$$\dot{x} = w\sin\theta, \;\; \dot{y} = w\cos\theta, \;\; \dot{\theta} = u; \quad |u| \leq 1, \; w \in [a, 1]. \tag{3}$$

Here $a \in [-1, 1]$ is the parameter of the problem. If $a = 1$, Dubins' car is obtained; if $a = -1$, Reeds and Shepp's car appears.

Finally, one can consider non-symmetric constraint $u \in [b,1]$ instead of the bound $|u| \leq 1$. Assume that the values of the parameter b satisfy the inclusion $b \in [-1,0)$. In this case, controlled system preserves essential properties inherent to the case $|u| \leq 1$ (since the value $u = 0$ remains the internal point of the restriction on u).

Let us write down the dynamics of the last system as the most general one among the above mentioned.

$$\dot{x} = w \sin\theta, \ \ \dot{y} = w \cos\theta, \ \ \dot{\theta} = u; \quad u \in [b,1], \ w \in [a,1]. \tag{4}$$

Here, a and b are fixed parameters with $a \in [-1,1]$, $b \in [-1,0)$.

The model (4) is kinematic, since it does not take into account forces acting on the body. The car is represented as a point mass. The control u determines the angular rate of the linear velocity vector; the control w changes the magnitude of the linear velocity.

In the paper, reachable sets at given time and those ones by given time for system (4) are studied using numerical constructions. The reachable set at time t_* is a collection of all states which can be obtained exactly at given time t_*. If such states are considered in the plane x, y, then we have a two-dimensional reachable set; if the consideration takes place in space x, y, θ, then a three-dimensional set is obtained. The peculiarity of the reachable set "by given time" is in accounting for the states reachable not only at time t_* but on the whole interval $[0, t_*]$.

3. Two-dimensional reachable sets

3.1 Reachable sets at given and by given time

Let $z_0 = (x_0, y_0)$ be the initial geometric position and θ_0 be the initial angle at time $t_0 = 0$. The reachable set $G^2(t_*; z_0, \theta_0)$ at given time t_* in the plane x, y is the set of all geometric positions which can be reached from the starting point z_0, θ_0 at time t_* by the trajectories of system (4) using admissible piecewise continuous controls $u(\cdot), w(\cdot)$.

Introducing the notation $z(t_*; z_0, \theta_0, u(\cdot), w(\cdot))$ for the solution of differential equation (4) with the initial point z_0, θ_0, one can write

$$G^2(t_*; z_0, \theta_0) := \bigcup_{u(\cdot), w(\cdot)} z(t_*; z_0, \theta_0, u(\cdot), w(\cdot)).$$

For $z_0 = 0, \theta_0 = 0$, the notation $G^2(t_*)$ will be used instead of $G^2(t_*; 0, 0)$.

Since the right-hand side of system (4) does not contain variables x, y, and due to the type of appearance of θ in the right-hand side of (4), the geometry of reachable sets does not depend on z_0, θ_0. Namely,

$$G^2(t_*; z_0, \theta_0) = \Pi_{\theta_0}(G^2(t_*)) + z_0.$$

Here, Π_θ is the operator of clockwise rotation of the plane x, y by the angle θ_0. Therefore, the study of reachable sets for point-wise initial conditions can be restricted to the case $z_0 = 0$, $\theta_0 = 0$.

The reachable sets by given time are defined in a similar way. Let

$$\mathcal{G}^2(t_*; z_0, \theta_0) := \bigcup_{t \in [0, t_*]} G^2(t; z_0, \theta_0).$$

Other formulas remain the same, we only replace G^2 by $\mathcal{G}^2$.

Classical and the best known is the construction of the reachable sets $G^2(t_*)$ and $\mathcal{G}^2(t_*)$ for system (1). It was established in paper (Cockayne & Hall, 1975) that any point of the boundary $\partial G^2(t_*)$ can be attained using piecewise control with at most one switch. Namely, the following variants of the control actions are possible: $(-1,0),(1,0),(-1,1),(1,-1)$. The first variant means that control $u \equiv 1$ acts on some interval $[0,\hat{t})$, and control $u \equiv 0$ works on the interval $[\hat{t},t_*]$. The similar is true for the rest of the four variants. Using this proposition and running the switch instant $\hat{t}$ from 0 to t_*, we can construct the boundary of the set $G^2(t_*)$. The form of the sets $G^2(t_*)$ for four instants of time $t_* = i\,0.5\pi, i = 1,2,3,4$, is shown in Fig. 1. The chosen values t_* correspond to the time needed to turn the velocity vector by the angle $i\,0.5\pi$ when moving with the control $u = 1$ ($u = -1$). For every t_*, the figure has its own scale. In Fig. 2, several trajectories arriving at the boundary of the reachable set are shown. The trajectories with the extreme controls $u = 1$ and $u = -1$ are the circles of radius 1. For $u = 0$, the motion along a straight line occurs.

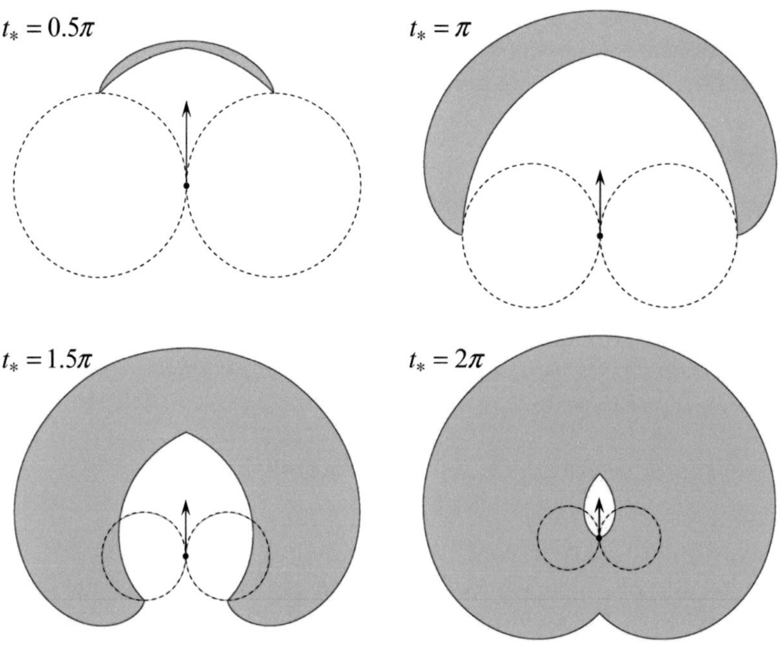

Fig. 1. Form of the reachable sets $G^2(t_*)$ for system (1)

Since the set $\mathcal{G}^2(t_*)$ is the union of the sets $G^2(t)$ over all $t \in [0,t_*]$, then any point of the boundary $\partial\mathcal{G}^2(t_*)$ is reachable with the control of the above mentioned structure (but this control is defined in general on $[0,\tilde{t}]$ where $\tilde{t} \in [0,t_*]$). The minimum time problem for system (1) with the initial point z_0 and given direction θ_0 of the velocity vector to the point z for which the angle θ is not specified is very popular. The structure of the solution to this problem was described by A. A. Markov. It was also known to R. Isaacs, since such problem is a degenerate case of the differential game "homicidal chauffeur". Nevertheless, accurate images of reachable sets $\mathcal{G}^2(t_*)$ have been appeared only in the beginning of 1990s (Boissonnat

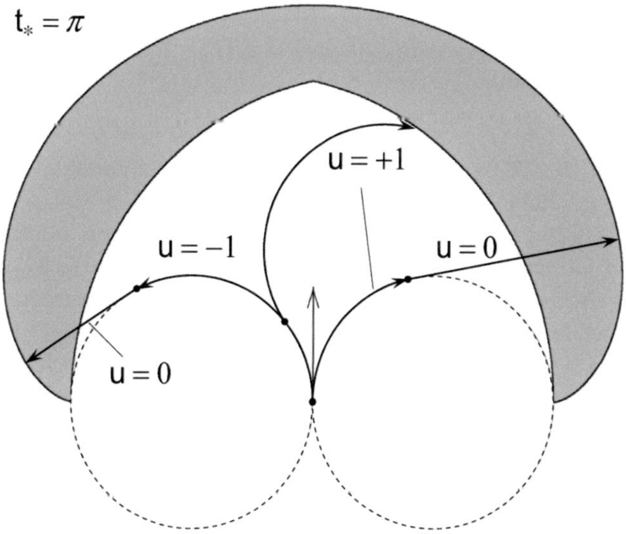

Fig. 2. Structure of the controls steering to the boundary of the reachable set $G^2(\pi)$ for system (1)

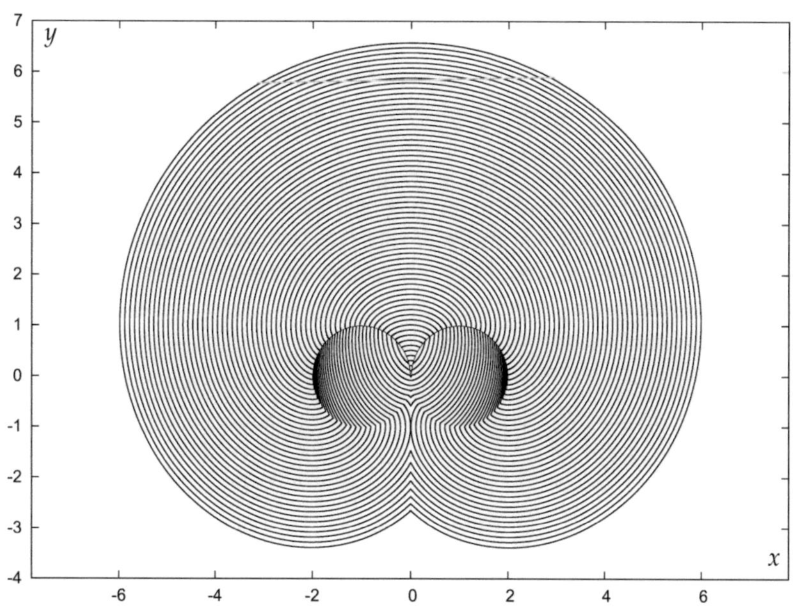

Fig. 3. Time-limited reachable sets $\mathcal{G}^2(t)$ for system (1), $\tau_f = 7.3$, $\delta = 0.1$

& Bui, 1994). Fig. 3 shows the form of the sets $\mathcal{G}^2(t)$ computed by the authors on the interval $[0, t_f]$, $t_f = 7.3$, the output step for the construction results is $\delta = 0.1$.

3.2 Reachable sets with regard to an orientedly added set

Let us now define the reachable set at time t_* with regard to an orientedly added set D:

$$G^2_D(t_*) := \bigcup_{u(\cdot),w(\cdot)} [z(t_*;0,0,u(\cdot),w(\cdot)) + \Pi_{\theta(t_*;0,0,u(\cdot),w(\cdot))}(D)].$$

Hence, when constructing $G^2_D(t_*)$, we add the set D which is rigidly rotated with respect to the origin by the angle $\theta(t_*;0,0,u(\cdot),w(\cdot))$ to each point (being attainable at the time t_* with $u(\cdot),w(\cdot)$) of the usual reachable set. It is assumed that $z_0 = 0$ and $\theta_0 = 0$.
The sense of the set $G^2_D(t_*)$ can be explained using the following example. Let us fix controls $u(\cdot),w(\cdot)$. At the time t_*, the geometric position $z(t_*;0,0,u(\cdot),w(\cdot))$ and the slope $\theta(t_*;0,0,u(\cdot),w(\cdot))$ of the velocity vector are realized. Suppose we are interested at this time instant in a point located at the distance d from the point $z(t_*;0,0,u(\cdot),w(\cdot))$ orthogonally to the velocity vector $\dot{z}(t_*) = (\dot{x}(t_*),\dot{y}(t_*))$ to the left from its direction. Such point is written as $z(t_*;0,0,u(\cdot),w(\cdot)) + \Pi_{\theta(t_*;0,0,u(\cdot),w(\cdot))}(D)$, if we take the set consisting from the point $(-d,0)$ in the plane x,y as the set D. The total collection of points at the time t_* with the property we are interested in is obtained by the enumeration of admissible controls $u(\cdot)$, $w(\cdot)$ and forms the set $G^2_D(t_*)$.
The reachable set $\mathcal{G}^2_D(t_*)$ by the time t_* with regard to an orientedly added set D is defined as

$$\mathcal{G}^2_D(t_*) := \bigcup_{t\in[0,t_*]} G^2_D(t).$$

3.3 Isaacs' transformation

System (4) is of the third order with respect to the state variables. Since we are interested in the construction of reachable sets in the plane of geometric coordinates, it is convenient to pass to a system of the second order with respect to the state variables. This can be done using Isaacs' transformation.
Isaacs utilized system (4) with $w \equiv 1$, $u \in [-1,1]$ (i.e. system (1)) for the formulation and solution of several pursuit-evasion differential games. The pursuit-evasion game "homicidal chauffeur" proposed by R. Isaacs in his report (Isaacs, 1951) was then published in his famous book "Differential games" (Isaacs, 1965) and became classical problem. We will apply the transformation, which R. Isaacs used in this game, for our purposes.
Let $h(t)$ be a unit vector in the direction of forward motion of system (4) at time t. The orthogonal to $h(t)$ unit vector is denoted by $k(t)$ (see Fig. 4). We have

$$h(t) = \begin{pmatrix} \sin\theta(t) \\ \cos\theta(t) \end{pmatrix}, \quad k(t) = \begin{pmatrix} \cos\theta(t) \\ -\sin\theta(t) \end{pmatrix}.$$

Axis $\mathbf{y}$ of the reference system is directed along the vector $h(t)$, axis $\mathbf{x}$ is directed along the vector $k(t)$.
Let $\tilde{z} = (\tilde{x},\tilde{y})$ be a fixed point in the plane x,y. The coordinates $\mathbf{x}(t),\mathbf{y}(t)$ of this point in the movable reference system whose origin coincides with the point $z(t)$ are

$$\mathbf{x}(t) = k'(t)(\tilde{z} - z(t)) = \cos\theta(t)(\tilde{x} - x(t)) - \sin\theta(t)(\tilde{y} - y(t)), \tag{5}$$

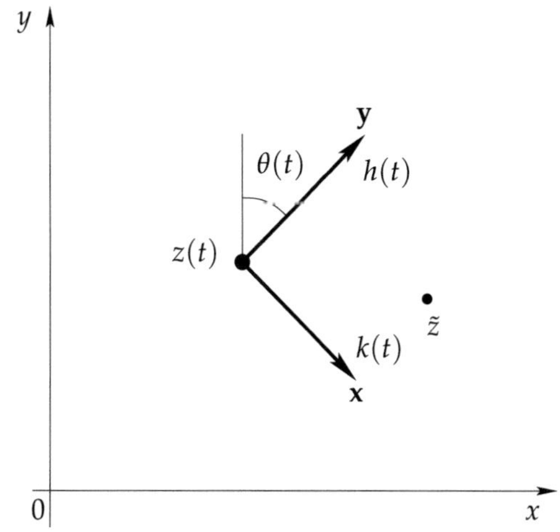

Fig. 4. Movable reference system

$$\mathbf{y}(t) = h'(t)(\tilde{z} - z(t)) = \sin\theta(t)(\tilde{x} - x(t)) + \cos\theta(t)(\tilde{y} - y(t)). \tag{6}$$

Here, the prime denotes the operation of transposition.
Taking into account (5) and (6), one obtains

$$\dot{\mathbf{x}}(t) = -\sin\theta(t)\dot{\theta}(t)(\tilde{x} - x(t)) - \cos\theta(t)\dot{x}(t) -$$
$$\cos\theta(t)\dot{\theta}(t)(\tilde{y} - y(t)) + \sin\theta(t)\dot{y}(t) = -\mathbf{y}(t)\dot{\theta}(t) - \cos\theta(t)\dot{x}(t) + \sin\theta(t)\dot{y}(t), \tag{7}$$

$$\dot{\mathbf{y}}(t) = \cos\theta(t)\dot{\theta}(t)(\tilde{x} - x(t)) - \sin\theta(t)\dot{x}(t) -$$
$$\sin\theta(t)\dot{\theta}(t)(\tilde{y} - y(t)) - \cos\theta(t)\dot{y}(t) = \mathbf{x}(t)\dot{\theta}(t) - \sin\theta(t)\dot{x}(t) - \cos\theta(t)\dot{y}(t). \tag{8}$$

The projection of the velocity vector of system (4) on the axis $\mathbf{x}$ at time t is zero, the projection on the axis $\mathbf{y}$ is $w(t)$. Therefore,

$$\cos\theta(t)\dot{x}(t) - \sin\theta(t)\dot{y}(t) = 0, \tag{9}$$

$$\sin\theta(t)\dot{x}(t) + \cos\theta(t)\dot{y}(t) = w(t). \tag{10}$$

Using (7) and (9), we obtain

$$\dot{\mathbf{x}}(t) = -\mathbf{y}(t)\dot{\theta}(t). \tag{11}$$

With (8) and (10), we get

$$\dot{\mathbf{y}}(t) = \mathbf{x}(t)\dot{\theta}(t) - w(t). \tag{12}$$

Equalities (11), (12), and $\dot{\theta} = u$ yield the system

$$\begin{aligned} \dot{\mathbf{x}} &= -\mathbf{y}u, \\ \dot{\mathbf{y}} &= \mathbf{x}u - w, \quad u \in [b, 1], \quad w \in [a, 1]. \end{aligned} \tag{13}$$

3.4 Solvability sets for problems of approach at given time and by given time

Let M be a closed set in the plane $\mathbf{x}, \mathbf{y}$. Denote by $W(\tau, M)$ (respectively, by $\mathcal{W}(\tau, M)$) the set of all points $\mathbf{z} = (\mathbf{x}, \mathbf{y})$ with the property: there exist piecewise continuous admissible controls $u(\cdot), w(\cdot)$ which bring system (13) from the initial point $\mathbf{z}$ to the set M at time τ (respectively, by time τ). The set $W(\tau, M)$ ($\mathcal{W}(\tau, M)$) is the solvability set in the problem of reaching the set M at time τ (by time τ).

Take now the same set M as a terminal set in the problem of reaching a given set for system (13) and as a set D in the problem of finding reachable set in geometric coordinates for system (4) with regard to an orientedly added set. It follows from the sense of Isaacs' transformation that the set $W(\tau, M)$ drawn in coordinates $\mathbf{x}, \mathbf{y}$ coincides with the set $G^2_M(\tau)$ depicted in coordinates x,y. Therefore,

$$W(\tau, M) = G^2_M(\tau). \tag{14}$$

Similarly,

$$\mathcal{W}(\tau, M) = \mathcal{G}^2_M(\tau). \tag{15}$$

In the following, we will utilize relations (14), (15) in order to obtain the reachable sets $G^2_D(t)$ and $\mathcal{G}^2_D(t)$ of system (4) using the sets $W(\tau, M)$ and $\mathcal{W}(\tau, M)$ computed for system (13) with $\tau = t$ and $M = D$. In addition, if M is a one-point set that coincides with the origin in the plane $\mathbf{x}, \mathbf{y}$, then

$$W(\tau, M) = G^2_M(\tau) = G^2(\tau), \quad \mathcal{W}(\tau, M) = \mathcal{G}^2_M(\tau) = \mathcal{G}^2(\tau).$$

By fixing some point $\bar{z} = (\bar{x}, \bar{y})$ in the plane and by increasing t, let us find the first instant $\bar{t}$ when $\bar{z} \in \mathcal{G}^2_M(\bar{t})$ (equivalently, $\bar{z} \in G^2_M(\bar{t})$). Such $\bar{t}$ be the optimal time $V(\bar{z})$ of passing from the point $z_0 = 0, \theta_0 = 0$ to the point $\bar{z}$ for system (4) with accounting for M. Hence, the Lebesgue set $\{z : V(z) \leq t\}$ of the optimal result function coincides with the set $\mathcal{G}^2_M(t) = \mathcal{W}(t, M)$. For the sets shown in Fig. 3 (where $M = \{0\}$), the function $z \to V(z)$ is discontinuous on the upper semi-circumferences of radius 1 with the centers at the points $(-1, 0)$, $(1, 0)$.

3.5 Backward procedure for construction of solvability sets

The algorithms developed by the authors for the numerical construction of the sets $W(\tau, M)$ and $\mathcal{W}(\tau, M)$ are based on the backward procedure (the parameter τ increases starting from $\tau = 0$) and being variants of the dynamic programming method for the considered class of problems. Backward procedures for the construction of the solvability sets at given time and by given time are intensively developed (Grigor'eva et al., 2005), (Mikhalev & Ushakov, 2007) for control problems and differential games. Elements of the backward constructions are included in one or another form (Sethian, 1999), (Cristiani & Falcone, 2009) into grid methods for solving differential games. For control problems, the backward procedures are simpler, since the second player whose actions should be accounted for in differential games is absent. Especially simple are the backward procedures for problems in the plane. In this case, one can storage the boundary of the sets $W(\tau, M)$ and $\mathcal{W}(\tau, M)$ in form of polygonal lines.

The general idea for the construction of the solvability sets $W(\tau, M)$ in the problem of approaching a set M by system (13) at time τ is the following. We deal with polygonal sets

which are considered as approximations of the ideal sets $W(\tau, M)$. Specify a time step Δ (it can also be variable) and define $\tau_0 = 0, ..., \tau_{i+1} = \tau_i + \Delta$.
Consider the vectogram $E(\mathbf{z}) = \bigcup_{\mathbf{u} \in P} f(\mathbf{z}, \mathbf{u})$ of the right hand side of the controlled system. For system (4), we have $\mathbf{u} = (u, w)$, $P = \{(u, w) : u \in [b, 1], w \in [a, 1]\}$. The set $\mathbf{z} - \Delta E(\mathbf{z})$ describes approximately the collection of points from which the point $\mathbf{z}$ can be approached at time Δ. Running over the boundary $\partial W(\tau_i, M)$ and attaching the set $\mathbf{z} - \Delta E(\mathbf{z})$ "to every" point of the boundary, we approximately construct the boundary $\partial W(\tau_{i+1}, M)$ of the set $W(\tau_{i+1}, M)$.
Theoretically, the solvability set $\mathcal{W}(\tau_*, M)$ for the problem of approaching the set M by system (13) by the time τ_* is defined through the union of the sets $W(\tau, M)$, $\tau \in [0, \tau_*]$. However, one can reject the explicit realization of the union operation and try to construct the boundary of the set $\mathcal{W}(\tau_{i+1}, M)$ directly on the base of the boundary of the set $\mathcal{W}(\tau_i, M)$. To this end, the notion of a front is introduced.
Using the optimal result function V, the front corresponding to the time τ is formally defined as

$$F(\tau) := \{\mathbf{z} \in \partial\mathcal{W}(\tau, M) : \; V(\mathbf{z}) = \tau\}.$$

If $\bar{\mathbf{z}} \notin \mathcal{W}(\tau, M)$, then every trajectory of system (13) starting from the point $\bar{\mathbf{z}}$ can approach $\mathcal{W}(\tau, M)$ through the front $F(\tau)$ only.
It is known (see e.g. Bardi & Capuzzo-Dolcetta (1997)) that if $F(\tau_*) = \partial\mathcal{W}(\tau_*, M)$ for some τ_*, then $F(\tau) = \partial\mathcal{W}(\tau, M)$ for all $\tau \geq \tau_*$.
Let $A(\tau) := \partial\mathcal{W}(\tau, M) \setminus F(\tau)$. It follows from results of optimal control theory and theory of differential games that the function $\mathbf{z} \to V(\mathbf{z})$ is discontinuous on the set $\mathcal{A} := \bigcup_{\tau \geq 0} A(\tau)$ and continuous in the remaining part of the plane.
Possible structure of the set $\mathcal{A}$ is not well explored for time-optimal problems. It is reasonable to assume that if the set $\mathcal{B} := \mathcal{A} \setminus M$ is not empty, then it consists of a collection of smooth arcs. Such arcs are called barriers (Isaacs, 1965).
By the definition of the front $F(\tau)$, we have $F(\tau) \subset W(\tau, M)$. From here, with accounting for the relations $F(\tau) \subset \partial W(\tau, M)$, $W(\tau, M) \subset \mathcal{W}(\tau, M)$, we obtain $F(\tau) \subset \partial\mathcal{W}(\tau, M)$.
The main idea of the algorithm of the backward construction of the sets $\mathcal{W}(\tau, M)$ is explained in Fig. 5. The next set $\mathcal{W}(\tau_{i+1}, M)$ for $\tau_{i+1} = \tau_i + \Delta$ is computed on the basis of the previous set $\mathcal{W}(\tau_i, M)$. The central role in this computation belongs to the front $F(\tau_i)$. As a result, the front $F(\tau_{i+1})$ is obtained, and a new set $A(\tau_{i+1})$ is formed via the extension or reduction of the set $A(\tau_i)$. The union $F(\tau_{i+1}) \bigcup A(\tau_{i+1})$ is the boundary of the next set $\mathcal{W}(\tau_{i+1}, M)$. The initial front $F(0)$ coincides with those part of the boundary ∂M that consists of the points from which the trajectories of system (13) (being written in backward time) leave M with increasing τ for at least one pair of admissible controls $u(\cdot)$, $w(\cdot)$. According to (Isaacs, 1965), such part of the boundary of M is called the usable part of M.
Therefore, the algorithm resembles the front propagation.
Several typical cases (local fragments) of the front propagation are presented in Figs. 6 and 7. Fig. 6a shows the case in which the left end of the front is moving from the endpoint c of the usable part of ∂M with increasing τ. In the algorithm, simultaneously with the computation of the next front $F(\tau_{i+1})$, the extension of the barrier is computed by means of connection of the left ends of the fronts $F(\tau_i)$ and $F(\tau_{i+1})$. In the case considered, this results in the local increase of the totality $A(\tau_{i+1})$ with respect to $A(\tau_i)$. The extension of the barrier forms a line on which the value function is discontinuous.

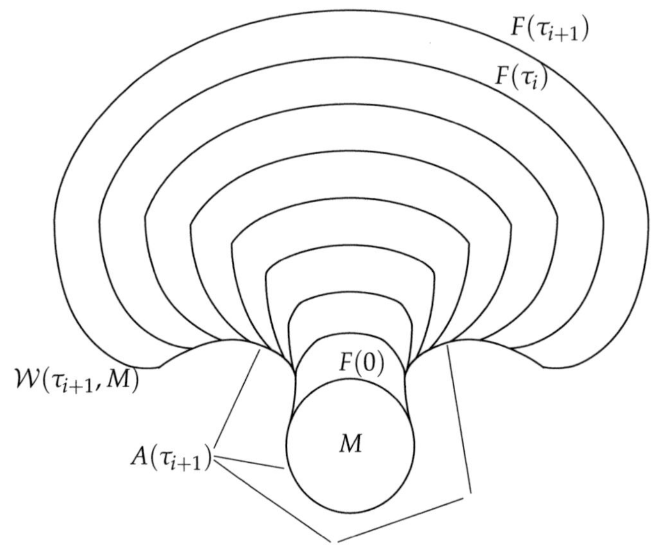

Fig. 5. Backward construction of the solvability sets $\mathcal{W}(\tau, M)$. The boundary of the set $\mathcal{W}(\tau_{i+1}, M)$ is $F(\tau_{i+1}) \cup A(\tau_{i+1})$

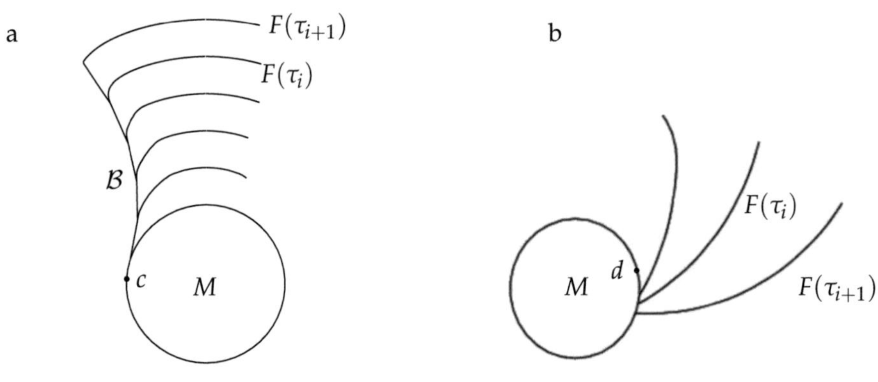

Fig. 6. a) The movement of the left end of the front generates the barrier line on which the value function is discontinuous; b) The right end of the front is moving along the boundary of the terminal set

In the case shown in Fig. 6b, the right end of the front starts to move along ∂M from the very beginning i.e. for small $\tau > 0$. Here, no barrier line emanates from the right endpoint d of the usable part. The value function near the point d outside the set M is continuous.

Fig. 7 represents the case where the left end of the front is running along the back side of the already constructed barrier. This results in the local decrease of the totality $A(\tau_{i+1})$ comparing to $A(\tau_i)$.

A more detailed description of the algorithm is given in (Patsko & Turova, 2009).

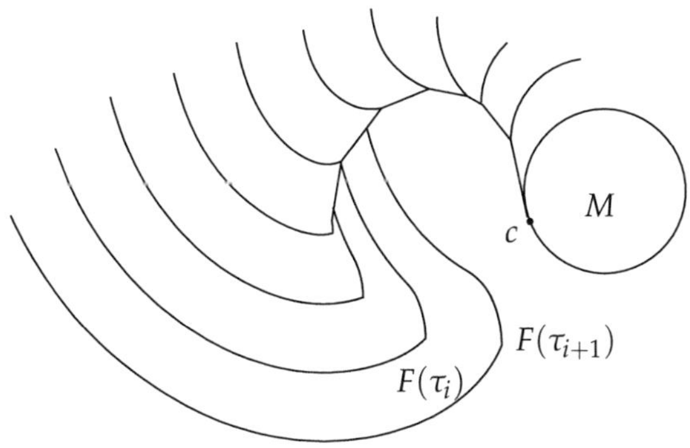

Fig. 7. The left end of the front is bending around the barrier line

3.6 Results of numerical construction of solvability sets (reachable sets)

Let M be a one-point set that coincides with the origin (in the numerical computations, a circle with a very small radius is taken). For system (3) with $a < 0$, the set $G^2(t)$ "swells" monotonically with increasing t, i.e. $G^2(t_2) \supset G^2(t_1)$ for $t_2 > t_1$, where the strict inclusion holds. This provides that the sets $G^2(t)$ and $\mathcal{G}^2(t)$ coincide. For $a = -1$ (i.e. for system (2)) the set $G^2(t) = \mathcal{G}^2(t)$ is symmetric with respect to the axes x, y.

After publishing the paper (Reeds & Shepp, 1990) related to the minimum time problem for system (2), the obtained results were refined and essentially propelled in the works (Sussmann & Tang, 1991), (Soueres et al., 1994), and (Soueres & Laumond, 1996) by using the Pontryagin maximum principle. The second paper describes in particular the construction of reachable sets $G^2(t)$ and give the structure of controls steering to the boundary of these sets. The properties of monotonic swelling of the sets $G^2(t)$ and the symmetry make system (2) very convenient for solving very complex problems of robot transportation (Laumond, 1998), pp. 23–43. For $a = -0.8$ and $a = -1$, the reachable sets are shown in Fig. 8. As before, the notation t_f means the end time of the construction interval. The symbol δ denotes the output step of the representation, which is not necessarily equal to the step Δ of the backward constructions. The latter is, as a rule, smaller.

For $a = 0.8$ and $a = 0.2$, the behavior of the reachable sets $G^2(t)$ for system (3) with increasing t is shown in Fig. 9. The dependency of the sets $G^2(t)$ and $\mathcal{G}^2(t)$ on the parameter a is presented for $t = 1.8$ in Fig. 10. Similar sets but for non-symmetric constraint $u \in [-0.6, 1]$ (i.e. for system (4)) are depicted in Fig. 11. Non-symmetry of the restriction on the control u results in the non-symmetry of the obtained sets with respect to the vertical axis. Note that from the theoretical point of view, the minimum time problem and the related problem of the construction of the sets $\mathcal{G}^2(t)$ for system (4) with $a = 1$ and $b \in [-1, 0)$ were studied in (Bakolas & Tsiotras, 2011).

In Figs. 12 and 13, the sets $\mathcal{G}^2_M(\tau)$ computed for $a = 0.2$ and $b = -0.6$ are shown. The set $\mathcal{G}^2_M(\tau) = \mathcal{W}(\tau, M)$ becomes for the first time non-simply-connected at $\tau_* = 4.452$ when the right part of the front collides with M. Here, one needs to fill a "hole" adjoining to the back side of initial part of the barrier. The hole is completely filled out by the time $\bar{\tau} = 4.522$. The second hole occurs at time $\tau^* = 5.062$ when left and right parts of the front meet. In this case, the filling out of the hole ends at $\bar{\bar{\tau}} = 5.21$. The function V of the optimal result is

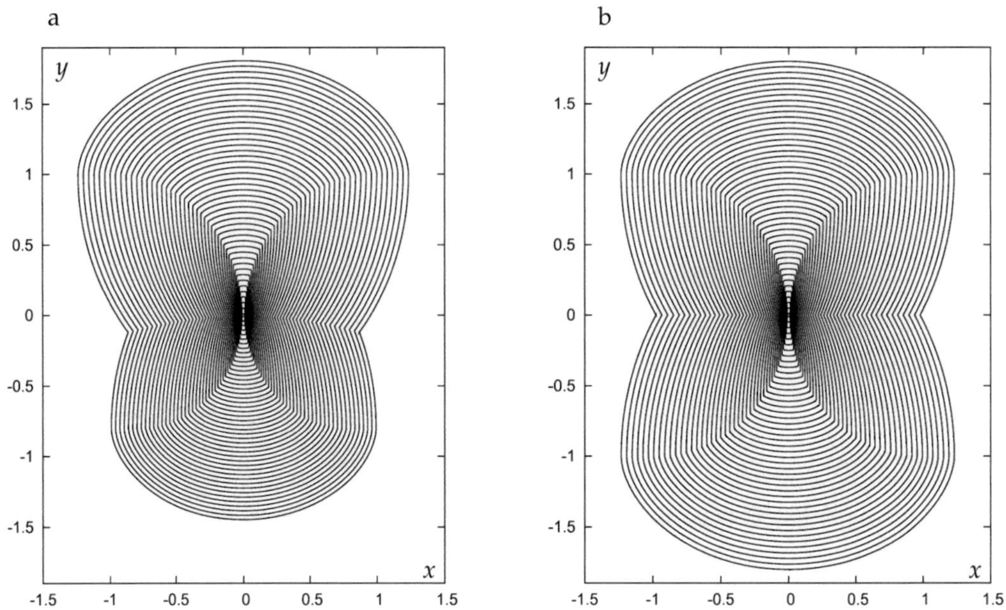

Fig. 8. Reachable sets $\mathcal{G}^2(t) = G^2(t)$ for system (3), $t_f = 1.8$, $\delta = 0.04$: a) $a = -0.8$; b) $a = -1$

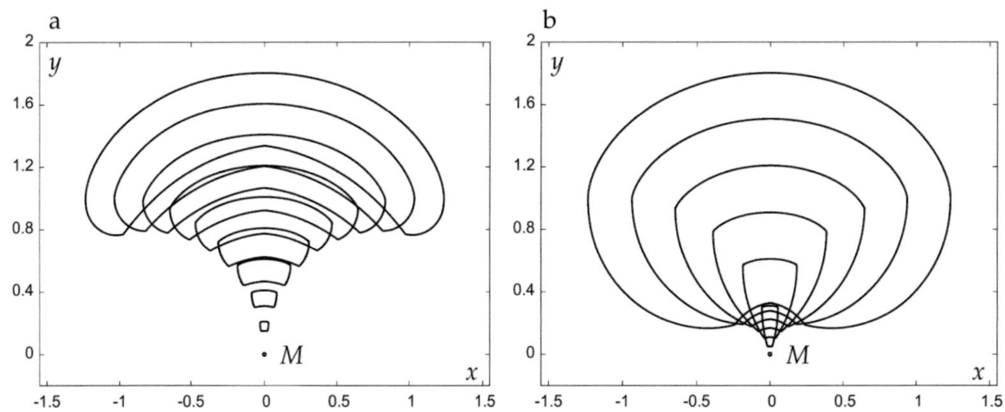

Fig. 9. The reachable sets $G^2(t)$ at given time for system (3), $t_f = 1.8$: a) $a = 0.8$, $\delta = 0.2$; b) $a = 0.2$, $\delta = 0.3$

discontinuous on the two barrier lines being the upper semi-circumferences with the centers at the points $(a/b, 0) = (-1/3, 0)$ and $(a, 0) = (0.2, 0)$ and the radiuses $\frac{1}{3} - r$ and $0.2 - r$, where $r = 0.01$ is the radius of the terminal circle M.

Let us consider an example where the set M is a circle of radius 0.3 centered at the point $(0.7, 0)$. Put $a = 0.8$. In Figs. 14 and 15, results of the construction of the sets $\mathcal{W}(\tau, M)$ are presented. We see which parts of the boundary of the reachable set $G^2_M(\tau) = \mathcal{W}(\tau, M)$

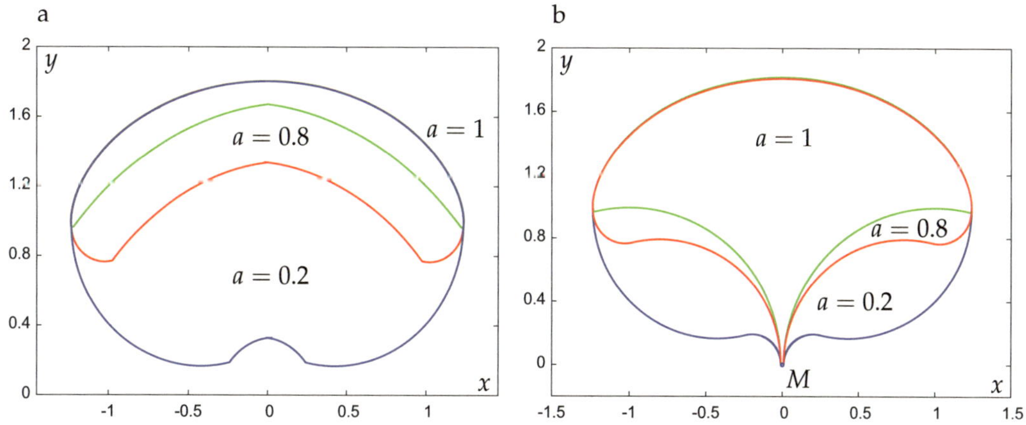

Fig. 10. Comparison of the reachable sets for system (3) for different values of a, $t = 1.8$: a) $G^2(t)$; b) $\mathcal{G}^2(t)$

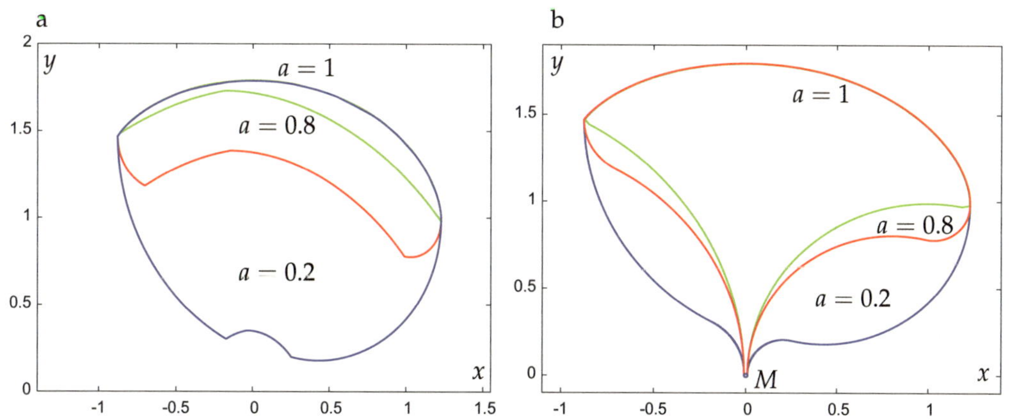

Fig. 11. Comparison of the reachable sets for system (4) for different values of a, $t = 1.8$: a) $G^2(t)$; b) $\mathcal{G}^2(t)$

propagate in a regular way with increasing τ and which ones (and from what time) are developed in a more complex manner. With increasing τ, the left end of the front moves along the barrier line (as in Fig. 6a). After passing the point d, the left end begins to move along the same barrier line but over its back side. The right end runs along the boundary of the terminal set with increasing τ (as in Fig. 6b), then changes over to the back side of the barrier. At time $\tau^* = 4.864$, the self-intersection of the front occurs. The front is divided into two parts: the inner and the outer ones. The ends of the inner front slide along the left barrier until they meet each other and a closed front is formed. The construction then is continued for the closed front. At time $\tau^* = 6.08$, the inner front shrinks into a point. The computation of the closed outer front is also continued till this time τ^*. The optimal result function V is discontinuous outside of M on the barrier ce.

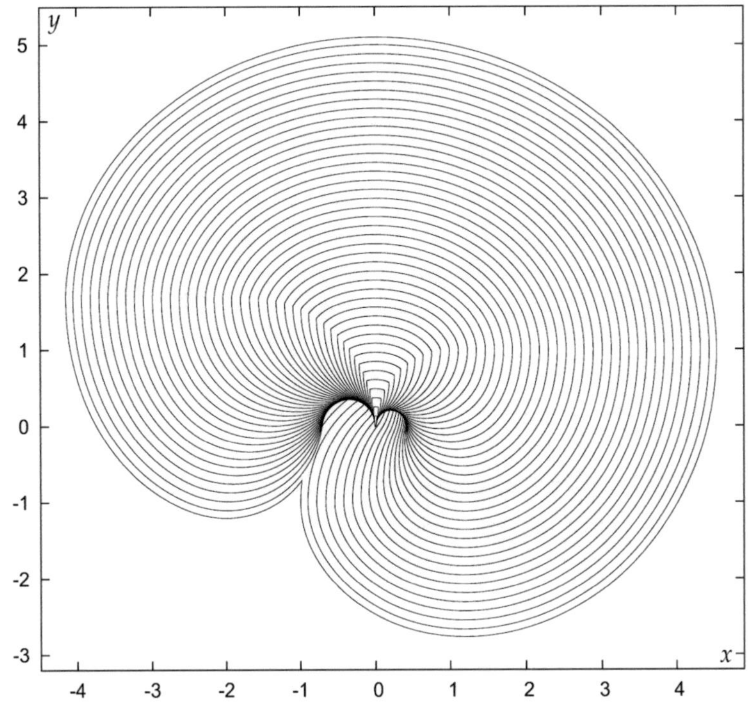

Fig. 12. Reachable sets $\mathcal{G}^2_M(\tau) = \mathcal{W}(\tau, M)$ for system (4) with $a = 0.2$, $b = -0.6$. The set M is the circle of radius 0.01 with the center at the origin. The output step of the representation is $\delta = 0.12$, the terminal time is $\tau_f = 5.21$

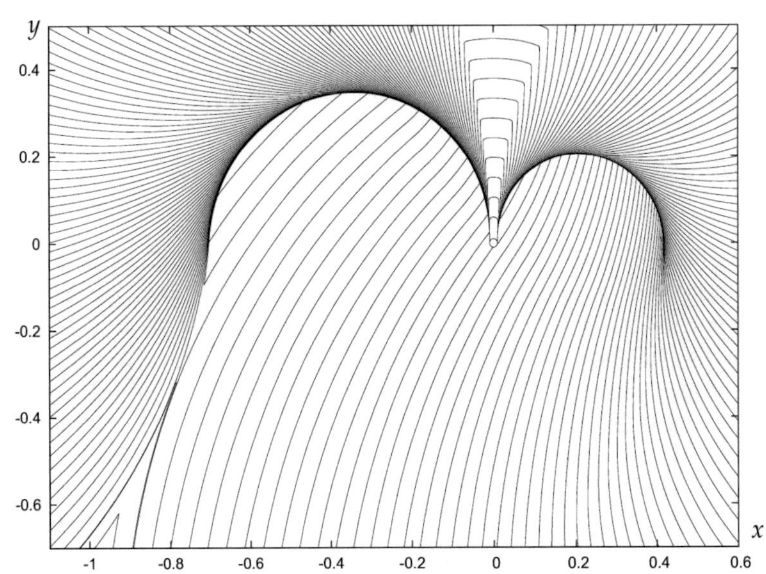

Fig. 13. Enlarged fragment of Fig. 12. The output step of the representation is $\delta = 0.046$

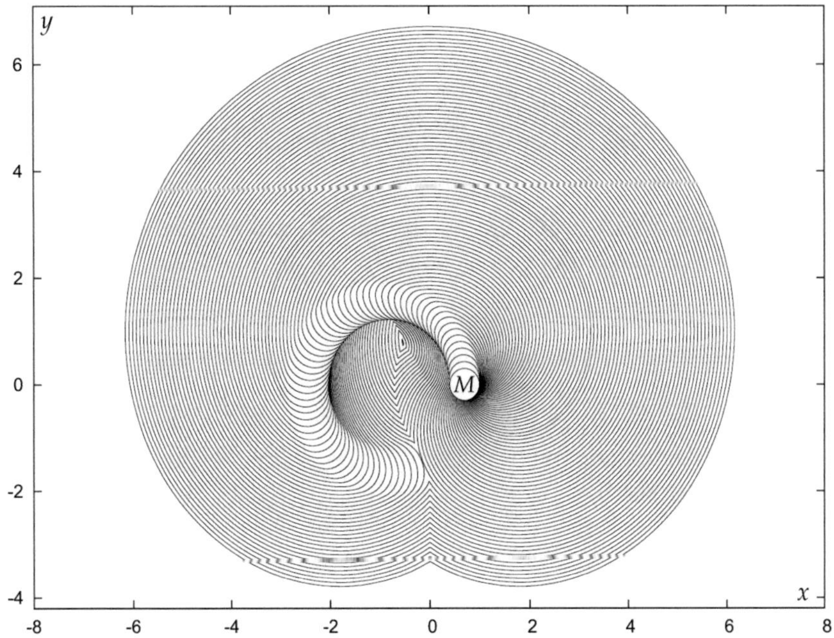

Fig. 14. Reachable sets $\mathcal{G}^2_M(\tau) = \mathcal{W}(\tau, M)$ for the circle M centered at $(0.7, 0)$; $a = 0.8$, $b = -1$, $\tau_f = 6.08$, $\delta = 0.076$

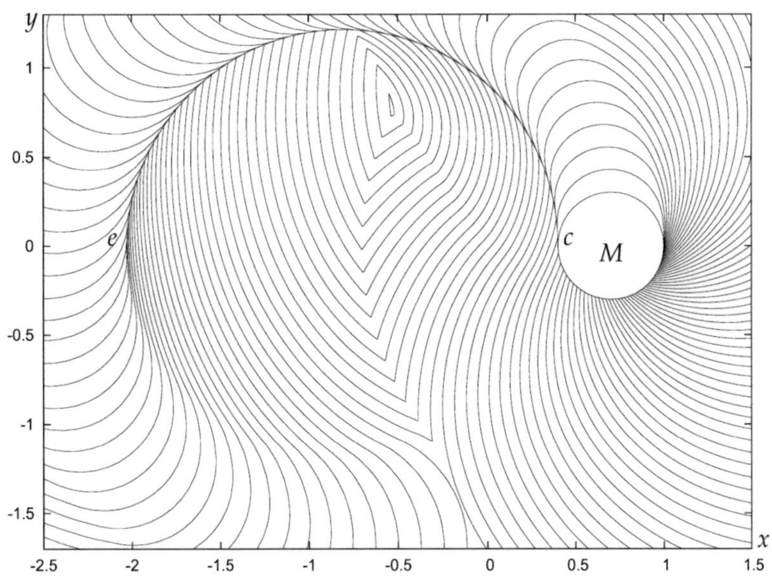

Fig. 15. Enlarged fragment of Fig. 14

4. Three-dimensional reachable sets

Let us describe the reachable sets

$$G^3(t_*) := \bigcup_{u(\cdot),w(\cdot)} (z(t_*;0,0,u(\cdot),w(\cdot)),\theta(t_*;0,0,u(\cdot),w(\cdot)), \quad \mathcal{G}^3(t_*) := \bigcup_{t\in[0,t_*]} G^3(t)$$

in the three-dimensional space x, y, θ. We restrict ourselves to the case of system (1) and a close to it system in which the control parameter u is restricted as $u \in [b,1]$, where $b \in [-1,0)$ is the fixed parameter.

4.1 Structure of controls steering to the boundary of reachable sets at given time

In paper (Patsko et al., 2003), it was established based on the application of the Pontryagin maximum principle to system (1) that for any point $(z,\theta) \in \partial G^3(t_*)$ the control steering to this point is piecewise continuous and has at most two switches. In addition, there are only 6 variants of changing the control:
1) 1, 0, 1; 2) −1, 0, 1; 3) 1, 0, −1; 4) −1, 0, −1; 5) 1, −1, 1; 6) −1, 1, −1.
The second variant means that the control $u \equiv -1$ acts on some interval $[0,t_1)$, the control $u \equiv 0$ works on an interval $[t_1,t_2)$, and the control $u \equiv 1$ operates on the interval $[t_2,t_*]$. If $t_1 = t_2$, then the second interval (where $u \equiv 0$) vanishes, and we obtain a single switch from $u = -1$ to $u = 1$. In the case $t_1 = 0$, the first interval where $u \equiv -1$ vanishes; in the case $t_2 = t_*$ the third interval with $u \equiv 1$ is absent. The control has constant value for all $t \in [0,t_*]$ if one of the following three conditions holds: $t_1 = t_*$, $t_2 = 0$, or both $t_1 = 0$ and $t_2 = t_*$. Similar is true for the other variants.
The proposition on six variants of the control $u(t)$ steering to the boundary of the reachable set $G^3(t_*)$ is similar in form to the Dubins theorem on the variants of the controls steering to the boundary of the reachable set $\mathcal{G}^3(t_*)$. The same variants are valid. However, due to the relation between the sets $G^3(t_*)$ and $\mathcal{G}^3(t_*)$ (the set $\mathcal{G}^3(t_*)$ is the union of the sets $G^3(t)$ over $t \in [0,t_*]$), the above mentioned properties of the controls leading to the boundary of the set $G^3(t_*)$ result in the analogous properties of the controls leading to the boundary of the set $\mathcal{G}^3(t_*)$, but the converse is false.

4.2 Numerical construction of three-dimensional reachable sets at given time

Let us apply the above formulated result on the structure of the control $u(t)$ steering to $\partial G^3(t_*)$ for the numerical construction of the boundary $\partial G^3(t_*)$.
To construct the boundary $\partial G^3(t_*)$ of the set $G^3(t_*)$, we search through all controls of the form 1–6 with two switches t_1, t_2. For every variant of switches, the parameter t_1 is chosen from the interval $[0,t_*]$, and the parameter t_2 from the interval $[t_1,t_*]$. In addition, controls with one switch and without switches are also considered. Taken a specific variant of switching and searching through the parameters t_1, t_2 on some sufficiently fine grid, we obtain a collection of points generating a surface in the three-dimensional space x, y, θ.
Therefore, each of the six variants yields its own surface in the three-dimensional space. The boundary of the reachable set $G^3(t_*)$ is composed of pieces of these surfaces. The six surfaces are loaded into the visualization program without any additional processing of data. Using this program, the boundary of the reachable sets is extracted. Some surfaces (in part or as a whole) find themselves inside of the reachable set. The visualization program does not plot such pieces.

The visualization of the three-dimensional sets is done with the program "Cortona VRML Client" utilizing the open standard format VRML/X3D for the demonstration of interactive vector graphics.

Fig. 16 shows the boundary of the set $G^3(t_*)$ at time $t_* = 1.5\pi$ from two perspectives. The initial values of x_0, y_0, and θ_0 are equal to zero. The different parts of the boundary are marked with different colors. For example, part 2 is reachable for the trajectories with the control $u(t)$ of the form $-1, 0, 1$ with two switches. The sections of the reachable set by the plane $\theta = \text{const}$ are depicted with some step along the axis θ. The points of junction lines of parts 1,2; 1,3; 2,4; 2,5; 2,6; 3,4; 3,5; 3,6 are obtained with a single-switch control. Any point of the common line of parts 5 and 6 is reachable for two trajectories with two switches each. Parts 5 and 6 have non-smooth junction along this line. The angle of the junction is not visible because it is rather small. The control $u(t) \equiv 0$ steers to the junction point of parts 1–4.

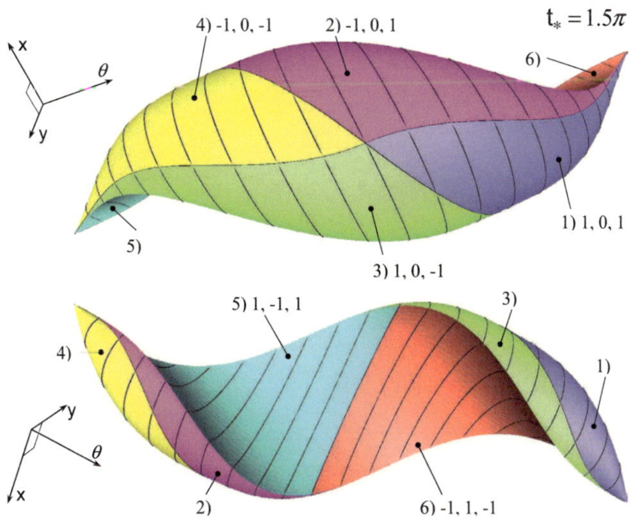

Fig. 16. The set $G^3(t_*)$ for $t_* = 1.5\pi$ shown from the two perspectives

Fig. 17 shows reachable sets $G^3(t_*)$ at the same perspective but with different scales for four time instants t_*. The transformation of the structure of the reachable set boundary is clearly seen. With increasing time, the forward part of the boundary covers the back part composed of patches 5, 6. Note that the angle θ is not restricted as $\theta \in [-\pi, \pi)$.

Passing from $t_* = 3\pi$ to $t_* = 4\pi$, one arrives at the time $t_* \approx 3.65\pi$ when the reachable set $G^3(t_*)$ becomes non-simply-connected for some small time interval. Namely, a cavity that does not belong to the reachable set arises. In Fig. 18, an origination of such a situation is shown. Here, a cut of two sets $G^3(t_*)$ corresponding to instants $t_* = 3\pi$ and $t_* = 3.65\pi$ is depicted. The cut is done using the plane $\theta = 0$. The set $G^3(3\pi)$ is simply connected and the set $G(3.65\pi)$ is not.

Fig. 19 shows the set $G^3(t_*)$ for $t_* = 1.6\pi$, $t_* = 2\pi$, and $t_* = 2.5\pi$. Here the angle θ is calculated modulo 2π.

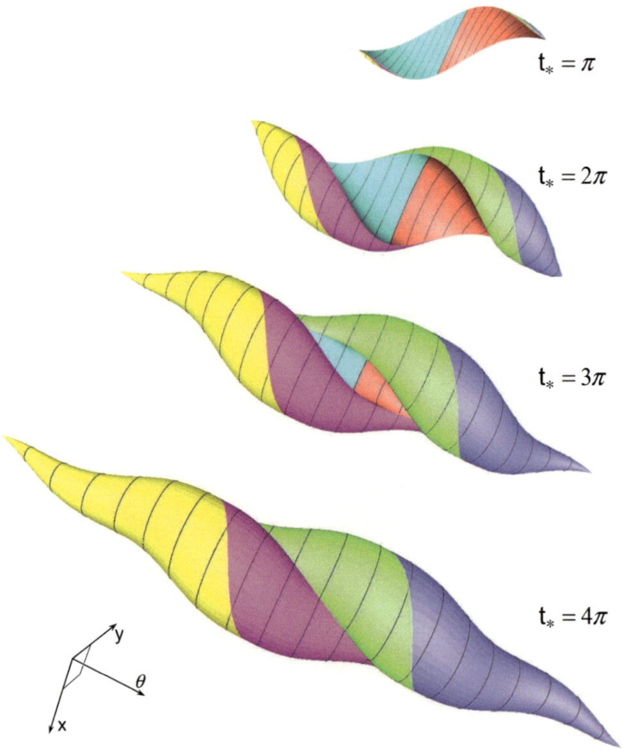

Fig. 17. Development of the reachable set $G^3(t_*)$

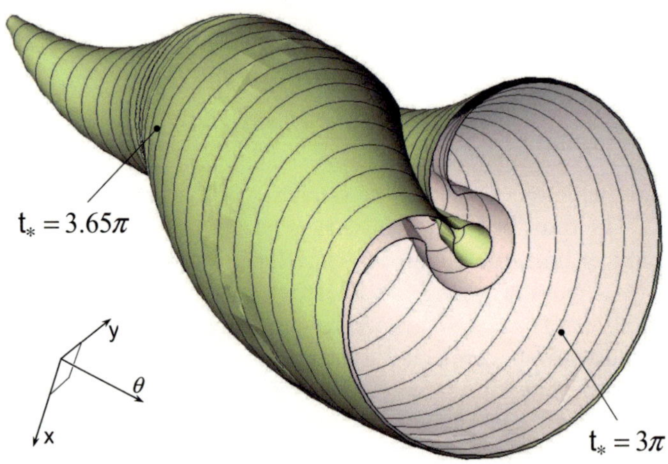

Fig. 18. Loss of simple connectivity

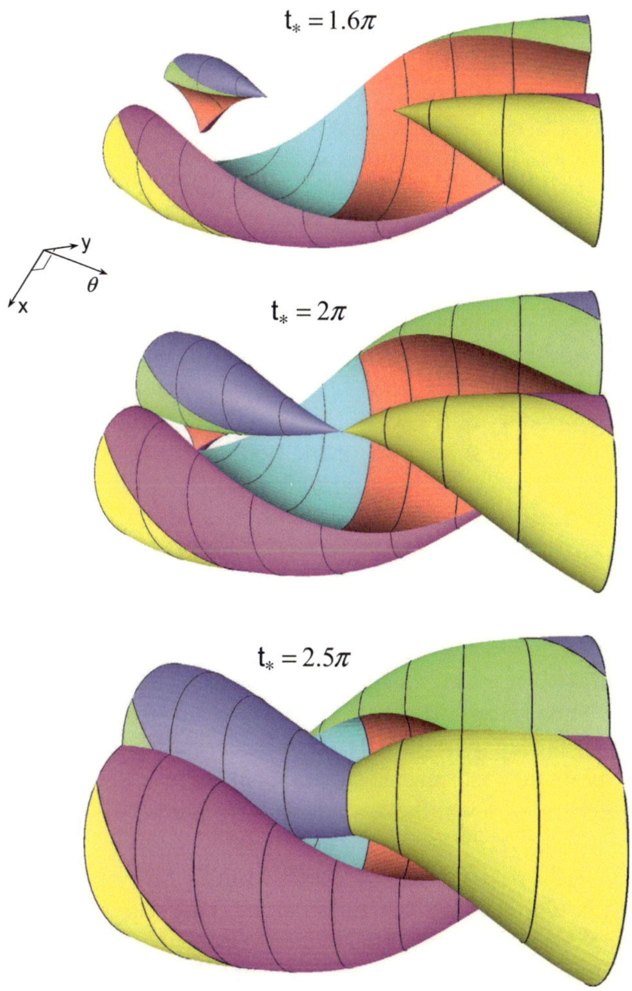

Fig. 19. The set $G^3(t_*)$ for three time instants with θ computed modulo 2π

4.3 Numerical construction of three-dimensional reachable sets by given time

Let us describe the reachable sets $\mathcal{G}^3(t_*)$ by given time. Theoretically, their construction can base on the definition $\mathcal{G}^3(t_*) = \bigcup\limits_{t \in [0, t_*]} G^3(t)$, and the boundary of the sets $G^3(t)$ can be obtained by running t on $[0, t_*]$ with a small step. However, this is very difficult approach for practical constructions. The analysis of the development of the sets $G^2(t_*)$ and $\mathcal{G}^2(t_*)$ in the plane x, y suggests a more thrifty method.

The observation of change of the set $G^3(t)$ gives the following.

For $t_* \in (0, \pi)$, any point inside of that part of the boundary $\partial G^3(t_*)$ which is generated by the controls of the kind 1–4 is strictly inside of the set $G^3(t_* + \Delta t)$ for any sufficiently small $\Delta t > 0$. Conversely, any point lying inside of that part of the boundary $\partial G^3(t_*)$ which is generated by the controls of the kind 5,6 is outside of the set $G^3(t_* + \Delta t)$.

Therefore, one can say that for $t_* \in (0, \pi)$, the front part of the boundary $\partial G^3(t_*)$ be the piece of the boundary generated by the controls 1–4, and the back part be the piece of the boundary $\partial G^3(t_*)$ corresponding to the controls 5,6. The junction of the front and back parts of the boundary $\partial G^3(t_*)$ occurs along two one-dimensional arcs in space x, y, θ, which correspond to the controls of the form (1,-1) and (-1,1) with one switch on $[0, t_*]$. Computing the collection of such arcs for every $t \in [0, t_*]$, we obtain a surface which forms the "barrier" part of the boundary of the set $\mathcal{G}^3(t_*)$. In total, the boundary of the set $\mathcal{G}^3(t_*)$ is composed of the front and barrier parts.

Thus, the construction of the boundary $\partial \mathcal{G}^3(t_*)$ for $t_* \in [0, \pi]$ requires loading of 4 surfaces corresponding to the controls 1–4 with two switches and 2 surfaces corresponding to the controls of the form (1,-1) (surface I) and (-1,1) (surface II) with one switch on the interval $[0, t]$, where $t \in [0, t_*]$, to the visualization program. The program constructs automatically the visible from the outside boundary of the set $\mathcal{G}^3(t_*)$.

Let now $t_* \in (\pi, 4\pi]$. In this case, some part of the boundary $\partial G^3(t_*)$ generated by the controls of the form 5 and 6 becomes the front one. For the construction of the boundary $\partial \mathcal{G}^3(t_*)$, 6 surfaces corresponding to the controls 1–6 with two switches and 2 surfaces I and II corresponding to the controls of the form (1,-1) and (-1,1) with one switch on $[0, t]$, where $t \in [0, t_*]$, are loaded into the visualization program. Note that for $t_* \in [2\pi, 4\pi]$ it is not necessary to increment two latter surfaces. It is sufficient to use their parts constructed up to time 2π. It should be emphasized that similarly to the case of the sets $G^3(t_*)$, there is a small time interval from $[3\pi, 4\pi]$ on which the set $\mathcal{G}^3(t_*)$ becomes non-simply-connected. For t_* from such an interval, the above described rule of the construction of the boundary using the visualization program gives only the external part of the boundary of the sets $\mathcal{G}^3(t_*)$. The detection of the "internal" boundary requires additional analysis and is not described here.

Starting from the instant of time $t_* = 4\pi$, the boundary of the set $G^3(t_*)$ becomes entirely a front. In this case, $\mathcal{G}^3(t_*) = G^3(t_*)$, $t_* \geq 4\pi$.

The set $\mathcal{G}^3(t_*)$ for $t_* = 1.5\pi$ is shown from two perspectives in Fig. 20; development of $\mathcal{G}^3(t_*)$ with increasing t_* is given in Fig. 21. These pictures can be compared with Figs. 16 and 17. The difference of the reachable sets by given time from the reachable sets at given time is in the presence of the barrier part formed by the smooth surfaces I and II. To understand better its arrangement, Fig. 22 gives the cut-off sets $\mathcal{G}^3(2\pi)$ and $\mathcal{G}^3(3\pi)$ (cutting plane is $\theta = 0$). The barrier part is developed from the initial point $(0, 0, 0)$ (white point in the pictures). Every of the shown level lines on the barrier part corresponds to its own instant of time. Till $t_* = \pi$, the level lines are closed curves. With t_* increasing, new level lines which are not anymore closed occur. In addition, the old level lines are reduced at their ends, among them those ones constructed until $t_* = \pi$. Starting from the instant $t_* = 2\pi$, the reduction of the constructed lines begins. It finishes at the time $t_* = 4\pi$ when the set $G^3(4\pi)$ captures the point $(0, 0, 0)$.

The set $\mathcal{G}^3(t_*)$ for $t_* = 2\pi$ with the angle θ computed modulo 2π is shown in Fig. 23.

Time-dependent construction of the reachable sets $\mathcal{G}^3(t_*)$ with the indication of which control from variants 1–6 corresponds to every particular piece on the front part of these sets is close to finding of optimal feedback control synthesis for the minimum time problem of steering to the point $(0, 0, 0)$. The optimal feedback control synthesis for system (1) was obtained in (Pecsvaradi, 1972) (θ is taken modulo 2π).

Note that some individual images of three-dimensional reachable sets by given time (with θ taken modulo 2π) obtained by other means are available in the works (Laumond, 1998), p. 7, and (Takei & Tsai, 2010), pp. 22, 23, 26, and 27.

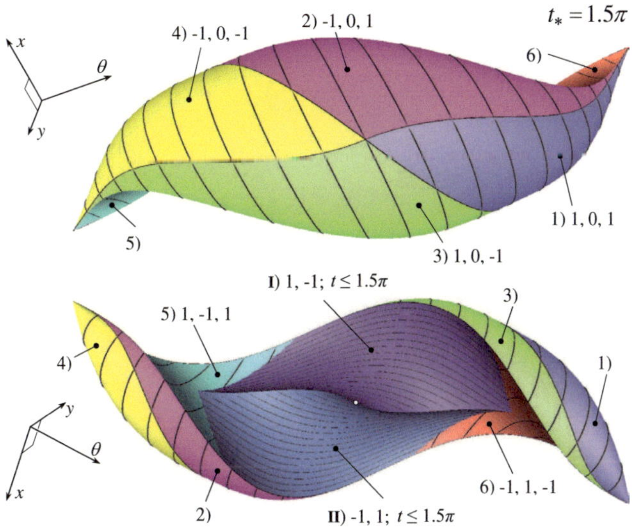

Fig. 20. The set $\mathcal{G}^3(t_*)$ for $t_* = 1.5\pi$ shown from the two perspectives

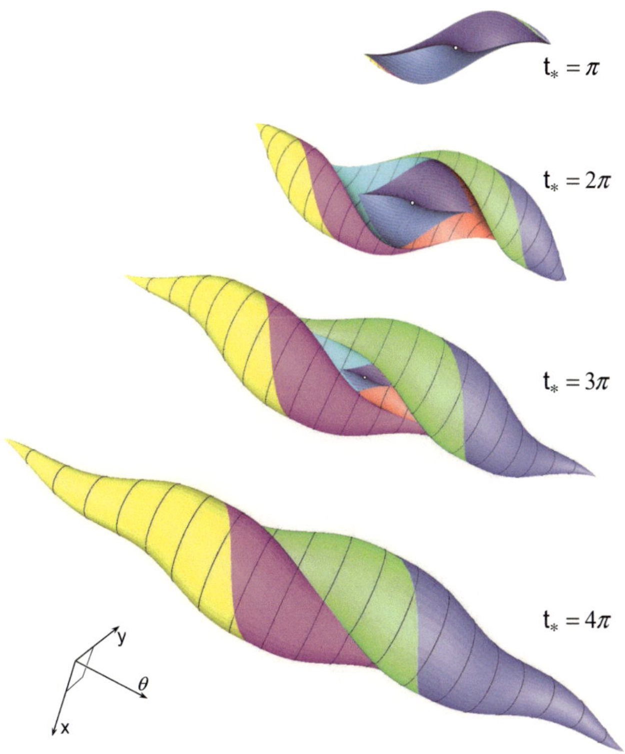

Fig. 21. Development of the reachable set $\mathcal{G}^3(t_*)$

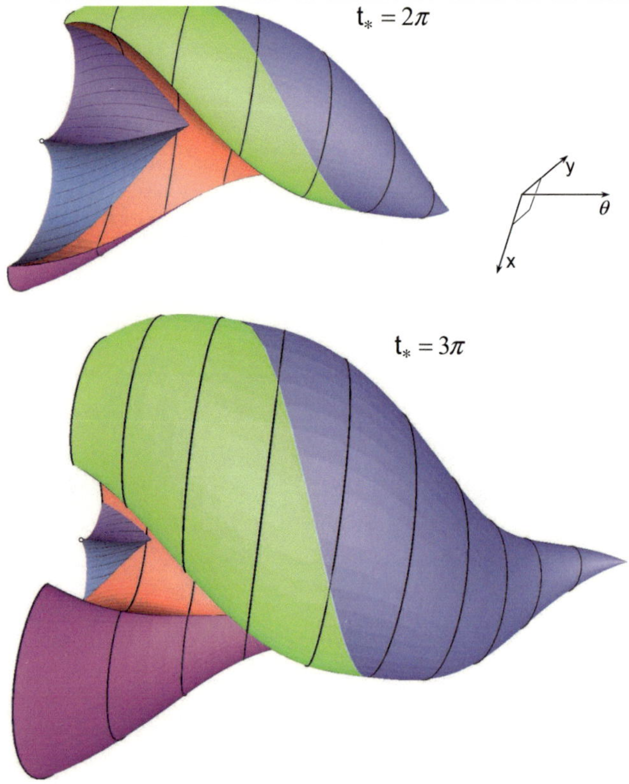

Fig. 22. The cut-off sets $\mathcal{G}^3(t_*)$ for $t_* = 2\pi$ and $t_* = 3\pi$, cutting plane $\theta = 0$

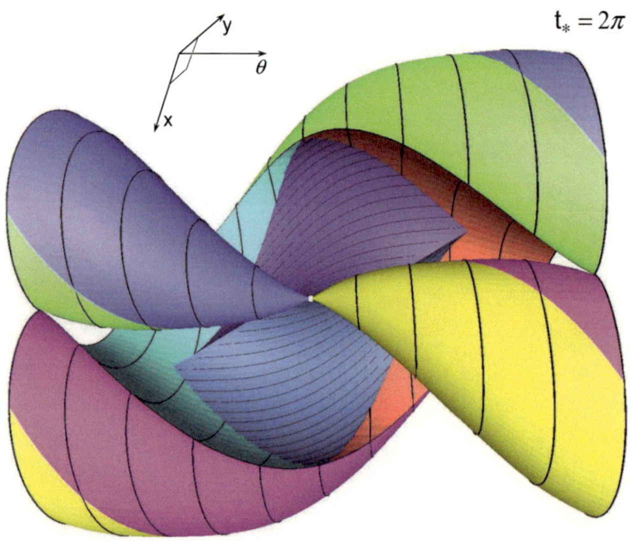

Fig. 23. The set $\mathcal{G}^3(2\pi)$ with the angle θ computed modulo 2π

4.4 Case of non-symmetric constraint on control u

The proposition on the structure of controls steering trajectories of system (1) to the boundary of the reachable set $G^3(t_*)$ is also preserved for the case of non-symmetric constraint $u \in [b, 1]$ with $b \in [-1, 0)$. One should only replace $u = -1$ by $u = b$. Results of the construction of the sets $G^3(t_*)$ are shown for $b = -0.25$ in Fig. 24. The sets $G^3(4\pi)$ and $G^3(6\pi)$ are depicted from the same perspective and have the same scale. Approximately the same perspective but a larger scale is used in Fig. 25 presenting the set $\mathcal{G}^3(4\pi)$. This set with the angle θ taken modulo 2π is shown in Fig. 26.

With a fixed point (x, y, θ), we can compute the first instant $V(x, y, \theta)$ when this point is on the boundary of the set $G^3(t)$ or, what is the same, on the boundary of $\mathcal{G}^3(t)$. The value $V(x, y, \theta)$ be the optimal steering time from the point $(0, 0, 0)$ to the point (x, y, θ). Paper (Bakolas & Tsiotras, 2011) gives results on the computation of level sets of the function $V(x, y, \theta)$ for fixed values θ (modulo 2π) and different values of the parameter b. This is equivalent to the computation of θ-sections of the sets $\mathcal{G}^3(t)$ for different values of b.

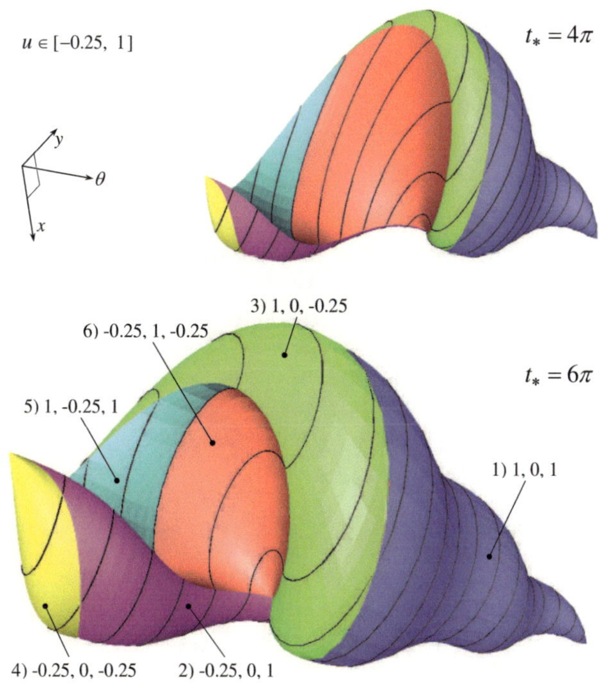

Fig. 24. The reachable sets $G^3(t_*)$ for the instants $t_* = 4\pi$ and $t_* = 6\pi$ for non-symmetric constraint on the control u

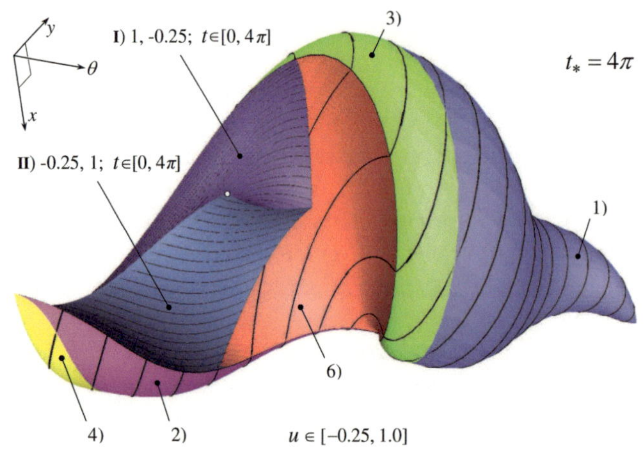

Fig. 25. The reachable set $\mathcal{G}^3(4\pi)$ for non-symmetric constraint on the control u

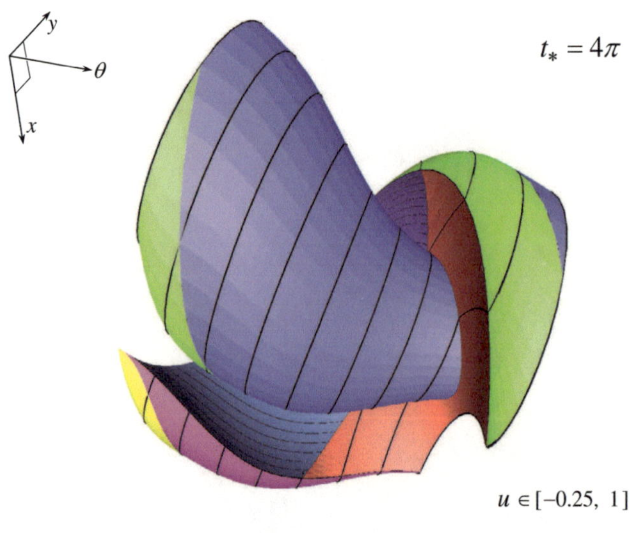

Fig. 26. The reachable set $\mathcal{G}^3(4\pi)$ with the angle θ computed modulo 2π

5. Conclusion

The paper considers reachable sets and inherent character of their development for simplest models of "car" motion used in the mathematical literature. Our investigation is restricted to the cases where reachable sets are constructed in two- and three-dimensional spaces. The understanding of the features and complexities that appear in low dimensional problems can be useful for the analysis of more complex models and for solving real practical problems (Laumond, 1998),(Lensky & Formal'sky, 2003), (LaValle, 2006), (Martynenko, 2007).

6. Acknowledgement

This work is partially supported by the Russian Foundation for Basic Research (project nos. 09-01-00436 and 10-01-96006) and by the Program "Mathematical control theory" of the Presidium of RAS (Ural Branch project 09-П-1-1015).

7. References

Agrachev, A. A. & Sachkov, Yu. (2004). *Control Theory from the Geometric Viewpoint*, Springer, Berlin.

Bakolas, E. & Tsiotras, P. (2011). Optimal synthesis of the asymmetric sinistral/dextral Markov–Dubins problem, *Journal of Optimization Theory and Applications*, Vol. 150 (No. 2), 233–250.

Bardi, M. & Capuzzo-Dolcetta, I. (1997). *Optimal Control and Viscosity Solutions of Hamilton–Jacobi–Bellman Equations. Systems & Control: Foundations and Applications.* Birkhäuser, Boston.

Boissonnat, J.-D. & Bui, X.-N. (1994). *Accessibility region for a car that only moves forwards along optimal paths*, Rapport de recherche N° 2181, INRIA.

Cockayne, E. J. & Hall, G. W. C. (1975). Plane motion of a particle subject to curvature constraints, *SIAM Journal on Control and Optimization*, Vol. 13 (No. 1), 197–220.

Cristiani, E. & Falcone, M. (2009). Fully-discrete schemes for the value function of pursuit-evasion games with state constraints, *Advances in Dynamic Games and Their Applications, Ann. Internat. Soc. Dynam. Games*, Vol. 10, Birkhäuser, Boston, MA, 177–206.

Dubins, L. E. (1957). On curves of minimal length with a constraint on average curvature and with prescribed initial and terminal positions and tangents, *American Journal of Mathematics*, Vol. 79, 497–516.

Grigor'eva, S. V.; Pakhotinskikh, V. Yu.; Uspenskii, A. A.; Ushakov, V. N. (2005). Construction of solutions in some differential games with phase constraints, *Sbornik: Mathematics*, Vol. 196 (No. 3-4), 513–539.

Isaacs, R. (1951). *Games of pursuit*, Scientific report of the RAND Corporation, Santa Monica.

Isaacs, R. (1965). *Differential Games*, John Wiley, New York.

Laumond, J.-P. (ed.) (1998). *Robot Motion Planning and Control*, Lecture Notes in Control and Information Sciences, Vol. 229, Springer, New York.

LaValle, S. M. (2006). *Planning Algorithms*, Chapters 13, 15, Cambridge University Press.

Lee, E. B. & Markus, L. (1967). *Foundations of Optimal Control Theory*, Wiley, New York.

Lensky, A. V. & Formal'sky, A. M. (2003). Two-wheel robot-bicycle with a gyroscopic stabilizer, *Journal of Computer and Systems Sciences International*, No. 3, 482–489.

Markov, A. A. (1889). Some examples of the solution of a special kind of problem on greatest and least quantities, *Soobscenija Charkovskogo Matematiceskogo Obscestva*, Vol. 2-1 (No. 5,6), 250–276 (in Russian).

Martynenko, Yu. G. (2007). Motion control of mobile wheeled robots, *Journal of Mathematical Sciences (New York)*, Vol. 147 (No. 2), 6569–6606.

Mikhalev, D. K. & Ushakov, V. N. (2007). On two algorithms for the approximate construction of a positional absorption set in a game-theoretic approach problem, *Automation & Remote Control*, Vol. 68 (No. 11), 2056–2070.

Patsko, V. S.; Pyatko, S. G.; Fedotov, A. A. (2003). Three-dimensional reachability set for a nonlinear control system, *Journal of Computer and Systems Sciences International*, Vol. 42 (No. 3), 320–328.

Patsko, V. S. & Turova, V. L. (2009). From Dubins' car to Reeds and Shepp's mobile robot, *Computing and Visualization in Science*, Vol. 12 (No. 7), 345–364.

Pecsvaradi, T. (1972). Optimal horizontal guidance law for aircraft in the terminal area, *IEEE Transactions on Automatic Control* Vol. 17 (No. 6), 763–772.

Pontryagin, L. S.; Boltyanskii, V. G.; Gamkrelidze, R. V.; Mischenko E. F. (1962). *The Mathematical Theory of Optimal Processes*, Interscience, New York.

Reeds, J. A. & Shepp, L. A. (1990). Optimal paths for a car that goes both forwards and backwards, *Pacific Journal of Mathematics*, Vol. 145, 367–393.

Sethian, J. A. (1999) *Level Set Methods and Fast Marching Methods*, Cambridge University Press, Cambridge, UK.

Souères, P.; Fourquet, J.-Y.; Laumond, J.-P. (1994). Set of reachable positions for a car, *IEEE Transactions on Automatic Control*, Vol. 39 (No. 8), 1626–1630.

Souères, P. & Laumond, J. P. (1996). Shortest paths synthesis for a car-like robot, *IEEE Transactions on Automatic Control*, Vol. 41 (No. 5), 672–688.

Sussmann, H.J. & Tang, W. (1991). *Shortest paths for the Reeds-Shepp car: a worked out example of the use of geometric techniques in nonlinear optimal control*, Report SYCON-91-lO, Rutgers University.

Takei, R. & Tsai, R. (2010). *Optimal trajectories of curvature constrained motion in the Hamilton-Jacobi formulation*, Report 10-67, University of California, Los Angeles.

9

Neural Networks Based Path Planning and Navigation of Mobile Robots

Valeri Kroumov[1] and Jianli Yu[2]
[1]*Department of Electrical & Electronic Engineering, Okayama University of Science, Okayama*
[2]*Department of Electronics and Information, Zhongyuan University of Technology, Zhengzhou*
[1]*Japan*
[2]*China*

1. Introduction

The path planning for mobile robots is a fundamental issue in the field of unmanned vehicles control. The purpose of the path planner is to compute a path from the start position of the vehicle to the goal to be reached. The primary concern of path planning is to compute *collision-free* paths. Another, equally important issue is to compute a *realizable* and, if possible, *optimal path*, bringing the vehicle to the final position.

Although humans have the superb capability to plan motions of their body and limbs effortlessly, the motion planning turns out to be a very complex problem. The best known algorithm has a complexity that is exponential to the number of degrees of freedom and polynomial in the geometric complexities of the robot and the obstacles in the environment (Chen & Hwang (1998)). Even for motion planning problems in the 2-dimensional space, existing complete algorithms that guarantee a solution often need large amount of memory and in some cases may take long computational time. On the other hand, fast heuristic algorithms may fail to find a solution even if it exists (Hwang & Ahuja (1992)).

In this paper we present a fast algorithm for solving the path planning problem for differential drive (holonomic)[1] robots. The algorithm can be applied to free-flying and snake type robots, too. Generally, we treat the two-dimensional known environment, where the obstacles are stationary polygons or ovals, but the algorithm can easily be extended for the three-dimensional case (Kroumov et al. (2010)). The proposed algorithm is, in general, based on the potential field methods. The algorithm solves the local minimum problem and generates optimal path in a relatively small number of calculations.

The paper is organized as follows. Previous work is presented in Section 2. In Section 3 we give a definition of the map representation and how it is used to describe various obstacles situated in the working environment. The path and the obstacle collisions are detected using artificial annealing algorithm. Also, the solution of the local minima problem is described there. In Section 4 we describe the theoretical background and the development of a motion

[1] In the mobile robotics, the term *holonomic* refers to the kinematic constraints of the robot chassis. The holonomic robot has zero nonholonomic kinematic constraints, while the nonholonomic one has one or more nonholonomic kinematic constraints.

planner for a differential drive vehicle. Section 5 presents the algorithm of the path planner. In Section 6 the calculation time is discussed and the effectiveness of the proposed algorithm is proven by presenting several simulation results. In the last section discussions, conclusions, and plans for further developments are presented.

2. Previous work

Comprehensive reviews on the work on the motion planning can be found in Latombe (1991). In this section we concentrate on motion planning for moving car-like robots in a known environment.
Motion planners can be classified in general as:

1) complete;

2) heuristic.

Complete motion planners can potentially require long computation times but they can either find a solution if there is one, or prove that there is none. Heuristic motion planners are fast but they often fail to find a solution even if it exists.
To date motion planners can be classified in four categories (Latombe (1991)):

1) skeleton;

2) cell decomposition;

3) subgoal graph;

4) potential field.

In the skeleton approach the free space is represented by a network of one-dimensional paths called a *skeleton* and the solution is found by first moving the robot onto a point on the skeleton from the start configuration and from the goal, and by connecting the two points via paths on the skeleton. The approach is intuitive for two-dimensional problems, but becomes harder to implement for higher degrees of freedom problems.
Algorithms based on the visibility graph (VGRAPH) (Lozano-Pèrez & Wesley (1979)), the Voronoi diagram (O'Dúnlaing & Yap (1982)), and the *silhouette* (Canny (1988)) (projection of obstacle boundaries) are examples of the skeleton approach. In the VGRAPH algorithm the path planning is accomplished by finding a path through a graph connecting vertices of the forbidden regions (obstacles) and the generated path is near optimal. The drawback of the VGRAPH algorithm is that the description of all the possible paths is quite complicated. Actually, the number of the edges of the VGRAPH is proportional to the squared total number of the obstacles vertices and when this number increases, the calculation time becomes longer. Another drawback is that the algorithm deals only with polygonal objects. Yamamoto et al. (1998) have proposed a near-time-optimal trajectory planning for car-like robots, where the connectivity graph is generated in a fashion very similar to that of Lozano-Pèrez & Wesley (1979). They have proposed optimization of the speed of the robot for the generated trajectory, but the optimization of the trajectory itself is not enough treated.
In the cell decomposition approach (Paden et al. (1989)) the free space is represented as a union of cells, and a sequence of cells comprises a solution path. For efficiency, hierarchical trees, e.g. octree, are often used. The path planning algorithm proposed by Zelinsky (1992) is quite reliable and combines some advantages of the above algorithms. It makes use of quadtree data structure to model the environment and uses the distance transform methodology to generate paths. The obstacles are polygonal-shaped which yields a quadtree

with minimum sized leaf quadrants along the edges of the polygon, but the quadtree is large and the number of leaves in the tree is proportional to the polygon's perimeter and this makes it memory consuming.

In the subgoal-graph approach, subgoals represent *key* configurations expected to be useful for finding collision-free paths. A graph of subgoals is generated and maintained by a global planner, and a simple local planner is used to determine the reachability among subgoals. This two level planning approach is first reported by Faverjon & Tournassoud (1987) and has turned out to be one of the most effective path planning methods.

The potential field methods and their applications to path planning for autonomous mobile robots have been extensively studied in the past two decades (Khatib (1986), Warren (1989), Rimon & Doditschek (1992), Hwang & Ahuja (1992), Lee & Kardaras (1997a), Lee & Kardaras (1997b), Chen & Hwang (1998), Ge & Cui (2000), Yu et al. (2002), Kroumov et al. (2004)). The basic concept of the potential field methods is to fill the workspace with an artificial potential field in which the robot is attracted to the goal position being at the same time repulsed away from the obstacles (Latombe (1991)). It is well known that the strength of potential field methods is that, with some limited engineering, it is possible to construct quite efficient and relatively reliable motion planners (Latombe (1991)). But the potential field methods are usually incomplete and may fail to find a free path, even if one exists, because they can get trapped in a local minimum (Khosla & Volpe (1988); Rimon & Doditschek (1992); Sun et al. (1997)). Another problem with the existing potential field methods is that they are not so suitable to generate optimal path: adding optimization elements in the algorithm, usually, makes it quite costly from computational point of view (Zelinsky (1992)). Ideally, the potential field should have the following properties (Khosla & Volpe (1988)):

1. The magnitude of the potential field should be unbounded near obstacle boundaries and should decrease with range.
2. The potential should have a spherical symmetry far away from the obstacle.
3. The equipotential surface near an obstacle should have a shape similar to that of the obstacle surface.
4. The potential, its gradient and their effects on the path must be spatially continuous.

The proposed in this paper algorithm is partially inspired by the results presented by Sun et al. (1997) and by Lee & Kardaras (1997a), but our planner has several advantages:

- the obstacle descriptions are implemented directly in the simulated annealing neural network—there is no need to use approximations by nonlinear equations (Lee & Kardaras (1997b));
- there is no need to perform a learning of the workspace *off-line* using backpropagation (Tsankova (2010));
- there is no need to perform additional calculations and processing when an obstacle is added or removed;
- a simple solution of the local minimum problem is developed and the generated paths are conditionally optimized in the sense that they are piecewise linear with directions changing at the corners of the obstacles.

The algorithm allows parallel calculation (Sun et al. (1997)) which improves the computational time. The experimental results show that the calculation time of the presented algorithm depends linearly on the total number of obstacles' vertices—a feature

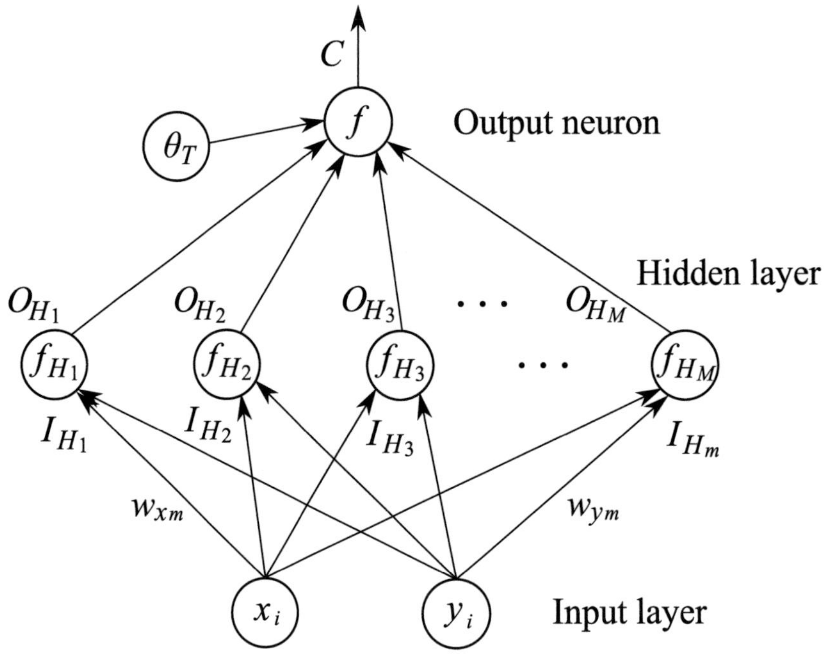

Fig. 1. Obstacle description neural network

which places it among the fastest ones. The only assumptions made in this work are that there are a finite number of stationary oval or polygonal obstacles with a finite number of vertices, and that the robot has a cylindrical shape. The obstacles can be any combination of polygons and ovals, as well. In order to reduce the problem of path planning to that of navigating a point, the obstacles are enlarged by the robot's polygon dimensions to yield a new set of polygonal obstacles. This "enlargement" of the obstacles is a well known method introduced formally by Lozano-Pèrez & Wesley (1979).

3. Environment description

3.1 Obstacles

Every obstacle is described by a neural network as shown in Fig. 1. The inputs of the networks are the coordinates of the points of the path. The output neuron is described by the following expression, which is called *a repulsive penalty function (RPF)* and has a role of repulsive potential:

$$C = f(I_0) = f\left(\sum_{m=1}^{M} O_{H_m} - \theta_T\right), \tag{1}$$

where I_0 takes a role of the induced local field of the neuron function $f(\cdot)$, θ_T is a bias, equal to the number of the vertices of the obstacle decreased by 0.5. The number of the neurons in the hidden layer is equal to the number of the vertices of the obstacle. O_{H_m} in eq. (1) is the output of the m-th neuron of the middle layer:

$$O_{H_m} = f(I_{H_m}), \qquad m = 1, \ldots, M, \tag{2}$$

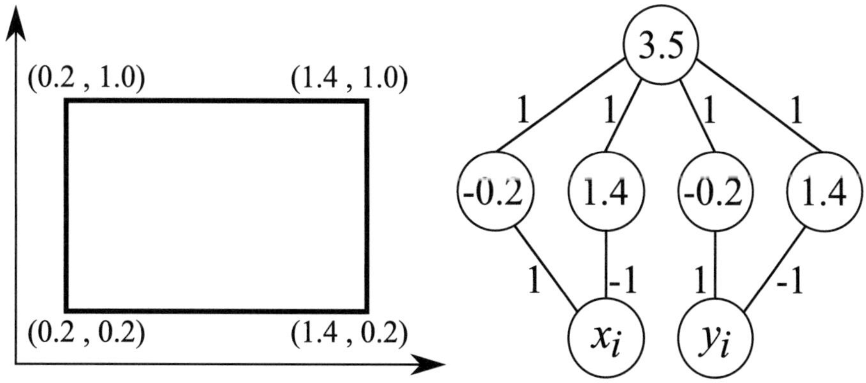

Fig. 2. The annealing network for a rectangular obstacle

where I_{H_m} is the weighted input of the m-th neuron of the middle layer and has a role of induced local field of the neuron function. The neuron activation function $f(\cdot)$ has the form:

$$f(x) = \frac{1}{1 + e^{-x/T}}, \tag{3}$$

where T is the pseudotemperature and the induced local field (x) of the neuron is equal to I_0 for eq. (1) or equal to I_{H_m} in the case of eq. (2). The pseudotemperature decreasing is given by:

$$T(t) = \frac{\beta_0}{\log(1+t)}. \tag{4}$$

Finally, I_{H_m} is given by the activating function

$$I_{H_m} = w_{xm}x_i + w_{ym}y_i + \theta_{H_m}, \tag{5}$$

where x_i and y_i are the coordinates of i-th point of the path, w_{xm} and w_{ym} are weights, and θ_{H_m} is a bias, which is equal to the free element in the equation expressing the shape of the obstacle.

The following examples show descriptions of simple objects.

Example: 1. *Description of polygonal obstacle (see Fig. 2).*

The area inside the obstacle can be described with the following equations:

$$x - 0.2 > 0; \quad -x + 1.4 > 0$$
$$y - 0.2 > 0; \quad -y + 1.0 > 0.$$

From these equations and eq. (5):

$$w_{x1} = 1 \quad w_{y1} = 0 \quad \theta_{H_1} = -0.2$$
$$w_{x2} = -1 \quad w_{y2} = 0 \quad \theta_{H_2} = 1.4$$
$$w_{x3} = 0 \quad w_{y3} = 1 \quad \theta_{H_3} = -0.2$$
$$w_{x4} = 0 \quad w_{y4} = -1 \quad \theta_{H_4} = 1$$

i.e. in the middle layer of the network, the activating functions become

$$\begin{aligned} I_{H_1} &= w_{x1}x_i + w_{y1}y_i + \theta_{H_1} = x_i - 0.2 \\ I_{H_2} &= w_{x2}x_i + w_{y2}y_i + \theta_{H_2} = -x_i + 1.4 \\ I_{H_3} &= w_{x3}x_i + w_{y3}y_i + \theta_{H_3} = y_i - 0.2 \\ I_{H_4} &= w_{x4}x_i + w_{y4}y_i + \theta_{H_4} = -y_i + 1.0. \end{aligned}$$

Hence, the free elements in the equations describing the obstacle are represented by the biases θ_{H_i} and the weights in the equations for I_{H_i} are the coefficients in the respective equations.

Example: 2. *Circular shape obstacle.*

When an obstacle has a circular shape, I_{H_m} is expressed as:

$$I_H = R^2 - (x_i - P)^2 - (y_i - Q)^2, \tag{6}$$

where R is the radius and (P, Q) are the coordinates of the centre.

Example: 3. *Description of elliptical obstacles.*

When the obstacle has elliptic (circular) shape, I_{H_m} is expressed as:

$$I_H = a^2b^2 - (X - x_i)^2b^2 + (Y - y_i)^2a^2, \tag{7}$$

which comes directly from the standard equation of the ellipse

$$\frac{(X - x_i)^2}{a^2} + \frac{(Y - y_i)^2}{b^2} = 1,$$

with xy-coordinates of the foci $(-\sqrt{a^2 - b^2} + X, Y)$ and $(\sqrt{a^2 - b^2} + X, Y)$ respectively, and the description network has two neurons in the middle layer.

Note 1. *The oval shaped obstacles are represented by neural network having two neurons (see the above Example 2 and Example 3) in the middle layer.*

It is obvious from the above that any shape which can be expressed mathematically can be represented by the description neural network. This, of course, includes configurations which are a combination of elementary shapes.
The description of the obstacles by the shown here network has the advantage that it can be used for parallel computation of the path, which can increase the speed of path generation. As it will be shown later, this description of the obstacles has the superiority that the calculation time depends only on the total number of the obstacles' vertices.

3.2 The local minima problem

The local minima remain an important cause of inefficiency for potential field methods. Hence, dealing with local minima is the major issue that one has to face in designing a planner based on this approach. This issue can be addressed at two levels (Latombe (1991)): (1) definition of the potential function, by attempting to specify a function with no or few local minima, and (2) in the design of the search algorithm, by including appropriate techniques for escaping from local minima. However, it is not easy to construct an "ideal" potential function with no local minima in a general configuration. The second level is more realistic and is

addressed by many researchers (see e.g. Latombe (1991); Lozano-Pèrez et al. (1994) and the references there).
In the proposed in this chapter algorithm, the local minima problem is addressed in a simple and efficient fashion:

1. After setting the coordinates of the start and the goal points respectively (Fig. 3(a)), the polygonal obstacles are scanned in order to detect concavity (Fig. 3(b)). In the process of scanning, every two other vertices of a given obstacle are connected by a straight line and if the line lies outside the obstacle, then a concavity exists.
2. If the goal (or the start) point lies inside a concavity, then a new, temporary, goal (start) lying outside the concavity is set (point g′ in Fig. 3(b)), and the initial path between the original goal (start) and the new one is set as a straight line. The temporary goal (start) is set at the nearest to the "original" start (goal) vertex.
3. Every detected concavity is temporarily filled, i.e. the obstacle shape is changed from a concave to a nonconcave one (see Fig. 3(c)). After finishing the current path-planning task, the original shapes are retained so that the next task could be planned correctly, and the temporary goal (start) is connected to the original one by a straight line (Fig. 3(d)).

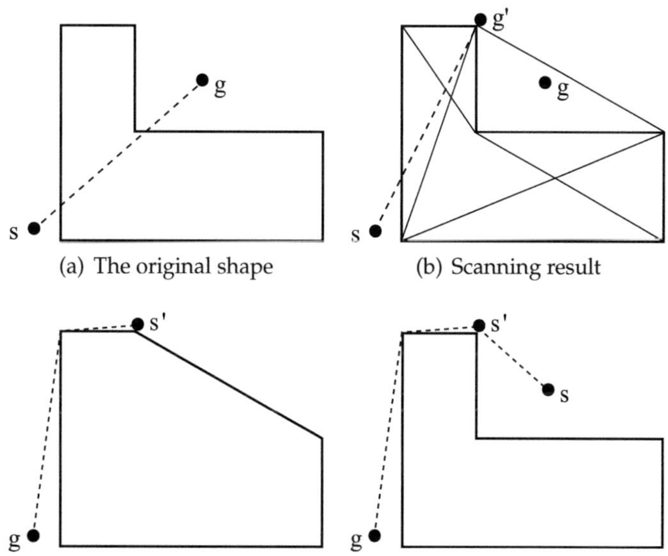

(a) The original shape (b) Scanning result

(c) The filled concavity (d) Final path

Fig. 3. Concave obstacle (the local minima solution)

The above allows isolating the local minima caused by concavities and increases the reliability of the path planning.

4. Optimal path planner

The state of the path is described by the following energy function.

$$E = w_l E_l + w_c E_c, \tag{8}$$

where w_l and w_c are weights ($w_l + w_c = 1$), E_l depicts the squared length of the path:

$$E_l = \sum_{i=1}^{N-1} L_i^2 = \sum_{i=1}^{N-1} [(x_{i+1} - x_i)^2 + (y_{i+1} - y_i)^2], \tag{9}$$

and E_c is given by the expression.

$$E_c = \sum_{i=1}^{N} \sum_{k=1}^{K} C_i^k, \tag{10}$$

where N is the number of the points between the start and the goal, K is the number of the obstacles, and C is obtained through eq. (1).

Minimizing eq. (8) will lead to obtaining an optimal in length path that does not collide with any of the obstacles. In order to minimize (8) the classical function analysis methods are applied. First, we find the time derivative of E:

$$\begin{aligned}\frac{dE}{dt} &= \sum_{i=1}^{N} \left[w_l \left(\frac{\partial L_i^2}{\partial x_i} + \frac{\partial L_{i-1}^2}{\partial x_i} \right) + w_c \sum_{k=1}^{K} \frac{\partial C_i^k}{\partial x_i} \right] \dot{x}_i \\ &+ \sum_{i=1}^{N} \left[w_l \left(\frac{\partial L_i^2}{\partial y_i} + \frac{\partial L_{i-1}^2}{\partial y_i} \right) + w_c \sum_{k=1}^{K} \frac{\partial C_i^k}{\partial y_i} \right] \dot{y}_i\end{aligned} \tag{11}$$

Setting

$$\begin{aligned}\dot{x}_i &= -\eta \left[w_l \left(\frac{\partial L_i^2}{\partial x_i} + \frac{\partial L_{i-1}^2}{\partial x_i} \right) + w_c \sum_{k=1}^{K} \frac{\partial C_i^k}{\partial x_i} \right], \\ \dot{y}_i &= -\eta \left[w_l \left(\frac{\partial L_i^2}{\partial y_i} + \frac{\partial L_{i-1}^2}{\partial y_i} \right) + w_c \sum_{k=1}^{K} \frac{\partial C_i^k}{\partial y_i} \right]\end{aligned} \tag{12}$$

we can rewrite the above equation as

$$\frac{dE}{dt} = -\frac{1}{\eta} \sum_{i=1}^{N} (\dot{x}_i^2 + \dot{y}_i^2) < 0 \tag{13}$$

where η is an adaptation gain. It is obvious that, when $\dot{x}_i \to 0$ and $\dot{y}_i \to 0$, E converges to its minimum. In other words, when all points of the path almost stop moving, there is no collision and the path is the optimal one, i.e. eq. (13) can be used as a condition for termination of the calculation iterations.

Now, from equations (12),

$$\begin{aligned}\frac{\partial L_i^2}{\partial x_i} + \frac{\partial L_{i-1}^2}{\partial x_i} &= -2x_{i+1} + 4x_i - 2x_{i-1}, \\ \frac{\partial L_i^2}{\partial y_i} + \frac{\partial L_{i-1}^2}{\partial y_i} &= -2y_{i+1} + 4y_i - 2y_{i-1}\end{aligned} \tag{14}$$

and

$$\frac{\partial C_i^k}{\partial x_i} = \frac{\partial C_i^k}{\partial (I_0)_i^k}\frac{\partial (I_0)_i^k}{\partial x_i} = \frac{\partial C_i^k}{\partial (I_0)_i^k}\left(\sum_{m=1}^{M}\frac{\partial (O_{H_m})_i^k}{\partial (I_{H_m})_i^k}\frac{\partial (I_{H_m})_i^k}{\partial x_i}\right)$$
$$= f'((I_0)_i^k)\left(\sum_{m=1}^{M} f'_{H_m}((I_{H_m})_i^k)w_{xm}^k\right);$$
$$\frac{\partial C_i^k}{\partial y_i} = \frac{\partial C_i^k}{\partial (I_0)_i^k}\frac{\partial (I_0)_i^k}{\partial y_i} = \frac{\partial C_i^k}{\partial (I_0)_i^k}\left(\sum_{m=1}^{M}\frac{\partial (O_{H_m})_i^k}{\partial (I_{H_m})_i^k}\frac{\partial (I_{H_m})_i^k}{\partial x_i}\right)$$
$$= f'((I_0)_i^k)\left(\sum_{m=1}^{M} f'_{H_m}((I_{H_m})_i^k)w_{ym}^k\right). \tag{15}$$

This leads to the final form of the function:

$$\dot{x}_i = -2\eta w_l(2x_i - x_{i-1} - x_{i+1}) - \eta w_c\sum_{k=1}^{K} f'((I_0)_i^k)\left(\sum_{m=1}^{M} f'_{H_m}((I_{H_m})_i^k)w_{xm}^k\right);$$
$$\dot{y}_i = -2\eta w_l(2y_i - y_{i-1} - y_{i+1}) - \eta w_c\sum_{k=1}^{K} f'((I_0)_i^k)\left(\sum_{m=1}^{M} f'_{H_m}((I_{H_m})_i^k)w_{ym}^k\right), \tag{16}$$

where f' is given by the following expressions:

$$f'(\cdot) = \frac{1}{T}f(\cdot)[1 - f(\cdot)]$$
$$f'_{H_m}(\cdot) = \frac{1}{T}f_{H_m}(\cdot)[1 - f_{H_m}(\cdot)]. \tag{17}$$

In eq. (16) the first member in the right side is for the path length optimization and the second one is for the obstacle avoidance.
One of the important advantages of the above path-planning is that it allows parallelism in the calculations of the neural network outputs (see Section 6), which leads to decreasing the computational time. The generated semi-optimal path can be optimized further by applying evolutionary methods (e.g. genetic algorithm). Such approach leads to an optimal solution but the computational cost increases dramatically (Yu et al. (2003)).

5. The path-planning algorithm

In this section an algorithm based on the background given in Sections 3 and 4 is proposed. The calculations for the path are conceptually composed by the following steps:

Step 1: Initial step

1. Let the start position of the robot is (x_1, y_1), and the goal position is denoted as (x_N, y_N).
2. Check for concavities (see Section 3.2.) and, if necessary, reassign the goal (start) position.
3. At $t = 0$ the coordinates of the points of the initial path (straight line) $(x_i, y_i;\ i = 2, 3, \ldots, N-1)$ are assigned as

$$x_i = x_0 + i(x_N - x_1)/(N-1),$$
$$y_i = (y_N - y_1)(x_i - x_1)/(x_N - x_1) + y_0, \tag{18}$$

i. e. the distance between the x and y coordinates of every two neighboring points of the path is equal.

Note 2. *It is assumed that the obstacles dimensions are enlarged by the robot's polygon dimensions (Lozano-Pèrez & Wesley (1979)).*

Step2: 1. For the points (x_i, y_i) of the path which lie inside some obstacle, the iterations are performed according to the following equations:

$$\dot{x}_i = -2\eta_1 w_l(2x_i - x_{i-1} - x_{i+1}) - \eta_1 w_c \sum_{k=1}^{K} f'((I_0)_i^k) \left(\sum_{m=1}^{M} f'_{H_m}((I_{H_m})_i^k) w_{xm}^k \right);$$
$$\dot{y}_i = -2\eta_1 w_l(2y_i - y_{i-1} - y_{i+1}) - \eta_1 w_c \sum_{k=1}^{K} f'((I_0)_i^k) \left(\sum_{m=1}^{M} f'_{H_m}((I_{H_m})_i^k) w_{ym}^k \right) \quad (19)$$
$$i = 2, 3, \ldots, N-1.$$

2. For the points (x_i, y_i) situated outside the obstacles, instead of eq. (19) use the following equations:

$$\dot{x}_i = -\eta_2 w_l(2x_i - x_{i-1} - x_{i+1});$$
$$\dot{y}_i = -\eta_2 w_l(2y_i - y_{i-1} - y_{i+1}), \quad (20)$$

i.e. for the points of the path lying outside obstacles, we continue the calculation with the goal to minimize only the length of the path.

Step 3: Perform p times the calculations of step 2, i.e. find $x_i(t+p), y_i(t+p)$ $(i = 2, 3, \ldots, N-1)$, where p is any suitable number, say $p = 100$.

Step 4: Test for convergence

Calculate the difference d of the path lengths at time t and time $(t+p)$, i.e.

$$d = \sum_{i=2}^{N-1} \{[x_i(t+p) - x_i(t)]^2 + [y_i(t+p) - y_i(t)]^2\}^{1/2}. \quad (21)$$

- If $d < \varepsilon$ then the algorithm terminates with the conclusion that the goal is reached via an "optimal" path. Here ε is a small constant, say $\varepsilon = 0.1$.
- If $d \geq \varepsilon$, then GO TO step 2.

Here eq. (19) is almost the same as eq. (16) with the difference that instead of only one RPF, different functions, as explained below, are used for every layer of the neural network. Every obstacle is described using a neural network as shown in Fig. 1. The output neuron is described by eq. (1), the neuron function $f(\cdot)$ is the same as in eq. (3) and the pseudotemperature is as in eq. (4).
The hidden layer inputs are as in eq. (1) but the outputs O_{H_m} now become

$$O_{H_m} = f_{H_m}(I_{H_m}), \quad m = 1, \ldots, M \quad (22)$$

with I_{H_m} becoming the induced local field of the neuron function f_{H_m}:

$$f_{H_m}(I_{H_m}) = \frac{1}{1 + e^{-I_{H_m}/T_{H_m}(t)}} \quad (23)$$

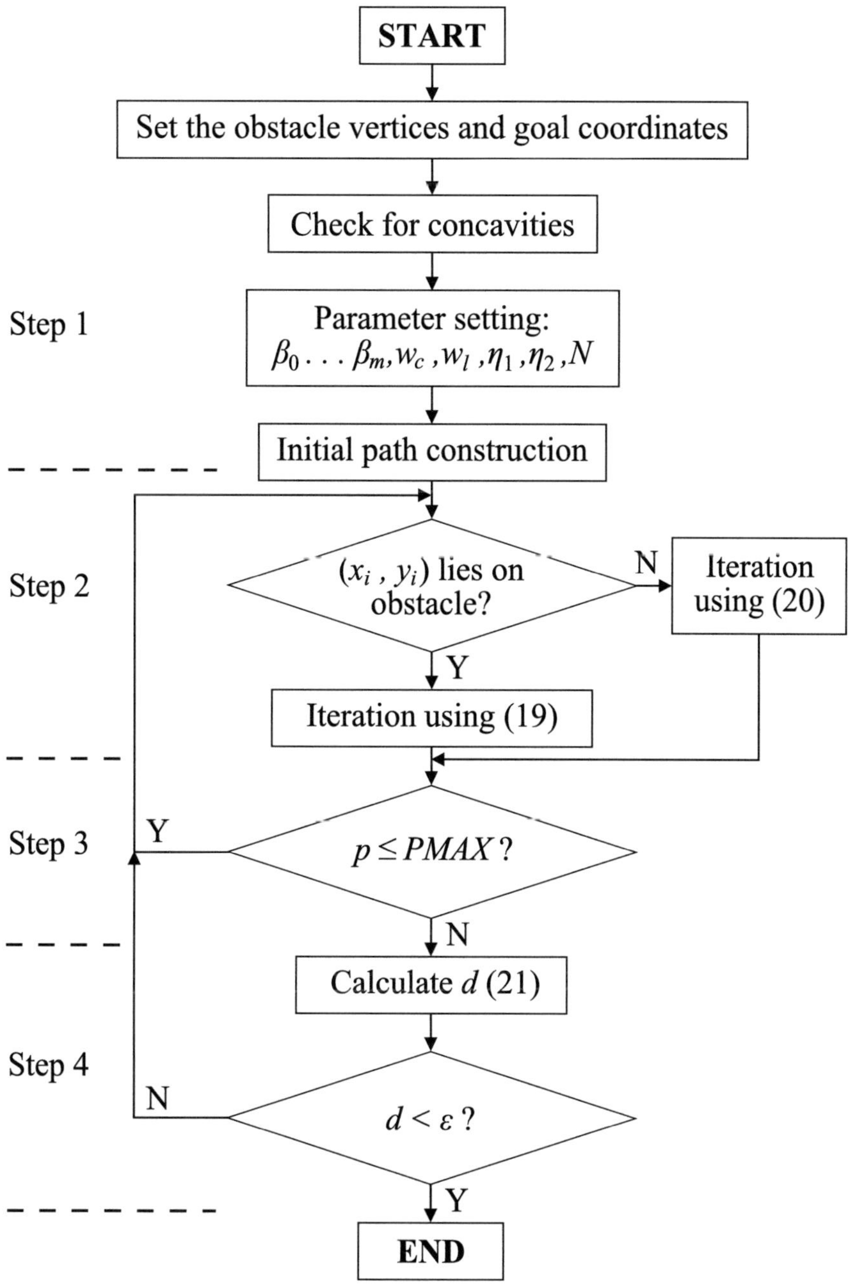

Fig. 4. Block diagram of the algorithm

and

$$T_{H_m}(t) = \frac{\beta_m}{\log(1+t)} \tag{24}$$

Finally, I_{H_m} is given by the activating function (5). Equations (3) and (23) include different values of pseudotemperatures, which is one of the important conditions for convergence of the algorithm and for generation of almost optimal path. The block diagram of the algorithm is shown in Fig. 4.

6. Experimental results

To show the effectiveness of the proposed in this paper algorithm, several simulation examples are given in this section.

6.1 Calculation speed evaluation

The simulations were run on ASUS A7V motherboards rated with 328 base (352 peak) SPEC CFP2000 and 409 base (458 peak) SPEC CINT2000 on the CPU2000 benchmark. The environment shown in Fig. 5 has been chosen for the benchmark tests. In the course of the speed measurements the number of the vertices was increased from 20 to 120 with step of 20 vertices. The final configuration of the obstacles inside the benchmark environment with three generated paths is shown in Fig. 5. To compare the run-times, 12 path generations have been performed randomly. After deleting the lowest and the highest path planning times, the average of the remaining 10 values has been adopted.

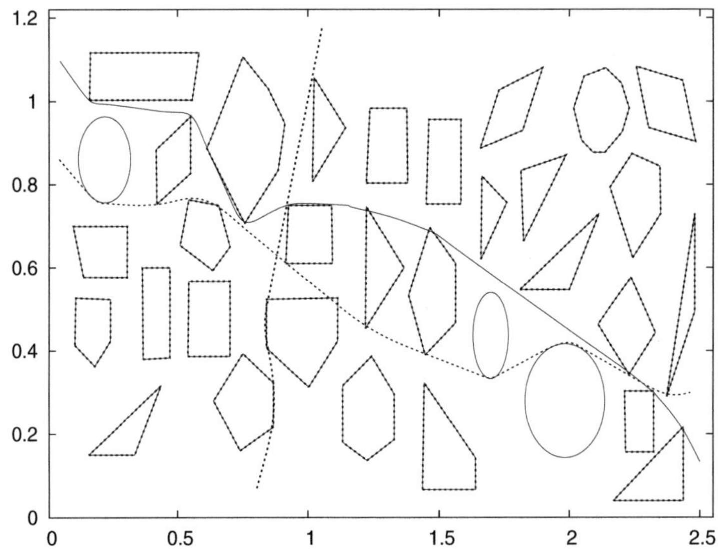

Fig. 5. The benchmark environment with three generated paths

Figure 6 shows the speed change depending on the total number of vertices of the obstacles. It is clear that the speed changes linearly with increasing the number of the vertices (cf. Lozano-Pèrez & Wesley (1979)).

As it was explained at the end of Section 4 the algorithm allows parallelizing the path computation. To calculate the speedups we ran parallel simulations for the same benchmark environment on one, two, and three processors. Figure 7 shows how the speed increases with increasing the number of the processors. For the case of two processors the simulation was run on two ASUS A7V boards, and for the case of three processors a Gigabyte 440BX board was added as a third processor. The communication during the parallel calculations was established by an Ethernet bus based network between the computers.

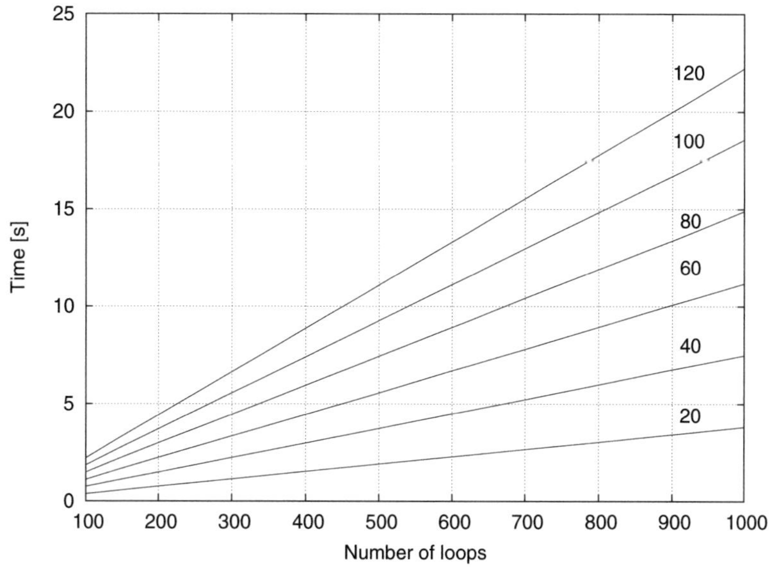

Fig. 6. Computation time depending on the number of the loops (100–1000) and number of the vertices (20–120)

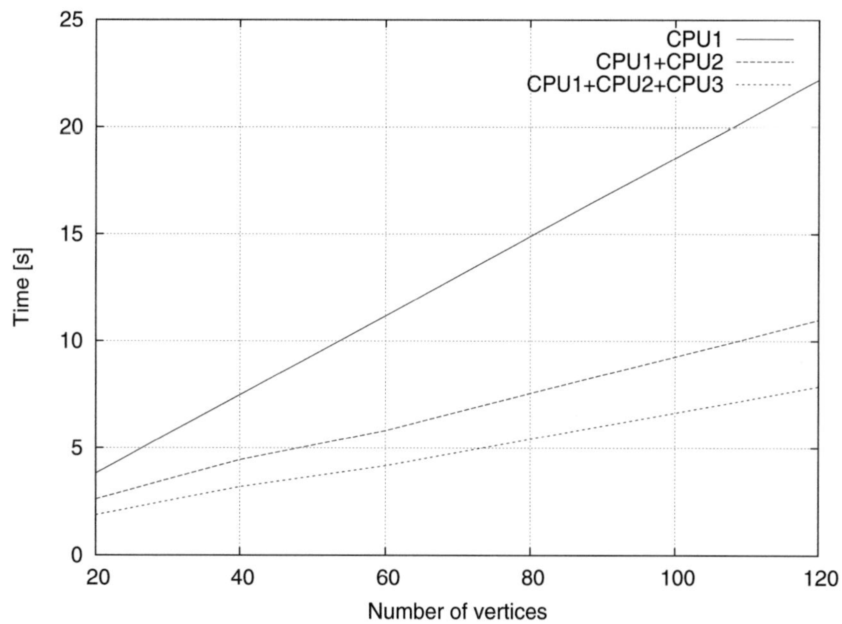

Fig. 7. Run-time t for different number of processors for 1000 loops depending on the number of vertices

It is possible to further increase the calculation speed by introducing adaptive adjustment of the parameters η_1 and η_2. For example, the following adaptive adjusting law increases the speed about three times (see Fig. 8):

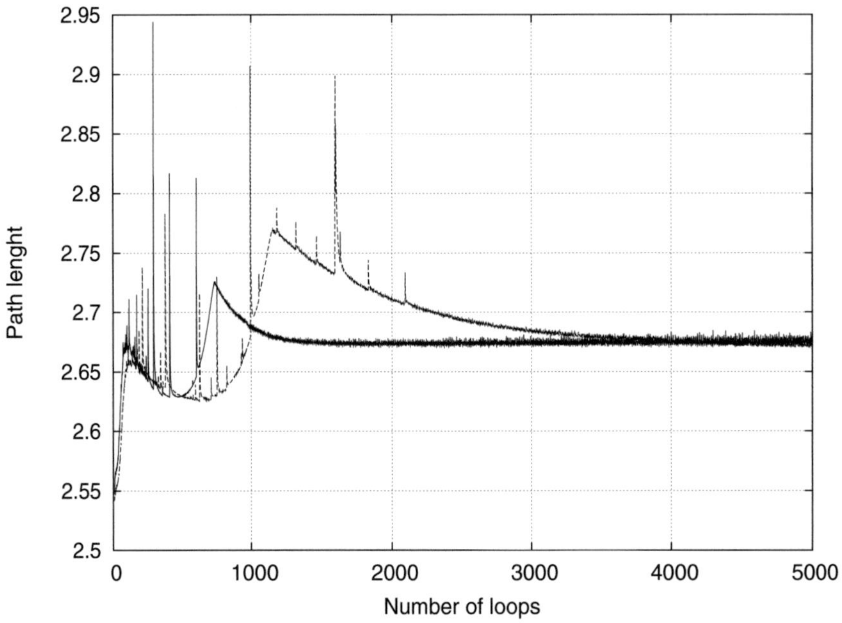

Fig. 8. Convergence speed of the algorithm depending on the way of setting η_1 and η_2: the slowly converging line is for the case when constant values are used, and the fast converging one depicts the convergence when adaptive adjustment is applied

$$\begin{aligned} \dot{\eta}_1(n+1) &= -\eta_1(n) + \delta\,\mathrm{sgn}[(d(n) - d(n-1))(d(n-1) - d(n-2))], \\ \dot{\eta}_2(n+1) &= -\eta_2(n) + \delta\,\mathrm{sgn}[(d(n) - d(n-1))(d(n-1) - d(n-2))] \end{aligned} \tag{25}$$

where δ is a small constant.

In the course of path calculations, a limited number of obstacles has impact on path generation—mostly the obstacles the initial, straight line path intersects with, contribute to path forming. Then, while performing *Step 1* of the algorithm, if all obstacles which do not collide with the initial path are ignored and excluded from the calculations, it is possible to drastically decrease the calculation time. A good example is the path in vertical direction in Fig. 5. When all obstacles are considered, the total number of vertices is 135 and RPF for all obstacles would be calculated. However, if only the four obstacles which collide with the initial straight line are used, the number of vertices decreases to only 17 and the expected calculation time will be approximately 8 times shorter. After finishing the path generation for the decreased number of obstacles, the original environment should be retained and a final correction of the path shape should be performed. Naturally, this final correction would require only few additional calculation loops.

6.2 Simulation results

Figure 9 shows additional results of path planning simulation in an environment with concaved obstacles. It can be seen that the local minimums do not cause problems in the path generation. In these simulations the number of the points between the start position and the goal was set to 200. In both examples the values of the coefficients in equations of the algorithm were not changed. In fact, the authors have performed many simulations using different environments, in which the obstacles were situated in many different ways. In all

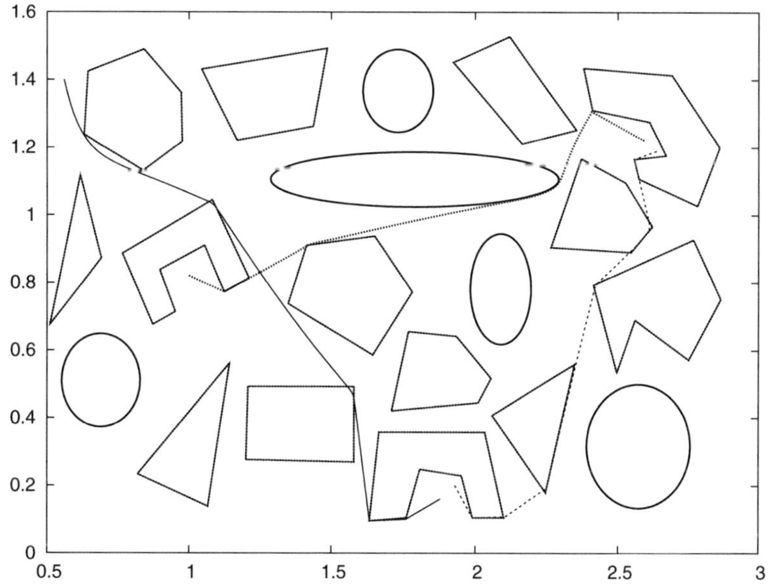

Fig. 9. Simulation example (environment including concave obstacles)

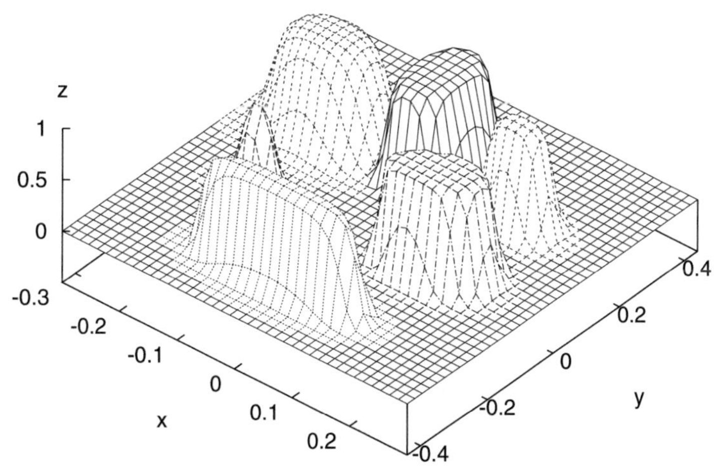

Fig. 10. Shape of the RPF in the course of simulation

the simulations the paths were successfully generated and the time for the calculations agreed with the calculation speed explained above. Figure 10 shows the potential fields of obstacles in the course of a simulation.

We have developed software with a graphic user interface (GUI) for the simulation and control of differential drive robots in a 2D environment (See Fig. 11). The environment allows the user to easily build configuration spaces and perform simulations. This environment is available at *http://shiwasu.ee.ous.ac.jp/planner/planner.zip*. The algorithm was successfully applied to the

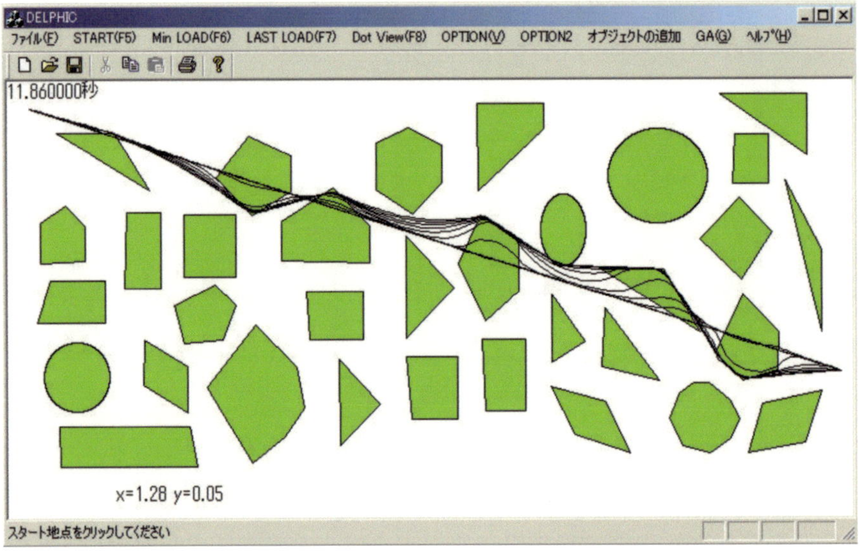

Fig. 11. GUI environment for simulation and control of differential drive robots

control and navigation of "Khepera" (Mondada et al. (1993)) and "Pioneer P3-DX" robots.

7. Discussion and conclusions

In this paper we have proposed an algorithm which guarantees planning of near optimal in length path for wheeled robots moving in an *a priori* known environment. We are considering wheel placements which impose *no* kinematic constraints on the robot chassis: castor wheel, Swedish wheel, and spherical wheel.

The proposed algorithm is a significant improvement of the potential field algorithms because it finds an almost optimal path without being trapped in local minima and the calculation speed for the proposed algorithm is reasonably fast. Because the only assumptions made are that the obstacles are stationary polygons and ovals, the algorithm can solve the navigation problem in very complex environments. For description of the obstacles the algorithm uses only the Cartesian coordinates of their vertices and does not need memory consuming obstacle transformations (cf. Zelinsky (1992); Zelinsky & Yuta (1993)). The consumed memory space is equal to twice the number of the coordinates of the path points and some additional memory for keeping the parameters of the description networks.

Because the generated paths are piecewise linear with changing directions at the corners of the obstacles, the inverse kinematics problems for the case of differential drive robots are simply solved: to drive the robot to some goal pose (x, y, θ), the robot can be spun in place until it is aimed at (x, y), then driven forward until it is at (x, y), and then spun in place until the required goal orientation θ is met.

It becomes clear from the course of the above explanations that the effectiveness of the proposed algorithm depends on the development of some mechanism for choosing the initial values of the pseudotemperatures (β_m) or properly adjusting the pseudotemperatures in order to generate potential fields satisfying the properties given in Section 2. Applying of an adaptive adjusting procedure to the coefficients β_m similar to that for adjusting of η_1 and η_2 (see. eq. (25)) is one possible solution. Moreover, there is no need to perform continuous adjustment throughout the whole path generation—adjustment in the first few steps gives good approximations for the initial values of the β_m coefficients. Strict solution of this problem is one of our next goals.

The proposed algorithm can be easily extended for the 3-D case (Kroumov et al. (2010)). The extension can be done by adding an additional input for the z (height of the obstacles) dimension to the obstacle description neural network. Adding of z coordinate and employing a technique for path planning, similar to that proposed in Henrich et al. (1998), would allow further adoption of the algorithm for planning of paths for industrial robots. Development of such application is another subject of our future work.

8. Acknowledgement

This work is partially supported by the Graduate School of Okayama University of Science, Okayama, Japan.

The authors would like to thank Mr. Yoshiro Kobayashi, a graduate student at the Graduate School of Okayama University of Science, for his help in preparing part of the figures. We are thankful to Ms Z. Krasteva for proofreading the final version of the manuscript.

9. References

Canny, J. F. (1988). *The Complexity of Robot Motion Planning*, The MIT Press, Cambridge, MA, U. S. A.

Chen, P. C. & Hwang, Y. K. (1998). Sandros: A dynamic graph search algorithm for motion planning, *IEEE Transactins on Robotics & Automation* 14(3): 390–403.

Faverjon, B. & Tournassoud, P. (1987). A local approach for path planning of manipulators with a high number of degrees of freedom, *Proceedings of IEEE International Conference on Robotics & Automation*, Raleigh, New Caledonia, U. S. A., pp. 1152–1159.

Ge, S. S. & Cui, Y. J. (2000). New potential functions for mobile robot path planning, *IEEE Transactins on Robotics & Automation* 16(5): 615–620.

Henrich, D., Wurll, C. & Wörn, H. (1998). 6 dof path planning in dynamic environments—a parallel on-line approach, *Proceedings of the 1998 IEEE Conference on Robotics & Automation*, Leuven, Belgium, pp. 330–335.

Hwang, Y. K. & Ahuja, K. (1992). Potential field approach to path planning, *IEEE Transactins on Robotics & Automation* Vol. 8(No. 1): 23–32.

Khatib, O. (1986). Real-time obstacle avoidance for manipulators and mobile robots, *International Journal of Robotics Research* 5(1): 90–98.

Khosla, P. & Volpe, R. (1988). Superquadratic artificial potentials for obstacle avoidance and approach, *Proceedings of IEEE International Conference on Robotics & Automation*, Leuven, Belgium, pp. 1778–1784.

Kroumov, V., Yu, J. & Negishi, H. (2004). Path planning algorithm for car-like robot and its application, *Journal of Systemics, Cybernetics and Informatics* Vol. 2(No. 3): 7.

Kroumov, V., Yu, J. & Shibayama, K. (2010). 3d path planning for mobile robots using simulated annealing neural network, *International Journal on Innovative Computing, Information and Control* 6(7): 2885–2899.

Latombe, J. C. (1991). *Robot Motion Planning*, Kluwer Academic Publishers, Norwell.

Lee, S. & Kardaras, G. (1997a). Collision-free path planning with neural networks, *Proceedings of 1997 IEEE International Conference on Robotics & Automation*, Vol. 4, pp. 3565–3570.

Lee, S. & Kardaras, G. (1997b). Elastic string based global path planning using neural networks, *Proceedings of 1997 IEEE International Symposium on Computational Intelligence in Robotics and Automation (CIRA'97)*, pp. 108–114.

Lozano-Pèrez, T., Mason, M. T. & Taylor, R. H. (1994). Automatic synthesis of fine motion strategies for robots, *International Journal of Robotics Research* 3(1): 3–24.

Lozano-Pèrez, T. & Wesley, M. A. (1979). An algorithm for planning collision-free paths among polyhedral obstacles, *Communications of the ICM* Vol. 22(No. 10): 560–570.

Mondada, F., Franzi, E. & Ienne, P. (1993). Mobile robot miniaturization: a tool for investigation in control algorithms, *Proceedings of ISER3*, Kyoto, Japan.

O'Dúnlaing, C. & Yap, C. K. (1982). A retraction method for planning the motion of a disk, *Journal of Algorithms* Vol. 6: 104–111.

Paden, B., Mees, A. & Fisher, M. (1989). Path planning using a jacobian-based freespace generation algorithm, *Proceedings of IEEE International Conference on Robotics & Automation*, Scottsdale, Arizona, U. S. A., pp. 1732–1737.

Rimon, E. & Doditschek, D. E. (1992). Exact robot navigation using artificial potential fields, *IEEE Transactins on Robotics & Automation* Vol. 8(No. 5): 501–518.

Sun, Z. Q., Zhang, Z. X. & Deng, Z. D. (1997). *Intelligent Control Systems*, Tsinghua University Press.

Tsankova, D. (2010). Neural networks based navigation and control of a mobile robot in a partially known environment, *in* A. Barrera (ed.), *Mobile Robots Navigation*, InTech, pp. 197–222.

Warren, C. W. (1989). Global path planning using artificial potential fields, *Proceedings of IEEE International Conference on Robotics & Automation*, pp. 316–321.

Yamamoto, M., Iwamura, M. & Mohri, A. (1998). Near-time-optimal trajectory planning for mobile robots with two independently driven wheels considering dynamical constraints and obstacles, *Journal of the Robotics Society of Japan* 16(8): 95–102.

Yu, J., Kroumov, V. & Narihisa, H. (2002). Path planning algorithm for car-like robot and its application, *Chinese Quarterly Journal of Mathematics* Vol. 17(No. 3): 98–104.

Yu, J., Kroumov, V. & Negishi, H. (2003). Optimal path planner for mobile robot in 2d environment, *Proceedings of the 7th World Multiconference on Systemics, Cybernetics and Informatics*, Vol. 3, Orlando, Florida, U. S. A., pp. 235–240.

Zelinsky, A. (1992). A mobile robot exploration algorithm, *IEEE Transactins on Robotics & Automation* Vol. 8(No. 6): 707–717.

Zelinsky, A. & Yuta, S. (1993). Reactive planning for mobile robots using numeric potential fields, *Proceedings of Intelligent Autonomous Systems 3 (IAS3)*, pp. 84–93.

10

Path Searching Algorithms of Multiple Robot System Applying in Chinese Chess Game

Jr-Hung Guo[1], Kuo-Lan Su[2] and Sheng-Ven Shiau[1]
[1]*Graduate school Engineering Science and technology, National Yunlin University of Science & Technology*
[2]*Department of Electrical Engineering, National Yunlin University of Science & Technology*
Taiwan

1. Introduction

Chinese chess game [1] is one of the most popular games. A two-player game with a complexity level is similar to Western chess. In the recent, the Chinese chess game has gradually attracted many researcher's attention. The most researchers of the fields are belong to expert knowledge and artificial intelligent. There are many evolutionary algorithms to be proposed. Darwen and Yao proposed the co-evolutionary algorithm to solve problems where an object measure to guide the searching process is extremely difficult to device [2]. Yong proposed multi-agent systems to share the rewards and penalties of successes and failures [3]. Almost all the chess game can be described by game tree. Game tree presents the possible movements and lists all situations for the Chinese chesses. We want to use the multi-robots system to present the scenario of the chess movement for the Chinese chess game, and play the Chinese chess game according to the real-time image feedback to the supervised computer via wireless image system.

The application of co-evolutionary models is to learn Chinese chess strategies, and uses alpha-beta search algorithm, quiescence searching and move ordering [4]. Wang used adaptive genetic algorithm (AGA) to solve the problems of computer Chinese chess [5]. Lee and Liu take such an approach to develop a software framework for rapidly online chess games [6]. Zhou and Zhang present the iterative sort searching techniques based on percentage evaluation and integrate percentage evaluation and iterative sort into problem of Chinese chess computer game [7]. Su and Shiau developed smart mobile robots to speak real-time status using voice module, and program the motion trajectories for multiple mobile robot system [8].

With the robotic technologies development with each passing day, robot system has been widely employed in many applications. Recently, more and more researchers are interest in the robot which can helps people in our daily life, such as entertaining robots, museum docent robots, educational robots, medical robots, service robots, office robots, security robots, home robots, and so on. In the future, we believe that intelligent robots will play an important role in our daily life. In the past literatures, many experts researched in the mobile robot, and proposed many methods to enhance the functions of the mobile robot [9]. So far, developing a big sized mobile robot to be equipped with many functions to become complex and huge, and the development period is too long. Thus, recently small-sized mobile robot system has been investigated for a specific task, and program the optimal motion path on the dynamic environment [18].

There is a growing in multi-robot cooperation research in recent year. Compare to single mobile robot, cooperation multiple mobile robots can lead to faster task completion, higher quality solution, as well as increase robustness owing its ability adjust to robot failure [10]. Grabowski and Navarro-serment [11] suggested multiple mobile robots in which each mobile platform has a specific sensor for some purpose and therefore the system's task can be distributed to each mobile platform during surveillance. The feature of this system is that each mobile robot has a common motion platform, but has different sensors. Chung et al. [12] composed of one ship and four small-sized search robots for team work in hazardous environment. Balch et al. [13] and Alami et al. [14] investigated cooperation algorithm of multiple robot system.

Some papers consider the problem of the multiple robot system working together. The multiple mobile robot system has more advantages than one single robot system [15]. The multiple mobile robots have the potential to finish some tasks faster than a single robot using ant colony system [16]. Multiple mobile robots therefore can be expected more fault tolerant than only one robot. Another advantage of multiple mobile robots is due to merging of overlapping information, which can help compensate for sensor uncertainty [17]. We have developed multiple small-size robot system to be applied in Chinese chess [8]. We extend the application field of the multiple mobile robot system, and program the shortest path moving on the chessboard platform using A* searching algorithm. The A* heuristic function are introduced to improve local searching ability and to estimate the forgotten value [19].

We present the searching algorithms based Chinese chess game, and use multiple mobile robots to present the scenario on the chessboard platform. The mobile robot has the shape of cylinder and its diameter, height and weight is 8cm, 15cm and 1.5kg. The controller of the mobile robot is MCS-51 chip, and acquires the detection signals from sensors through I/O pins, and receives the command from the supervised compute via wireless RF interface. The chesses (mobile robots) can speak Chinese language for real-time status using voice module. We develop the user interface of multiple mobile robots according to the basic rules of Chinese chess game on the supervised computer. The mobile robot receives the command from the supervised computer, and calculates the displacement using the encoder module The supervised computer program the motion trajectories using evaluation algorithm and A* searching algorithm for the mobile robot to play the Chinese chess game. The A* searching algorithm can solve shortest path problem of mobile robots from the start point to the target point on the chessboard platform.

The simulation results can found the shortest motion path for mobile robots (chesses) moving to target points from start points in a collision-free environment. The two points are selected by two players according to the rules of the Chinese chess game. In the experimental results, we test some functions of the mobile robot to obey the moving rules of the game on the chessboard, and implement the simulation results on the chessboard platform using mobile robots. Users can play the Chinese chess game on the supervised computer using the mouse. Mobile robots can receive the command from the supervised computer, and move the next position according to the attribute of the chess. The chess (mobile robot) moving scenario of the Chinese chess game feedback to the user interface using wireless image system.

2. System architecture

The system architecture of the multiple mobile robots based Chinese chess game system is shown in Fig 1. The system contains a supervised computer, a image system, some wireless RF modules, a remote supervised computer and thirty-two mobile robots. Mobile robots are classified two sides (red side and black side) belong two players. There are sixteen mobile robots in each side. The supervised computer can transmit the command signals to control mobile robots, and receives the status of the mobile robots via wireless RF interface. These signals contain ID code, orientation and displacement of mobile robots from the supervised computer to mobile robots. The feedback signals of mobile robots contain ID code, position (X axis and Y axis) to the supervised computer. Each robot is arranged an ID code to classify the attribute of the chess. Users move the chess on the user interface using the mouse. The supervised computer can controls the mobile robot moving the target position according the command. The mobile robot transmits the ending code to the supervised computer after finishing the task via wireless RF interface.

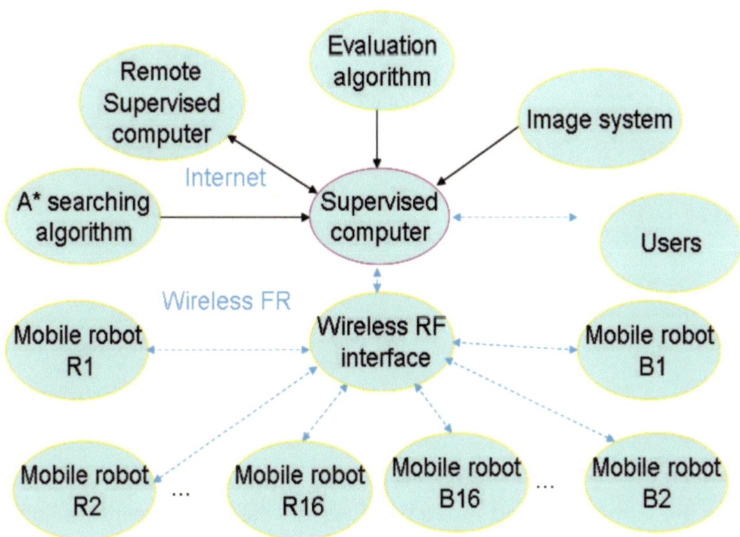

Fig. 1. The architecture of the mobile robots based Chinese chess game system.

The Chinese chess game system can transmit the real-time image to the supervised computer via image system. Users can play the Chinese chess game with others on the supervised computer using the mouse, or play the game on the remote supervised computer via wireless Internet. The player can move the chess piece using the mouse according to the rules of the Chinese chess game. Mobile robots receive the status from the supervised computer via wireless RF interface. There are two sides of the game system. One is the red side; the other is black side. Each side has sixteen chesses. There are one "king" chess, two "advisor" chesses, two "elephant" chesses, two "horse" chesses, two "rook" chesses, two "cannon" chesses and five "pawn" chesses. The definition of the chessboard is shown in Fig. 2. Then we define the initial position all chesses. Such as the position of "red king" is (4,0), and "black king" is (4,9)...etc.

The basic rules of the Chinese chess are found easily on the Internet. Before we want to control the multiple mobile robots based Chinese chess game. A chess engine needs to be designed and tested to ensure that all chess displacement and game rules are strictly adhered. We proposed the engine to be simple programmed basic rules of the games. The chessboard, denoted a 9 by 10 matrix, is the most important information that determine a players next moving position. We use axis position to define all chesses. First, the chessboard, denoted a axis position (x,y) from (0,0) to (8,9).

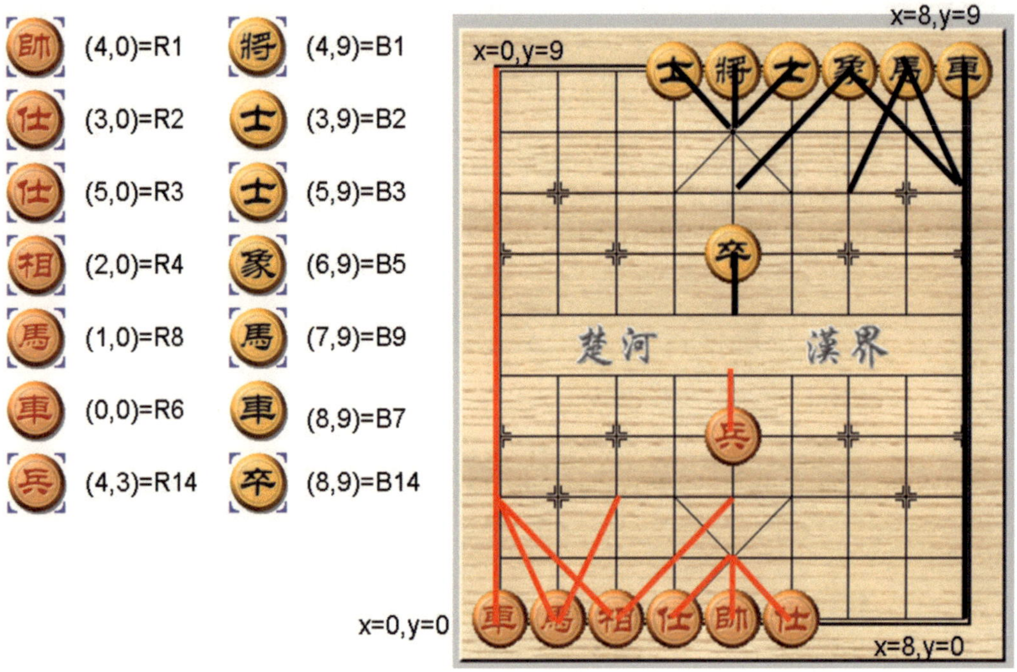

Fig. 2. The positions of the chesses.

We plot the possible motion trajectories using thick lines for the chesses on the chessboard. Then we define the game tree, and move the chess to the target position. Such as the chess "black horse" can moves to the position (8,7), (6,7) or (5,8). But the chess can't moves to the position (5,8) according the rules of the Chinese chess game. The chess "black horse" has an obstacle (black elephant) on the right side, and can't walk over the chessboard. We plot the possible motion trajectory for the chess piece on the chessboard. Then we define the game tree, and move the chess to the assigned position. Such as the chess "red horse" can moves to the position (0,2) or (2,2). But the chess can't moves to the position (3,1) according to the rules of the Chinese chess game.

The communication protocol of the multiple mobile robot system is 10 bytes. It contains one start byte, eight data bytes and one checksum byte. The supervised computer transmits 10 bytes to control the mobile robot, and the communication protocol of the control command is listed in Table 1. The mobile robot receives the command to discriminate the robot ID code to be right, and moves to the target point step by step, and transmits the environment status to the supervised computer on real-time. The communication protocol of the feedback data from the mobile robot is listed in Table 2. The byte 3 and 4 represents the positions of the mobile robot on X axis and Y axis.

Byte	0	1	2	3	4
Definition	Start byte	ID code	Robot ID code	X axis of start position	Y axis of start position
5	6		7	8	9
X axis of target position	Y axis of target position		Orientation	No use	Checksum

Table 1. The communication protocol of the supervised computer.

Byte	0	1	2	3	4
Definition	Start byte	ID code	Robot ID code	X axis of robot	Y axis of robot
5	6		7	8	9
Obstacle status	Orientation of robot		No use	No use	Checksum

Table 2. The communication protocol of the mobile robot.

3. Mobile robot

We develop two module based mobile robot (MBR-I and MBR-II) for the Chinese chess game. The mobile robot (module based robot-I, MBR-I) has the shape of cylinder, and it's equipped with a microchip (MCS-51) as the main controller, two DC motors and driver circuits, a reflect IR sensor circuits, a voice module, an encoder module, three Li batteries and some wireless RF modules. Meanwhile, the mobile robot has four wheels to provide the capability of autonomous mobility.

The mobile robot has some hardware circuits to be shown in Fig. 3. The encoder module uses two reflect IR sensors to calculate the pulse signals from the two wheels of the mobile robot, and calculates the moving distance on the chessboard platform. The power of the mobile robot is three Li batteries, and connects with parallel arrangement. The driver circuits can control two DC motors to execute the displacement command through PWM signal from the controller. The controller of the mobile robot can acquires the detection signals from sensors through I/O pins, and receives the control command via wireless RF interface. The switch input can turns on the power of the mobile robot, and selects power input to be Li batteries or adapter. The voice module can speak the real-time status of the environment for mobile robot moving on the chessboard. In the encoder module, we plot the white line and black line on the wheel of the mobile robot, and use two reflect IR sensors to calculate the pulse signals from the two wheels of the mobile robot. We can set the pulse number for per revolution to be P, and the mobile robot moves pulse number to be B. We can calculate the movement displacement D of the mobile robot using the equation.

$$D = 4.25 \times \pi \frac{B}{P} \tag{1}$$

The diameter of the wheel is 4.25 cm. We design the new chessboard using grid based platform to be shown in Fig. 4. The arrangement of the chess board is 11 grids on the horizontal direction (X axis), and is 12 grids on the vertical direction (Y axis). The distance is 30cm between the center of corridor on the X axis and Y axis of the chessboard. The width of the corridor is 12cm. The mobile robot uses IR sensors to detect obstacles, and decides the cross points of the chess board. The game only uses 9 grids (horizontal) by 10 grids (vertical). We release two grids on each direction (horizontal and vertical) to arrange the leaving chesses of red side and black side. We put the mobile robots that are eaten moving around the platform. We improve the mobile robot (MBR-I) to be implemented in the new chessboard platform, and design new driver device using DC servomotor.

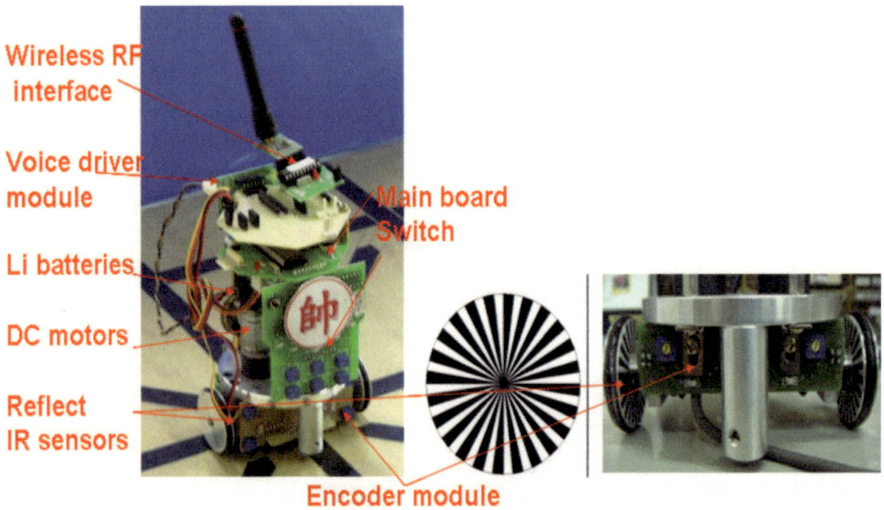

Fig. 3. The structure of the MBR-I.

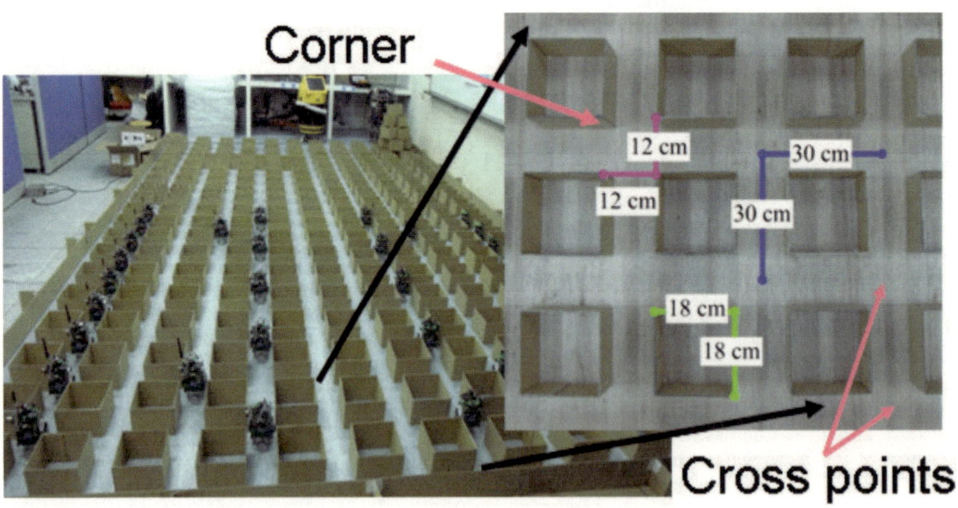

Fig. 4. The chessboard of the Chinese chess game.

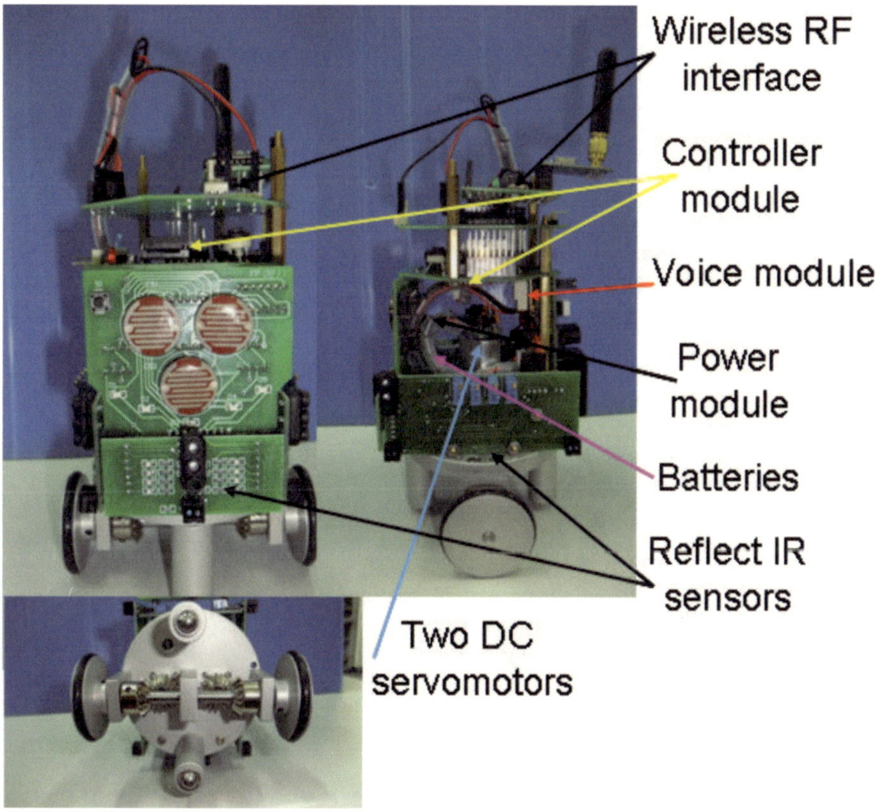

Fig. 5. The structure of the MBR-II.

The mobile robot (module based robot-II, MBR-II) has the shape of cylinder to be equipped with a microchip (MCS-51) as the main controller, and contains two DC servomotors and driver devices, some sensor circuits, a voice module, a compass module, three Li batteries, a wireless RF interface and three reflect IR sensor modules. Meanwhile, the mobile robot has four wheels to provide the capability of autonomous mobility. The structure of the mobile robot is shown in Fig. 5.

The controller calculates the orientation of the mobile robot from the compass module. The compass module has the error range to be 5^0, and transmits the measured values $(x \text{ and } y)$ in X axis and Y axis, and transmits to the controller of the mobile robot. The mobile robot can calculates the orientation angle θ is

$$\theta = Tan^{-1}(-y / x) \tag{2}$$

The distance is 30cm between the center of corridor on the X axis and Y axis, and the width of the corridor is 12cm. The mobile robot uses three IR sensors to detect the signals on the front side, right side and left side of the mobile robot, and decides the position of the obstacles, and detect the cross points. The definition of the orientation is east (X- axis), west (X+ Axis), south (Y+ axis) and north (Y- axis) on the motion platform. We set measurement range on the orientation of the mobile robot. For example, the measurement value of the orientation between 226^0 and 315^0, and define the direction of the mobile robot is west.

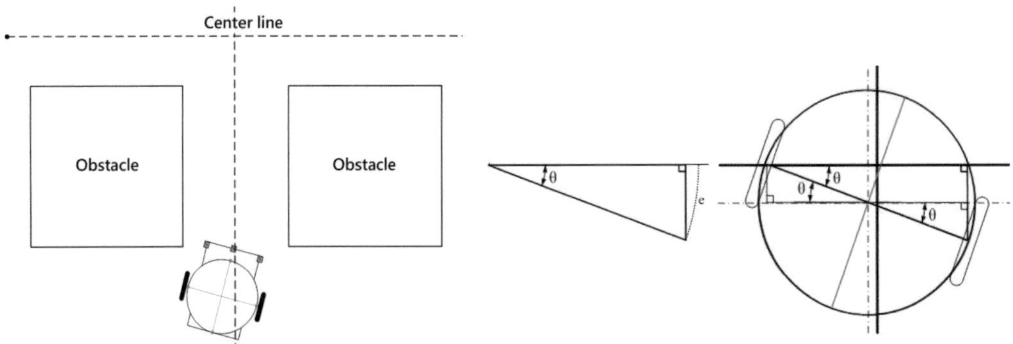

Fig. 6. The mobile robot tuning method.

The mobile robot moves on the corridor of the motion platform, and detects the location to be leaving the center line using the three reflect IR sensors. It tunes the motion path moving to the center line of the corridor. For example, the mobile robot leaves the center line, and moves on the left side of the center line. The mobile robot turns right θ (degree) to be set by users, and is calculated by the encoder module. Then it moves the displacement S according to Eq. (3), and moves to the center line. Finally, the mobile robot turns left θ to face the center line. The overview of the tuning schedule is shown in Fig. 6.

$$S = \frac{R\pi\theta}{180} \tag{3}$$

The parameter R (=4.25cm) is the diameter of the wheel. Furthermore, the mobile robot moves on the right side of the centre line. The tuning processing is the same as the previous method.

4. User interface

The operation interface of the multiple robot based Chinese chess game system is shown in Fig. 7. There are two regions in the operation interface. The chessboard of the Chinese chess is the main monitor of the user interface. Next we explain the chessboard of the Chinese chess game, and describe how to use the interface. It can displays "communication port", "communication protocol" and "axis position" on the bottom side of the operation interface. We set the communication port is 2, and Baud rate is 9600. The start position and target position of each chess displays on the left-bottom side of the operation interface. We make an example to explain the operation interface to be shown in Fig. 7. Players can move "black elephant"to the target position (5,8) from the start position (3,10). The bottom side of the interface can displays the start position (3,10) and the target position (5,8). We develop the operation interface using Visual Basic language for the Chinese chess game system.
We make other example to explain the Chinese chess game interface to be shown in Fig. 8. We can move the chess "red horse" from the start position (2,1) to the next position (3,3) using the mouse. The supervised computer control "red horse" chess moving to the next position (3,3). The scenario of the operation interface is shown in Fig. 8. The bottom side of the operation interface can displays the start position (2,1) and the next position (3,3).

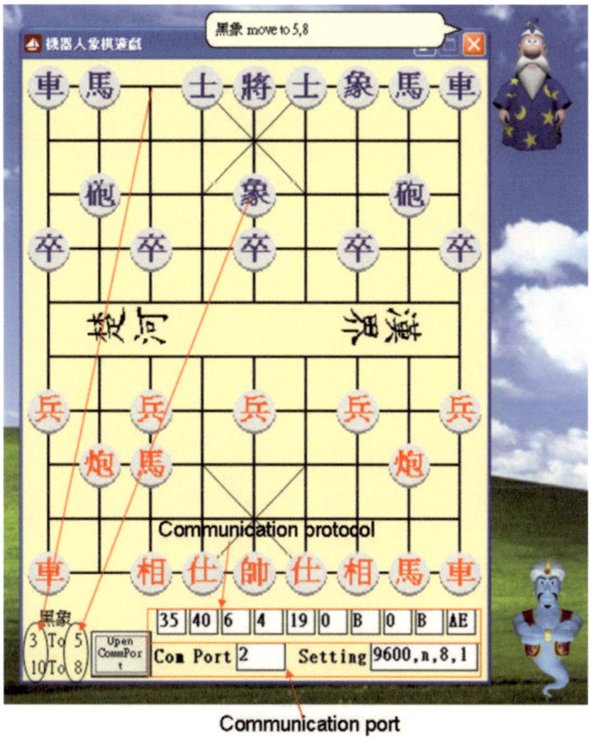

Fig. 7. The operation interface of moving the chess "black elephant".

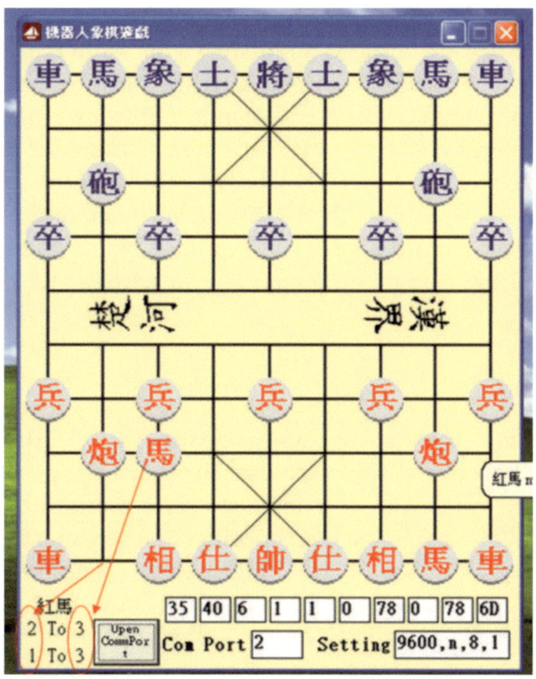

Fig. 8. The operation interface of moving the chess "red horse".

The user interface of the multiple mobile robot based Chinese chess game system has four parts to be shown in left side of the Fig. 9. Players can start or close the game in the part "I". The part "II" can displays the communication protocol on the bottom of the monitor, and the communication port is 2, and Baud rate setting is 9600. The supervised computer transmits the control command to the mobile robot, and receives the status of the mobile robot, and displays the position of the mobile robot moving on the chessboard platform simultaneously to be shown in the part "III".

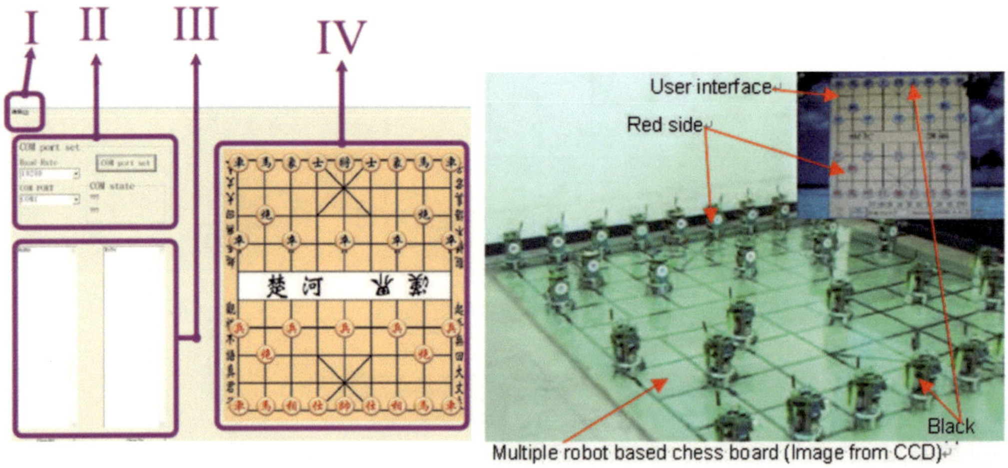

Fig. 9. The user interface of the game.

The part "IV" is chessboard platform. It can displays the motion path that is computed by evaluation algorithm and A* searching algorithm. The operation interface of the multiple robot based Chinese chess game system is shown in right side of the Fig. 9. There are two regions in the user interface. One is real-time image from the image system, and is equipped in the mobile robot based chessboard. The other is the chessboard of the Chinese chess game, and displays on the right side or the monitor. The players can move the chess using the mouse according the rules of the Chinese chess game. Mobile robots receive the status from the supervised computer via wireless RF interface, and move to the next position.

5. Evaluation method

The evaluation algorithm of the Chinese chess game uses the rule based method. Players can move the chess to the target position on the user interface. The position of the chess is not abided by the proposed rules. The user interface can't execute the command, and can't control the mobile robot moving to the target position. Players must renew the operation on the step, and obey the exact rules of the Chinese chess game. We can define the initial position of the chess piece is (x,y), and define the movement rules of all chesses as following. n is movement displacement on the x axis, and m is movement displacement on the y axis. n and m of the target position must be plus integer. The positions of all chesses must obey the rules that are listed in the Table 3 at any time.

Chess	Target position	Rules
Red king	$(x\pm n, y\pm m),\ 0\le n\le 1,\ 0\le m\le 1$	$3\le x\le 5,\ 3\le x\pm n\le 5,$ $0\le y\le 2,\ 0\le y\pm m\le 2$
Black king	$(x\pm n, y\pm m),\ 0\le n\le 1,\ 0\le m\le 1$	$3\le x\le 5,\ 0\le x\pm n\le 5,$ $7\le y\le 9,\ 0\le y\pm m\le 9$
Red advisor	$(x\pm n, y\pm n),\ n=1$	$3\le x\le 5,\ 0\le x\pm n\le 5,$ $0\le y\le 2,\ 0\le y\pm n\le 2$
Black advisor	$(x\pm n, y\pm n),\ n=1$	$3\le x\le 5,\ 0\le x\pm n\le 5,$ $7\le y\le 9,\ 0\le y\pm n\le 9$
Red elephant	$(x\pm n, y\pm n),\ n=2$	$0\le x\le 8,\ 0\le x\pm n\le 8,$ $0\le y\le 4,\ 0\le y\pm n\le 4$
Black elephant	$(x\pm n, y\pm n),\ n=2$	$0\le x, x\pm n\le 8,\ 5\le y, y\pm n\le 9$
Red rook, Red cannon Black rook, Black cannon	$(x\pm n, y\pm m),\ n=0$ or $m=0$	$0\le x, x\pm n\le 8,\ 5\le y, y\pm m\le 9$
Red horse, Black horse	$(x\pm n, y\pm m),\ n+m=3$	$0\le x\le 8,\ 0\le x\pm n\le 8,$ $0\le y\le 9,\ 0\le y\pm m\le 9$
Red pawn	$(x, y+1)$	$0\le x\le 8,$ $3\le y\le 4,\ 3\le y\pm m\le 4$
	$(x\pm n, y+m),\ n+m=1$	$0\le x\le 8,\ 0\le x\pm n\le 8,$ $5\le y\le 9,\ 5\le y+m\le 9$
Black pawn	$(x, y-1)$	$0\le x\le 8,$ $5\le y\le 6,\ 5\le y\pm m\le 6$
	$(x\pm n, y-m),\ n+m=1$	$0\le x\le 8,\ x\pm n\le 8,$ $0\le y\le 4,\ 0\le y-m\le 4$

Table 3. The rules of Chinese chesses.

Mobile robots move on the chessboard to obey the rules of Chinese chess game. The user interface knows the start position and the target position, and programs the shortest path of the chess (mobile robot) moving to the target position from the start position. We use A* searching algorithm to find the shortest path on the Chinese chess game. A^* searching algorithm is proposed by Hart in 1968, and solved the shortest path problem of multiple nodes travel system. The formula of A^* searching algorithm is following

$$f(n)=g(n)+h(n) \tag{4}$$

The core part of an intelligent searching algorithm is the definition of a proper heuristic function $f(n)$. $g(n)$ is the exact cost at sample time n from start point to the target point. $h(n)$ is an estimate of the minimum cost from the start point to the target point. In this study, n is reschedules as n' to generate an approximate minimum cost schedule for the next point. The equation (5) can be rewritten as follows:

$$f(n) = g(n) + h(n') \tag{5}$$

We make an example to explain algorithm. Such as a mobile robot move to the target point "T" from the start point "S". The position of the start point is (2,6), and the target position is (2,9). We set some obstacle on the platform. The white rectangle is unknown obstacle. The black rectangle (obstacle) is detected by the mobile robot using A* searching algorithm. We construct two labels (Open list and Close list) in the right side of Fig. 8. The neighbour points of the start point fill in the "Open list". The "Close list" fills the start point and evaluation points. We construct label on the first searching result to be shown in Fig. 8. We calculate the values of $f(n)$, $g(n)$ and $h(n)$ function according to the pulse numbers by the encoder of the DC servomotor, and use the proposed method to compare the values of the function. We select the minimum value of the function $f(n)$ to be stored in "Close list". We can find the target point on the final searching result to be shown in Fig. 10, and we can decide a shortest path to control the mobile robot moving to the target point.
The total distance of the shortest path C_{st} can be calculated as

$$C_{st} = \sum_{n=1}^{m} G_n (m = t - 1) = 2043 \tag{6}$$

In the other condition, we rebuild the positions of the obstacles on the platform to be shown in Fig. 11. The mobile robot can't find the path C_{st} moving to the target position using the proposed method. The total distance C_{st} as

$$C_{st} = \infty \tag{7}$$

Open	Closed	Point	F_{12}	G_{12}	H_{12}	D
(3,6)	S	(3,6)	1380	732	648	→
(1,6)	(2,7)	(1,6)	1380	732	648	↑
(1,7)	(3,7)	(1,7)	1056	570	486	↑
(6,8)	(3,8)	(6,8)	1704	864	810	↑
(6,9)	(4,8)	(6,9)	1338	690	648	↑
(6,10)	(5,8)	(6,10)	1662	852	810	↑
(4,11)	(5,9)	(4,11)	1338	690	648	↑
(3,11)	(5,10)	(3,11)	1014	528	486	↑
(2,11)	(4,10)	(2,11)	648	162	324	↑
	(3,10)					
	(2,10)					
	T					

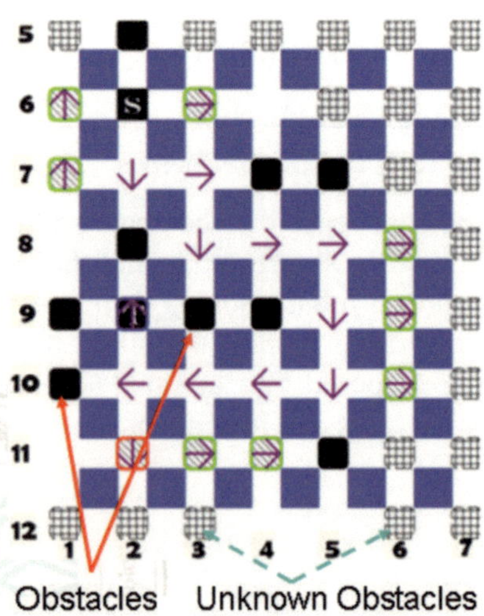

Fig. 10. The searching result for case 1.

Open	Closed	Point	F_{1i}	G_{1i}	H_{1i}	D
	S					
	(2,7)					
	(3,7)					
	(3,8)					
	(4,8)					
	(5,8)					
	(5,9)					
	(5,10)					
	(6,10)					
	(6,9)					
	(6,8)					
	(3,6)					
	(4,6)					
	(1,6)					
	(1,7)					
	(1,8)					

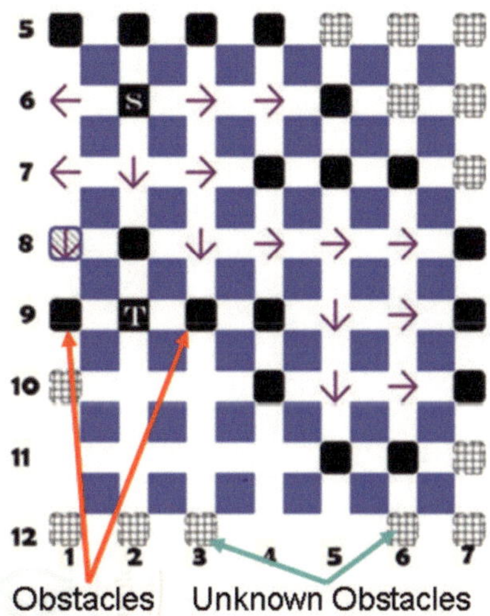

Fig. 11. The searching result for case 2.

6. Experimental results

We execute some experimental scenarios using the MBR-I robot on the chessboard for the Chinese chess game system, and test the movement functions of some chesses (mobile robots). The two players are located by two sides, and move the chess one time each other. One moves red chesses; the other moves black chesses. The first experimental scenario is "red king". The player moves forward the chess "red king" using the mouse to be shown in Fig. 12 (a). The chess "red king" moves one grid to the target position on the user interface to be shown in Fig. 12 (b). The supervised computer orders the command to the mobile robot "red king" moving forward to the target position via wireless RF interface. The mobile robot can calculates the movement displacement using the encoder module, and speaks the movement status using the voice module. The experimental result is shown in Fig. 12 (c).

The second experimental scenario is "red rook". The player moves forward chess "red rook" using the mouse to be shown in Fig. 13 (a). The supervised computer orders the command

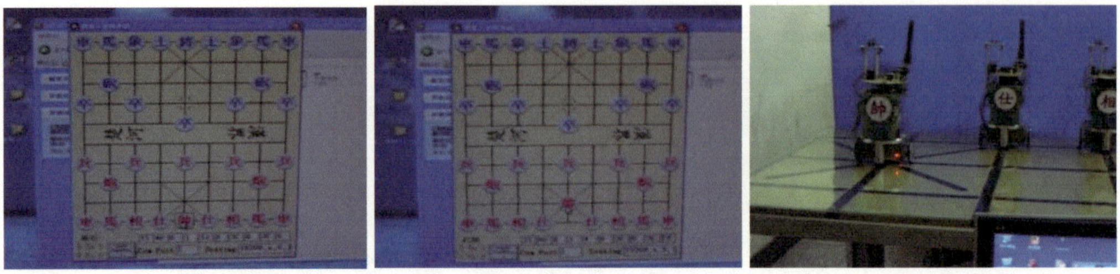

(a) (b) (c)

Fig. 12. The experimental result for "red king".

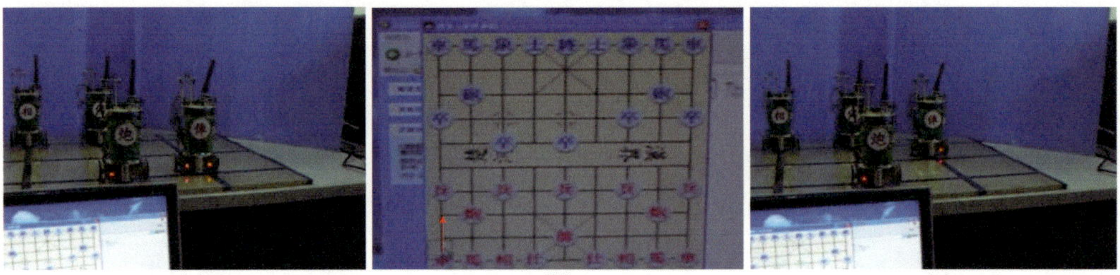

(a) (b) (c)

Fig. 13. The experimental result for "red rook".

to the mobile robot "red rook". The mobile robot "red rook" moves to the target position (8,2) from the start position (8,0). The mobile robot can calculates the movement displacement using the encoder module, and stops at the target position. The experimental result is shown in Fig. 13 (b) and (c).erwe

The third experimental scenario is "red cannon". The player moves the chess "red cannon" to right side using the mouse. The chess "red cannon" moves to the target position (4,2) from the start position (1,2) to be shown in Fig. 14 (a) and (b). The supervised computer orders the command to the mobile robot "red cannon" moving to the target position. The mobile robot receives the command to turn right, and moves to the target position. Then it can turn left and face to the black chess. The mobile robot calculates the movement displacement and the orientation using the encoder module, and speaks the movement status using the voice module. The experimental results are shown in Fig. 14 (c) and (d).

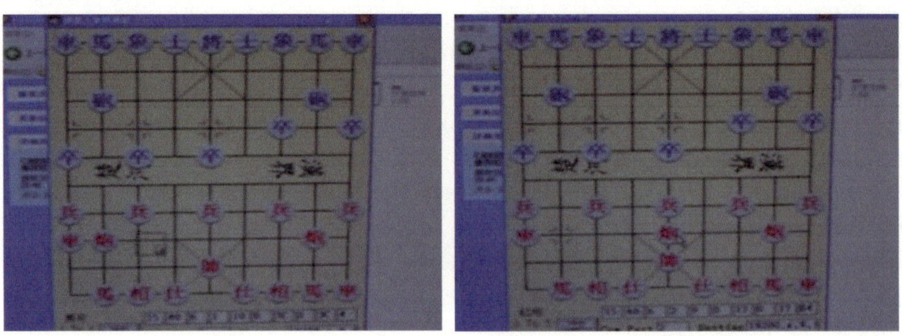

(a) (b)

Fig. 14. (Continued)

(c) (d)

Fig. 14. The experimental result for "red cannon".

We implement the functions of the mobile robot to obey the movement rules of Chinese chess game. Then we execute experimental scenarios of the multiple mobile robots based Chinese chess game system. The first experimental scenario is "red elephant" that is moved by the player (red side). The player moves the chess "red elephant" to the target position (4,2) using the mouse to be shown in Fig. 15 (a) and (b). The supervised computer orders the command to the mobile robot "red elephant" via wireless RF interface. The start position of the mobile robot is (6,0). The movement orientation of the mobile robot is west-north. The mobile robot receives the command to turn left on 90^0, and moves forward to the target position according to the Chinese chess rules. The mobile robot speaks the movement status using voice module, and calculates the movement displacement using the encoder module, and moves to the target position (4,2). The experimental results are shown in Fig. 15 (c) and (d).

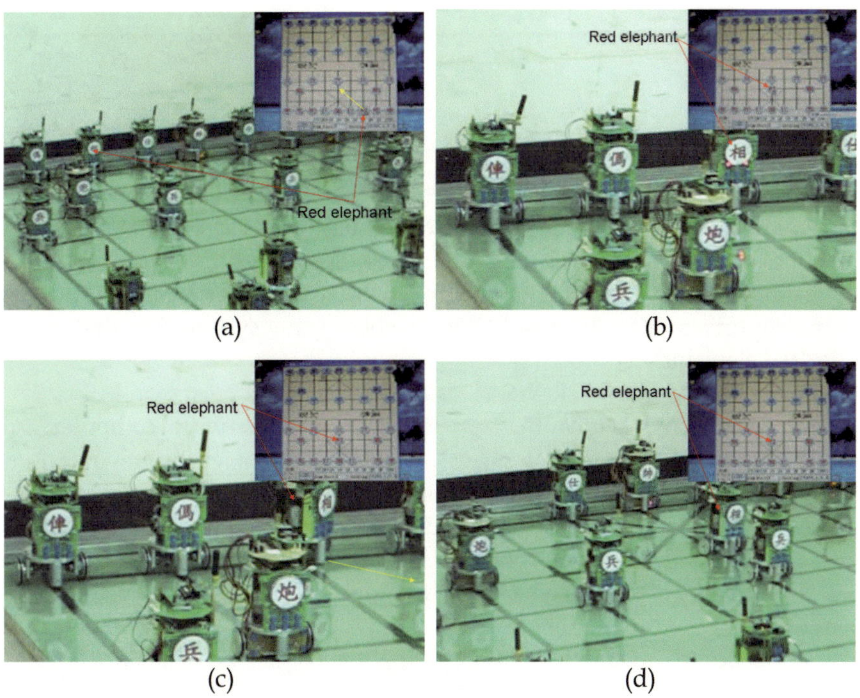

Fig. 15. The experimental result for "red elephant".

The second experimental scenario is "black advisor" that is moved by the other player (black side). The player moves the chess "black advisor" using the mouse to be shown in Fig. 16 (a). The start position of the "black advisor" chess is (3,9). The supervised computer orders the command to the mobile robot "black advisor". The chess turn left on 90^0, and moves to the target position (4,8) and stop. The mobile robot can calculates the movement displacement using the encoder module, and speak the movement status using voice module, too. The experimental result is shown in Fig. 16 (b). The next step must moves red chess by the player (red side). The players (red side and black side) move one chess step by step on the chessboard each other. Finally, the user interface can decides which player to be a winner, and closes the game.

(a) (b)

Fig. 16. The experimental result for "black advisor".

The controller of the mobile robots calculates the movement displacement to be error, or loses pulse numbers from the encoder module. Mobile robots can't stay at the cross point of two black lines on the chessboard for a long time. We design the grid based chessboard platform to improve the weakness, and implement some experimental results using the multiple mobile robots (MBR-II) based Chinese chess game system. The first experimental scenario is "red elephant".

The user moves the chess "red elephant" using the mouse to be shown in Fig. 17 (a). The start position of the "red elephant" chess is (6,0). The supervised computer orders the command to the mobile robot "red elephant". The chess moves forward two grids, and turn left angle to be 90^0. Then the mobile robot moves two grids to the next position (4,2), and turn right angle 90^0 to face the black side and stop. The user interface of the Chinese chess game system uses multiple mobile robots to be shown in Fig. 17(a). The mobile robot can calculates the movement displacement via encoder of DC servomotor, and speaks the movement status of the mobile robot using voice module, too. The experimental scenarios use the mobile robot to execute the motion path of the chess "red elephant" to be shown in Fig. 17 (b)-(e). There is not obstacle on the motion path of the chess. We program the shortest path using A* searching algorithm.

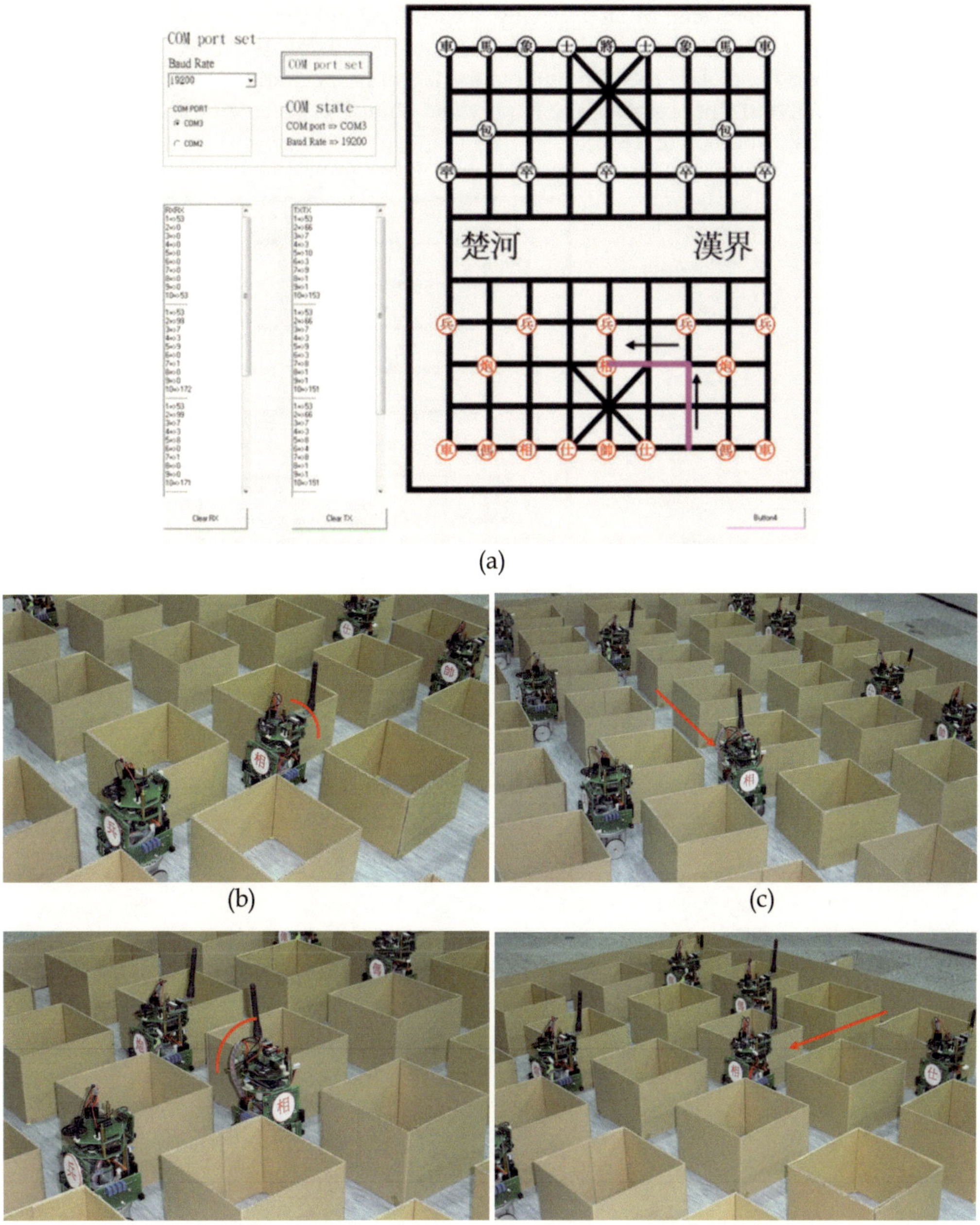

(a) (b) (c) (d) (e)

Fig. 17. The experimental results for the chess "red elephant".

The second experimental scenario is "red cannon". The user moves the chess "red cannon" using the mouse to be shown in Fig. 18 (a). The start position of the "red cannon" chess is (1,2), and the target point is (1,6). The mobile robot "red cannon" can't moves forward to the target position (6,1). The motion path of the mobile robot has an obstacle (chess) at the position (1,5). The supervised computer reconstructs all possible paths using A* searching algorithm for red cannon step by step. Finally, we can find out the shortest path to avoid the obstacle (chess). The path can displays on the interface using red line. The supervised computer controls the mobile robot moving to the target point from the start point using the shortest path to avoid the obstacle. The supervised computer calculates the cost *f(n)* to be listed on the left side of the Fig. 18 (a).

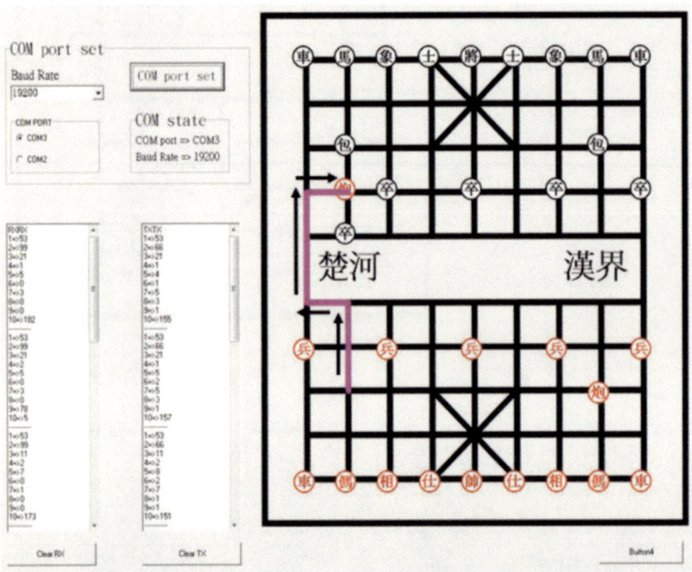

(a)

(b)

(c)

(d)

(e)

Fig. 18. The experimental results for the chess "red cannon".

The supervised computer orders the command to the mobile robot "red cannon". The chess piece moves forward two grids, and turn left angle to be 90^0. Then the mobile robot moves one grid to, and turn right angle 90^0. Then the mobile robot moves two grids, and turn right angle 90^0. Finally the mobile robot moves one grid to the target position (1,6), and turn left angle 90^0 to face the black side and stop. Finally, the experiment scenarios are shown in Fig. 18(b)-(e).

7. Conclusion

We have presented the experimental scenario of the Chinese chess game system using multiple mobile robots. The system contains a supervised computer, a image system, some wireless RF modules and thirty-two mobile robots. Mobile robots are classified red side and black side. We design two types' mobile robots (MBR-I and MBR-II) for Chinese chess game. The two mobile robots have the shape of cylinder and its diameter, height and weights is 8cm, 15cm and 1.5kg, and execute the chess attribute using two interfaces. One is wireless RF interface, and the other is voice interface. We develop the user interface on the supervised computer for the Chinese chess game. The supervised computer can programs the motion paths of the mobile robots using evaluation algorithm according to the rule of Chinese chess game, and receive the status of mobile robots via wireless RF interface. The MBR-I calculates the movement displacement using the pulse numbers from reflect IR sensors of the encoder module. Then we develop the MBR-II robot to implement scenario of the Chinese chess game on the grid based chessboard platform. The MBR-II robot calculates the movement displacement using encoder of DC servomotor. The chess (mobile robot) uses A* searching algorithm to program the shortest path moving to the target point. The supervised computer can controls mobile robots, and receives the status of mobile robots via wireless RF interface. Players can move the chess piece using the mouse on the supervised computer, and obey the game rules. The supervised computer can controls the mobile robot moving to the target position. The mobile robot speaks Chinese language for the movement status.

8. Acknowledgment

This work was supported by the project "Development of an education robot", under National Science Council of Taiwan, (NSC 99-2622-E-224-012-CC3).

9. References

S. J. Yen, J. C. Chen, T. N. Yang and S. C. Hsu, "Computer Chinese Chess," ICGA Journal, Vol.27, No. 1, pp.3-18, Mar, 2004.

P. Darwen and X. Yao, "Coevolution in Iterated Prisoner's Dilemma with Intermediate Levels of Cooperation: Application to Missile Defense," International Journal of Computational Intelligence Applications, Vol. 2, No. 1, pp.83-107, 2002.

C. H. Yong and R. Miikkulainen, "Cooperative Coevolution of Multi-agent Systems," University of Texas, Austin, USA, Tech. Rep. AI01-287,2001.

C. S. Ong, H. Y. Quek, K. C. Tan and A. Tay, "Discovering Chinese Chess Strategies Through Coevolutionary Approaches," IEEE Symposium on Computational Intelligent and Games, pp.360-367, 2007.

J. Wang, Y. H. Luo, D. N. Qiu and X. H. Xu, "Adaptive Genetic Algorithm's Implement on Evaluation Function in Computer Chinese Chess," Proceeding of ISCIT 2005, pp.1206-1209.

W. P. Lee, L. J. Liu and J. A. Chiou, "A Component-Based Framework to Rapidly Prototype Online Chess Game for Home Entertainment," IEEE International Conference on System, Man and Cybernetics, pp.4011-4016, 2006.

W. Zhou, J. L. Zhang and Y. Z. Wang, "The Iterative Sort Search Techniques Based on Percentage Evaluation," Chinese Control and Decision Conference (CCDC 2008), pp.5263-5266.

Kuo-Lan Su, Sheng-Ven Shiau, Jr-Hom Guo and Chih-Wei Shiau, "Mobile Robot Based Onlin Chinese Chess Game," The Fourth International Conference on Innovative Computing, Information and Control, pp.63, 2009.

C. Buiu and N. Popescu, Aesthetic emotions in human-robot interaction implications on interaction design of robotic artists, International Journal of Innovative Computing, Information and Control, Vol.7, No. 3, pp.1097-1107, 2011.

T. Song, X. Yan, A Liang and K. Chen, A distributed bidirectional auction algorithm for multi-robot coordination, IEEE International Conference on Research Challenges in Computer Science, pp.145-148, 2009.

R. Grabowski and L. Navarro-Serment, C. Paredis, and P. Khosla, Heterogeneous teams of modular robots for mapping and exploration, Autonomous Robots, Vol.8, No. 3, pp.293-308, 2000.

E. J. Chung, Y. S. Kwon, J. T. Seo, J. J. Jeon, andH. Y. Lee, Development of a multiple mobile robotic system for team work, SICE-ICASE International Joint Conference, pp.4291-4296, 2006.

T. Balch, and R. Arkin, Behavior-based formation control for multirobot teams, IEEE Trans. on Robotics and Automation, Vol.14, No. 6, pp.926-939, 1998.

R. Alami, S. Fleury, M. Herrb, F. Ingrand, and F. Robert, Multi-robot cooperation in the MARTHA project, IEEE Robotics and Automation Magazine, Vol. 5, No. 1, pp.36-47, 1998.

Y. Cao *et al*, Cooperative mobile robotics: antecedents and directions, Autonomous Robots, Vol.4, No. 1, pp.7-27, 1997.

J. H. Guo and K. L. Su, Ant system based multi-robot path planning, ICIC Express Letters, Part B: Applications, Vol. 2, No. 2, pp. 493-498, 2011.

W. Burgard, M. Moors *et al*, Coordinated Multi-Robot Exploration, IEEE Transaction on Robotics, Vol. 21, No. 3, pp.376-386, 2005.

K. L. Su, C. Y. Chung, Y. L. Liuao and J. H. Guo, A* searching algorithm based path planning of mobile robots, ICIC Express Letters, Part B: Applications, Vol. 2, No. 1, pp. 273-278, 2011.

Y. Saber and T. Senjyu (2007) Memory-bounded ant colony optimization with dynamic programming and A* local search for generator planning, IEEE Trans. on Power System, Vol.22, No. 4, pp.1965-1973

11

Motion Planning for Mobile Robots Via Sampling-Based Model Predictive Optimization

Damion D. Dunlap[1], Charmane V. Caldwell[2], Emmanuel G. Collins[2], Jr. and Oscar Chuy[2]

[1]*Naval Surface Warfare Center - Panama City Division*
[2]*Center for Intelligent Systems, Control and Robotics (CISCOR)*
FAMU-FSU College of Engineering
U.S.A

1. Introduction

Path planning is a method that determines a path, consecutive states, between a start state and goal state, LaValle (2006). However, in motion planning that path must be parameterized by time to create a trajectory. Consequently, not only is the path determined, but the time the vehicle moves along the path. To be successful at motion planning, a vehicle model must be incorporated into the trajectory computation. The motivation in utilizing a vehicle model is to provide the opportunity to predict the vehicle's motion resulting from a variety of system inputs. The kinematic model enforces the vehicle kinematic constraints (i.e. turn radius, etc.), on the vehicle that limit the output space (state space). However, the kinematic model is limited because it does not take into account the forces acting on the vehicle. The dynamic model incorporates more useful information about the vehicle's motion than the kinematic model. It describes the feasible control inputs, velocities, acceleration and vehicle/terrain interaction phenomena. Motion planning that will require the vehicle to perform close to its limits (i.e. extreme terrains, frequent acceleration, etc.) will need the dynamic model. Examples of missions that would benefit from using a dynamic model in the planning are time optimal motion planning, energy efficient motion planning and planning in the presence of faults, Yu et al. (2010).

Sampling-based methods represent a type of model based motion planning algorithm. These methods incorporate the system model. There are current sampling-based planners that should be discussed: The Rapidly-Exploring Random Tree (RRT) Planner, Randomized $A^\star$ ($RA^\star$) algorithm, and the Synergistic Combination of Layers of Planning (SyCLoP) multi-layered planning framework. The Rapidly-Exploring Random Tree Planner was one of the first single-query sampling based planners and serves as a foundation upon which many current algorithms are developed. The RRT Planner is very efficient and has been used in many applications including manipulator path planning, Kuffner & LaValle. (2000), and robot trajectory planning, LaValle & Kuffner (2001). However, the RRT Planner has the major drawback of lacking any sort of optimization other than a bias towards exploring the search space. The $RA^\star$ algorithm, which was designed based on the RRT Planner, addresses this drawback by combining the RRT Planner with an $A^\star$ algorithm. The SyCLoP framework is

presented because it not only represents a very current sampling-based planning approach, but the framework is also one of the few algorithms to directly sample the control inputs.
Originally, this research began by applying nonlinear model predictive control (NMPC), implemented with sequential programming, to generate a path for an autonomous underwater vehicle (AUV), Caldwell et al. (2006). As depicted in Fig. 1, NMPC was attractive because it is an online optimal control method that incorporates the system model, optimizes a cost function and includes current and future constraints all in the design process. These benefits made planning with NMPC promising, but there were weaknesses of NMPC that had to be addressed. Since MPC must solve the optimization problem online in real-time, the method was limited to slow systems. Additionally, even though models were used in the design process, linear models where typically used in order to avoid the local minima problem that accompany the use of nonlinear models. In order to exploit the benefits of MPC these issues had to be addressed.

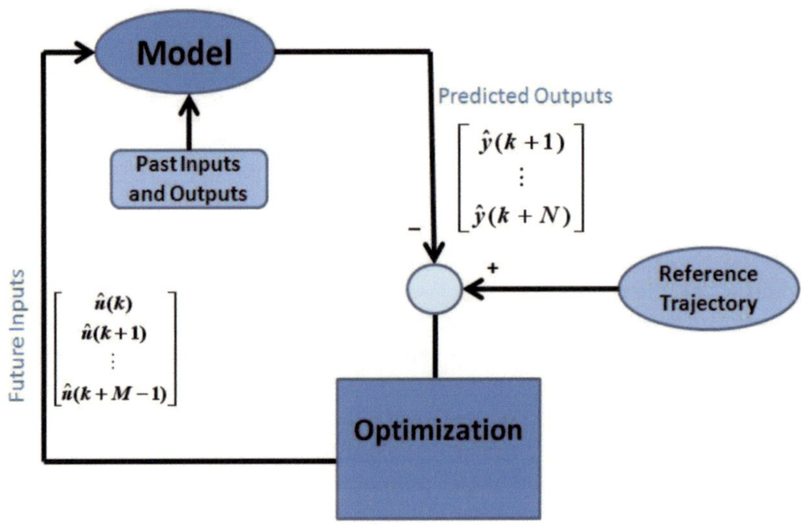

Fig. 1. The stages of the MPC algorithm.

Since the robotics and AI communities had the same goal for planning but have different approaches that tend to yield computationally efficient algorithms, it was decided to integrate these various concepts to produce a new enhanced planner called Sampling Based Model Predictive Control (SBMPC). The concept behind SBMPC was first presented in Dunlap et al. (2008). Instead of utilizing traditional numerical methods in the NMPC optimization phase in Fig. 1, Sampling Based Model Predictive Optimization (SBMPO) uses A^* type optimization from the AI community. This type of graph search algorithm results in paths that do not become stuck in local minima. In addition, the idea of using sampling to consider only a finite number of solutions comes from robotic motion planning community. Sampling is the mechanism used to trade performance for computational efficiency. Instead of sampling in the output space as traditional sampling based planning methods, SBMPC follows the view of traditional MPC and SyCLoP, which samples the input space. Thus, SBMPC draws from the control theory, robotics and AI communities.
Section 2 of this chapter will present the novel SBMPC algorithm in detail and compare Sampling Based Model Predictive Optimization and traditional Sampling based methods. Section 3 provides simulation results utilized on an AUV kinematic model. Section 4 presents

results of both an AUV and an unmanned ground vehicle (UGV) that perform steep hill climbing. An evaluation of SBMPO tuning parameters on the computation time and cost is presented in Section 5. Finally, Section 6 concludes and presents future work.

2. SBMPC algorithm

This section provides the SBMPC Algorithm and the comparison of SBMPO to other traditional sampling based methods. However, first the variables used in the algorithm are described. The SBMPO algorithm and terms follow closely with Lifelong Planning $A^\star$ ($LPA^\star$), Koenig et al. (2004). However, the variation is in the Generate Neighbor algorithm which generates the next state by integrating the system model and considering constraint violations. All the components of SBMPC are described in this Section, but the later simulation results in Section 3 and 4 utilize only the SBMPO and Generate Neighbors algorithms.

2.1 SBMPC variables

SBMPC operates on a dynamic directed graph G which is a set of all nodes and edges currently in the graph. $SUCC(n)$ represents the set of successors (children) of node $n \in G$ while $PRED(n)$ denotes the set of all predecessors (parents) of node $v \in G$. The cost of traversing from node n' to node $n \in SUCC(n')$ is denoted by $c(n',n)$, where $0 < c(n',n) < \infty$. The optimization component is called Sampling Based Model Predictive Optimization and is an algorithm that determines the optimal cost (i.e. shortest path, shortest time, least energy, etc.) from a start node $n_{start} \in G$ to a goal node $n_{goal} \in G$.
The start distance of node $v \in G$ is given by $g^\star(v)$ which is the cost of the optimal path from the given start node v_{start} to the current node v. SBMPC maintains two estimates of $g^\star(v)$. The first estimate $g(v)$ is essentially the current cost from v_{start} to the node v while the second estimate, $rhs(v)$, is a one-step lookahead estimate based on $g(v')$ for $v' \in PRED(v)$ and provides more information than the estimate $g(v)$. The $rhs(v)$ value satisfies

$$rhs(v) = \begin{cases} 0, & \text{if } v = v_{start} \\ \min_{v' \in PRED(v)} (g(v') + c(v',v)), & \text{otherwise.} \end{cases} \tag{1}$$

A node v is locally consistent iff $g(v) = rhs(v)$ and locally inconsistent iff $g(v) \neq rhs(v)$. If all nodes are locally consistent, then $g(v)$ satisfies (1) for all $v \in G$ and is therefore equal to the start distance. This enables the ability to trace the shortest path from v_{start} to any node v by starting at v and traversing to any predecessor v' that minimizes $g(v') + c(v',v)$ until v_{start} is reached.
To facilitate fast re-planning, SBMPC does not make every node locally consistent after an edge cost change and instead uses a heuristic function $h(v, v_{goal})$ to focus the search so that it only updates $g(v)$ for nodes necessary to obtain the optimal cost. The heuristic is used to approximate the goal distances and must follow the triangle inequality: $h(v_{goal}, v_{goal}) = 0$ and $h(v, v_{goal}) \leq c(v,v') + h(v', v_{goal})$ for all nodes $v \in G$ and $v' \in SUCC(s)$. SBMPO employs the heuristic function along with the start distance estimates to rank the priority queue containing the locally inconsistent nodes and thus all the nodes that need to be updated in order for them to be locally consistent. The priority of a node is determined by a two component key vector:

$$key(v) = \begin{pmatrix} k_1(v) \\ k_2(v) \end{pmatrix} = \begin{pmatrix} \min(g(v), rhs(v)) + h(v, v_{goal}) \\ \min(g(v), rhs(v)) \end{pmatrix} \tag{2}$$

where the keys are ordered lexicographically with the smaller key values having a higher priority.

2.2 SBMPC algorithm

The SBMPC algorithm is comprised of three primary methods: Sampling Based Model Predictive Control, Sampling Based Model Predictive Optimization and Generate Neighbor. The main SBMPC algorithm follows the general structure of MPC where SBMPC repeatedly computes the optimal path between the current state $x_{current}$ and the goal state x_{goal}. After a single path is generated, $x_{current}$ is updated to reflect the implementation of the first control input and the graph G is updated to reflect any system changes. These steps are repeated until the goal state is reached.

The second algorithm SBMPO is the optimization phase of SBMPC that provides the prediction paths. SBMPO repeatedly generates the neighbors of locally inconsistent nodes until v_{goal} is locally consistent or the key of the next node in the priority que is not smaller than $key(v_{goal})$. This follows closely with the ComputeShortestPath algorithm of LPA^* Koenig et al. (2004). The node, v_{best}, with the highest priority (lowest key value) is on top of the priority que. The algorithm then deals with two potential cases based on the consistency of the expanded node v_{best}. If the node is locally overconsistent, $g(v) > rhs(v)$, the g-value is set to $rhs(v)$ making the node locally consistent. The successors of v are then updated. The update node process includes recalculating $rhs(v)$ and key values, checking for local consistency and either adding or removing the node from the priority que accordingly. For the case when the node is locally underconsistent, $g(v) < rhs(v)$, the g-value is set to ∞ making the node either locally consistent or overconsistent. This change can affect the node along with its successors which then go through the node update process.

The Generate Neighbor algorithm determines the successor nodes of the current node. In the input space, a set of quasi-random samples are generated that are then used with a model of the system to predict a set of paths to a new set of outputs (nodes) with $x_{current}$ being the initial condition. The branching factor B (sampling number) determines the number of paths that will be generated and new successor nodes. The path is represented by a sequence of states $x(t)$ for $t = t_1, t_1 + \Delta t, \cdots, t_2$, where Δt is the model step size. The set of states that do not violate any state or obstacle constraints is called $\mathbf{X}_{free}$. If $x(t) \in \mathbf{X}_{free}$, then the new neighbor node x_{new} and the connecting edge can be added to the directed graph, G. If $x_{new} \in STATE\_GRID$, then the node currently exists in the graph and only the new path to get to the existing node needs to be added.

Algorithm 1 Sampling Based Model Predictive Control

```
x_current ⇐ start
repeat
    SBMPO ()
    Update system state, x_current
    Update graph, G
until the goal state is achieved
```

2.3 Comparison of SBMPO and traditional sampling based methods

This section discusses the conceptual comparison between SBMPO and traditional Sampling-based methods. Similar to many other planning methods, there have been many variants of the sampling based methods that seek to improve various aspects of their

Algorithm 2 SBMPO ( )

while $PRIORITY.TopKey() < v_{goal}.key \parallel v_{goal}.rhs \neq v_{goal}.g$ **do**
  $v_{best} \Leftarrow PRIORITY.Top()$
  Generate_Neighbors (v_{best}, **B**)
  if $v_{best}.g > v_{best}.rhs$ **then**
    $v_{best}.g = v_{best}.rhs$
    for all $v \in SUCC_{v_{best}}$ **do**
      Update the node, v
    end for
  else
    $v_{best}.g = \infty$
    for all $v \in SUCC(v_{best}) \cup v_{best}$ **do**
      Update the node, v
    end for
  end if
end while

Algorithm 3 Generate_Neighbors (Vertex v, Branching **B**)

for $i = 0$ to **B** **do**
  Generate sampled input, $u \in \mathbb{R}^u \cap \mathbf{U}_{free}$
  for $t = t_1 : dt_{integ} : t_2$ **do**
    Evaluate model: $x(t) = f(v.x, u)$
    if $x(t) \notin \mathbf{X}_{free}(t)$ **then**
      Break
    end if
  end for
  $x_{new} = x(t_2)$
  if $x_{new} \in STATE\_GRID$ and $x_{new} \in \mathbf{X}_{free}$ **then**
    Add $Edge(v.x, x_{new})$ to graph, **G**
  else if $x_{new} \in \mathbf{X}_{free}$ **then**
    Add $Vertex(x_{new})$ to graph, **G**
    Add $Edge(v.x, x_{new})$ to graph, **G**
  end if
end for

performance. It is not possible to cover every variant, but the purpose of this section is to put in perspective how SBMPO is a variant of the traditional sampling-based method.

2.3.1 Traditional sampling based methods

Examples of traditional sampling based motion planning algorithms include RRTs, LaValle (1998), and probability roadmaps,. A common feature of each of these algorithms is they work in the output space of the robot and employ various strategies for generating samples (i.e., random or pseudo-random points). In essence, as shown in Fig. 2, sampling based motion planning methods work by using sampling to construct a tree that connects the root (initial state) with a goal region.
Most online sampling based planning algorithms follow this general framework:

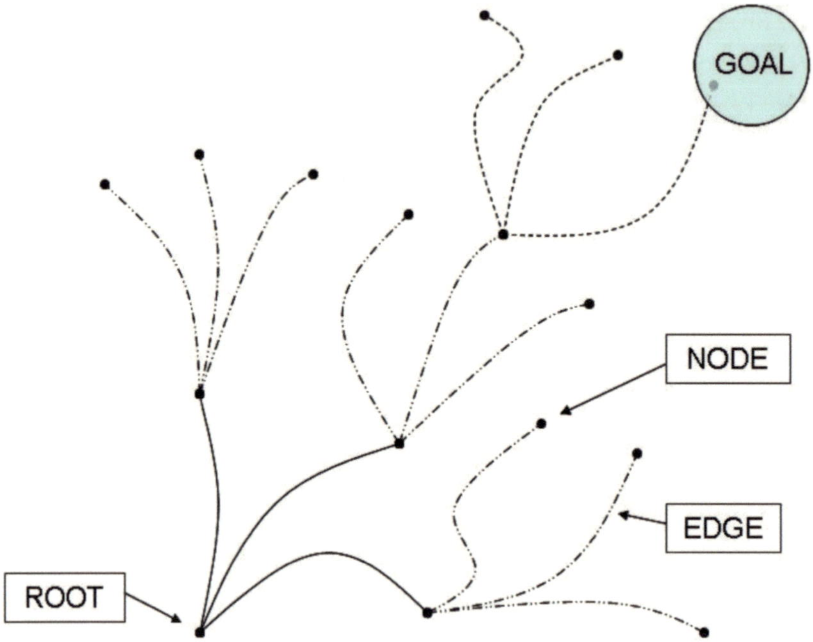

Fig. 2. A tree that connects the root with a goal region.

1. **Initialize:** Let $G(V;E)$ represent a search graph where V contains at least one vertex (i.e., node), typically the start vertex and E does not contain any edges.
2. **Vertex Selection Method (VSM):** Select a vertex u in V for expansion.
3. **Local Planning Method (LPM):** For some $u_{new} \in C_{free}$ (free states in the configuration space) and attempt to generate a path $\tau_s : [0,1] \rightarrow: \tau(0) = u$ and $\tau(1) = u_{new}$. The path must be checked to ensure that no constraints are violated. If the LPM fails, then go back to Step 2.
4. **Insert an Edge in the Graph:** Insert τ_s into E, as an edge from u to u_{new}. Insert u_{new} into V if it does not already exist.
5. **Check for a Solution:** Check G for a solution path.
6. **Return to Step 2:** Repeat unless a solution has been found or a failure condition has been met.

The model is part of the local planning method (LPM), which determines the connection between the newly generated sample and the existing graph. Essentially, it is a two-point value boundary problem.

2.3.2 Similarities of SBMPO and traditional sampling based methods

There are some similarities that both SBMPO and traditional sampling methods share.

2.3.2.1 Sampling

As its name implies, SBMPC is dependent upon the concept of sampling, which has arisen as one of the major paradigms for robotic motion planning community, LaValle (2006). Sampling is the mechanism used to trade performance for computational efficiency. SBMPO employs quasi-random samples of the input space. Properly designed sampling algorithms provide theoretical assurances that if the sampling is dense enough, the sampling algorithm will find a solution when it exists (i.e. it has some type of completeness).

2.3.3 Differences of SBMPO and traditional sampling based methods

Since SBMPO is the outgrowth of both MPC and graph search algorithms, there are some fundamental differences in SBMPO and traditional sampling based methods.

2.3.3.1 Input sampling

There are two primary disadvantages to using output (i.e., configuration space) sampling as is commonly done in traditional sampling based methods. The first limitation lies within the VSM, where the algorithm must determine the most ideal node to expand. This selection is typically made based on the proximity of nodes in the graph to a sampled output node and involves a potentially costly nearest neighbor search. The LPM presents the second and perhaps more troublesome problem, which is determining an input that connects a newly sampled node to the current node. This problem is essentially a two-point boundary value problem (BVP) that connects one output or state to another. There is no guarantee that such an input exists. Also, for systems with complex dynamics, the search itself can be computationally expensive, which leads to a computationally inefficient planner. A solution to the problem is to introduce input sampling. The concept of input sampling is not new and has been integrated into methods like the SyCLoP algorithm, Plaku et al. (2010). When the input space is sampled as proposed in this chapter, the need for a nearest-neighbor search is eliminated, and the LPM is reduced to the integration of a system model, and therefore, only generates outputs that are achievable by the system. Sampling the control inputs directly also prevents the need to determine where to connect new samples to the current graph and therefore avoid costly nearest-neighbor searches.

In order to visualize this concept, consider an Ackerman steered vehicle at rest that has position (x, y) and orientation θ, which are the outputs of the kinematic model. The model restricts the attainable outputs. All the dots in Fig. 3 are output nodes obtained from sampling the output space even though only the dots on the mesh surface can physically be obtained by the vehicle. There are a larger number of dots (sampled outputs) in the output space that do not lie in the achievable region (mesh surface). This means those sampled outputs are not physically possible, so traditional sample based methods would have to start the search over. This leads to an inefficient search that can substantially increase the computational time of the planner. The intersection of the grid lines in Fig. 3 correspond to the points in output space generated by a uniform sampling of the model inputs, the left and right wheel velocities. In essence, sampling in the input space leads to more efficient results since each of the corresponding dots in the output space is allowed by the model.

2.3.3.2 Implicit state grid

Although input sampling avoids two of the primary computational bottle-necks of sampling-based motion planning, there is also a downside of input sampling. Input sampling has not been used in most planning research, because it is seen as being inefficient. This type of sampling can result in highly dense samples in the output space since input sampling does not inherently lead to a uniformly discretized output space, such as a uniform grid. This problem is especially evident when encountering a local minimum problem associated with the $A^\star$ algorithm, which can occur when planning in the presence of a large concave obstacle while the goal is on the other side of the obstacle. This situation is considered in depth for discretized 2D path planning in the work of Likhachev & Stentz (2008), which discusses that

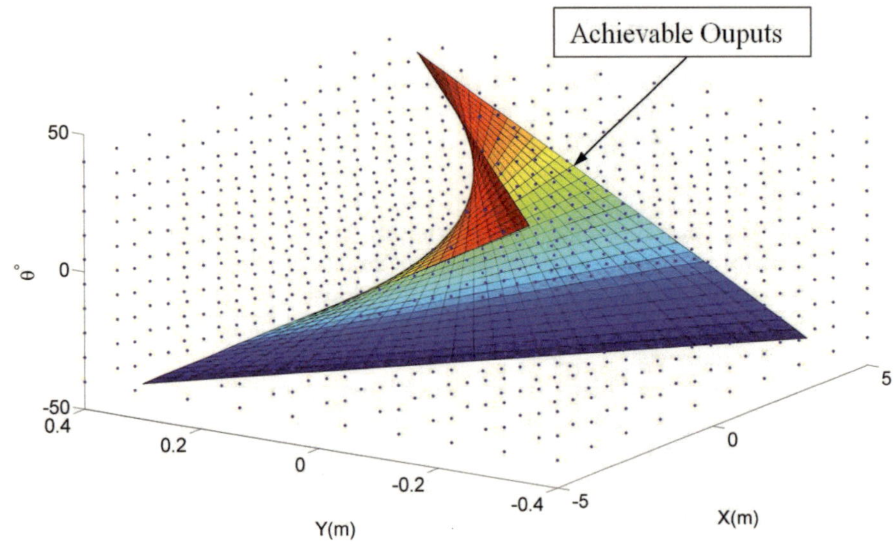

Fig. 3. Illustration of the potential inefficiency of sampling in the output space.

the $A^{\star}$ algorithm must explore all the states in the neighborhood of the local minimum, shown as the shaded region of Fig. 4, before progressing to the final solution. The issue that this presents to input sampling methods is that the number of states within the local minimum is infinite because of the lack of a discretized output space.

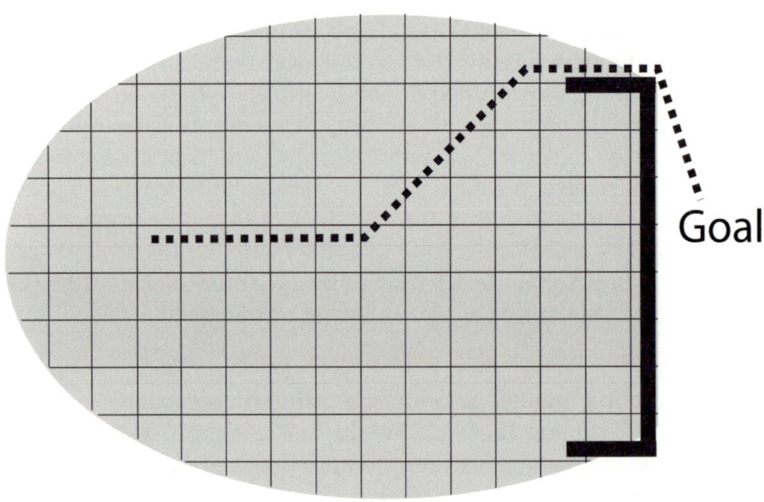

Fig. 4. Illustration of the necessity of an implicit state grid.

The second challenge resulting from the nature of input sampling as well as the lack of a grid is that the likelihood of two outputs (states) being identical is extremely small. All $A^{\star}$-like algorithms utilize Bellman's optimality principle to improve the path to a particular output by updating the paths through that output when a lower cost alternative is found. This feature is essential to the proper functioning of the algorithm and requires a mechanism to identify when outputs (states) are close enough to be considered the same. The scenario presented in Fig. 5 is a situation for which the lack of this mechanism would generate an inefficient path. In this situation, node v_1 is selected for expansion after which the lowest cost node is v_3. The *implicit state grid* then recognizes that v_2 and v_3 are close enough to be considered the same and updates the path to their grid cell to be path c since $c < a + b$.

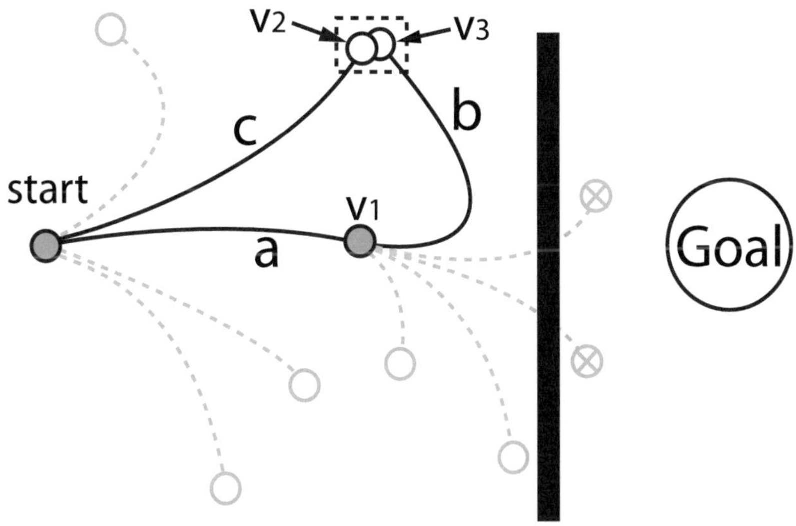

Fig. 5. Illustration of the necessity of an implicit state grid.

The concept of an *implicit state grid*, Ericson (2005), is introduced as a solution to both of the challenges generated by input sampling. The implicit grid ensures that the graph generated by the SBMPC algorithm is constructed such that only one active output (state) exists in each grid cell, limiting the number of nodes that can exist within any finite region of the output space. In essence, the *implicit state grid* provides a discretized output space. It also allows for the efficient storage of potentially infinite grids by only storing the grid cells that contain nodes, which is increasingly important for higher dimensional problems, Ericson (2005). The resolution of the grid is a significant factor in determining the performance of the algorithm with more fine grids in general requiring more computation time, due to the increased number of outputs, with the benefit being a more optimal solution. Therefore, the grid resolution is a useful tuning tool that enables SBMPC to effectively make the trade off between solution quality and computational performance.

2.3.3.3 Goal directed optimization

There is a class of discrete optimization techniques that have their origin in graph theory and have been further developed in the path planning literature. In this study these techniques will be called *goal-directed optimization* and refer to graph search algorithms such as Dijkstra's algorithm and the $A^{\star}$, $D^{\star}$, and $LPA^{\star}$ algorithms Koenig et al. (2004); LaValle (2006). Given a graph, these algorithms find a path that optimizes some cost of moving from a start node to some given goal. In contrast to discrete optimization algorithms such as branch-and-bound optimization Nocedal & Wright (1999), which "relaxes" continuous optimization problems, the goal-directed optimization methods are inherently discrete, and have often been used for real-time path planning.

Generally, sampling based methods such as RRTs do not incorporate any optimization and terminate when an initial feasible solution is determined. In essence, instead of determining an optimal trajectory, traditional sampling based methods only attempt to find feasible trajectories. To remedy these problems, the Randomized $A^\star$ ($RA^\star$) algorithm was introduced in Diankov & Kuffner. (2007), as a hybrid between RRTs and the $A^\star$ search algorithm. Similar to $RA^\star$, SBMPO incorporates a goal directed optimization to ensure the trajectory is optimal subject to the sampling.

Although not commonly recognized, goal-directed optimization approaches are capable of solving control theory problems for which the ultimate objective is to plan an optimal trajectory and control inputs to reach a goal (or set point) while optimizing a cost function. Hence, graph search algorithms can be applied to terminal constraint optimization problems and set point control problems. To observe this, consider the tree graph of Fig. 2. Each node of this tree can correspond to a system state, and the entire tree may be generated by integrating sampled inputs to a system model. Assume that the cost of a trajectory is given by the sum of the cost of the corresponding edges (i.e., branches), where the cost of each edge is dependent not only on the states it connects but also the inputs that are used to connect those states. The use of the system model can be viewed simply as a means to generate the directed graph and associated edge costs.

3. 3D motion planning with kinematic model

In order to demonstrate SBMPO capabilities, two local minima scenarios will be considered: 1) a concave obstacle and 2) a highly cluttered area. The purpose is to test how SBMPO handles these types of local minima environments. In this section, the kinematic model of an AUV is used for the motion planning simulations.

$$\begin{bmatrix} \dot{x} \\ \dot{y} \\ \dot{z} \end{bmatrix} = \begin{bmatrix} c\theta c\psi & s\phi s\theta c\psi - c\phi s\psi & c\phi s\theta c\psi - s\phi s\psi \\ c\theta s\psi & s\phi s\theta s\psi - c\phi c\psi & c\phi s\theta s\psi - s\phi c\psi \\ s\theta & s\phi c\theta & c\phi c\theta \end{bmatrix} \begin{bmatrix} u \\ v \\ w \end{bmatrix}$$

$$\begin{bmatrix} \dot{\phi} \\ \dot{\theta} \\ \dot{\psi} \end{bmatrix} = \begin{bmatrix} 1 & s\phi t\theta & c\phi t\theta \\ 0 & c\phi & -s\phi \\ 0 & s\phi s\theta & c\phi s\theta \end{bmatrix} \begin{bmatrix} p \\ q \\ r \end{bmatrix}, \tag{3}$$

where u, v, w are linear velocities in the local body fixed frame along the x, y, z axes, respectively and p, q, r are the angular velocities in the local body fixed frame along the x, y, z axes, respectively. The AUV posture can be defined by six coordinates, three representing the position $x_1 = (x, y, z)^T$ and three corresponding to the orientation $x_2 = (\phi, \theta, \psi)^T$, all with respect to the world frame. The constraints for the vehicle is given in Table 1.

The basic problem in each of these scenarios is to use the kinematic model to plan a minimum distance trajectory for the AUV from a start posture to a goal position while avoiding the obstacles. A 2.93 GHz Intel Core 2 Duo desktop was used for simulations in this Section.

3.1 AUV concave obstacle

As previously stated, SBMPO can handle local minimum problems that other path planning methods have difficulties handling. A local minima problem is a possible scenario a vehicle

Inputs	min	max	States	min	max
u	0 m/s	2 m/s	x	-5 m	30 m
v	-0.1 m/s	0.1 m/s	y	-5 m	30 m
w	-0.1 m/s	0.1 m/s	z	-20 m	0 m
p	$-5^\circ/s$	$5^\circ/s$	ϕ	-15°	15°
q	$-5^\circ/s$	$5^\circ/s$	θ	-15°	15°
r	$-15^\circ/s$	$15^\circ/s$	ψ	-360°	360°

Table 1. Simulation Constraints for the 3D Kinematic Model

can be presented with that has a set of concave obstacles in front of the goal. Note that whenever a vehicle is behind an obstacle or group of obstacles and has to increase its distance from the goal to achieve the goal, it is in a local minimum position.

The simulations were run with a sampling number of 25 and grid resolution of 0.1m. The vehicle has a start posture of $(5m, 0m, -10m, 0^\circ)$ and a goal position of $(5m, 10m, -10m)$. As shown in Fig. 6, SBMPO does not get stuck behind the obstacles, but successfully determines a trajectory in 0.59s. The successful traversal is largely due to the optimization method used in SBMPO. The goal-directed optimization allows a more promising node (lower cost) to replace a higher cost node as shown in the the example in Fig 7. Goal-directed optimization can accomplish this because they compute each predicted control input separately and backs up when needed as illustrated by the iterations corresponding to the 3rd, 4th, and 5th arrays. This feature enables it to avoid local minima. It converges to the optimal predicted control sequence $\{u^*(k)\}$ (denoted by the right-most array), which is the optimal solution subject to the sampling, whereas a nonlinear programming method may get stuck at a local minimum. In addition, these results show that SBMPO's implementation of the implicit state grid helps prevent the issues with input sampling discussed in Section 2.3.3.2. Since the implicit grid is applied, it does not require significant time to explore the area around the concave obstacles.

3.2 AUV cluttered multiple obstacles

In Section 3.1, there was one local minimum in the scenario. However, some underwater environments will require the AUV to navigate around multiple cluttered obstacles. This can produce a more complex situation because now there are multiple local minima . The simulations in this section assume there were random start, goal and obstacle locations in order to represent 100 different multiple obstacle underwater environment configurations. The start locations X, Y, Z and ψ were chosen randomly in the respective ranges $[0\ 20]m$, $[0\ 1]m$, $[-12\ \ -8]m$, $[30^\circ\ 150^\circ]$, and the goal was chosen randomly in the respective ranges $[0\ 20]m$, $[19\ 20]m$, $[-12\ \ -8]m$. In addition, there were 40 randomly generated obstacles. The 100 simulation runs had a sampling number of 25. Fig. 8 exemplifies one random scenarios generated. In the scenarios SBMPO was capable of allowing the AUV to maneuver in the cluttered environment successfully reaching the goal.

For a vehicle to be truly autonomous, it must be capable of determining a trajectory that will allow it to successfully reach the goal without colliding with an obstacle. In these simulations SBMPO was 100% successful in assuring the vehicle accomplished this task. It is important to consider both SBMPO's mean computation time of 0.43s and median computation time of 0.15s to compute these trajectories. Since the scenarios generated were random, there were a few scenarios created that were more cluttered which caused a larger CPU time. This is the reason for the discrepancy between the mean and median computation times.

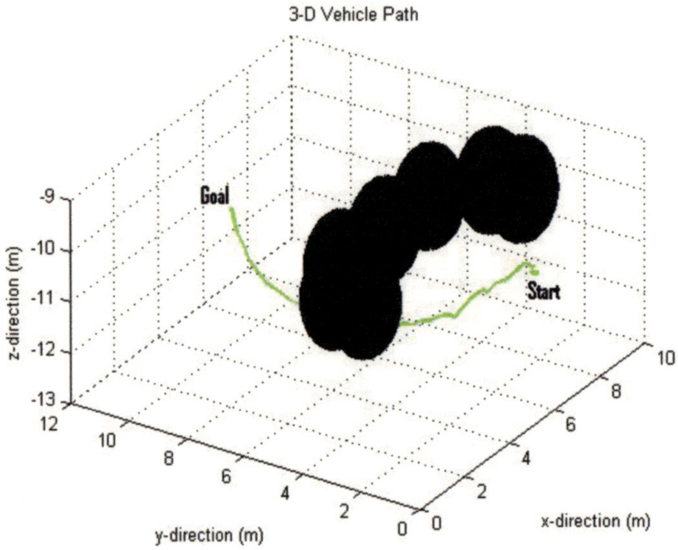

Fig. 6. A local minima scenario.

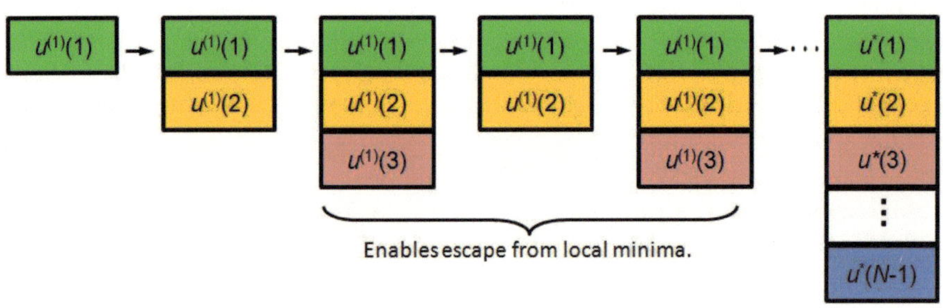

Fig. 7. Example of goal-directed optimization.

4. 3D motion planning with dynamic model

In some scenarios it is sufficient to plan using the kinematic model. However, in cases where the vehicle is pushed to an extreme, it is necessary to consider the vehicles dynamic model when planning.

4.1 Steep hill climbing

In this section, two different types of vehicles, an AUV and an UGV, consider steep hill motion planning using their respective dynamic model. The vehicle must be capable of acquiring a certain amount of momentum to successfully traverse the hill. In order to determine if the vehicle can produce the correct amount of momentum, a dynamic model is not physically capable of traversing.

4.1.1 AUV

The AUV dynamic model used for these simulations can be found in Healey & Lienard (1993). The model constraints are given in Table 2. The SBMPO parameters are in Table 3. The steep

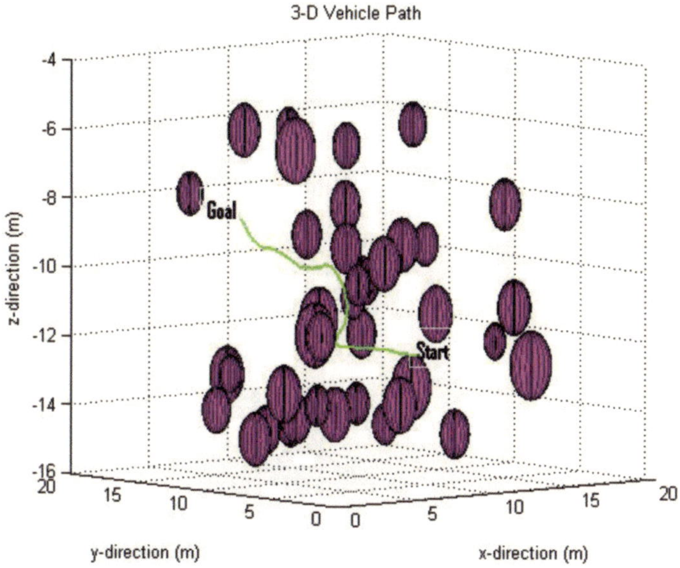

Fig. 8. A random start, goal and obstacle scenario.

hill was constructed utilizing 6 sphere obstacles constraints stacked to give a peak at $8m$. The AUV's start location was $(-4m\ 17m\ \ -18m\ 0°\ 0°\ 0°)$ and goal was $(19m\ 7m\ \ -14m)$.

States	min	max	States	min	max	Inputs	min	max
x	-30 m	130 m	u	0 m/s	2 m/s	δ_r	$-22°$	$22°$
y	-30 m/s	130 m	v	-2 m/s	2 m/s	δ_s	$-22°$	$22°$
z	-20 m	5 m	w	-2 m/s	2 m/s	δ_b	$-22°$	$22°$
ϕ	$-15°$	$15°$	p	$-5°/s$	$5°/s$	δ_{bp}	$-22°$	$22°$
θ	$-85°$	$85°$	q	$-5°/s$	$5°/s$	δ_{bs}	$-22°$	$22°$
ψ	$-360°$	$360°$	r	$-15°/s$	$15°/s$	n	0 rpm	1500 rpm

Table 2. The simulation constraints for the AUV dynamic model.

Model Time Steps	0.5s
Control updates	10s
No. of Input Samples	20
Grid Resolution	0.5

Table 3. The simulation parameters for the AUV dynamic model.

A dynamic model was utilized to determine the path over a steep hill. The AUV was not able to determine a path because the vehicle starts too close to the steep hill to gain momentum. As depicted in Fig. 9, the dynamic model was able to predict that there was not enough momentum to overcome the hill in such a short distance. Note the vehicle constraints do not allow this type of AUV to have a negative velocity which would allow the vehicle to be able to reverse in order to acquire enough momentum. As a result of the vehicle constraint, Fig 9 shows the unsuccessful path. It is not the SBMPO algorithm that cannot successfully determine a path, but the vehicle constraint (dynamic model) that predicts there was not enough momentum to overcome the hill in such a short distance. In order to demonstrate this consider the same scenario using the kinematic model in Fig 10. SBMPO does determine a path, but this is only because the kinematic model utilized does not provide all the vehicle information to correctly predict the vehicle motion. This further shows the importance of using the proper model when motion planning. The trajectory determined by the planner is only as accurate as the model used.

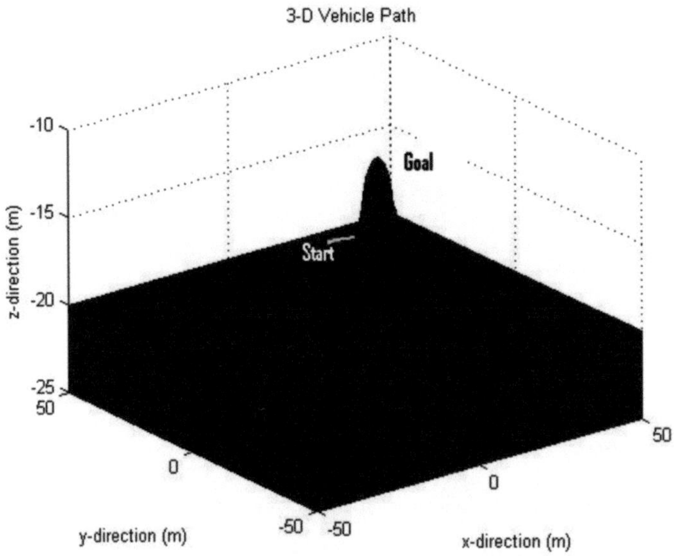

Fig. 9. The AUV dynamic model steep hill scenario.

4.1.2 UGV

This section discusses momentum-based motion planning applied to UGV steep hill climbing. Note that steep hill climbing capability of UGVs is very important to aid the completion of assigned tasks or missions. As an additional requirement, the motion planning is constrained such that the UGV has a zero velocity at the goal (e.g. top of the hill) and this has a unique application, such as reconnaissance, where UGV needs to climb and stop at the top of the hill to gather information. In this section, the momentum-based motion planning is implemented using SBMPO with UGV's dynamic model and a minimum time cost function. The minimum time cost function is employed to achieve zero velocity at the goal.

Figure 11 shows a scenario where a UGV is at the bottom of a steep hill and the task is to climb to the top of the hill. The general approach is to rush to the top of the hill. However, if the torque of the UGV and the momentum are not enough, it is highly possible that the UGV will fail to climb as shown in Fig. 11(a). An alternative approach for the UGV is to back up to gain more momentum and rush to the top of the hill as shown in Fig 11(b). The aforementioned approaches can be done using SBMPO with UGV's dynamic model. SBMPO can generate a trajectory for successful steep hill climbing, and it can also determine if the UGV needs to back up or how far the UGV needs to back up to successfully climb the hill.

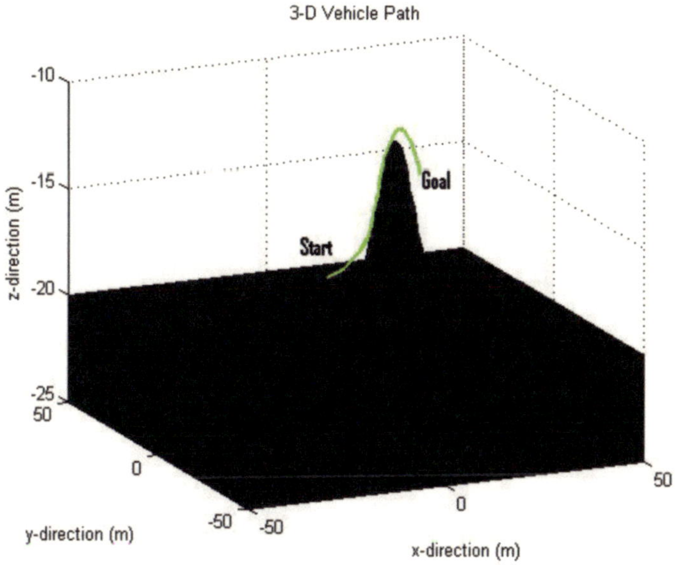

Fig. 10. The AUV kinematic model steep hill scenario.

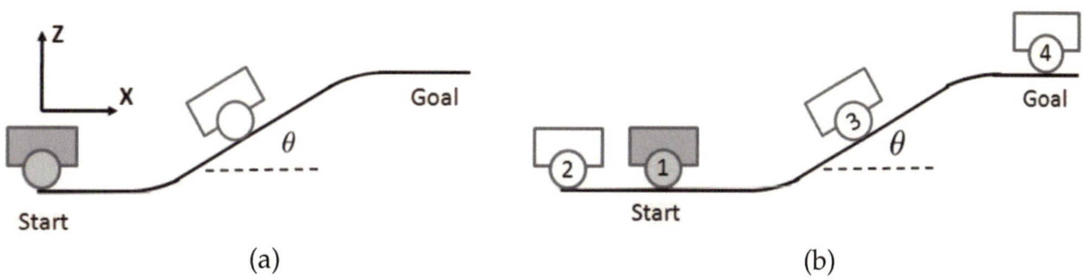

Fig. 11. A steep hill climbing scenario for a UGV (a) the UGV rushes to the top of the hill without enough momentum and torque and leads to unsuccessful climb (b) the UGV backs up to gain momentum and leads to a successful climb.

A minimum time cost function is used to implement steep hill climbing with zero velocity at the goal. To formulate the minimum time, consider a system described by

$$\ddot{q} = u; \;\; q(0) = q_0, \;\; \dot{q}(0) = \omega_0, \tag{4}$$

where u is bounded by $-a \leq u \leq b$. The state space description of (4) is given by

$$\dot{q}_1 = q_2, \;\; \dot{q}_2 = u; \; q_1(0) = q_0 \stackrel{\Delta}{=} q_{1,0}, \; q_2(0) = \omega_0 \stackrel{\Delta}{=} q_{2,0}, \tag{5}$$

where $q_1 = q$ and $q_2 = \dot{q}$. It is desired to find the minimum time needed to transfer the system from the original state $(q_{1,0}, q_{2,0})$ to the final state $(q_{1,f}, 0)$, where $q_{1,f} \stackrel{\Delta}{=} q_f$. Since the solution for transferring the system from $(q_{1,0}, q_{2,0})$ to the origin $(0, 0)$ is easily extended to the more general case by a simple change of variable, for ease of exposition it is assumed that

$$q_{1,f} = 0. \tag{6}$$

The minimum time control problem described above can be solved by forming the Hamiltonian and applying the "Minimum Principle" (often referred to as "Pontryagin's Maximum Principle") as described in Bryson & Ho (1975). In fact, the above problem is solved in Bryson & Ho (1975) for the case when the parameters a and b are given by $a = b = 1$. Generalizing these results yields that the minimum time is the solution $\underline{t}_f$ of

$$\begin{aligned} \underline{t}_f^2 - \tfrac{2q_{2,0}}{a}\underline{t}_f &= \tfrac{q_{2,0}^2+2(a+b)q_{1,0}}{ab}, \text{ if } q_{1,0} + \tfrac{q_{2,0}|q_{2,0}|}{2b} < 0, \\ \underline{t}_f^2 + \tfrac{2q_{2,0}}{b}\underline{t}_f &= \tfrac{q_{2,0}^2-2(a+b)q_{1,0}}{ab}, \text{ if } q_{1,0} + \tfrac{q_{2,0}|q_{2,0}|}{2a} > 0. \end{aligned} \tag{7}$$

The minimum time ($\underline{t}_f$) computed using (7) corresponds to a "bang-bang" optimal controller illustrated by Fig. 12, which shows switching curves that take the system to the origin using either the minimum or maximum control input (i.e., $u = -a$ or $u = b$). Depending on the initial conditions, the system uses either the minimum or maximum control input to take the system to the appropriate switching curve. For example, if $(q_{1,0}, q_{2,0})$ corresponds to point p_1 in Fig. 12, then the control input should be $u = -a$ until the system reaches point p_2 on the switching curve corresponding to $u = b$. At this point the control is switched to $u = b$, which will take the system to the origin.

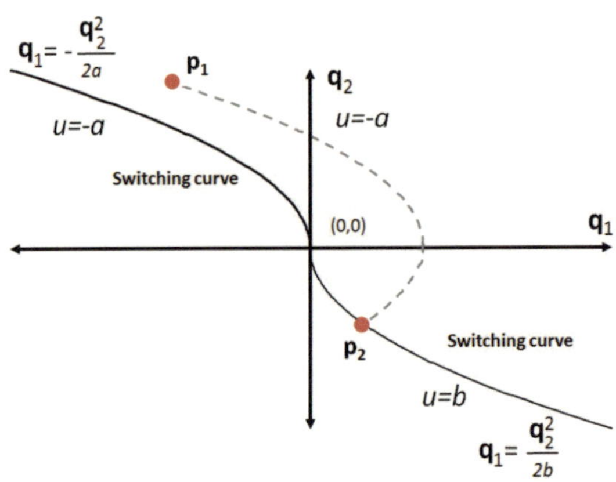

Fig. 12. Illustration of bang-bang minimum time optimal control which yields the minimum time solution $\underline{t}_f$ of (7).

To demonstrate steep hill climbing, the UGV starts at (0,0,0)[m] and the goal is located at (2.5,0,0.76)[m]. As shown in Fig. 13, the hill is described with the following parameters: $R = 1m$, $l = 0.75m$, $d = 0.4m$ and $\theta = 30^o$. A UGV dynamic model discussed in Yu et al. (2010) is used and it is given by

$$M\ddot{q} + C(\dot{q}, q) + G(q) = \tau, \tag{8}$$

where

$$-\tau_{max} < \tau < \tau_{max} \tag{9}$$

and $\tau_{max} = 10Nm$. $\ddot{q}$, $\dot{q}$, and q are respectively the wheel angular acceleration, velocity, and position, M is the inertia, $C(\dot{q}, q)$ is the friction term, and $G(q)$ is the gravity term. Based on the parameters of the hill and the UGV, the maximum required torque to climb quasi-statically the hill is 14.95Nm. This clearly shows that the UGV cannot climb without using momentum.

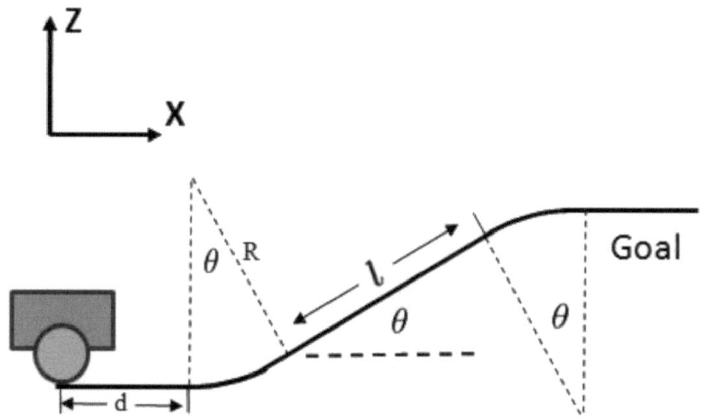

Fig. 13. The UGV steep hill parameters.

The results of the motion planning using SBMPO with the UGV's dynamic model and minimum time cost function are shown in Fig. 14. Fig. 14(a) shows the desired X-Z position of the UGV and Figs. 14(b)-(d) show respectively the desired wheel angular position, velocity, and acceleration, which are the trajectory components of the UGV's. In practice, the resulting trajectory is fed to the UGV's low-level controller for tracking. In Fig. 14(b), the desired wheel angular position starts at zero, and it goes negative (UGV backs up) before it proceeds to the goal. Fig. 14(c) shows the desired angular velocity of the wheel, and it is negative before the UGV accelerates to climb the hill. It also shows that the angular velocity at the goal is zero. The results clearly show that the UGV backs up to increase momentum, which is automatically done by SBMPO.

5. Tuning parameters

Similar to other algorithms, SBMPO has parameters that have to be tuned to guarantee optimal results. SBMPO has two main tuning parameters, the sampling number (branching factor) and grid resolution (size). Each tuning parameter has an effect on the computation time and cost. In this Section, one of the random scenarios from Section 3.2 was investigated.

5.1 Sampling number

The sample number is the number of samples that are selected to span the input space. In order to determine how the sample number effects the computation time of SBMPO the grid resolution was held constant at 0.1m, and the sample number was varied from 10 to 40 by increments of 2. Fig. 15 shows that the effect of the sampling number is nonlinear, so there is no direct relationship between the sample number and computation time. Originally it was thought that an increase in sample number would cause an increase in computation time because there would be more nodes to evaluate. However, as shown in Fig. 15 this is not completely true.

The reason for the nonlinear trend is threefold. First as shown in Fig. 15 by samples 10 and 12 , when there are not enough samples (the sample number is too low) to span the space, it can

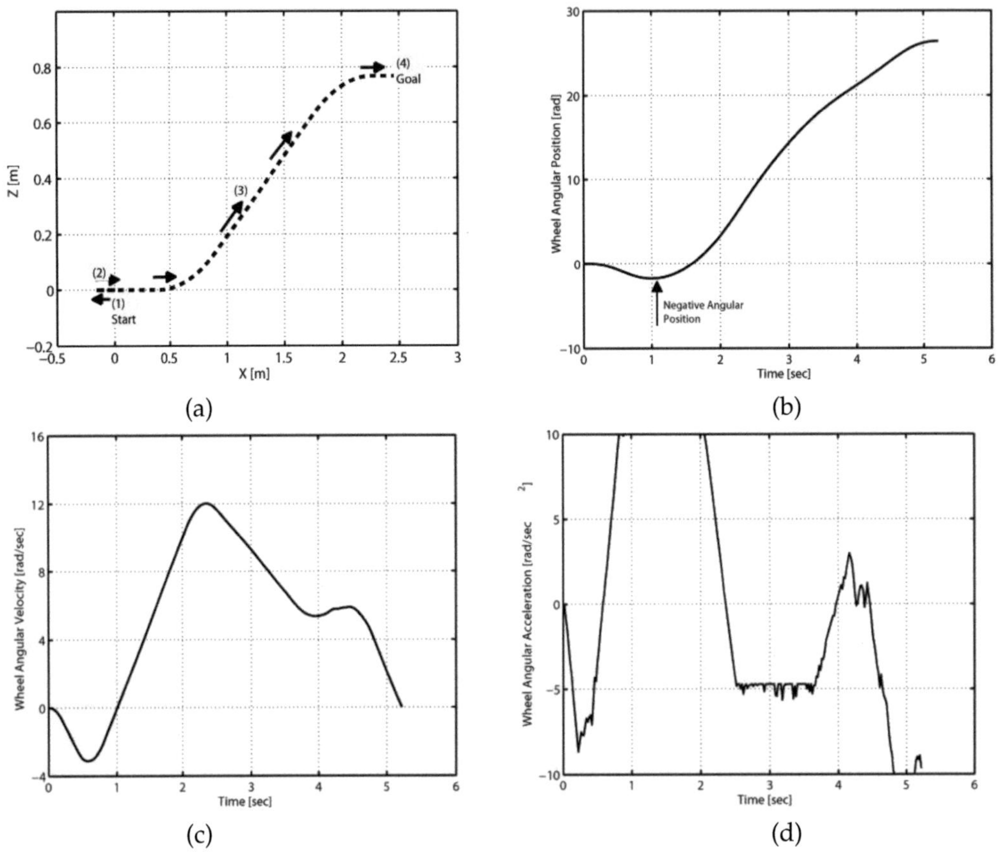

Fig. 14. (a) X-Z position of the UGV (b) wheel angular position (c) wheel angular velocity (d) wheel angular acceleration.

also increase the CPU time, because it takes more iterations (i.e. steps in SBMPO) to determine the solution. A good tuning of the parameter occurs at 14 samples which results in a smaller computation time. The second trend, as shown in Fig. 15 between samples 14 and 22, is that after a good tuning of the parameter, increasing the number of samples also increases the computation times which corresponds to the original hypothesis that an increase in sample number will result in an increase in CPU time. Lastly, a factor that contributes to the effect of the sample number on the computation time is the path produced by SBMPO. It is possible for a larger sample number to have a lower computation time when the path SBMPO generates to the goal encounters a smaller cluster of obstacles. Figs. 16a and 16b show the paths generated respectively using 26 and 36 samples. The path of Fig. 16a which has the lower sampling number takes the AUV through a cluster of obstacles, whereas the path of Fig. 16b which has the larger sample number takes a path that largely avoids the obstacles. Even though Fig. 16b corresponds to a sample number of 36, referring to Fig. 15, its computation time of 0.29s is smaller than that for Fig. 16a, which corresponds to a sample number of 26 and has a computation time of 0.5s.

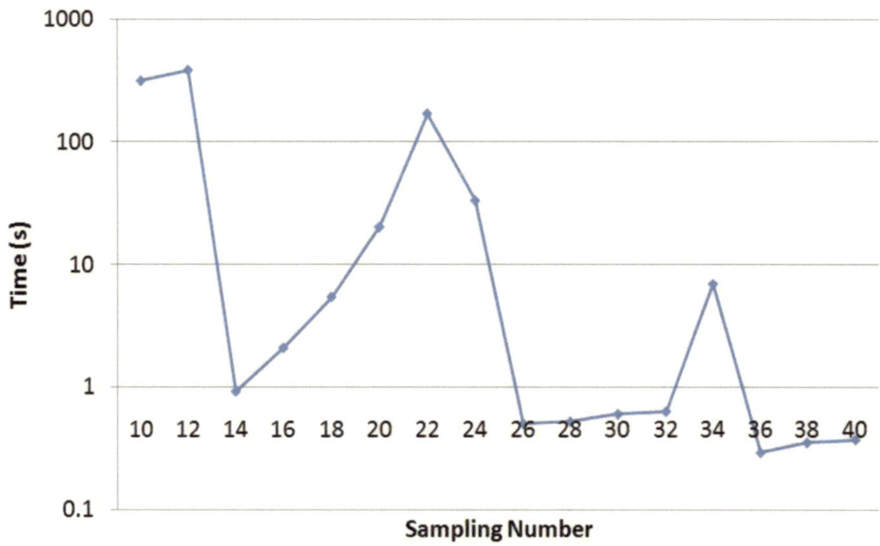

Fig. 15. The effect of sample size on computation time.

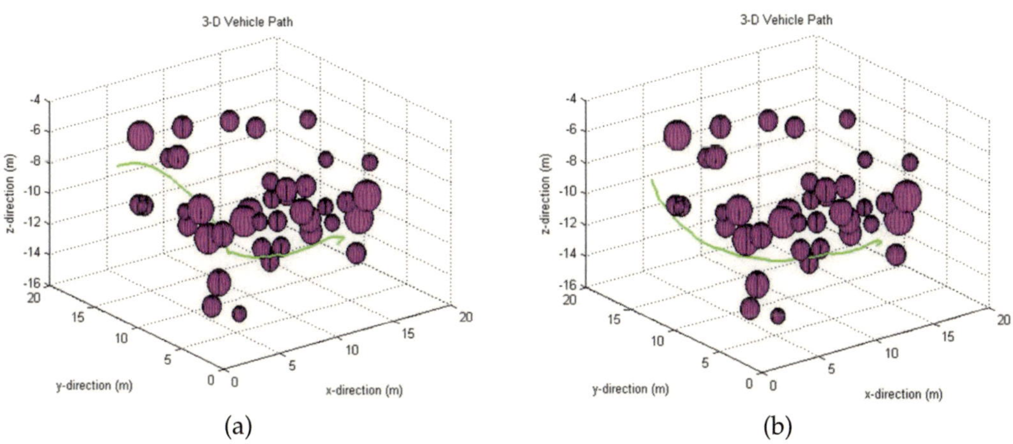

Fig. 16. (a) Scenario with sample number = 26, (b) Scenario with sample number = 36.

Fig. 17 depicts how varying the sample number effects the cost (i.e. distance). The cost is larger in the smaller sample numbers 10 and 12. Afterwards, the variation in the cost is small, which leads to more of an optimal solution.

5.2 Grid size

The grid size is the resolution of the implicit state grid. To evaluate how this tuning parameter effects the computation time, the sampling number was held constant at 25, and the grid resolution was varied between 0.02 to 0.5. Again this tuning parameter is not monotonic with respect to the computation time as depicted in Fig. 18. This shows the importance of properly tuning the algorithm. It may be thought that increasing the grid size would cause

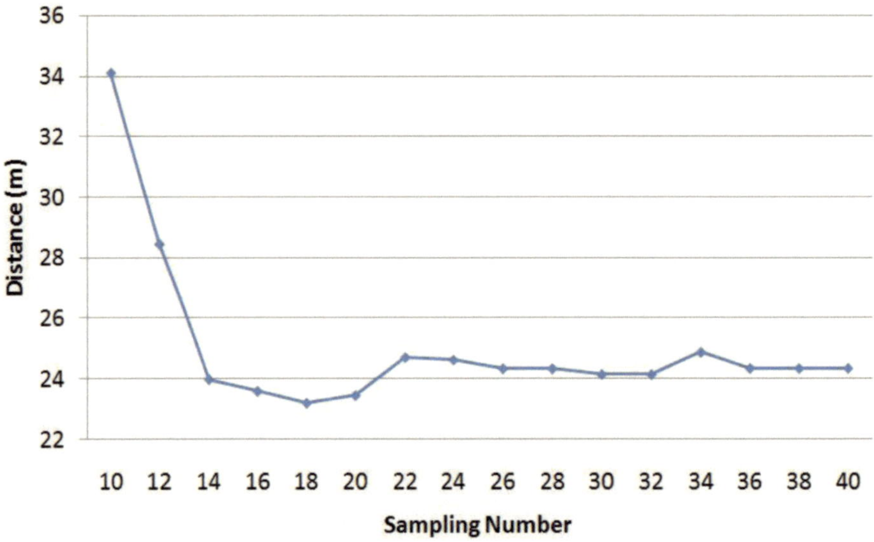

Fig. 17. The effect of sample size on path cost.

less computation. However, the opposite is true. The larger the grid size, the higher the possibility that two nodes are considered as the same state, which leads to the need for more sampling of the input space and an increased computation time. When choosing the grid resolution, it is important to recognize that increasing the grid size tends to lead to higher cost solutions as depicted in Fig. 19.

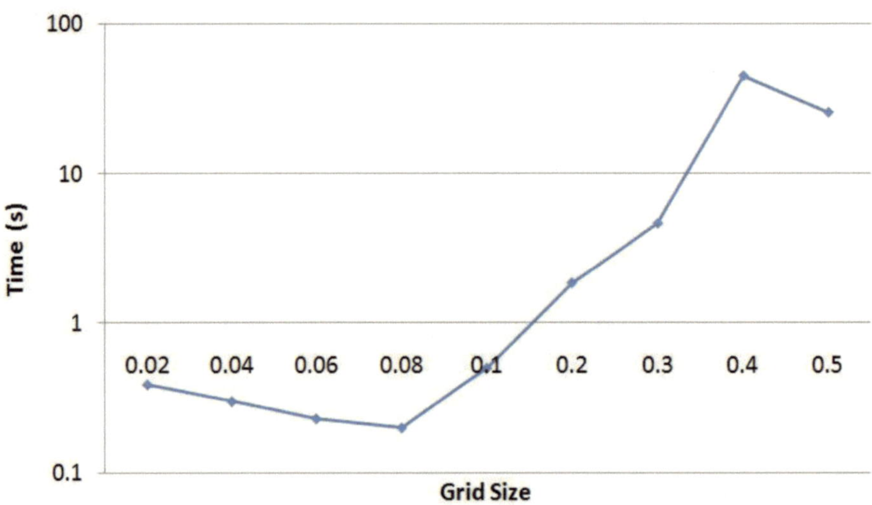

Fig. 18. The effect of grid size on computation time.

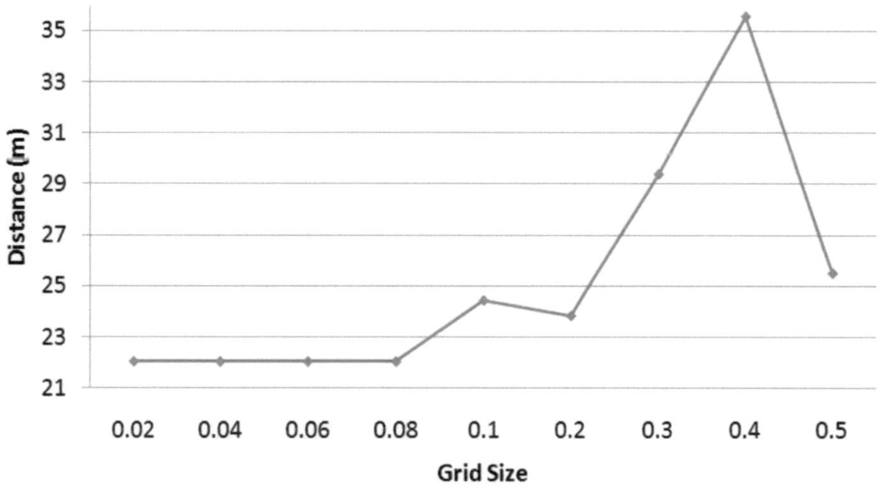

Fig. 19. The effect of grid size on path cost.

6. Conclusion

SBMPO is a NMPC method that exploits sampling-based concepts from the robotics literature along with the LPA^* incremental optimization algorithm from the AI literature to achieve the goal of quickly and simultaneously determining the control updates and paths while avoiding local minima. The SBMPO solution is globally optimal *subject to the sampling*. Sampling Based Model Predictive Optimization has been shown to effectively generate paths in the presence of nonlinear constraints and when vehicles are pushed to extreme limits. It was determined that SBMPO is only as good as the model supplied to predict the vehicle's movement. Selecting the correct model is important.

The future work is to develop the replanning feature of SBMPC. Currently, SBMPO is applied, but the ability to replan is essential to the SBMPC algorithm. SBMPC utilizes LPA^* which allows quick replanning of the path without having to completely restart the planning process when new information is obtained or changes in the environment occur. Only the nodes that are affected by a change in the environment must be reevaluated. This reduces the computation time and aids the method in achieving fast computation times. Once the replanning feature of SBMPC is in place, scenarios that include disturbance, model mismatch, unknown obstacles and moving obstacles can be examined to test more realistic situations. The algorithm will also be in a framework that is more comparable to traditional NMPC that only takes the first input and replans at every time step to create a more robust controller. Then SBMPC can be considered a general fast NMPC method that is useful for any nonlinear system or systems subject to nonlinear constraints.

7. References

Bryson, A. & Ho, Y. (1975). *Applied Optimal Control Optimization, Estimation, and Control*, HPC, New York.

Caldwell, C., Collins, E. & Palanki, S. (2006). Integrated guidance and control of AUVs using shrinking horizon model predictive control, *OCEANS Conference* .

Diankov, R. & Kuffner., J. (2007). Randomized statistical path planning, *Conference on Intelligent Robots and Systems* .

Dunlap, D. D., E. G. Collins, J. & Caldwell, C. V. (2008). Sampling based model predictive control with application to autonomous vehicle guidance, *Florida Conference on Recent Advances in Robotics* .

Ericson, C. (2005). *Real–Time Collision Detection*, Elsevier.

Healey, A. & Lienard, D. (1993). Multivariable sliding-mode control for autonomous diving and steering for unmanned underwater vehicle, *IEEE Journal of Oceanic Engineering* 18(3): 327–338.

Koenig, S., Likhachev, M. & Furcy, D. (2004). Lifelong planning A*, *Artificial Intelligence* .

Kuffner, J. J. & LaValle., S. M. (2000). Rrt-connect: An efficient approach to single-query path planning, *IEEE International Conference on Robotics and Automation* p. 995ï£¡1001.

LaValle, S. (1998). Rapidly-exploring random trees: A new tool for path planning, *Technical report*, Iowa State University.

LaValle, S. M. (2006). *Planning Algorithms*, Cambridge University Press.

LaValle, S. M. & Kuffner, J. J. (2001). Randomized kinodynamic planning, *International Journal of Robotics Research* 20(8): 378–400.

Likhachev, M. & Stentz, A. (2008). R* search, *Proceedings of the National Conference on Artificial Intelligence (AAAI)* pp. 1–7.

Nocedal, J. & Wright, S. (1999). *Numerical Optimization*, Springer, New York.

Plaku, E., Kavraki, L. & Vardi, M. (2010). Motion planning with dynamics by synergistic combination of layers of planning, *IEEE Transaction on Robotics* pp. 469–482.

Yu, W., Jr., O. C., Jr., E. C. & Hollis, P. (2010). Analysis and experimental verification for dynamic modeling of a skid-steered wheeled vehicle, *IEEE Transactions on Robotics* pp. 340 – 353.

Part 3

Mobile Robots Navigation

12

Fictitious Fuzzy-Magnet Concept in Solving Mobile–Robot Target Navigation, Obstacle Avoidance and Garaging Problems

Srđan T. Mitrović[1] and Željko M. Đurović[2]
[1]*Defense University, Military Academy*
[2]*University of Belgrade, School of Electrical Engineering*
Serbia

1. Introduction

A human driver can successfully execute different vehicle navigation tasks, such as: forward garaging, reverse garaging, parallel parking, diagonal parking, following a given trajectory, stopping at a given point, avoiding one or more obstacles, and the like. These processes are either too difficult to model accurately or too complex to model at all, while experienced drivers are able to successfully execute the tasks without any knowledge of the vehicle's mathematical model, even when they switch vehicles. Existing knowledge about how such problems are solved and the ability to express this knowledge, are benefits which can be put to good use when designing controllers for wheeled mobile robot (WMR) navigation tasks. It is well known that fuzzy logic is the most effective tool for solving a problem for which there is a human solution (Zadeh, 2001), because it offers a mathematical apparatus in the background of an inaccurate, qualitative description, which projects input variables into output variables through several stages. However, describing the tasks executed by a driver in spoken language is not a simple process. When we drive a car, we rely on our driving skills and prefer to use our hands and feet, rather than our brain. It is safe to assume that compiling rules based on driver skills will not be an easy task.

In view of the above, this chapter presents a new approach to the modeling of driver skills, based on a fuzzy model and an original virtual fuzzy-magnet concept. Driver experience is used to solve simple tasks (turn left or right, slow down or accelerate), while the entire problem is solved by specifying an appropriate number of fictitious fuzzy magnets and defining their tasks. Each fictitious fuzzy magnet is defined by its position and a set (or subset) of fuzzy rules which determine its action. For example, to solve a target navigation problem in open space requires only one fuzzy magnet which attracts the mobile robot to the target, while a more complex problem, such as bidirectional garaging of a mobile robot, requires two fuzzy magnets: one immediately in front of the garage and the other inside the garage. The first point is used to approach the garage and while the vehicle is entering the garage, it serves as a reference point for proper orientation, similar to human driving skills. The second point is both a target and a reference point. Compared to other algorithms which address this type of problem, the proposed algorithms are very simple; they are not based on the WMR model. The proposed fuzzy controllers are of the Takagi-Sugeno (T-S) type; they are manually

generated (in a manner similar to human driving skills), and have clearly-defined physical parameters.
The first section of this chapter describes the fictitious fuzzy-magnet concept and its application in solving target navigation problems in an obstacle-free environment. It further analyzes the application of a pair of fuzzy magnets in solving a garaging problem involving a differential-drive mobile robot. The fictitious fuzzy magnet concept allows navigation to the target in a single maneuver, without changing the direction of WMR travel. The symmetry of the differential-drive WMR is utilized fully, such that the algorithm provides a bidirectional solution to the WMR garaging problem. The robot is automatically parked from the end of the robot that is closer to the garage entrance. The algorithm can be applied when the control variable is of the discrete type and where there are relatively few quantization levels. The efficiency and shortfalls of the proposed algorithm are analyzed by means of both detailed simulations and multiple re-runs of a real experiment. Special attention is devoted to the analysis of different initial robot configurations and the effect of an error in the estimation of the current position of the robot on garaging efficiency.
The third section of this chapter analyzes the possibility of applying the fictitious fuzzy-magnet concept to navigate a mobile robot to a target in a workspace which contains one or more obstacles. At the preprocessing stage, the left and right side of the obstacle in the robot's line of sight are determined based on the mutual positions of the robot and the obstacle. Since controller inputs are based on the relative dimensions of the obstacle, the controller can be applied to obstacles of different shapes and sizes. Following the design of the above-described controllers, and the successful testing of navigation efficiency to the goal in a workspace which includes a single stationary obstacle, the algorithm was applied, without modification, to avoid an obstacle moving along a straight line at a constant speed equal to one-third of the maximum speed of the robot. Here the proposed algorithm demonstrated limited efficiency. The final section of the chapter proposes a method for a simple but efficient generalization of the algorithm to support a group of obstacles, using parallel data processing, illustrated by several computer simulations of WMR navigation to the target point through a group of obstacles. The disadvantages of the proposed algorithm when applied in complex environments are discussed at the end of the chapter.

2. Fictitious fuzzy magnets concept

A given point $A(x_A, y_A)$, is assumed to be in the Cartesian coordinate system and to represent the position of the fictitious fuzzy magnet FM. The FM is defined as an arranged pair comprised of its position A and an added sub-set of fuzzy rules, $FRsS$ (Fuzzy Rules subSet):

$$FM = (A, FRsS) \tag{1}$$

The $FRsS$ enables the determination of the zone of influence of the fuzzy magnet, as well as the definition of the action it causes. To establish the structure of the fuzzy rules, it is necessary to adopt input and output variables (i.e., premise variables and conclusion variables). Figure 1 shows the position of the WMR relative to the point A, which represents the position of the fictitious fuzzy magnet, FM.
Point (x_R, y_R) is the center of the WMR, and its orientation relative to the x axis is denoted by θ_R, such that the configuration of the WMR q_R is unequivocally defined by three coordinates: $q_R(x_R, y_R, \theta_R)$. The bold line denotes the front end of the robot. The angle formed by the

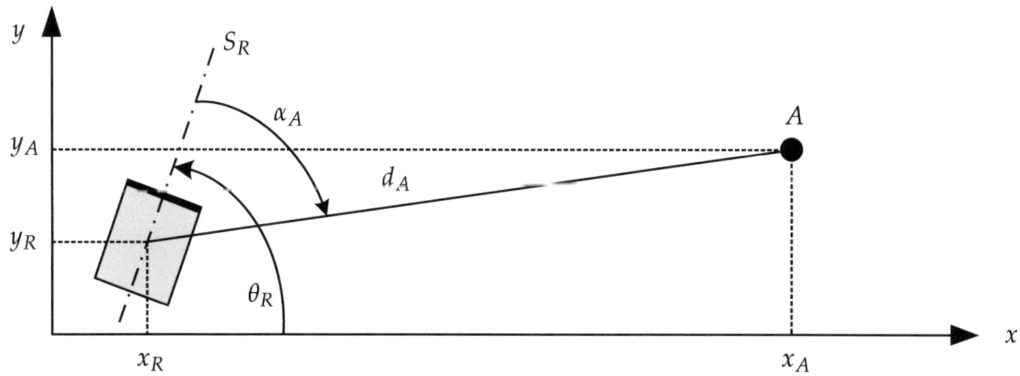

Fig. 1. Mutual spatial positions of wheeled mobile robot and fictitious fuzzy magnet.

longitudinal line of symmetry of the WMR, S_R, and the segment which connects the points (x_R, y_R) and A is denoted by α_A. If the coordinates of the WMR center are denoted by $(x_R(k), y_R(k))$ at the sample time k, the distance between the WMR and the fictitious fuzzy magnet, $d_A(k)$, is:

$$d_A(k) = \sqrt{(x_A - x_R(k))^2 + (y_A - y_R(k))^2}. \tag{2}$$

The position of the WMR at the sample time k, relative to the point A, is unequivocally determined by its orientation relative to the point A – angle $\alpha_A(k)$ and the distance $d_A(k)$, which will be adopted as input variables of the *FRsS*. In the general case, we will state that the output velocity variables are v_1 and v_2. The *FRsS* of r rules for the discrete T-S fuzzy system is defined as:

$$\begin{array}{ll} \textit{Control Rule } i: \text{If } d_A(k) \text{ is } \mu_{i1} \text{ and } \alpha_A(k) \text{ is } \mu_{i2} & \\ \qquad \text{then } v_1(k) = C_{i1} \text{ and } v_2(k) = C_{i2} & i = 1, 2, \ldots, r \end{array} \tag{3}$$

where $d_A(k)$ and $\alpha_A(k)$ are premise variables. The membership function which corresponds to the ith control rule and the jth premise variable is denoted by μ_{ij}, and C_{ij} are constants. System outputs are $v_1(k)$ and $v_2(k)$, obtained from:

$$\begin{bmatrix} v_1(k) \\ v_2(k) \end{bmatrix} = \frac{\sum\limits_{i=1}^{r} \mu_{i,1}(d_A(k)) \cdot \mu_{i,2}(\alpha_A(k)) \cdot [C_{i,1}\ C_{i,2}]^T}{\sum\limits_{i=1}^{r} \mu_{i,1}(d_A(k)) \cdot \mu_{i,2}(\alpha_A(k))} \tag{4}$$

where $\mu_{ij}(x_j(k))$ is the degree of membership of $x_j(k)$ in μ_{ij}. As is well known, the control of a differential drive mobile robot can be designed in two ways: using the speed of the left and right wheels, or using the linear and angular velocities of the robot. Consequently, the general definition of the *FRsS* can be written more precisely in two ways. The navigation of a differential drive mobile robot was designed in (Mitrović & Đurović, 2010a) based on wheel speed control, such that the output variables of the *FRsS* are the speed of the left wheel – v_L and the speed of the right wheel – v_R, such that (3-4) become:

$$\begin{array}{ll} \textit{Control Rule } i: \text{If } d_A(k) \text{ is } \mu_{i1} \text{ and } \alpha_A(k) \text{ is } \mu_{i2} & \\ \qquad \text{then } v_L(k) = C_{i1} \text{ and } v_R(k) = C_{i2} & i = 1, 2, \ldots, r \end{array} \tag{5}$$

$$\begin{bmatrix} v_L(k) \\ v_R(k) \end{bmatrix} = \frac{\sum\limits_{i=1}^{r} \mu_{i,1}(d_A(k)) \cdot \mu_{i,2}(\alpha_A(k)) \cdot [C_{i,1}\ C_{i,2}]^T}{\sum\limits_{i=1}^{r} \mu_{i,1}(d_A(k)) \cdot \mu_{i,2}(\alpha_A(k))} \tag{6}$$

while in (Mitrović & Đurović, 2010b) the output variables of the *FRsS* are linear and angular velocities of the robot v_R and ω_R, and in this case (3-4), can be written as:

$$\begin{array}{l} \textit{Control Rule } i: \text{If } d_A(k) \text{ is } \mu_{i1} \text{ and } \alpha_A(k) \text{ is } \mu_{i2} \\ \qquad\qquad\qquad \text{then } v_R(k) = C_{i1} \text{ and } \omega_R(k) = C_{i2} \end{array} \quad i = 1,\ 2, \ldots, r \tag{7}$$

$$\begin{bmatrix} v_R(k) \\ \omega_R(k) \end{bmatrix} = \frac{\sum\limits_{i=1}^{r} \mu_{i,1}(d_A(k)) \cdot \mu_{i,2}(\alpha_A(k)) \cdot [C_{i,1}\ C_{i,2}]^T}{\sum\limits_{i=1}^{r} \mu_{i,1}(d_A(k)) \cdot \mu_{i,2}(\alpha_A(k))} \tag{8}$$

2.1 Navigation in an obstacle-free environment

The application of the proposed concept will be illustrated by an example involving WMR navigation to a target in an obstacle-free environment. The WMR configuration (x_R, y_R, θ_R) and the target position (x_A, y_A) are shown in Fig. 1. We will assume the robot control inputs (v_R, ω_R) to be the outputs of the fuzzy logic controller for navigation in an obstacle free environment (FLC_{OFE}). The WMR angle relative to the target α_A and the distance to the target d_A are fuzzy controller inputs, and the corresponding membership functions are shown in Fig. 2.

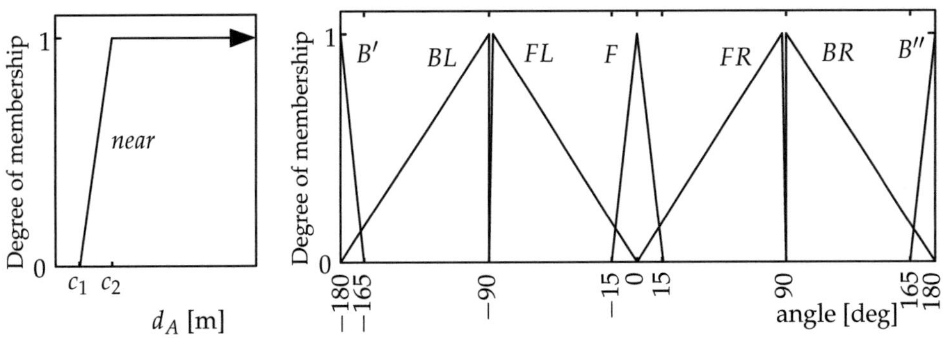

(a) Distance to the target (b) Orientation relative to the target

Fig. 2. Membership functions of input linguistic variables: (a) Orientation relative to the target – α_A, (b) Distance relative to the target d_A.

The linguistic variable α_A is defined through the following membership functions:

$$\alpha_a \,\{Front, FrontLeft, FrontRight, Back, BackLeft, Back, Right\} \tag{9}$$

or abbreviated as:

$$\alpha_A \,\{F, FL, FR, B, BL, BR\} \tag{10}$$

as shown in Figure 2(b). In order to enable bidirectional navigation, the variable α_A is strictly divided into two groups – the angles related to orientation at the front $\{F, FL, FR\}$, and the angles related to orientation at the rear $\{B, BL, BR\}$. The method proposed in this chapter analyzes mobile objects with an equal ability of maneuvering by front and rear pace; therefore, the objective of defined membership functions and fuzzy rules is to provide identical performance in both cases. For this reason, the membership function B (*Back*) is divided into two sub-functions B' and B'', in the following manner:

$$B = S(B', B'') \tag{11}$$

where S represents the operator of $S - norm$, which corresponds to the fact that the union of sets B' and B'' produces set B. Since the sets B' and B'' are disjunctive, the calculation of $S - norm$ is not important (Mitrović & Đurović, 2010a).
The distance-to-target linguistic variable d_A is described by means of a single membership function – *near* (Fig. 2(a)), and its task is to stop the WMR in close proximity to the target. This is ensured by adjustment of free parameters c_1 and c_2. The designed FLC is of the T-S type, with two inputs: the distance between the robot and the target – d_A and the orientation of the robot relative to the target – α_A, and two outputs: linear velocity v_R and angular velocity ω_R of the robot. The fuzzy controller rules for robot navigation to the target in an obstacle-free environment are shown in Table 1.

	B	BL	FL	F	FR	BR
v_R	$-V_m$	$-V_m/2$	$V_m/2$	V_m	$V_m/2$	$-V_m/2$
ω_R	0	ω_m	$-\omega_m$	0	ω_m	$-\omega_m$

Table 1. Fuzzy Logic Controller for navigation in an Obstacle–Free Environment (FLC_{OFE}) – fuzzy rules base.

The maximum linear velocity of the analyzed robot is $V_m = 0.5\,\text{m/s}$ and its maximum angular velocity is $\omega_m = 7.14\,\text{rad/s}$. The approximate dynamic model of a mobile robot (Mitrović, 2006) was used for the simulations, and the distance between robot wheels was 7 cm. Figure 3 shows WMR trajectories to the target (coordinate origin), from three initial configurations. Figure 4 shows controller outputs for the navigation scenarios depicted in Fig. 3. It is apparent from Fig. 4 that during navigation, one of the controller outputs assumes maximum values and this reduces the duration of the navigation process.

3. Bidirectional garage parking

The WMR garage parking (garaging) problem can also be viewed as a navigation problem in an environment with an obstacle, where the garage is the obstacle of a specific shape, but is also the target. When navigation problems with obstacles are solved, the goal is to achieve the ultimate configuration without colliding with the obstacle, which involves obstacle avoidance. Numerous obstacle avoidance and mobile robot navigation methods are discussed in literature. The potential field method, originally proposed by Khatib (1986) for obstacle avoidance in real-time by manipulators and mobile robots, plays a major role in solving this type of problem. A robot is navigated in a potential field which represents the sum of attractive forces originating from the target, and repulsive forces originating from the obstacles, while the analytical solution is the negative gradient of a potential function. Additionally, the resulting potential field may contain local minima where navigation ends without reaching the target, which is the greatest disadvantage of this method (Koren & Borenstein, 1991). The most frequent example of a local minimum is the case where obstacles

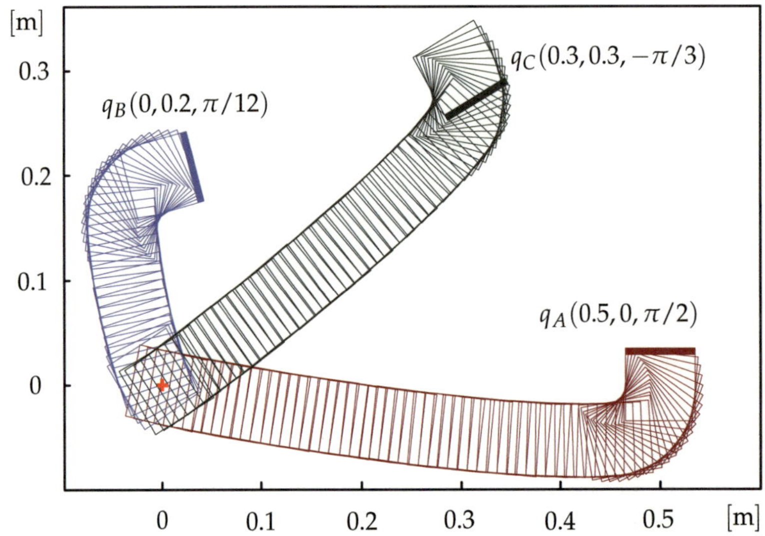

Fig. 3. Robot navigation trajectories from three initial postures (x_R, y_R, θ_R) to the target $(0, 0)$.

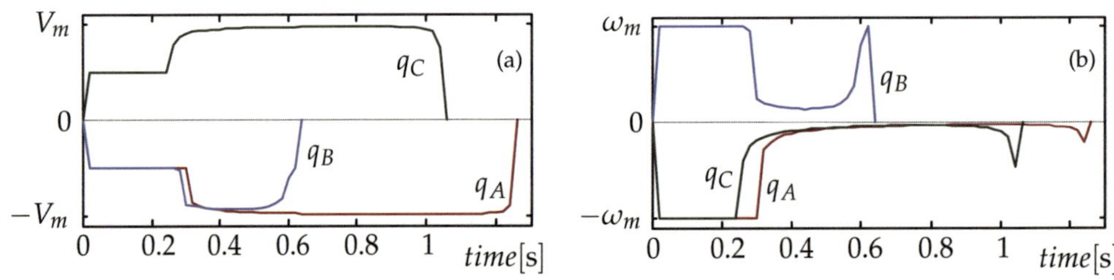

Fig. 4. Graphs for scenarios depicted in Fig. 3: (a) WMR linear velocity, (b) WMR angular velocity.

are deployed in such a way that they form a concave **U** barrier, such that the algorithm enters into an infinite loop or halts at the local minimum (Krishna & Kalra, 2001; Motlagh et al., 2009). Since the garage is a concave obstacle, the main shortfall of the potential field method could also be a solution to the garaging problem; however, analyzed literature does not provide an example of the use of the potential field method to solve mobile robot garaging problems. A probable reason for this is that the goal of garaging is to reach the target configuration in a controlled manner, with limited deviations, while the completion of navigation at the local minimum of the potential field is accompanied by oscillatory behavior of the object being parked, or the entry into an infinite loop (Koren & Borenstein, 1991), which is certainly not a desirable conclusion of the process.

The garaging/parking problem implies that a mobile robot is guided from an initial configuration $(x_{R(0)}, y_{R(0)}, \theta_{R(0)})$ to a desired one (x_G, y_G, θ_G), such that it does not collide with the garage. Figure 5 shows the garage and robot parameters; the position of the garage is defined by the coordinates of its center – C_m, while its orientation is defined by the angle of the axis of symmetry of the garage – S_G relative to the x axis, identified as β in the figure. The width of the garage is denoted by W_G and the length by L_G, and it is understood that garage dimensions enable the garaging of the observed object. The implication is that the goal

(x_G, y_G) is inside the garage and is set to coincide with the center of the garage C_m, such that $C_m(x_G, y_G)$. In the case of bidirectional garaging, the targeted configuration is not uniquely defined because the objective is for the longitudinal axis of symmetry of the robot to coincide with that of the garage, such that $\delta = 0$ or $\delta = \pi$, meaning that there are two solutions for the angle $\theta_G : \theta_{G_1} = \beta$, $\theta_{G_2} = \beta + \pi$. For reasons of efficiency, the choice between these two possibilities should provide the shortest travel distance of the mobile robot. The controller proposed in this section does not require *a priori* setting of the angle θ_G, because it has been designed in such a way that the mobile robot initiates the garaging process from the end closer to the garage.

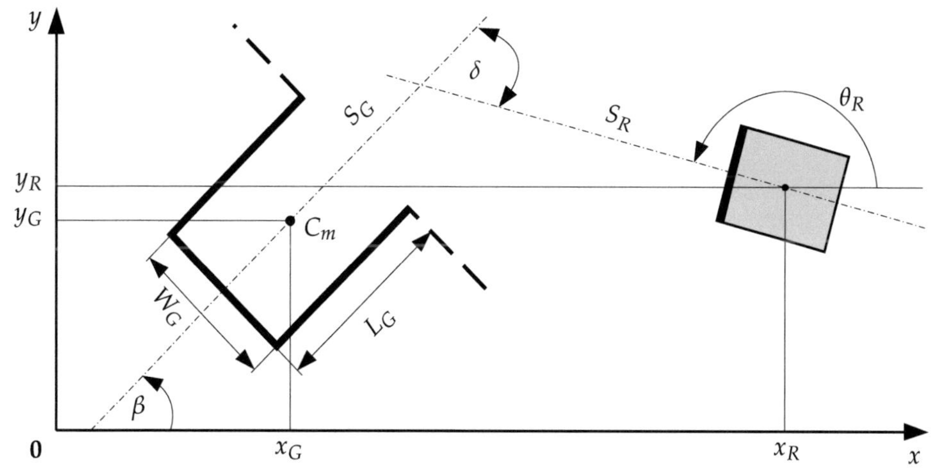

Fig. 5. Mutual spatial positions of wheeled mobile robot and garage.

The bidirectional garaging problem is similar to the stabilization problem, also known as the "parking problem", formulated by Oriolo et al. (2002): "the robot must reach a desired configuration $q_G(x_G, y_G, \theta_G)$ starting from a given initial configuration $q_{(0)}(x_{R(0)}, y_{R(0)}, \theta_{R(0)})$". The differences are that there are two solutions for the desired configuration in bidirectional garaging problems and that the stabilization problem need not necessarily involve constrains imposed by the presence of the garage.
Although the bidirectional garaging problem is addressed in this chapter as a stabilization problem, the proposed solution can also be used as a garaging trajectory generator because it includes non-holonomic constraints of the mobile robot.

3.1 System of two fictitious fuzzy magnets

From a human control skill perspective, the garaging of a vehicle is comprised of at least two stages: approach to the garage and entry into the garage. The adjustment of the vehicle's position inside the garage might be the third stage, but the need for this stage depends on the success of the second stage (i.e., driver skills). Since there is much less experience in differential drive mobile vehicle control than in car and car-like mobile robot control, and based on recommendations in (Sugeno & Nishida, 1985), the proposed WMR garaging system is a model of operator control actions, designed by means of two fictitious fuzzy magnets, one of which is located immediately in front of the garage and the other inside the garage. The first point is used to approach the garage and while the vehicle is entering the garage, it serves as a reference point for proper orientation, similar to human driving skills. The second point is both the target and a reference point for the WMR when it enters the garage and when it completes the garaging process. Contrary to the well-known point-to-point navigation approach, navigation aided by a system of two fuzzy magnets is executed in a single stage since the garaging process requires the robot to approach the area in front of the garage, but not to also reach the point which is the center of this area. The fictitious fuzzy magnets are denoted by FM_{Fm} (*Forward magnet*) and FM_{Cm} (*Central magnet*), and created according to (1):

$$FM_{Fm} = (F_m, FRsS_{Fm}), \tag{12}$$

$$FM_{C_m} = (C_m, FRsS_{Cm}). \tag{13}$$

In Fig. 6, their positions are denoted by C_m and F_m. The point F_m lies on the garage axis of symmetry S_G at a distance d_F from the front line. Let us imagine that these points are fictitious fuzzy magnets with attraction regions around them and if the mobile object/vehicle finds itself in that region, the attraction force will act on it. The activity of these fictitious fuzzy magnets will be neutralized in point C_m, thus finishing the garaging process.

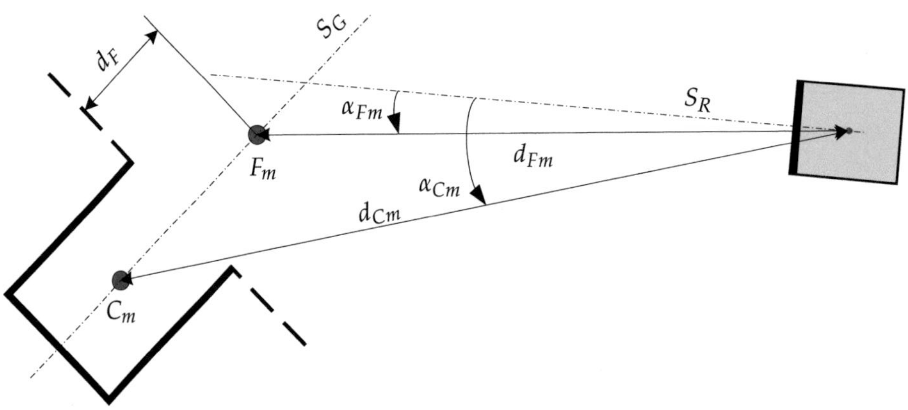

Fig. 6. Mutual spatial positions of wheeled mobile robot and fictitious fuzzy magnet.

The implementation of this concept requires the definition of the interaction between the fictitious fuzzy magnets and the vehicle, consistent with the definition of the fictitious fuzzy magnets. Figure 6 also shows the parameters which define the relationship between the fictitious fuzzy magnets and the vehicle. The orientation of the WMR relative to the point F_m is denoted by α_{Fm}, and relative to the point C_m by α_{Cm}. The distance between the WMR and the fictitious fuzzy magnet C_m is d_{Cm}, while the distance to the fictitious magnet F_m is d_{Fm}. The key step in the design of such an approach is the observation of these two distances as fuzzy sets which need to be attributed appropriate membership functions. The first task of the FM_{Fm} is to guide the WMR to the area immediately in front of the garage, such that the linguistic variable d_{Fm} is assigned a single membership function, denoted by far in Fig. 7(a). The membership function was not assigned to the linguistic variable $\mu_{far}(d_{Fm})$ in the standard manner because the entire universe of discourse is not covered (since $F_1 > 0$). This practically means that FM_{Fm} will have no effect on the robot when $d_{Fm} < F_1$, but FM_{Cm} will; the membership function for the distance to its position C_m is shown in Fig. 7(b). The membership function $\mu_{near}(d_{Cm})$ allows the action of fuzzy magnet C_m to affect the WMR only when the WMR is near FM_{Cm}, or when the distance of d_{Cm} is less than C_4. Even though no linguistic variable related to distance covers the entire set of possible values, the WMR is in the region

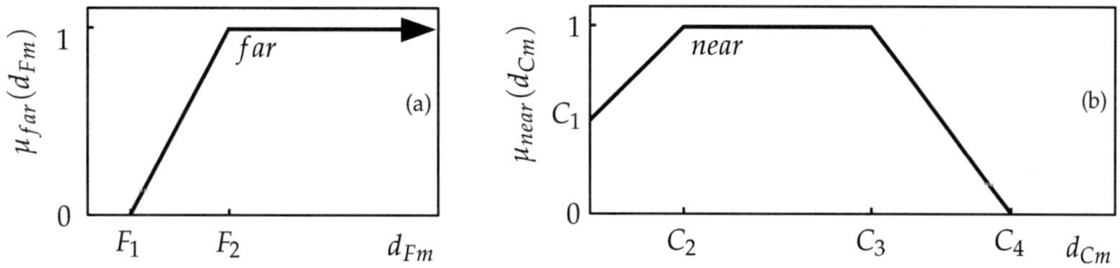

Fig. 7. Definition of membership functions for distances: (a) d_{Fm}, (b) d_{Cm}.

of attraction of at least one fictitious fuzzy magnet at all times. A more detailed procedure for the selection of parameters and fuzzy rules will be presented later in the text.

The next step in the definition of fuzzy rules subsets is the definition of the second input variable, associated with the WMR orientation relative to the fictitious fuzzy magnets. The angles are denoted by α_{Fm} and α_{Cm} in Fig. 6. If angle α_{Fm} is close to zero, then the vehicle's front end is oriented toward the fictitious fuzzy magnet F_m, and if the value of angle α_{Fm} is close to π, the vehicle's rear end is oriented toward F_m. This fact is important, as will be shown later in the text, from the perspective that the objective of vehicle control will be to reduce the angles α_{Fm} and α_{Cm} to a zero or π level, depending on initial positioning conditions. Linguistic variables α_{Fm} and α_{Cm} are together identified as *Direction* and are described by membership functions which should point to the orientation of the WMR relative to the fictitious fuzzy magnets according to (9) and (10), whose form is shown in Fig. 8.

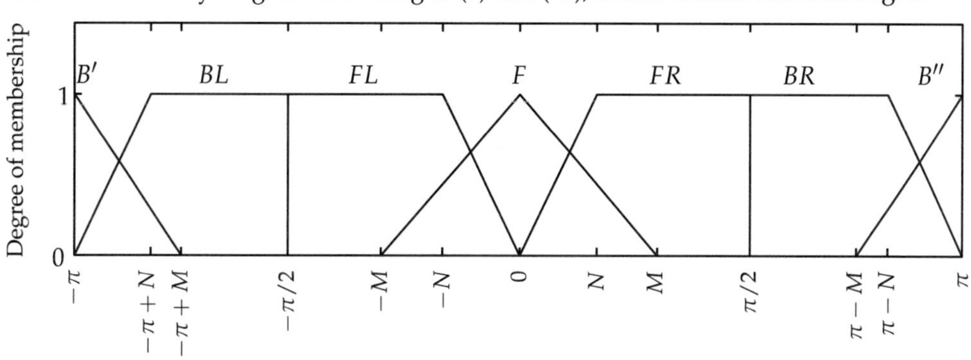

Fig. 8. Membership functions of linguistic variable *Direction*.

3.2 Definition of fuzzy rules

Since the WMR bidirectional garaging system is comprised of two fictitious fuzzy magnets, FM_{F_m} and FM_{C_m}, the fuzzy rules base is created from the rules subsets of each fuzzy magnet. It follows from the definition of the concept (3) that a fuzzy rules subset is determined by two inputs (WMR distance and orientation relative to fuzzy magnet position), meaning that the number of fuzzy controller inputs is equal to $K \times 2$, where K is the number of fuzzy magnets which make up the system, such that the number of inputs into a fuzzy system comprised of two fuzzy magnets is four. This concept is convenient because inherence rules for each fuzzy magnet are generated from inputs related to that magnet, such that the addition of a fuzzy magnet increases the number of rules in the controller base only in its subset of fuzzy rules. We will now specifically define two fuzzy rules subsets, $FRsS_{F_m}$ and $FRsS_{C_m}$. Since the

fuzzy rules subsets are created for the same object, the output variables are defined equally, regardless of whether concept (5-6) or (7-8) is applied.
The definition of output variables of the FLC requires a clear definition of the object of control. The algorithm described in this chapter is primarily intended for the control of two-wheeled, differential drive mobile robots. The wheels of this type of vehicle are excited independently and their speeds are defined autonomously. Guidance is controlled by the wheel speed difference, and the fuzzy rules subsets will be generated in the form of (5-6). We will assume that the physical characteristics of the actuators of both wheels are identical, and that the object has identical front- and back-drive capabilities. Since the inherence rules will be defined in general, we will assume certain general characteristics of the object of control, which will not affect the practical application of the concept. The maximum wheel speed will be denoted by the positive constant V_m, where $-V_m$ means that the wheel is rotating at maximum velocity which induces the object to move in reverse. We will also assume a nonnegative constant V_1, which satisfies the condition:

$$V_1 \leq V_m. \tag{14}$$

Fuzzy rules are defined based on the following principles (Mitrović & Đurović, 2010a):

- As the distance from the vehicle to the garage grows, the total speed of the vehicle should increase;
- As the distance from the vehicle to the garage shrinks, the total speed of the vehicle should drop;
- The difference between the wheel speeds causes turning, which must depend on the robot's orientation toward the garage;
- In the case of good orientation, the robot speed may be maximal.

We will identify the fuzzy logic controller for bidirectional garaging using the fictitious fuzzy magnet concept as FLC_{BG}. The fuzzy rules base of the FLC_{BG} is comprised of two subsets: the $FRsS_{Fm}$ subset, which corresponds to the fictitious fuzzy magnet FM_{Fm}, shown in Table 2(a), and the $FRsS_{C_m}$ subset shown in Table 2(b).

Rule	d_{Fm}	α_{Fm}	v_L	v_R
1.	*far*	*B*	$-V_m$	$-V_m$
2.	*far*	*BL*	0	$-V_m$
3.	*far*	*FL*	0	V_m
4.	*far*	*F*	V_m	V_m
5.	*far*	*FR*	V_m	0
6.	*far*	*BR*	$-V_m$	0

(a) Fuzzy rules subset $FRsS_{Fm}$

Rule	d_{Cm}	α_{Cm}	v_L	v_R
7.	*near*	*B*	$-V_m$	$-V_m$
8.	*near*	*BL*	0	$-V_1$
9.	*near*	*FL*	0	V_1
10.	*near*	*F*	V_m	V_m
11.	*near*	*FR*	V_1	0
12.	*near*	*BR*	$-V_1$	0

(b) Fuzzy rules subset $FRsS_{Cm}$

Table 2. Fuzzy rules base for bidirectional garaging fuzzy controller – FLC_{BG}.

Membership functions of input variables d_{Fm} and d_{Cm} independently activate particular rules (Table 2), an action which results in a considerable reduction in the number of rules. The rule which produces zero commands on both wheels does not exist; this might lead to the wrong conclusion that the vehicle never stops and does not take a final position. Since the points C_m and F_m are on the garage axis of symmetry, when the vehicle finds itself near its final position the difference between orientation angles α_{Fm} and α_{Cm} is close to $\pm\pi$. In this case, at least two rules generating opposite commands are activated, and their influence becomes annulled (Mitrović & Đurović, 2010a).

3.3 Selection of fuzzy controller parameters

However, with regard to the endless set of possible combinations of input variables, the selection of fuzzy rules does not guarantee that the vehicle will stop at the desired position. It is therefore, necessary to pay close attention to the selection of parameters d_F, M, N, F_1, $F2$, C_1, C_2, C_3, and C_4.

3.3.1 Parameter d_F

FM_{Fm} is described by (12) and, before the parameters which define the membership functions of the linguistic variables of $FRsS_{Fm}$ are determined, the position F_m of FM_{Fm} needs to be established. In Fig. 6, the point F_m is located on the longitudinal axis of symmetry of the garage and its distance from the garage door is d_F. The parameter d_F should be adopted in the first iteration, observing the following limitation:

$$d_F > \sqrt{(W_R/2)^2 + (L_R/2)^2}, \tag{15}$$

where W_R is the robot width, and L_R is the robot length, with the recommendation $d_F \approx W_R$.

3.3.2 Parameters M and N

The selection of parameters M and N, in their geometrical sense as presented in Fig. 8, influences the nature of the maneuver (curve and speed of turning) performed by the vehicle during garaging, especially at the beginning of the garaging process when the robot is at some distance from the garage. When the values of M and N are low, the vehicle rotates with a very small curve diameter; as these values increase, the arches circumscribed by the vehicle also increase (Mitrović & Đurović, 2010a). During the selection of these values, a compromise must be made to maintain the maneuvering capabilities of the vehicle, as well as the constraints imposed by geometry, namely, the ratio between the vehicle dimensions and the width of the garage. Generally, these parameters are adjusted in such a way that the vehicle circumscribes larger arches when distant from the garage, whereas more vivid maneuvers are needed in the vicinity of the garage. As such, in the second iteration, parameters M and N related to input α_{Fm} are adjusted, followed by those related to the input α_{Cm}, with the following constraints:

$$\pi/2 > M \geq N > 0. \tag{16}$$

3.3.3 Coefficients C_3, C_4, F_1, and F_2

The trapezoidal membership function *far* of the linguistic variable d_{Fm} (Fig. 7(a)) is determined by the coefficients F_1 and F_2, while the membership function *near* of the linguistic variable d_{Cm} is determined by the coefficients C_1, C_2, C_3 and C_4 (Fig. 7(a)). The coefficients C_3, C_4, F_1 and F_2 enable gradual activation of fictitious magnet C_m and the deactivation of fictitious magnet F_m, as the robot approaches its destination. The selection of these parameters has a critical impact on the performance of the entire fuzzy controller. Inequalities (17) and (18) guarantee the form of the membership functions shown in Fig. 7(a) and Fig. 7(b), respectively.

$$C_4 > C_3 > 0 \tag{17}$$

$$F_2 > F_1 \geq 0 \tag{18}$$

Figure 9(a) illustrates the action of FM_{F_m}. Outside the circle of radius F_2 , with its center at F_m (dark gray in the figure), this fuzzy magnet exhibits maximum activity. When the WMR is within this region, the distance between the robot and the point F_m, which was earlier denoted by d_{Fm} is greater than F_2 and $\mu_{far}(d_{F_m} > F_2) = 1$, as shown in Fig. 7(a).

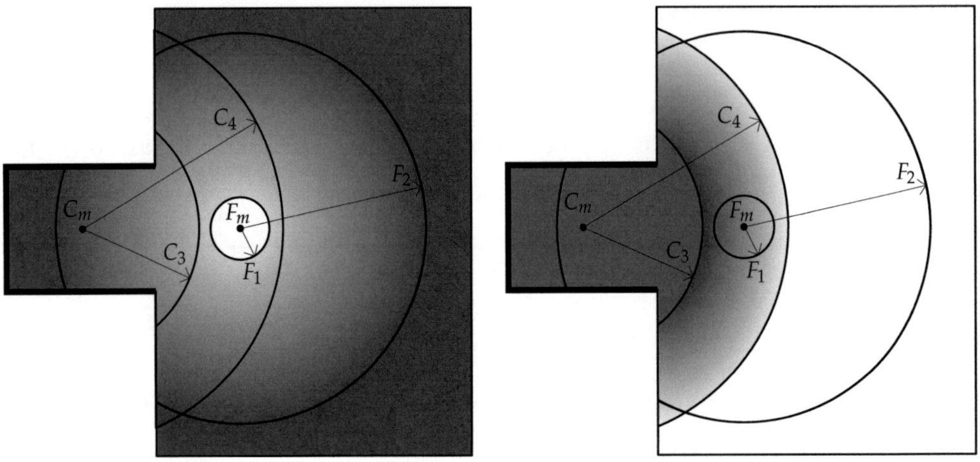

(a) Activity of fictitious fuzzy magnet FM_{Fm}. (b) Activity of fictitious fuzzy magnet FM_{Cm}.

Fig. 9. Illustration of fictitious fuzzy magnet activity.

The activity of FM_{Fm} declines as the position of the robot approaches the point F_m, through to the edge of the circle of radius F_1, inside which the activity of this fuzzy magnet equals zero (i.e., $\mu_{far}(d_{Fm} \leq F_1) = 0$). To facilitate adjustment of the overlap region of the two fuzzy magnets, it is convenient for the point C_m, which determines both the position of FM_{Cm} and the position of the center of the garage, to lie within the circle of radius F_2, which can be expressed, based on Fig. 9(a), by the following inequality:

$$F_2 \geq L_G/2 + d_F. \tag{19}$$

The coefficients C_1 and C_2 (Fig. 7(b)) should initially be set at $C_1 = 1$ and $C_2 = 0$. Figure 9(b) illustrates the activity of FM_{Cm}. Outside the circle of radius C_4, whose center is at C_m, the activity of this fuzzy magnet is equal to zero (i.e., $\mu_{near}(d_{Cm} > C_4) = 0$). The activity of FM_{Cm} increases as the robot approaches the point C_m, through to the edge of the circle of radius C_3, within which the activity of this fuzzy magnet is maximal (i.e., $\mu_{near}(d_{Cm} \leq C_3) = 1$), shown in dark gray in the figure). To ensure that at least one fuzzy magnet is active at any point in time, the following constraint must strictly be observed:

$$C_4 > L_G/2 + d_F + F_1. \tag{20}$$

This constraint ensures that the circle of radius F_1 is included in the circle of radius C_4, which practically means that the vehicle is within the region of attraction of at least one fuzzy magnet at all times. The dominant impact on vehicle garaging is obtained by overlapping fictitious fuzzy magnet attraction regions, and proper vehicle garaging is achieved through the adjustment of the region in which both fictitious fuzzy magnets are active.

3.3.4 Coefficients C_1 and C_2

The selection of the above parameters might enable vehicle stopping near point C_m. The simultaneous adjustment of the garaging path and stopping at the target position, due to a large number of free parameters, requires a compromise which may have a considerable impact on the quality of the controller. The introduction of parameters C_1 and C_2 enables unharnessed adjustment of vehicle stopping. The selection of parameters C_1 and C_2, enabled by theoretically exact stopping of vehicles at point C_m, will cause oscillations in movement around point C_m due to the dynamics of the vehicle, a factor which was neglected in the process of controller design. Accordingly, the selected desired final point of garaging is in the immediate vicinity of the fictitious magnet C_m location. Coefficients C_1 and C_2 take into account the neglected dynamics of the vehicle, and their adjustment is performed experimentally, observing the following constraints:

$$d_{Cm} < L_G/2 \Rightarrow d_{Fm} \leq L_G/2 + d_F. \tag{21}$$

The above relation ensures that at least two opposed rule-generating commands are always activated around the point C_m. If the variable d_{Fm} assumes a value larger than $L_G/2 + d_F$, meaning that the vehicle has practically surpassed the target point, the angles α_{Cm} and α_{Fm} will become nearly equal, and results in the activation of rules that generate commands of the same sign, which in turn causes oscillations around the point C_m.

3.4 Simulation results

An approximate dynamic model of the *Hemisson* robot (Mitrović, 2006) was used for the simulations. This is a mobile vehicle of symmetrical shape, with two wheels and a differential drive. Wheels are independently excited, and their speeds may be defined independently from each other. Guidance is controlled by the difference in wheel speeds. Each of these speeds can be set as one of the integer values in the interval $[-9, 9]$, (Table 3), where a negative value means a change in the direction of wheel rotation. The dependence of the command on wheel speed is not exactly linear causing certain additional problems in the process of vehicle control (Mišković et al., 2002). Specifically, for the subject case, the selected output variables of the FLC are the speed commands of the left and right wheels: VL and VR, respectively. Accordingly, this discrete set of integer values from the interval $[-9, 9]$ is selected as the domain of membership functions attributed to the output variables.

Wheel Speed Command	0	1	2	3	4	5	6	7	8	9
Wheel Speed $[\mathrm{mm/s}]$	0	3	18	42	64	86	109	134	156	184

Table 3. Wheel speed of the Hemisson robot as a function of wheel speed command.

The sampling period of the FLC was $T_s = 0.2\,\mathrm{s}$. The size of the garage was $16\,\mathrm{cm} \times 20\,\mathrm{cm}$, and the size of the robot was $10\,\mathrm{cm} \times 12\,\mathrm{cm}$. The FLC_{BG} was designed to conform to the constraints defined in the previous section, with the objective of ensuring successful garaging. The experimentally-determined values of these parameters, for the examined case of the Hemisson robot, are presented in Table 4.

Figure 10 shows WMR garaging trajectories for a set of 17 different initial conditions. Depending on the initial configuration, the WMR is guided to the target by forward or back drive, which illustrates the bidirectional feature of the proposed fictitious fuzzy magnets concept.

Coeff.	Value	Coeff.	Value
$M(\alpha 1)$	$7\pi/36$	C_1	20/63
$N(\alpha 1)$	$\pi/12$	C_2	4.3 cm
$M(\alpha 2)$	$\pi/4$	C_3	16 cm
$N(\alpha 2)$	$\pi/18$	C_4	40.0 cm
F_1	4.0 cm	d_F	11.0 cm
F_2	44.5 cm	V_1	Command 3

Table 4. FLC_{BG} – coefficient values.

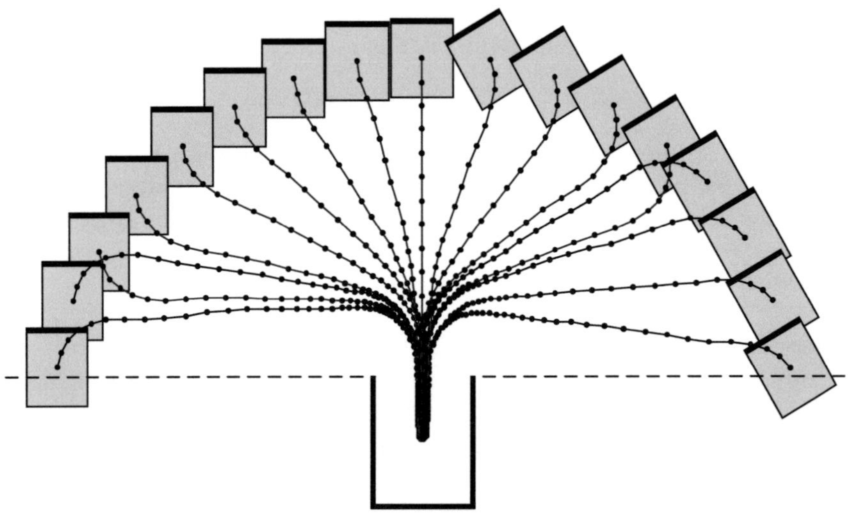

Fig. 10. Illustrative trajectories for front- and back-drive garaging

To test the efficiency and applicability of the proposed method using an extended set of initial configurations, a new test set of N initial conditions was created. This set of initial configurations ($x_{Ri,(t=0)}, y_{Ri,(t=0)}$, and $\theta_{Ri,(t=0)}$), $i = 1 : N$, was formed such that the robot is placed at equidistant points on the x and y axes spaced 1 cm apart, while the angle $\theta_{Ri,(t=0)}$ was a uniform-distribution random variable in the interval $[-\pi/2, \pi/2]$. The test configuration set was divided into two sub-sets: the first satisfies the condition that the robot in its initial configuration is at an adequate distance from the garage and contains N_d elements, while the other includes initial configurations near the garage and contains $N_c = N - N_d$ elements. Garaging performance was measured by the distance of the robot from the target position d_{Cm} (Fig. 6), and the angle δ (Fig. 5), which showed that the robot and garage axes of symmetry did not coincide:

$$\delta = |(\beta - \theta_R) \bmod \pi| \tag{22}$$

The results of simulations with the FLC_{BG}, for the set of initial conditions N_d, are shown in Fig. 11. In all N_d cases, the garaging process was completed with no collision occurring between the robot and the garage. In view of the non-linear setting of the speeds and the long sampling period, the conclusion is that the results of garaging were satisfactory. The average deviation in N_d cases was $\overline{d}_{Cm} = 1.1$ cm, while the average angle error was $\overline{\delta} = 1.4°$.

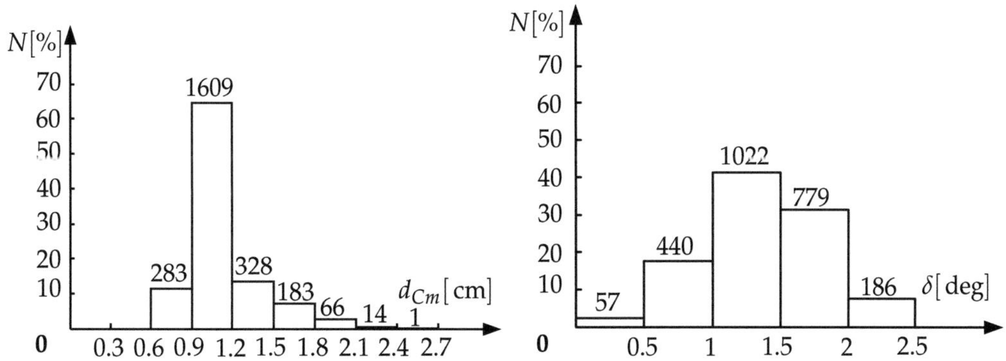

(a) Histograms of WMR distances from final position (b) Histograms of WMR deviations from final orientation

Fig. 11. Results of simulations with FLC_{BG}, for the set of initial conditions N_d.

3.5 Robustness of the system

Since the system was designed for use with a real vehicle, its efficiency needed to be examined under less-than-ideal conditions. To operate the system, the positions and orientations of the robot for each selection period must be determined, whereby the robot position and orientation need not necessarily coincide with the real position and orientation. Since inputs into the system consist of two distances and two angles whose accuracy depends directly on the accuracy of the determination of the position of the robot during the garaging process, the effect of inaccuracies of the vehicle coordinates on the efficiency of the garaging process were analyzed. The set of experiments was repeated for the FLC_{BG} with N_d initial conditions, and the WMR coordinate determination error was modeled by noise with uniform distribution within the range $[-1\,\text{cm}, 1\,\text{cm}]$. Figure 12 shows histograms of distance and orientation deviations from the targeted configuration, under the conditions of simulated sensor noise. It was found that the system retained its functionality but that the deviations were greater than those seen in the experiment illustrated in Fig. 11.

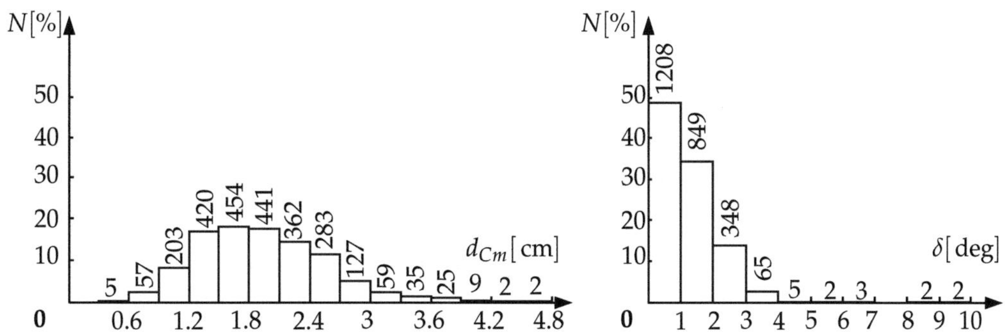

(a) Histograms of WMR distances from final position (b) Histograms of WMR deviations from final orientation

Fig. 12. Results of simulations with FLC_{BG} with simulated sensor noise, for the N_d initial conditions.

Table 5 shows extreme and mean values derived from the garaging experiments. The subscript N denotes that the simulations were conducted under simulated sensor noise conditions. The sensor noise mostly affected the maximum distances from the target position $d_{Cm\,\max}$; the mean values of angle error $\overline{\delta}$ were unexpectedly lower under simulated sensor noise conditions.

	$\overline{\delta}$ [°]	$\delta_{\max}$ [°]	$\overline{d}_{Cm}$ [cm]	$d_{Cm\max}$ [cm]
FLC_{BG}	1.37	2.35	1.10	2.41
FLC_{BG_N}	1.19	9.52	1.93	4.80

Table 5. Extreme and mean values derived from garaging experiments.

3.6 Limitations of the proposed FLC

Simulations were conducted for the sub-set N_c of initial conditions, under which the robot was not at a sufficient distance from the garage and was placed at the points of an equidistant grid, spaced 0.25 cm apart, where the initial orientation of the robot $\theta_{Ri,(t=0)}$ was a uniform-distribution random variable in the interval $[-\pi/2, \pi/2]$. Figure 13 shows the regions of initial conditions where the probability of collision of the robot with the garage is greater than zero. Full lines identify the limits of the region under ideal conditions, while dotted lines denote simulated sensor noise conditions. Figure 13 shows that sensor noise has little effect on the restricted area of initial conditions, which is an indicator of the robustness of the system to sensor noise.

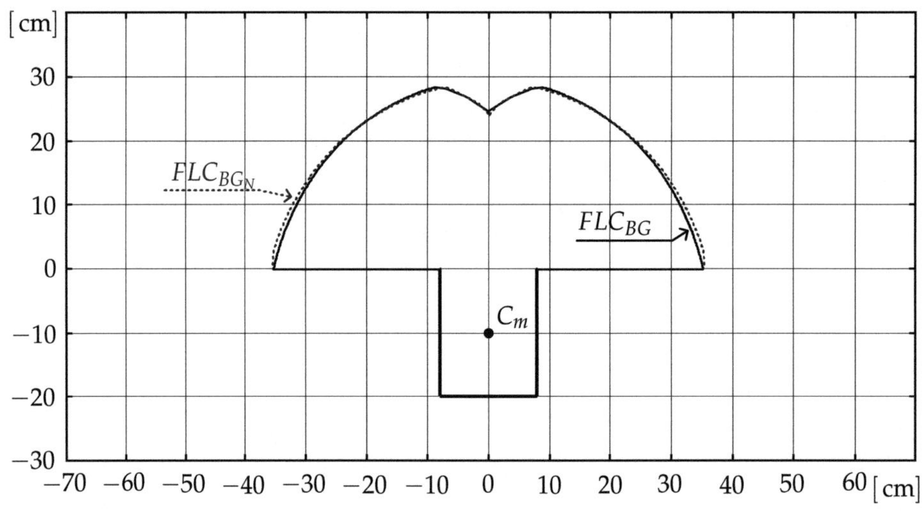

Fig. 13. Limits of the system – region of initial configurations with probable collision of the WMR with the garage: full line – ideal conditions (FLC_{BG}); dotted line – with sensor noise conditions (FLC_{BG_N}).

3.7 Experimental results

An analogous experiment was performed with a real mobile vehicle (Hemisson mobile robot) and a real garage whose dimensions are 16 cm × 20 cm. A block diagram of the garaging experiment is shown in Fig. 14.

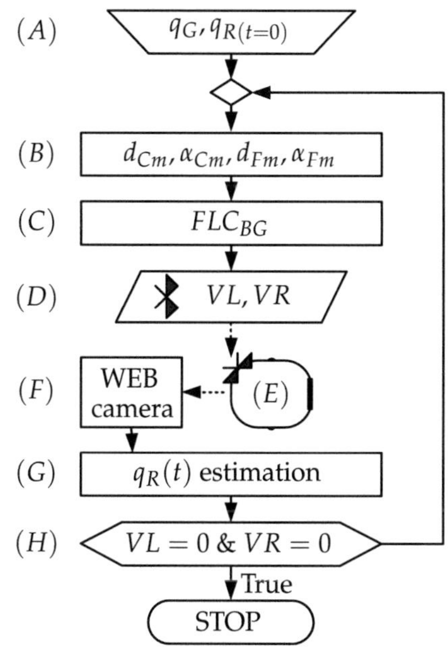

Fig. 14. Block diagram of real-time experiment: (A) Determining of initial conditions; (B) Calculation of: $d_{Cm}, d_{Fm}, \alpha_{Cm}$ and α_{Fm}; (C) Fuzzy logic controller for bidirectional garaging; (D) Bluetooth interface; (E) Hemisson mobile robot; (F) Web camera; (G) Estimation of the WMR posture; (H) Examination of VL and VR values.

A personal computer with a Bluetooth interface (D) and web camera (F) – resolution 640×480 pixels, were used during the experiment. The initial conditions were determined prior to the initiation of the garaging process, namely (A): the position and orientation of the garage q_G, and the posture (position and orientation) of the robot $q_{R(t=0)}$. Based on the posture of the robot and the position and orientation of the garage, in block (B), input variables $d_{Cm}, d_{Fm}, \alpha_{Cm}$ and α_{Fm} were calculated for the FLC_{BG} (C). FLC outputs were wheel speed commands for the Hemisson robot (VL and VR), which were issued to the robot via the Bluetooth interface. The web camera performed a successive acquisition of frames in real time, with a 0.2 s repetition time, which dictated the sampling period for the entire fuzzy controller. In block (G), web camera frames were used to estimate the position of the robot and compute its orientation (Đurović & Kovačević, 1995; Kovačević et al., 1992). During the garaging process, wheel speed commands were different from zero. When both commands became equal to zero, garaging was completed and block (H) halted the execution of the algorithm.

Figures 15(a) and 15(b) show 12 typical trajectories obtained during the process of garaging of a Hemisson robot in a real experiment. The illustrative trajectories shown in Fig. 15(a) deviate from expected trajectories and are mildly oscillatory in nature, which is a result of robot wheel eccentricity and the ultimate resolution of the camera used to estimate the position and orientation of the robot. The system which tracks the WMR to the target can be upgraded by a non-linear algorithm (Đurović & Kovačević, 2008) or instantaneous acquisition by two sensors (Đurović et al., 2009). Figure 15(b) shows trajectories obtained during the course of garaging from "difficult" initial positions (initial angle between the axis of symmetry of the robot and the axis of symmetry of the garage near $\pi/2$). These trajectories were obtained equally under front- and back-drive conditions (the front of the robot is identified by a thicker line). All of the illustrated trajectories are indicative of good performance of the proposed garaging algorithm.

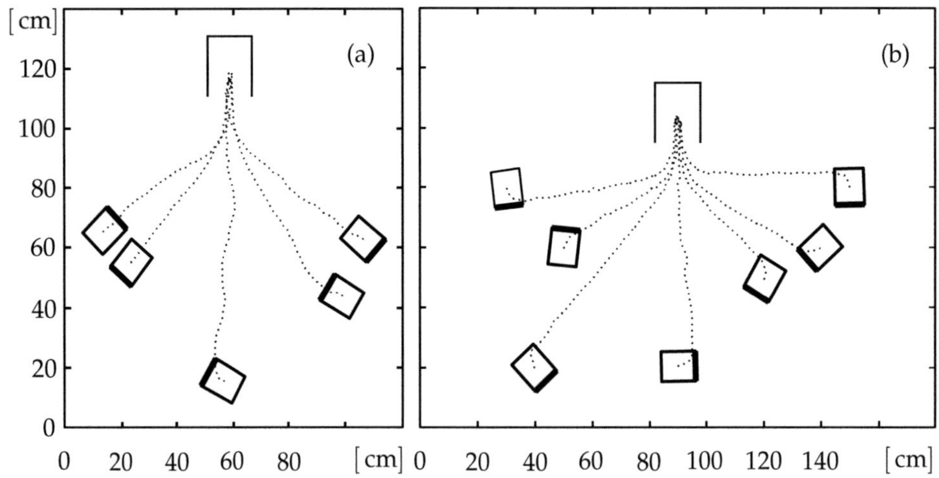

Fig. 15. Real-time garaging experiment results.

4. Obstacle avoidance

The navigation of a mobile robot in a workspace with obstacles is a current problem which has been extensively researched over the past few decades. Depending on the definition of the problem, the literature offers several approaches based on different methodologies. Relative to the environment, these approaches can be classified depending on whether the vehicle moves in a known (Kermiche et al., 2006), partially known (Rusu et al., 2003), or totally unknown environment (Pradhan et al., 2009). Further, the environment can be stationary or dynamic – with moving obstacles (Hui & Pratihar, 2009), or a moving target (Glasius et al., 1995). With regard to the ultimate target, the mobile robot is navigated to a given point (Mitrović et al., 2009), or a given configuration – a point and a given orientation at that point (Mitrović & Đurović, 2010a), or navigated to reach and follow a given trajectory (Tanaka & Sano, 1995). The problems are solved both in real-time and off-line.

Numerous obstacle avoidance methods are discussed in literature. The potential field method (Khatib, 1986) plays a major role in solving this type of problem. The potential function has to be derived for each individual obstacle and this can be a drawback from a real-time perspective if there are a large number of obstacles. Additionally, the resulting potential field may contain local minima where navigation ends without reaching the goal (Koren & Borenstein, 1991).

Fuzzy-logic-based controllers are widely applied to address obstacle avoidance and mobile robot navigation problems, and they range from reactive fuzzy controllers (which do not include a planning stage and generate outputs based on real-time sensor signals) to controllers which resolve the trajectory planning problem of a WMR through the synthesis of fuzzy logic with genetic algorithms, neural networks, or the potential field approach. Reignier (1994) proposed a reactive FLC for WMR navigation through a workspace with obstacles,

based on data from 24 ultrasonic sensors. Abdessemed et al. (2004) proposed a reactive controller inspired by the potential field method, with a rule base optimized by evolutionary techniques. Hui & Pratihar (2009) compared the potential field method with soft computing methods used to solve moving obstacle avoidance problems.
This section proposes a new methodology for robot navigation in a workspace which contains one or more obstacles. The approach is based on the fictitious fuzzy magnet concept but, contrary to the system described in the previous section, the rules are generated such that the action of the fuzzy magnet is to repel the vehicle. It should be noted that the inputs into the controller based on the fuzzy magnet concept are the distance and orientation relative to the fuzzy magnet position, which is a point, while the obstacles have dimensions which must be taken into account at the controller design stage. The controller proposed in this section uses information about the relative dimensions of the obstacle. At the preprocessing stage, the left and right side of the obstacle in the robot's line of sight are determined based on the mutual positions of the robot and the obstacle. Since controller inputs are based on the relative dimensions of the obstacle, the controller can be applied to obstacles of different shapes and sizes.

4.1 Navigation in a single-obstacle environment

In general, WMR navigation to a target in a workspace with obstacles is comprised of two tasks: avoidance of obstacles and navigation to the target. The controller may be designed to integrate these two tasks, or the tasks can be performed independently but in parallel. Parallel processing of target and obstacle data, contrary to integrated approaches, offers a number of advantages, ranging from fewer FLC inputs to parallel distribute compensation discussed by Wang et al. (1996). To allow for the algorithm to be used for navigation in an environment which contains an unknown number of obstacles, the parallel fuzzy controller data processing approach was selected. The FLC_{OFE} controller discussed in Section 2.1 was used to execute a navigation task to a target in an obstacle-free environment. Instead of v_R and ω_R, in this section its outputs will be denoted by v_{OFE} and ω_{OFE}, respectively, since notations with R are reserved for outputs from the entire navigation system.
Variables which will reflect the mutual positions of the robot and the obstacle need to be defined and efficient navigation in real-time enabled, regardless of the position and size of the obstacle relative to the target and the current configuration of the robot. Figure 16 shows the WMR and obstacle parameters. Since we address the WMR navigation problem in planar form, we will assume that the obstacle in the robot's workspace is a circle defined by its center coordinates and radius: $O(x_O, y_O, r_O)$. Figure 16 shows a real obstacle whose radius is r_r, in light gray. Because of the finite dimensions of the robot, it is extended by a circular ring of width r_{cl} (dark gray). Hereafter we will use r_O for the radius of the obstacle, computed from the equation:

$$r_O = r_r + r_{cl}, \qquad r_{cl} > W_G/2. \tag{23}$$

In Fig. 16, the distance between the WMR center and the obstacle is denoted by d_{RO}. Point P_C lies at the intersection of the WMR course line and the obstacle's radius which form a right angle and represent a virtual collision point. When this intersection is an empty set, the obstacle does not lie on the WMR trajectory and point P_C is not defined. The portion of the radius which is located to the left of the point P_C relative to the WMR is denoted by l_L, while the portion of the radius to the right of P_C is denoted by l_R. We will use variables l_L and l_R to define the obstacle avoidance algorithm. Since l_L and l_R are dimensional lengths and since

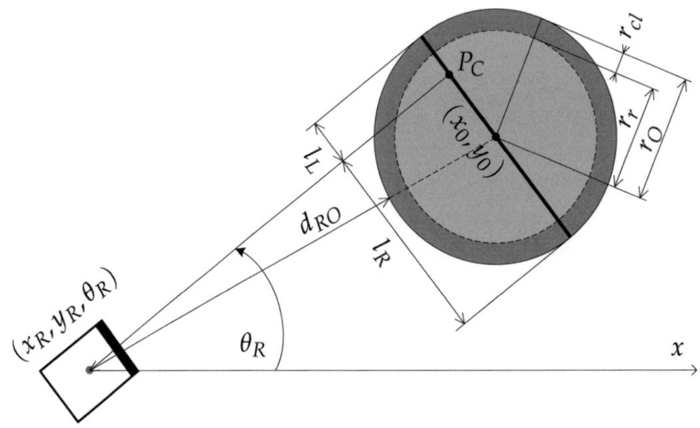

Fig. 16. Relative positions of obstacle and mobile robot.

their values depend on the WMR position and obstacle dimensions, we will normalize them based on the equation:

$$\begin{aligned} L_L &= l_L/(2 \cdot r_O), \\ L_R &= l_R/(2 \cdot r_O), \end{aligned} \tag{24}$$

where L_L and L_R are the left and right side of the obstacle, respectively, and can assume values from the interval $[0, 1]$, regardless of obstacle dimensions, since $L_L + L_R = 1$. If L_L or L_R is equal to 1, the robot will follow a course which will bypass the obstacle. If point P_C is not defined, we will assume that L_L and L_R are equal to zero, meaning that the obstacle does not lie on the robot's course. Based on the position of the virtual collision point P_C and the linear velocity of the robot v_R, we will generate the obstacle avoidance controller input using:

$$Side = \begin{cases} [-1, -0.5) & \text{if } L_L \in (0.5, 1] \text{ and } v_R \leq 0 \\ [-0.5, 0) & \text{if } L_R \in [0.5, 1] \text{ and } v_R \leq 0 \\ (0, 0.5] & \text{if } L_R \in [0.5, 1] \text{ and } v_R > 0 \\ (0.5, 1] & \text{if } L_L \in (0.5, 1] \text{ and } v_R > 0 \\ 0 & \text{if } L_R = 0 \quad \text{ and } L_L = 0\,. \end{cases} \tag{25}$$

The variable *Side* provides information about the position of the obstacle relative to the robot and is equal to zero if the obstacle does not lie on the robot's course. The corresponding membership functions are shown in Fig. 17(a). In the name of the membership function, subscript B stands for *Backward* and F for *Forward*. To ensure efficient bypassing of the obstacle, in addition to the position of the obstacle relative to the WMR, the controller has to be provided with the distance between the WMR and the obstacle, denoted by d_{RO} in Fig. 16, whose membership functions are shown in Fig. 17(b).
The outputs of the obstacle avoidance controller are the linear velocity v_{OA} and the angular velocity ω_{OA} of the robot. The respective fuzzy rules are shown in Table 6.
It is important to note that the outputs of the obstacle avoidance controller are equal to zero when the obstacle does not lie on the robot's course. For this reason the algorithm for WMR navigation to the target point in a single-obstacle environment must integrate FLC_{OFE} and FLC_{OA} solutions. The linear velocity v_R and the angular velocity ω_R of the robot are computed

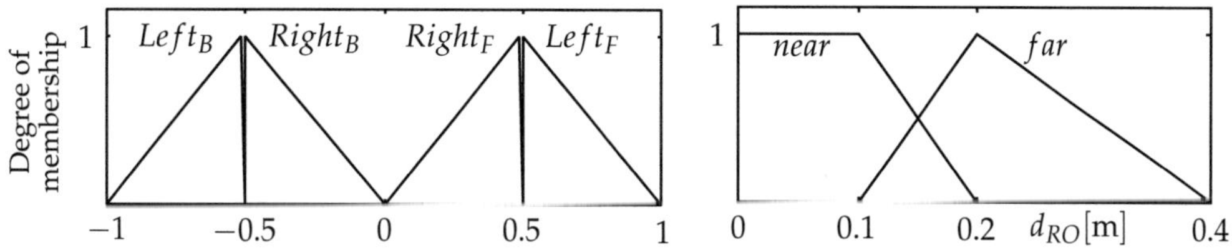

(a) Membership functions of linguistic variable *Side*.

(b) Membership functions of linguistic variable distance to obstacle – d_{RO}.

Fig. 17. Membership functions of obstacle avoidance fuzzy controller FLC_{OA}.

Rule	$Side$	d_{OR}	v_{OA}	ω_{OA}
1.	$Left_F$	$near$	$V_m/2$	$-\omega_m$
2.	$Left_B$	$near$	$-V_m/2$	ω_m
3.	$Right_F$	$near$	$V_m/2$	ω_m
4.	$Right_B$	$near$	$-V_m/2$	$-\omega_m$
5.	$Left_F$	far	V_m	$-\omega_m/2$
6.	$Left_B$	far	$-V_m$	$\omega_m/2$
7.	$Right_F$	far	V_m	$\omega_m/2$
8.	$Right_B$	far	$-V_m$	$-\omega_m/2$

Table 6. FLC_{OA} – fuzzy rules base.

from the following relation:

$$\omega_R = \begin{cases} \omega_{OFE} & \text{if } \omega_{OA} = 0 \\ \omega_{OA} & \text{if } \omega_{OA} \neq 0 \end{cases}; \quad v_R = \begin{cases} v_{OFE} & \text{if } v_{OA} = 0 \\ v_{OA} & \text{if } v_{OA} \neq 0 \end{cases}. \tag{26}$$

To decrease the duration of the navigation process, the controller output v_R was post-processed to ensure maximum linear velocity v_{Rmax} of the robot based on (Mitrović & Đurović, 2010b):

$$v_{Rmax}(\omega_R, v_R) = sgn(v_R) \cdot (V_m - |\omega_R| \cdot D/2), \tag{27}$$

where D is the distance between the robot wheels. The maximum linear velocity of the robot is a function of the angular velocity ω_R and the physical constraints of the robot. Figure 18 illustrates the trajectories followed by the WMR to avoid two obstacles: $O_1(-0.4, 0, 0.15)$ and $O_2(0.3, 0.1, 0.05)$, from four initial positions: $q_A(-0.8, 0, \pi)$, $q_B(-0.7, 0.2, \pi/4)$, $q_C(0.45, 0.25, 0)$, and $q_D(0.6, 0.2, -2\pi/3)$.

4.1.1 Moving obstacle avoidance

The proposed algorithm, without modification, was applied in a single moving obstacle avoidance scenario. It was assumed that the obstacle travels along a straight line at a constant speed v_{obs}, which is equal to one-third of the maximum speed of the robot. The outcome was successful when the robot overtook or directly passed by the obstacle. However, if it attempted to pass by at an angle, the robot collided with the obstacle. This was as expected, since the controller does not take the speed of the obstacle into account. Figure 19 illustrates the scenarios in which the robot overtakes, directly passes by, and meets with the moving obstacle at an angle.

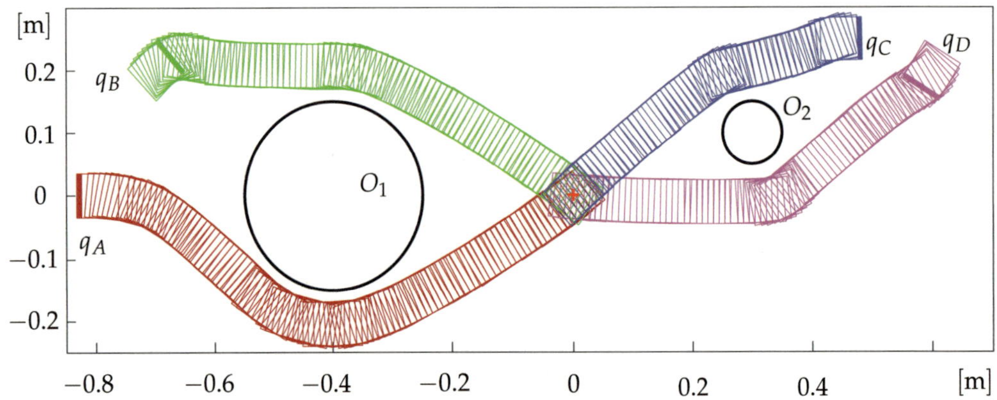

Fig. 18. Bypass trajectories of the robot from four initial positions around two obstacles of different dimensions.

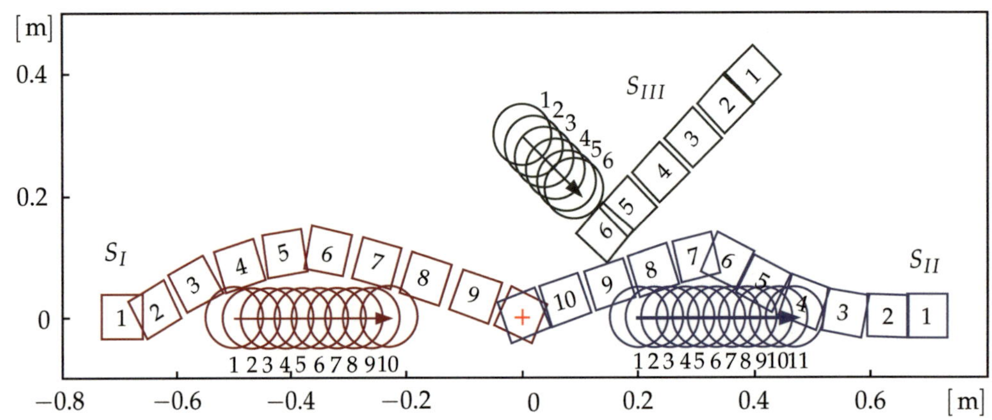

Fig. 19. Avoidance of moving obstacles: S_I – overtaking, S_{II} – directly passing, S_{III} – meeting at an angle.

4.2 Navigation trough a group of obstacles

We will now address the scenario in which there is a group of N obstacles in the workspace of the robot, where $N > 1$. To benefit from the approach described in the previous section, we will use the single-obstacle FLC_{OA} for each individual obstacle and thus obtain N values for the robot's linear and angular velocities. We will then compute the final angular velocity from the following equation:

$$\omega_{OA} = \left(\sum_{k=1}^{N} C_k \cdot \omega_{OAk} \cdot d_{ROk}^{-1} \right) \cdot \left(\sum_{k=1}^{N} C_k \cdot d_{ROk}^{-1} \right)^{-1}, \tag{28}$$

where ω_{OAk} is the FLC_{OA} output for the kth obstacle and d_{ROk} is the distance between the robot and the kth obstacle. As the distance from the vehicle to the kth obstacle shrinks, its influence to the controller outputs should increase. Using the coefficient C_k, we will take into account only those obstacles which lie on the robot's course. It is computed using (25), as follows:

$$C_k = \begin{cases} 0 & \text{if } Side_k = 0 \\ 1 & \text{if } Side_k \neq 0 \end{cases} . \tag{29}$$

We will derive the speed for the system of N obstacle avoidance controllers in a manner similar to (28), from:

$$v_{OA} = \left(\sum_{k=1}^{N} C_k \cdot v_{OAk} \cdot d_{ROk}^{-1} \right) \cdot \left(\sum_{k=1}^{N} C_k \cdot d_{ROk}^{-1} \right)^{-1} , \tag{30}$$

where v_{OAk} is the FLC_{OA} output for the kth obstacle. Outputs from the system of N FLC_{OA} and FLC_{OFE} controllers are integrated by means of (26), and the speed of the robot is maximized based on (27). Figure 20 shows a scenario with several randomly-distributed obstacles. Initial positions and the target (coordinate origin) are beyond the obstacle area. It is apparent from the figure that the proposed methodology efficiently solves the problem of navigation through an area which includes obstacles.

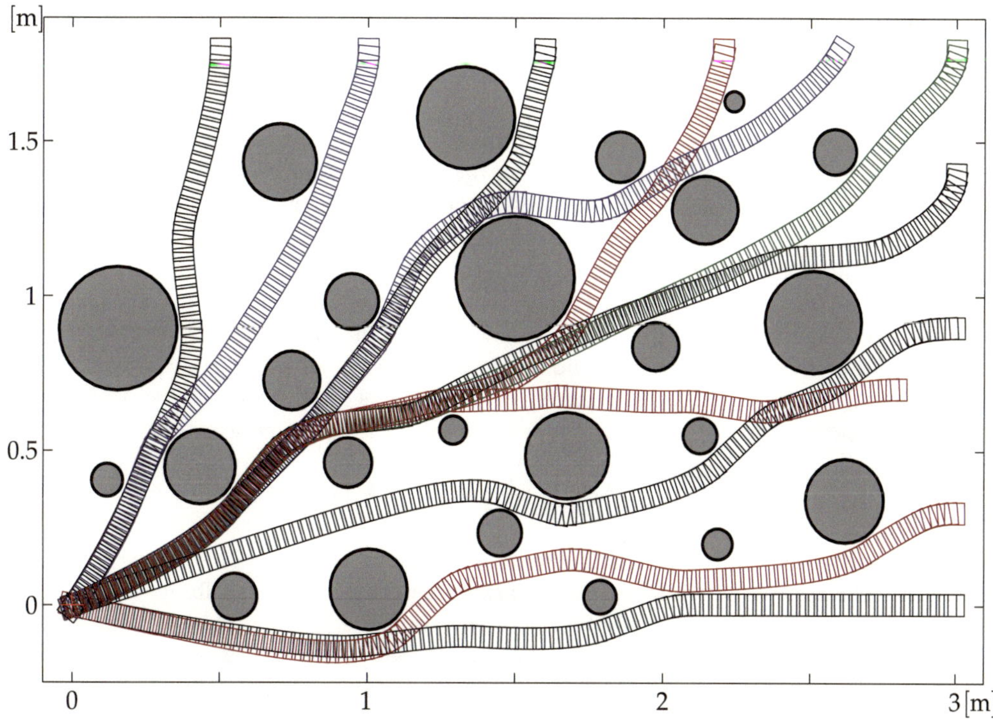

Fig. 20. Navigation to the goal (coordinate origin) through a group of obstacles.

4.2.1 Constraints of the system

Although the positions of the obstacles and the target are known, a closer analysis of the proposed approach leads to the conclusion that navigation to the target takes place in an unknown environment. The position of the target is known in all of the illustrated scenarios, but obstacles are taken into account only after they have triggered appropriate FLC rules, that is, when the distance between the robot and the obstacle is not greater than 40 cm (universe of

discourse of the linguistic variable d_{RO}, Fig. 17(b)). This basically means that the robot takes into account obstacles within a 40 cm radius.
It is well-known that a basic FLC cannot solve navigation through a workspace with arbitrarily distributed obstacles, if prior information about them is not available. A special problem is encountered when the obstacles are deployed such that they create a concave, **U**, barrier, in which case the robot enters an infinite loop or local minima (Krishna & Kalra, 2001; Motlagh et al., 2009).

5. Conclusion

A fictitious fuzzy magnet concept was introduced and its application analyzed in this chapter, using different navigation problems of differential drive mobile robots. The concept allows for navigation to the target in a single maneuver, without changing the direction of travel of the mobile robot. The symmetricity of the differential drive mobile robot is fully utilized, such that approaches based on the fictitious fuzzy magnet concept provide bidirectional solutions to mobile robot navigation problems. For example, during garaging, the mobile robot is automatically parked from the end of the robot which is closer to the garage entrance. The algorithm can be applied when the control variable is of the discrete type and where there are relatively few quantization levels.
When applied to a mobile robot garaging problem, a detailed analysis of simulation and experimental results illustrates the efficiency of the proposed algorithm and its robustness in the case of a random or systematic WMR position estimation error, as well as its limitations. The most significant shortfall of the proposed algorithm is that it does not provide a solution which will ensure that regardless of initial conditions the garage parking process is completed with no collision with the garage. The geometrical position of the initial conditions which lead to a collision of the robot with the garage is a compact, finite area which is discussed in the chapter. Some of the constraints mentioned here could be overcome in further research aimed at improving the proposed algorithm through a higher level of FLC complexity. Additionally, if a larger number of fictitious fuzzy magnets are introduced, the concept could be used to perform more intricate garage parking tasks. Re-configuration of controller outputs would render the proposed algorithm applicable for garage parking of a broader class of car-like mobile robots.
Additionally, it was shown that a larger number of fictitious fuzzy magnets can be introduced to execute tasks more complex than garage parking. Specifically, the feasibility was explored of applying the fictitious fuzzy magnet concept to execute the task of mobile robot navigation to a target in an environment which contains one or more obstacles. It was shown that the algorithm designed to avoid a single stationary obstacle can be successfully generalized to a multiple obstacles avoidance problem, preserving the bidirectional capability. The proposed algorithm can be extended by fuzzification of the coefficients C_k, taking into account the probability of collision with obstacles which lie on the robot's course.

6. References

Abdessemed, F., Benmahammed, K. & Monacelli, E. (2004). A fuzzy-based reactive controller for a non-holonomic mobile robot, *Robotics and Autonomous Systems* 47(1): 31–46.
Đurović, Ž. & Kovačević, B. (1995). QQ–plot approach to robust Kalman filtering, *International Journal of Control* 61(4): 837–857.

Đurović, Ž. & Kovačević, B. (2008). A sequential LQG approach to nonlinear tracking problem, *International Journal of Systems Science* 39(4): 371–382.

Đurović, Ž., Kovačević, B. & Dikić, G. (2009). Target tracking with two passive infrared non-imaging sensors, *IET Signal Processing* 3(3): 177–188.

Glasius, R., Komoda, A. & Gielen, S. (1995). Neural network dynamics for path planning and obstacle avoidance, *Neural Networks* 8(1): 125–133.

Hui, N. & Pratihar, D. (2009). A comparative study on some navigation schemes of a real robot tackling moving obstacles, *Robotics and Computer Integrated Manufacturing* 25(4-5): 810–828.

Kermiche, S., Saidi, M., Abbassi, H. & Ghodbane, H. (2006). Takagi–Sugeno based controller for mobile robot navigation, *Journal of Applied Sciences* 6(8): 1838–1844.

Khatib, O. (1986). Real-time obstacle avoidance for manipulators and mobile robots, *The International Journal of Robotics Research* 5(1): 90–98.

Koren, Y. & Borenstein, J. (1991). Potential field methods and their inherent limitations for mobile robot navigation, *Proceedings of the IEEE Conference on Robotics and Automation*, Sacramento, pp. 1398–404.

Kovačević, B., Đurović, Ž. & Glavaški, S. (1992). On robust Kalman filtering, *International Journal of Control* 56(3): 547–562.

Krishna, K. & Kalra, P. (2001). Perception and remembrance of the environment during real-time navigation of a mobile robot, *Robotics and Autonomous Systems* 37: 25–51.

Mišković, Lj., Đurović, Ž. & Kovačević, B. (2002). Application of the minimum state error variance approach to nonlinear system control, *International Journal of Systems Science* 33(5): 359–368.

Mitrović, S. (2006). Design of fuzzy logic controller for autonomous garaging of mobile robot, *Journal of Automatic Control* 16: 13–16.

Mitrović, S. & Đurović, Ž. (2010a). Fuzzy logic controller for bidirectional garaging of a differential drive mobile robot, *Advanced Robotics* 24(8–9): 1291–1311.

Mitrović, S. & Đurović, Ž. (2010b). Fuzzy–based controller for differential drive mobile robot obstacle avoidance, *in* G. Indiveri & A. M. Pascoal (eds), *7th IFAC Symposium on Intelligent Autonomous Vehicle*, Vol. 7, IFAC–PapersOnLine, Lecce, Italy.

Mitrović, S., Đurović, Ž. & Aleksić, M. (2009). One approach for mobile robot navigation throw area with obstacles, *eProc. of LIII ETRAN Conference*, Vol. 53, Vrnjačka Banja, Serbia, pp. AU3.5 1–4.

Motlagh, O., Hong, T. & Ismail, N. (2009). Development of a new minimum avoidance system for a behavior-based mobile robot, *Fuzzy Sets and Systems* 160: 1929–1945.

Oriolo, G., Luca, A. & Vendittell, M. (2002). WMR control via dynamic feedback linearization: Design, implementation, and experimental validation, *IEEE Trans. Control Systems Technology* 10: 835–852.

Pradhan, S., Parhi, D. & Panda, A. (2009). Fuzzy logic techniques for navigation of several mobile robots, *Applied Soft Computing Journal* 9(1): 290–304.

Reignier, P. (1994). Fuzzy logic techniques for mobile robot obstacle avoidance, *Robotics and Autonomous Systems* 12(3-4): 143–153.

Rusu, P., Petriu, E., Whalen, T., Cornell, A. & Spoelder, H. (2003). Behavior–based neuro–fuzzy controller for mobile robot navigation, *IEEE Transactions on Instrumentation and Measurement* 52(4): 1335–1340.

Sugeno, M. & Nishida, M. (1985). Fuzzy control of model car, *Fuzzy Sets and Systems* 16(2): 103–113.

Tanaka, K. & Sano, M. (1995). Trajectory stabilization of a model car via fuzzy control, *Fuzzy Sets and Systems* 70(2-3): 155–170.

Wang, H., Tanaka, K. & Griffin, M. (1996). An approach to fuzzy control of nonlinear systems: Stability and design issues, *IEEE Transactions on Fuzzy Systems* 4: 14 – 23.

Zadeh, L. (2001). Outline of a computational theory of perceptions based on computing with words, *in* N. Sinha, M. Gupta & L. Zadeh (eds), *Soft Computing and Intelligent Systems: Theory and Applications*, Academic Press, London, pp. 2–33.

13

Reliable Long-Term Navigation in Indoor Environments

Mattias Wahde, David Sandberg and Krister Wolff
Department of Applied Mechanics, Chalmers University of Technology, Göteborg Sweden

1. Introduction

Long-term robustness is a crucial property of robots intended for real-world tasks such as, for example, transportation in indoor environments (e.g. warehouses, industries, hospitals, airports etc.). In order to be useful, such robots must be able to operate over long distances without much human supervision, something that sets high demands both on the actual robot (hardware) and its artificial brain[1] (software). This paper is focused on the development of artificial robotic brains for reliable long-term navigation (for use in, for example, transportation) in indoor environments.

Reliable decision-making is an essential tool for achieving long-term robustness. The ability to make correct decisions, in real-time and often based on incomplete and noisy information, is important not only for mobile robots but also for animals, including humans (McFarland, 1998; Prescott et al., 2007). One may argue that the entire sub-field of behavior-based robotics emerged, at least in part, as a result of the perceived failure of classical artificial intelligence to address real-time decision-making based on a robot's imperfect knowledge of the world. Starting with the *subsumption method* (Brooks, 1986), many different methods for decision-making, often referred to as methods for *behavior selection* or *action selection*, have been suggested in the literature on behavior-based robotics (see, for example, Bryson (2007); Pirjanian (1999) for reviews of such methods).

In actual applications, a common approach is to combine a *reactive layer* of decision-making, using mainly behavior-based concepts, with a *deliberative layer* using, for example, concepts from classical artificial intelligence, such as high-level reasoning) (Arkin, 1998). Several approaches of this kind have been suggested (see, for example, Arkin (1987) and Gat (1991)) and applied in different robots (see, for example, Sakagami et al. (2002)).

In the *utility function* (UF) method for decision-making (Wahde, 2003; 2009), which will be used here, a somewhat different approach is taken in which, for the purposes of decision-making, no distinction is made between reactive and deliberative aspects of the robotic brain. In this method, an artificial robotic brain is built from a repertoire of *brain processes* as well as a single decision-making system responsible for activating and de-activating brain processes

[1] The term *artificial (robotic) brain* is here used instead of the more common *control system* since the latter term, in the authors' view, signifies the low-level parts (such as motor control to achieve a certain speed) of robot intelligence, whereas more high-level parts, such as decision-making are better described by the term used here.

(several of which may run in parallel). Two kinds of brain processes are defined, namely *(motor) behaviors*, which make use of the robot's motors and *cognitive processes*, which do not. Both kinds of processes may include both reactive and deliberative aspects. Note also that, regardless of content, the brain processes used in the UF method are all written in a unified manner, described in Sect. 4 below.

As its name implies, the UF method, described in Sect. 3 below, is based on the concept of utility, formalized by von Neumann & Morgenstern (1944). The method is mainly intended for complex motor tasks (e.g. transportation of objects in arenas with moving obstacles) requiring both reactivity and deliberation. While the method has already been tested (both in simulation and in real robots) for applications involving navigation over distances of 10-20 m or less (Wahde, 2009), this paper will present a more challenging test of the method, involving operation over long distances. The task considered here will be a delivery task, in which a robot moves to a sequence of target points in a large arena containing both stationary and moving obstacles. The approach considered here involves long-term simulations, preceded by extensive validation of the simulator, using a real robot.

2. Robot

The differentially steered, two-wheeled robot (developed in the authors' group) used in this investigation is shown in Fig. 1. The robot has a circular shape with a radius of 0.20 m. Note that an indentation has been made for the wheels. The height (from the ground to the top of the laser range finder (LRF)) is around 0.84 m. The weight is around 14.5 kg. The robot's two actuators are Faulhaber 3863A024C DC motors, each with a Faulhaber 38/1S-14:1 planetary gearhead. Each motor is controlled by a Parallax HB-25 motor controller. The robot is equipped with two 8 Ah power sources, with voltages of 12 V and 7.2 V, respectively.

In addition to the two drive wheels, the robot has two Castor wheels, one in the back and one in the front. The front Castor wheel is equipped with a suspension system, in order to avoid situations where the drive wheels lose contact with the ground, due to bumps in the surface on which the robot moves. The wheels are perforated by 24 holes which, together with a Boe-Bot Digital Encoder Kit detector on each wheel, are used for measuring the rotation of the wheels for use in odometry. In addition, the robot is equipped with a Hokuyo URG-04LX LRF with a range of 4 m and a 240 degree sweep angle. For proximity detection, the robot has two forward-pointing Sharp GP2D12 IR sensors (oriented towards ± 30 degrees, respectively, from the robot's front direction), and one IR sensor, of the same type, pointing straight backwards. The robot is equipped with two Basic Stamp II microcontrollers (which, although slow, are sufficient for the task considered here), one that reads the wheel encoder signals, and one that (i) sends signals to the motor controllers and (ii) receives signals from the three IR sensors. The two Basic Stamps are, in turn, connected via a USB hub to a laptop that can be placed on top of the robot (under the beam that holds the LRF). The LRF is also connected to the laptop, via the same USB hub.

3. Decision-making structure

The decision-making structure is based on the UF method, the most recent version of which is described by Wahde (2009). The method is mainly intended for use in tasks such as navigation, transportation, or mapping. In this method, the artificial brain is built from (i) a set of brain processes and (ii) a decision-making system based on the concept of utility, allowing the robot

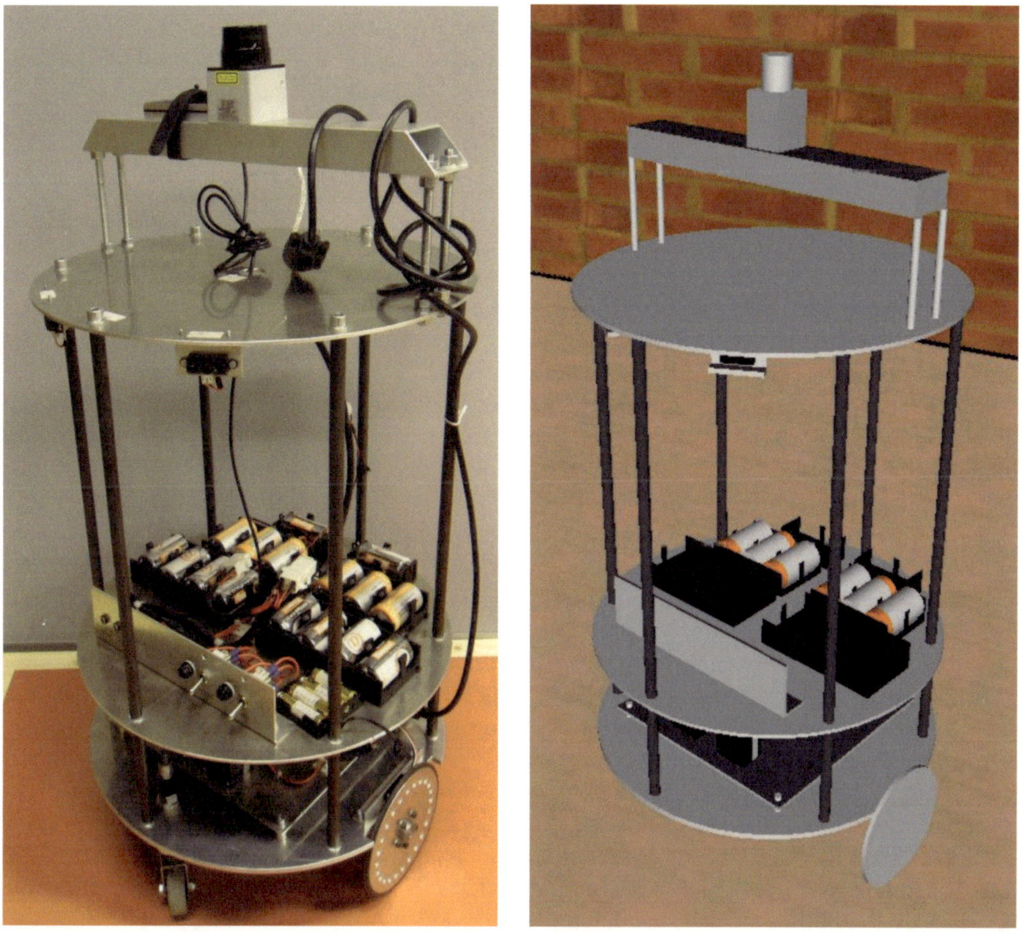

Fig. 1. Left panel: The differentially steered robot considered here. The laser range finder is mounted on top. Right panel: The simulated version of the robot.

to activate and de-activate the various brain processes. Here, only a brief description will be given. For a more detailed description, see Wahde (2009).

Brain processes are divided into several types. The two categories used here are *cognitive processes* (that do not make use of the robot's motors) and *motor behaviors* that *do* make use of the motors. In the UF method, any number of cognitive process can be active simultaneously, whereas exactly one motor behavior is active at any time, the rationale being that a robot of the kind considered here (with two wheels as its only actuators) can only carry out one motor action at any given time.

Each brain process (regardless of its type) is associated with a *utility function* whose task it is to determine the merit of running the brain process, in any given situation. The utility functions, in turn, depend on scalar *state variables*. A simple example of a state variable z may be the reading of an IR sensor mounted at the front of a robot. The value of z is an (admittedly incomplete) measure of the risks involved in moving forward. A more complex example is the *moving obstacle danger level* (see Subsect. 4.4 below), which uses consecutive LRF scans to obtain a scalar value measuring the risk of collision with moving obstacles.

State variable values are measured continuously, but asynchronously (since not all sensors are updated with the same frequency), and the most recent values available are used as inputs to the utility functions from which the utility value u_i for brain process i is obtained using the equation

$$\tau_i \dot{u}_i + u_i = \sigma_i \left(\sum_{k=1}^{m} a_{ik} z_k + b_i + \Gamma_i \right), \quad i = 1, \ldots, n, \tag{1}$$

where n is the number of brain processes, τ_i are time constants determining the reaction time of the robot (typically set to around 0.1 s), m is the number of state variables (denoted z_k), and a_{ik} and b_i are tunable parameters. The squashing function $\sigma_i(x)$ is taken as $\tanh(c_i x)$ where c_i is a positive constant (typically set to 1). Thus, the squashing functions σ_i serve to keep the utility values in the range $[-1, 1]$.

Once the values of the utility functions have been computed, decision-making is simple in the UF method, and works as follows: (i) The motor behavior with largest utility (among all motor behaviors) is active and (ii) any cognitive process with positive utility is active. All other brain processes are inactive.

The parameters Γ_i (referred to as *gamma parameters*), which are normally equal to zero, allow direct activation or de-activation of a brain process. Ideally, the state variables z_k should provide the robot will the information needed to make an informed decision regarding which brain processes to use in any situation encountered. However, in practice, the parameters Γ_i are needed in certain situations. Consider, for example, a simple case in which the utility function for an obstacle avoidance behavior depends on a single state variable z (multiplied by a positive constant a) that measures obstacle proximity (using, for example, IR or sonar sensors). Now, if the obstacle avoidance behavior is activated, the robot's first action is commonly to turn away from the obstacle. When this happens, the value z will then drop, so that u also drops, albeit with a slight delay depending on the value of the time constant in Eq. (1). At this point, the obstacle avoidance behavior is likely to be de-activated again, before actually having properly avoided the obstacle. The gamma parameters, which can be set directly by the brain processes, have been introduced to prevent such problems. Thus, for example, when the obstacle avoidance behavior is activated, it can set its *own* gamma parameter to a positive value, thus normally avoiding de-activation when the state variable z drops as described above. Whenever the obstacle avoidance behavior has run its course, it can set the gamma parameter to a large negative value, thus effectively de-activating itself. Note that the decision-making system is active continuously, so that, in the example above, obstacle avoidance *can* be de-activated (even after raising its own gamma parameter) should another brain process reach an even higher utility value. Once a gamma parameter has been set to any value (positive or negative) other than zero, its magnitude falls of exponentially with time, with a time constant (τ_i^{Γ}) specific to the brain process at hand. This time constant typically takes a larger value than the time constant in the utility function.

As mentioned above, the UF method allows several (cognitive) brain processes to run in parallel with the (single) active motor behavior. Thus, for example, while moving (using a motor behavior), a robot using the UF method would simultaneously be able to run two cognitive processes, one for generating odometric pose estimates from encoder readings and one for processing laser scans in order to recalibrate its odometric readings, if needed.

As is evident from Eq. (1), for a given set of state variables, the actual decisions taken by the robot will depend on the parameters τ_i, a_{ij}, b_i, and c_i as well as the values of the gamma parameters (if used). Thus, in order to make a robot carry out a specific task correctly, the user must set these parameters to appropriate values. Note that, for many cognitive processes (e.g. odometry, which normally should be running continuously), all that is needed is for the utility values to be positive at all times, which can easily be arranged by setting all the a-parameters for that process to zero, and the b-parameter to any positive value.

In many cases, the remaining parameters (especially those pertaining to the motor behaviors) can be set by hand using trial-and-error. In more complex situations, one can use a stochastic optimization method such as, for example, a genetic algorithm or particle swarm optimization to find appropriate parameters values. In those situations, one must first define a suitable objective function, for example the distance travelled in a given amount of time, subject to the condition that collisions should be avoided. Optimization runs of this kind are normally carried out in the simulator (described in Subsect. 5.1 below) rather than a real robot, since the number of evaluations needed during optimization can become quite large.

4. Brain processes

Just as the decision-making system, brain processes used in the UF method also have a unified structure; the main part of each brain process is a finite-state machine (FSM) consisting of *states* (in which various computations are carried out and, in the case of motor behaviors, actions are taken) and *conditional transitions* between states. In any brain process, the FSM executes the *current state* until a transition condition forces it to jump to another state, which is then executed etc. An illustration of a brain process FSM is shown in Fig. 8 below.

For the task considered here, namely reliable long-term navigation, six brain processes have been used, namely *Grid navigation* (denoted B_1), *Odometry* (B_2), *Odometric calibration* (B_3), *Moving obstacle detection* (B_4), *Moving obstacle avoidance* (B_5), and *Long-term memory* (B_6). The six brain processes will be described next.

4.1 Grid navigation

The navigation method used here is contained in a brain process called *Grid navigation*. As the name implies, the navigation method (which is intended for indoor, planar environments) relies on a grid. Dividing a given arena into a set of convex cells is a common problem in robotics, and it is typically approached using *Voronoi diagrams* (or their dual, *Delaunay triangulation*) (Okabe et al., 2000) or *Meadow maps* (Singh & Agarwal, 2010). However, these tessellations tend to generate jagged paths, with unnecessary turns that increase the length of the robot's path (a problem that can be overcome using *path relaxation*; see Thorpe (1984)). Nevertheless, in this paper, a different method for automatic grid generation will be used, in which the map of the arena is first contracted to generate a margin for the robot's movement. Next, the arena is divided into convex cells in a preprocessing step similar to that described by Singh & Wagh (1987). Note, however, that the method described in this paper also can handle non-rectangular arenas (and obstacles). Finally, the cells are joined in order to generate larger convex cells in which the robot can move freely. Once such a grid is available, a path is generated using Dijkstra's algorithm (see Sect. 4.1.2 below), optimizing the placement of waypoints (on the edges of the cells) in order to minimize the path length. The method for grid generation, which is summarized in Table 1, will now be described.

4.1.1 Grid generation method

It is assumed that a map is available, in the form of a set of closed *map curves* consisting, in turn, of a sequence of connected (i.e. sharing a common point) *map curve segments*, each

Step 1	Generate the contracted map.
Step 2	Generate the preliminary map curve intersection grid.
Step 3	Process non-axis parallel map curve segments.
Step 4	Generate the map curve intersection grid by removing cells outside the grid.
Step 5	Generate the convex navigation grid by joining cells.

Table 1. The algorithm for generating the convex navigation grid. See the main text for a detailed description of the various steps.

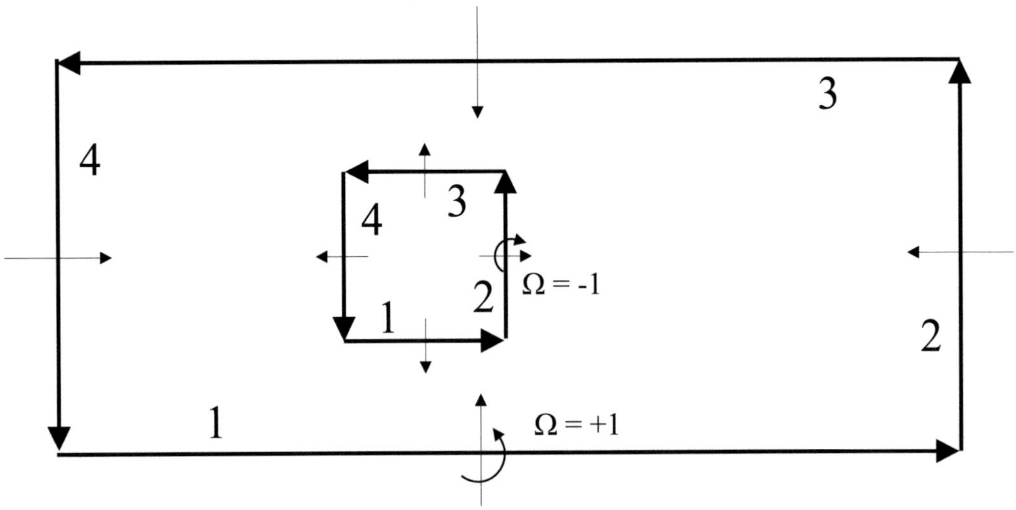

Fig. 2. The definition of in-directions in a rectangular arena with a square obstacle. In this simple arena, there are two map curves, each with four enumerated map curve segments. The thick arrow on each map curve segment points from the starting point to the end point of the segment. The thin arrows point towards the in-direction. The corresponding value of Ω (see Eq. (2)) is equal to $+1$ for the outer map curve (wall) and -1 for the inner map curve (obstacle). Note that the sign of Ω depends on the (arbitrary) choice of enumeration (clockwise or counterclockwise) for the map curve segments.

defined by a starting point $\mathbf{P}_a = (x_a, y_a)$ and an end point $\mathbf{P}_b = (x_b, y_b)$. Thus, each map curve is a simple (i.e. without self-intersections) polygon. In addition, for each map curve segment, an *in-direction* is defined. The in-direction Ω is given by the sign (either -1 or 1) of the z-component of the cross product between the vector $\mathbf{p} = \mathbf{P}_b - \mathbf{P}_a$ and a unit vector $\mathbf{d}$ (orthogonal to $\mathbf{p}$) pointing towards the side of the map curve segment which is accessible to the robot. Thus,

$$\Omega = \text{sgn}\left((\mathbf{p} \times \mathbf{d}) \cdot \hat{z}\right). \tag{2}$$

The in-direction concept is illustrated in Fig. 2. Note that, for any given map curve, all segments will have the same in-direction.

The method divides the accessible area of a robot's environment into a set of convex grid cells. Due to the convexity of the grid cells the robot can move freely within any grid cell, a property that facilitates path optimization, as described below. The grid generation method operates as follows: First (Step 1 in the algorithm described in Table 1), since the robot has a certain size, the actual map is shrunk to generated a *contracted map*. The margin μ (a tunable parameter) used when generating the contracted map should be such that, if the robot is positioned on a

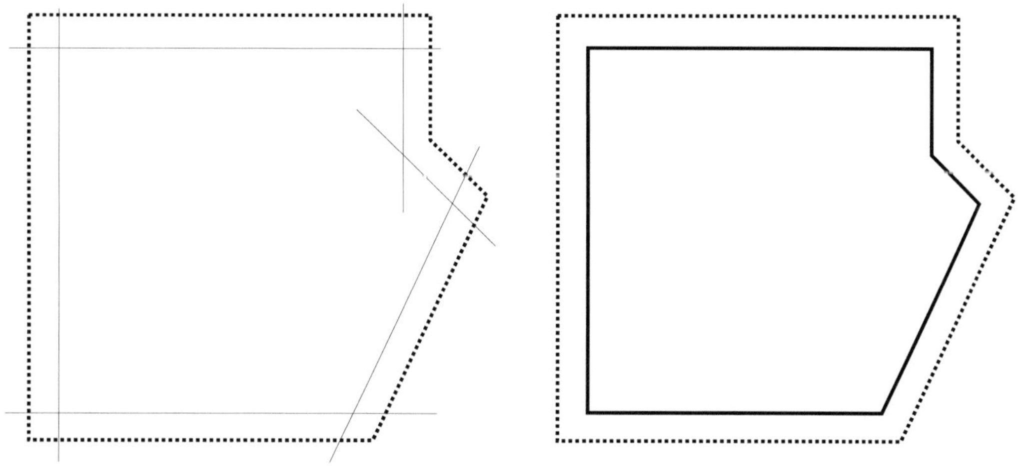

Fig. 3. The map contraction procedure. The left panel shows the original map (dotted line), along with the line segments obtained using the procedure described in the main text. In the right panel, the intersection points between those lines have been determined, resulting in the contracted map (solid line).

line in the contracted map, it should not touch any wall or other fixed obstacle. The procedure is illustrated in Fig. 3. The algorithm for generating the contracted map is straightforward: Consider a given line segment $\mathbf{p} = (p_x, p_y) = \mathbf{P}_b - \mathbf{P}_b = (x_b - x_a, y_b - y_a)$. Let $\mathbf{c}$ be a vector of length μ orthogonal to $\mathbf{p}$ and pointing towards the in-direction. Thus

$$\mathbf{c} = \mu\Omega\frac{(-p_y, p_x)}{\sqrt{p_x^2 + p_y^2}}. \tag{3}$$

It is now possible to obtain two points on the contracted map curve segment as

$$\mathbf{Q}_a = \mathbf{P}_a + \mathbf{c} \tag{4}$$

and

$$\mathbf{Q}_b = \mathbf{P}_b + \mathbf{c}. \tag{5}$$

Note, however, that these two points are normally not the end points of the contracted map curve segment. In order to determine the end points, one must first carry out the process described above for all map curve segments in a given map curve. Once that process has been completed, the end points of the contracted map curve segments can be determined as the intersection points between the lines passing through the point pairs generated during the contraction process just described. If these lines are extended to infinity, any given line may intersect many other lines. In that case, the points chosen as end points of the contracted map curve segments are taken as those intersection points that are nearest to the points $\mathbf{Q}_a$ and $\mathbf{Q}_b$, respectively. The process is illustrated in Fig. 3, where a map consisting of a single map curve, with six map curve segments, is contracted.

Note that, in its simplest form, the map contraction method requires that no contracted map *curve* should intersect any of the other contracted map curves. The cases in which this happens

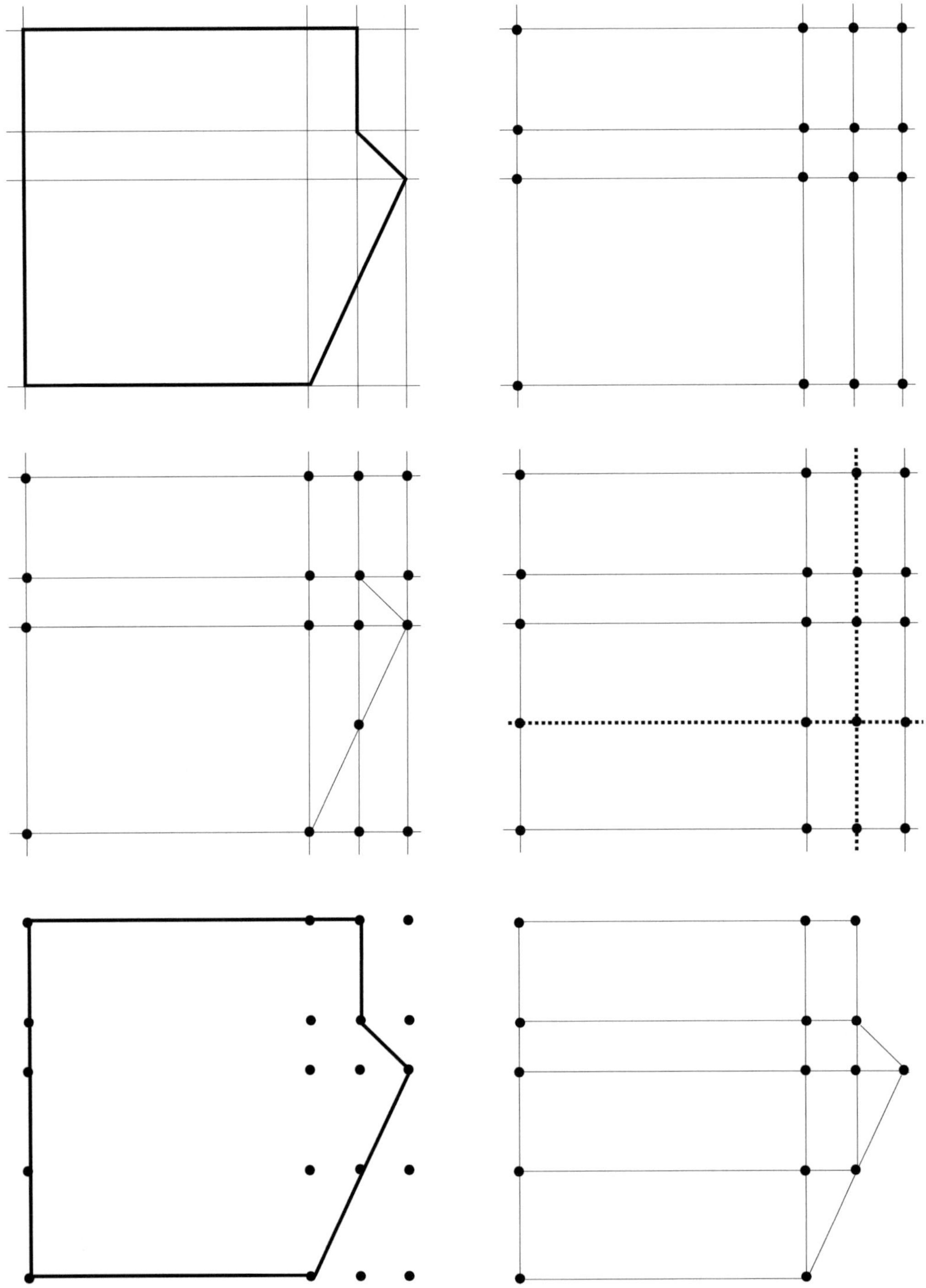

Fig. 4. The grid generation algorithm, applied to the contracted map from Fig. 3. See the main text for a detailed description of the various steps.

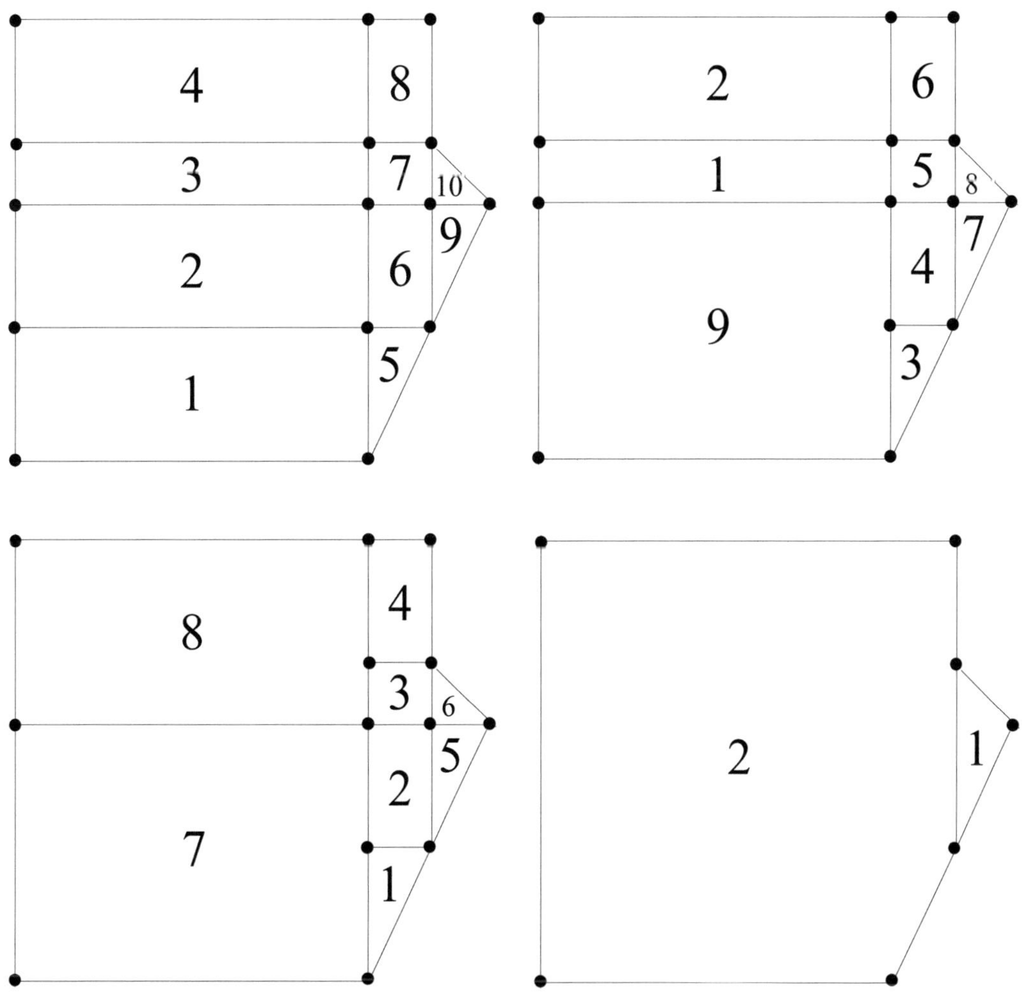

Fig. 5. The grid simplification algorithm, applied to the grid shown in Fig. 4. During grid simplification, polygonal cells are combined to form larger (but still convex) cells. Note that, in each step, the enumeration changes after the removal of the two cells that are joined. The upper left panel shows the initial grid, whereas the upper right panel shows the grid after the first step. In the bottom row, the left panel shows the grid after the second step, and the right panel shows the final result, after several steps.

can be handled, but the procedure for doing so will not be described here. Instead, we shall limit ourselves to considering maps in which the contracted map curves do not intersect.

Once the contracted map has been found, the algorithm (Step 2) proceeds to generate a temporary grid (referred to as the *preliminary map curve intersection grid*), consisting of the intersections between all axis-parallel (horizontal or vertical) map curve segments expanded to infinite length, as illustrated in the first row of Fig. 4. Next (Step 3), all non-axis parallel map curve segments are processed to check for intersections between those lines and the sides of the cells in the preliminary map curve intersection grid. For any intersection found, the corresponding point is added to the grid, as illustrated in the middle row (left panel) of Fig. 4. Lines are then drawn horizontally and vertically from the newly added points, and the points corresponding to any intersections between those two lines and the cells of the preliminary map curve intersection grid are added as well; see the right panel in the middle row of Fig. 4. Some of the cells will be located outside the (contracted) map, as shown in the left panel in the bottom row of Fig. 4. Those cells are removed (Step 4), resulting in the *map curve intersection grid*, shown in the final panel of Fig. 4.

Note that the cells in the map curve intersection grid are convex by construction. In principle, one could just remove the inaccessible cells and use the remaining cells during navigation. However, in a realistic arena (more complex than the simple arena used for illustrating the procedure) the map curve intersection grid typically contains very many cells, thus slowing down the path generation. Furthermore, with a large number of cells, the paths generated will typically have an unwanted zig-zag nature. Thus, in Step 5 of the algorithm, cells in the map curve intersection grid are joined to form larger cells, with the condition that two cells are only joined if the result is again a convex polygon, something that can easily be checked since, in a convex polygon, all cross products between consecutive sides (i.e. sides that share a common point) have the same sign. In the current version of the algorithm, no attempt has been made to generate a minimal (in number) set of convex polygons. Instead, the cells are simply processed in order, removing joined cells and adding newly formed cells, until there are no cells left such that their combination would be convex. The first steps of this process, as well as the final result, are shown in Fig. 5. The result is referred to as the *convex navigation grid*.

4.1.2 Path planning

The path planning procedure uses the convex navigation grid (which can be generated once and for all for any given (fixed) arena). Given a starting point (i.e. the robot's estimate of its current location) and a desired goal point, a path is generated using Dijkstra's algorithm (Dijkstra, 1959), augmented with a process for selecting navigation waypoints on the edges of the grid cells through which the path passes. As in the standard Dijkstra algorithm, cells are considered in an expanding pattern from the starting cell (i.e. the cell in which the robot is currently located). For any considered cell c, all neighbors a_i of c (i.e. those cells that, partially or fully, share a common side with c) are considered. For each a_i, a randomly generated waypoint q_{c,a_i} is selected (and stored) along the common side shared by the two cells (see Fig. 6). The distance from a_i to the starting point can then be computed and stored.

However, the path planning algorithm should also be able to handle situations in which a moving obstacle (which, of course, would not be included in the convex navigation grid) is found. If, for a line segment l connecting c to a_i, an intersection is found between l and a circle (with an added margin) around a detected moving obstacle (see below), a large penalty is added in the computation of the distance from a_i to the starting point, thus effectively rendering that path useless. Note that the robot, of course, has no global knowledge of the positions of moving obstacles: Only those moving obstacles that it can actually *detect* are included in the analysis. Note also that the path planning method will ignore any detected moving obstacle at a distance of at least Δ (a tunable parameter) from the robot. This is so, since there is no point to account for a faraway moving obstacle (which may not even be headed in the robot's direction) while planning the path.

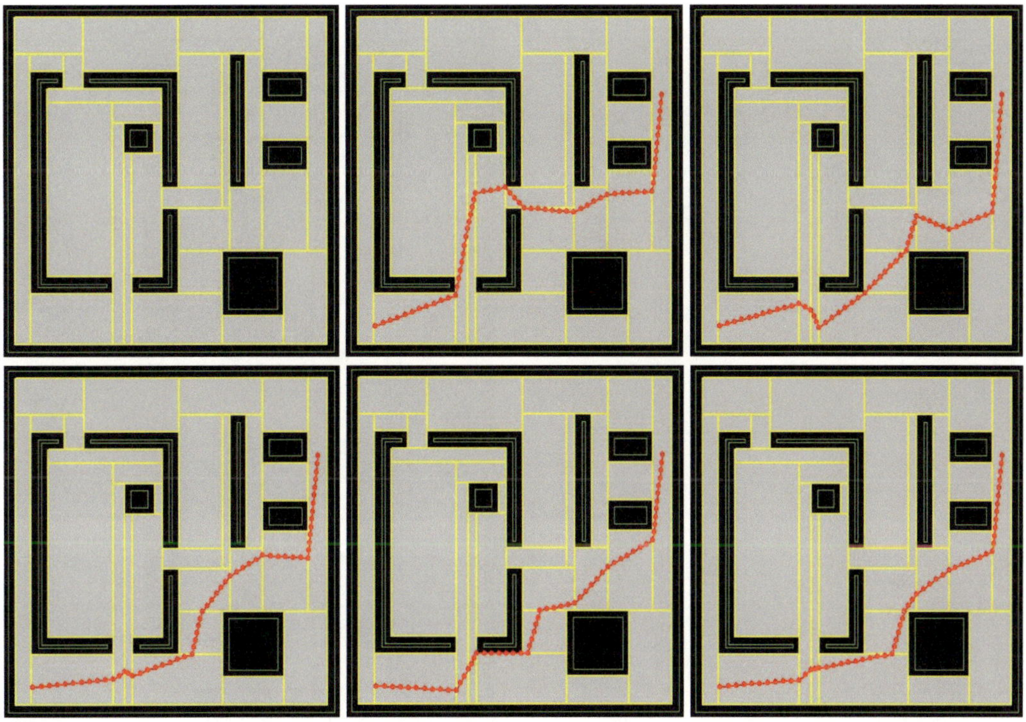

Fig. 6. Path planning: The upper left panel shows the convex navigation grid for the arena described in Sect. 6 and the remaining panels show the paths considered during path planning. In this case, the robot was located in the lower left corner of the arena, and its navigation target was located near the upper right corner. The first path considered is shown in the upper middle panel. After some time, the optimization procedure found a shorter path (upper right panel), along a different route. In the remaining steps (lower panels) the path planning algorithm further optimized the path (length minimization) by adjusting the placement of the waypoints (on the edges of the convex grid cells). Note that, in this figure, the actual navigation path is shown, i.e. not only the waypoints on the cell edges, but also all the intermediate (interpolated) waypoints.

The process is carried out for all grid cells, since the geometrically shortest path may involve a potential collision with a moving obstacle, as described above, in which case another path must be found, as shown in Fig. 7. A navigation path candidate is obtained once all grid cells have been considered.

However, as noted above, the exact locations of the waypoints q_{c,a_i} (on the edges of the navigation grid cells) are randomly chosen. Thus, an optimization procedure is applied, during which the entire path generation process is iterated n times, using different randomly selected waypoints along the cell sides (see also Fig. 6), and the shortest path thus found is taken as the final path. Once this path has been obtained, the actual *navigation path* is formed by adding interpolated waypoints between the waypoints obtained during path generation, so that the distance between consecutive waypoints is, at most, d_{wp} (a user-specified parameter).

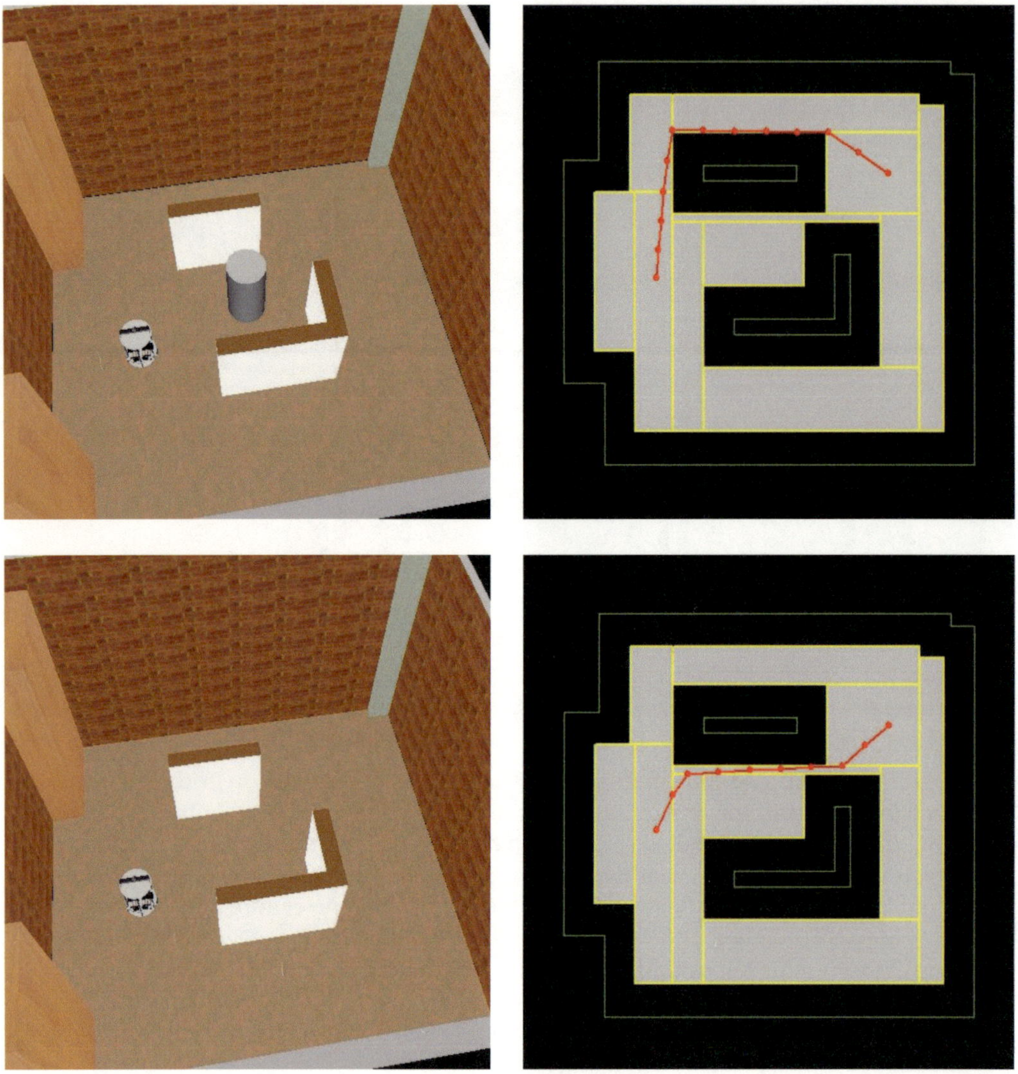

Fig. 7. Path planning with (upper panels) and without (lower panels) moving obstacles present. The moving obstacle is represented as a gray cylinder.

4.1.3 Navigation

The navigation motor behavior, illustrated in Fig. 8, consists of an FSM with 9 states. In State 1, the robot stops (path planning takes place at standstill), and jumps to State 2 in which the robot determines whether or not (based on its odometric position estimate) it is actually located in the convex navigation grid. This check is necessary, since the robot may have left the grid (slightly) as a result of carrying out moving obstacle avoidance; see below. If the robot finds that it is located *outside* the grid, it jumps to State 3 in which it checks, using the LRF[2], whether a certain sector (angular width γ_c, radius r_c) in front of the robot is clear, i.e. does

[2] It is assumed that the height of the moving obstacles is such that they are detectable by the (two-dimensional) LRF.

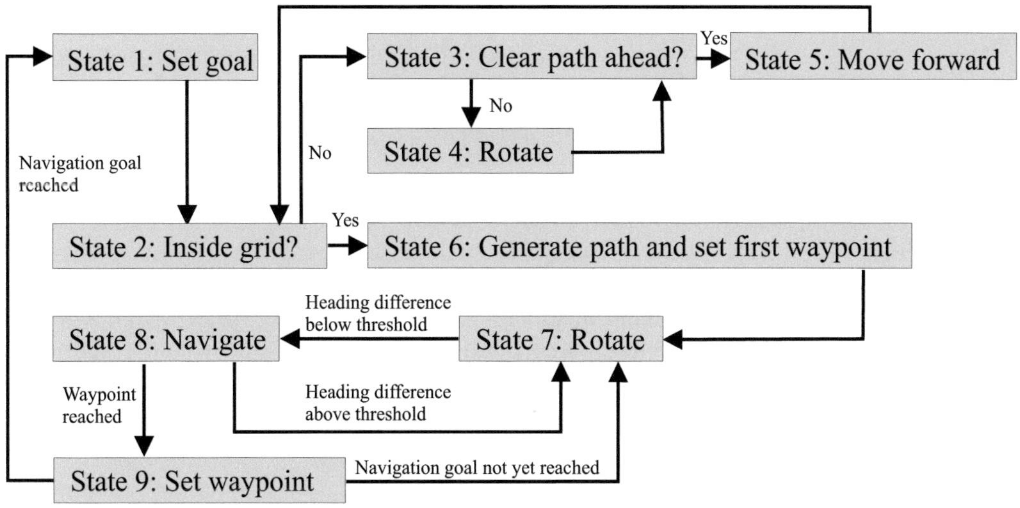

Fig. 8. The FSM implementing grid navigation.

not contain any obstacle. If this is so, the robot jumps to State 5 of the FSM, in which it sets the motor signals to equal, positive values, making the robot move forward, essentially in a straight line[3]. It then jumps back to State 2, again checking if it has reached the grid etc. If instead the robot finds that the path ahead is not clear, it jumps to State 4, in which it sets the motor signals to equal values but with opposite signs, so that it begins rotating without moving its center of mass. It then jumps back to State 3, to check whether the path ahead is clear.

If, in State 2, the robot finds that it is located *inside* the grid, it jumps to State 6, in which it determines the starting point (from the current odometric estimate) and the navigation goal (provided by the user), then generates an optimized path using the procedure described in Subsect. 4.1.2 and, finally, sets the current waypoint as the second waypoint in the path (the first point being its current location). It then jumps to State 7, in which it checks whether (the absolute value of) the difference[4] in estimated heading (using odometry) and desired heading (obtained using the direction to the current waypoint) exceeds a threshold δ. If so, it begins a pure rotation (as in State 4, see above). Once the difference in heading drops below the threshold, the robot jumps to State 8, in which the left and right motor signals are set as

$$v_{\mathrm{L}} = V_{\mathrm{nav}} - \Delta V \tag{6}$$

and

$$v_{\mathrm{R}} = V_{\mathrm{nav}} + \Delta V, \tag{7}$$

respectively, where

$$\Delta V = K_p V_{\mathrm{nav}} \Delta\varphi, \tag{8}$$

where $\Delta\varphi$ is the difference in heading angle, and K_p is a constant. While in State 8, the robot continuously checks the difference between estimated and desired heading. If it should

[3] Even though the motor settings are equal, the actual path may differ slightly from a straight line, due to actuator noise and other similar sources of error.

[4] The difference $\Delta\phi$ between two heading angles φ_1 and φ_2 is defined as $\varphi_1 - \varphi_2 \bmod 2\pi$.

exceed the threshold δ, the robot returns to State 7. If instead the robot reaches within a distance d_p (around 0.1 - 0.2 m) of the current waypoint, the robot jumps to state 9 where it checks whether it has reached the navigation goal (i.e. the final waypoint). If this is the case, the robot picks the next navigation goal from the list, and returns to State 1.
If not, the robot selects the next waypoint in the current path. Now, the simplest way to do so would be to just increase the index of the waypoint by one. Indeed, this is normally what happens. However, in certain situations, it may be favorable to skip one waypoint, namely in cases where the path changes direction abruptly. In such cases, passing all waypoints will lead to an unnecessarily jerky motion. Thus, in State 9, before selecting the next waypoint, the robot considers the angle between the vectors $\mathbf{v}_1 = \mathbf{p}_{i+1} - \mathbf{p}_i$ and $\mathbf{v}_2 = \mathbf{p}_{i+2} - \mathbf{p}_{i+1}$, where $\mathbf{p}_i$ is the waypoint that the robot just has reached, and $\mathbf{p}_{i+1}$ and $\mathbf{p}_{i+2}$ are the two next waypoints (this procedure is of course skipped if $\mathbf{p}_i$ or $\mathbf{p}_{i+1}$ is the final waypoint along the path). If this angle exceeds a certain threshold *and* the path ahead of the robot is clear (measured as in State 3, but possibly with different thresholds defining the sector in front of the robot), the robot skips waypoint $i+1$ and instead targets waypoint $i+2$.

4.2 Odometry

The *Odometry* brain process is straightforward: Given the estimated wheel rotations since the last update (based on the pulse counts from the wheel encoders), the robot uses the simple kinematic model of a differentially steered two-wheeled robot (with two additional supports for balance) to obtain an estimate of its velocity (v_x, v_y). Using standard forward kinematics, the robot also obtains estimates of its position (x, y) and angle of heading φ.

4.3 Odometric calibration

The purpose of the *odometric calibration* brain process is to correct the inevitable errors in the pose obtained from the *odometry* brain process. Odometric calibration relies on scan matching and builds upon an earlier version (called *laser localization*) described by Sandberg et al. (2009). However, whereas the earlier version required the robot to stop before correcting the odometric errors, the new version operates continuously and in concert with other brain processes. Thus, using the brain process taxonomy described in Sect. 3, the new version is a cognitive process rather than a motor behavior.
The FSM of the odometric calibration brain process operates as follows: In State 1, the robot checks whether the (modulus of the) difference in motor signals (between the left and right motor) exceeds a given threshold $T_{\Delta v}$. If it does, the robot remains in State 1, the rationale being that scan matching is less likely to give a reliable result if the robot is currently executing a sharp turn rather than moving forward (or backward). This condition is less of a restriction than one might think: Since the grid navigation method attempts to find the shortest path from the starting point to the goal point, the resulting path typically consists of long straight lines, with occasional turns where necessary.
When the wheel speed difference drops below the threshold, the robot jumps to State 2 of the FSM, where it stores the most recent laser scan and then carries out a laser scan match. The procedure for the scan match has been described in full detail by Sandberg et al. (2009), and will not be given here. Suffice it to say that the scan matching procedure compares the stored laser scan to a virtual scan taken in the map, by directly computing (essentially) the root mean square error between the laser ray readings. Thus, unlike many other methods, this method does *not* rely on the (brittle and error-prone) identification of specific landmarks. Furthermore, unlike the version presented by Sandberg et al. (2009), the current version is able to filter out

moving obstacles. It does so by noting that there is an upper limit on the pose errors, since the odometric calibration is executed at least a few times per second. Thus, laser rays for which the discrepancy between the expected and actual reading is too large to be caused by pose errors are simply ignored in the scan matching procedure; see also Subsect. 5.3 below.
If the scan match error is below a threshold T_{ϵ}, the robot concludes that its pose is correct and jumps back to State 1. However, if the error exceeds the threshold, a search procedure is initiated in State 3 of the FSM. Here, the robot carries out a search in a region of size $L_x \times L_y \times L_{\varphi}$ in pose space, around the estimated pose (obtained from odometry) recorded when the stored laser scan was taken.
Since the search procedure takes some time (though less than one second) to carry out, once it has been completed the robot must correct for the movement that has occurred since the stored laser scan was taken. This correction is carried out using odometry, which is sufficiently accurate to provide a good estimate of the *change* in the robot's pose during the search procedure (even though the *absolute* pose may be incorrect, which is the reason for running odometric calibration in the first place!)

4.4 Moving obstacle detection

As mentioned above, in the case considered here, the only long-range sensor is an LRF mounted on top of the robot. Thus, a method for detecting moving obstacles, i.e. any obstacle that is not part of the map, will have to rely on differences between the expected and actual LRF readings at the robot's current pose. Needless to say, this requires, in turn, that the robot's pose estimate, generated by the *Odometry* and *Odometric calibration* brain processes, is rather accurate. Assuming that this is the case, the FSM implementing the *Moving obstacle detection* (MOD) brain process works as follows: The FSM cycles between Steps 1 and 2, where Step 2 is a simple waiting state (the duration of which is typically set to around 0.02 to 0.10 s). In State 1, the robot compares the current LRF readings with the expected readings (given its pose in the map). Essentially, those readings for which the difference between the actual and expected readings exceed a certain threshold T_{mo} are considered to represent moving obstacles. Thus, even in cases where the robot's pose is slightly incorrect (due to odometric drift), it will still be able reliably to detect moving obstacles, provided that T_{mo} is set to a sufficiently large value. On the other hand, the threshold cannot be set *too* large, since the robot would not be able to detect, for example, moving obstacles just in front of a wall. In the investigation considered here, the value of T_{mo} was equal to 0.30 m.
Normally, when a moving obstacle is present within the range of the LRF, a set of consecutive LRF rays will be found to impinge on the obstacle. Given the first and last of those rays, estimates can be obtained for the position and radius of the obstacle. However, in order to minimize the risk of spurious detections, a list of moving obstacle *candidates* is maintained, such that only those candidates that fulfill a number of conditions are eventually placed in the actual list of moving obstacles. To be specific, detected obstacle candidates are removed if (i) they are located beyond a certain distance $d_{\mathrm{mo}}^{\mathrm{max}}$ or (ii) the inferred radius is smaller than a threshold $r_{\mathrm{mo}}^{\mathrm{min}}$. An obstacle candidate that survives these two checks is artificially increased in size by a given margin (typically 50%). On the other hand, for a detected obstacle candidate such that the inferred radius (after adding the margin) exceeds a threshold $r_{\mathrm{mo}}^{\mathrm{max}}$ (typically as a result of pose errors), the actual radius is set equal to $r_{\mathrm{mo}}^{\mathrm{max}}$.
Despite the precautions listed above, the robot will, from time to time, make spurious detections. Hence, a probability measure p_i for each moving obstacle i is introduced as well. A newly detected moving obstacle candidate is given probability $p_i = p_0$. If the

obstacle candidate is found again (in the next time step), its probability is modified as $p_i \leftarrow \max(\alpha p_i, 1)$, where $\alpha > 1$ is a constant. If, on the other hand, the previously detected obstacle candidate is not found, the probability is modified as $p_i \leftarrow \beta p_i$, where $\beta < 1$ is a constant. Whenever the probability p_i drops below a threshold $p_{\min}$, the corresponding obstacle is removed from the list of candidates. Finally, the remaining candidates are added to the list of actual moving obstacles, when they have been detected (i.e. with probability above the minimum threshold) for at least ΔT_{mo} s (a user-specified parameter, typically set to around 0.5 s).

In order to avoid multiple detections of the same moving obstacle (remembering that the MOD method runs many times per second), detected moving obstacle candidates that are within a distance $\delta_{\text{mo}}^{\max}$ of a previously detected obstacle candidate are identified with that obstacle candidate. Obstacle candidates that do not fulfill this condition are added as separate obstacle candidates. In addition to generating a list of moving obstacles, the brain process also determines which of the moving obstacles is closest to the robot.

Furthermore, a scalar variable, referred to as the *moving obstacle danger level* (denoted μ) is computed (also in State 1 of the FSM). This variable is used as a state variable for the decision-making system, and influences the activation (or de-activation) of the moving obstacle avoidance motor behavior, which will be described below. Needless to say, a scalar variable for assessing the threat posed by a moving obstacle can be generated in a variety of ways, without any limit on the potential complexity of the computations involved. One may also, of course, consider using additional long-range sensors (e.g. cameras) in this context. Here, however, a rather simple definition has been used, in which the robot only considers moving obstacles in the frontal[5] half-plane (relative direction from $-\pi/2$ to $\pi/2$). The robot first computes the change in distance to the nearest obstacle between two consecutive executions of State 1. If the distance is increasing, the current danger level m is set to zero. If not, the current danger level is determined (details will not be given here) based on the distance and angle to the moving obstacle. Next, the actual danger level μ is obtained as $\mu \leftarrow (1-\alpha)\mu + \alpha m$, where α is a constant in the range $]0,1]$.

4.5 Moving obstacle avoidance

The *Moving obstacle avoidance* (MOA) motor behavior is responsible for handling emergency collision avoidance[6]. Once the MOA motor behavior has been activated[7], it devotes all its computational capacity to avoiding the nearest moving obstacle; it makes no attempt to assess the danger posed by the obstacle, a task that is instead handled by the decision-making system, using the moving obstacle danger level computed in the MOD brain process. Just as in the case of the MOD brain process, MOA can also be implemented in complex and elaborate ways.

Here, however, a rather simple MOA method has been used, in which the region in front of the robot is divided into four zones, as illustrated in Fig. 9. Should the MOA motor behavior be activated, the actions taken by the robot depend on the zone in which the moving obstacle

---

[5] The robot thus only handles frontal collisions; it is assumed that a moving obstacle, for example a person, will not actively try to collide with the robot from behind.

[6] Note that the grid navigation motor behavior also handles *some* collision avoidance, in the sense that it will not deliberately plan a path that intersects a nearby moving obstacle.

[7] Note that, whenever the MOA motor behavior is activated, the grid navigation motor behavior is turned off, and vice versa. This is so since, in the UF method, only one motor behavior is active at any given time; see Sect. 3.

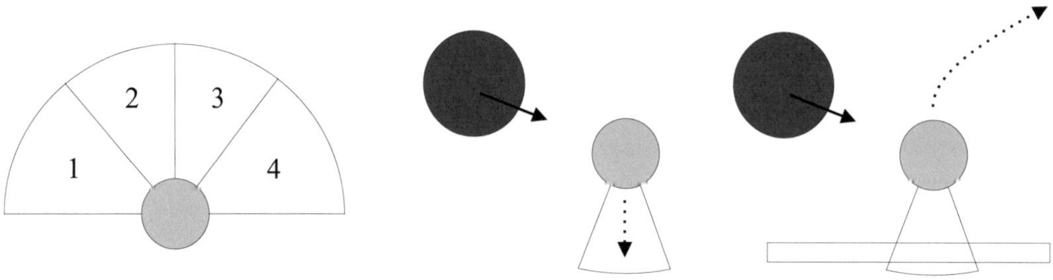

Fig. 9. Left panel: The zones used in the moving obstacle avoidance motor behavior. The robot is shown as a light gray disc. Middle and right panels: The two evasive maneuvers defined for cases in which the moving obstacle (dark gray disc) is located in Zone 1. If the robot is able to move backwards (based on the readings from the rear IR sensor) it does so (middle panel). On the other hand, if an obstacle behind the robot prevents it from moving backwards (right panel), it attempts instead to evade the obstacle by moving forward and to the right.

is located. For example, if the moving obstacle is located in Zone 1, the robot checks (using the rear IR sensor) whether it can move backwards without collisions. If so, it sets the motor signals to equal, negative values, thus (asymptotically) moving straight backwards, until it has moved a certain, user-specified distance, at which point the robot stops as the MOA motor behavior is awaiting de-activation.

If the robot is unable to move backwards (either initially or after some time), it instead attempts to circumnavigate the moving obstacle by turning right, setting the left motor signal to a large positive value, and the right motor signal to a small negative value. Once the robot has turned a certain angle (relative to the initial angle of heading), it instead sets the motor signals to equal positive values, thus (asymptotically) moving forward in a straight line, until it has moved a certain distance. At some point, the obstacle will disappear from the field of view of the LRF (either because it has passed the robot, or as a result of the robot's evasive action), at which point the obstacle danger level (μ) drops, and the MOA motor behavior is de-activated shortly thereafter, so that grid navigation can be resumed. A similar, but somewhat more elaborate, procedure is used if the obstacle is located in Zone 2. Zones 3 and 4 are handled as Zones 2 and 1, respectively, but with the motor signal signs etc. reversed.

The utility function for the MOA motor behavior uses the obstacle danger level (see Subsect. 4.4) as a state variable. Hence, when the perceived obstacle danger level rises above a threshold, which is dependent on the parameter values used in Eq. (1), the MOA motor behavior is activated. At this point, the robot attempts to avoid the moving obstacle, as described above. In cases where it does so by turning, the perceived obstacle danger level typically drops. Thus, in order to avoid immediate de-activation, the MOA motor behavior uses a gamma parameter to raise its own utility once activated (in State 1); see also Sect. 3 above.

4.6 Long-term memory

The *Long-term memory* (LTM) brain process simply stores (and makes available to the other brain processes) the resources necessary for the robot to carry out its navigation task. In the case considered here, the LTM stores the map of the arena (needed during odometric calibration) and the convex navigation grid (needed for navigation).

5. Implementation and validation

5.1 The general-purpose robotics simulator

A simulator referred to as the *General-purpose robotics simulator* (GPRSim) has been written with the main purpose of evaluating the UF decision-making method. The simulator allows a user to set up an arbitrary arena, with or without moving obstacles, and a robot, including both its physical properties (e.g. actuators, sensors, physical parameters such as mass, moment of inertia etc.) and its artificial brain (i.e. a set of brain process, as described above, and the decision-making system implementing the UF method, also described above). In order to make the simulations as realistic as possible, the simulator introduces noise at all relevant levels (e.g. in sensors and actuators). Moreover, simulated sensor readings are taken only with the frequency allowed by the physical counterpart.

5.2 Simulations vs. reality

Simulations are becoming increasingly important in many branches of science, and robotics is no exception. However, in the early days of behavior-based robotics many researchers expressed doubts as to the possibility of transferring results obtained in simulations onto real robots. In particular, Brooks took an especially dim view of the possibility of such transfer, as discussed in Brooks (1992). Similar critique of simulations has been voiced by, among others, Jakobi et al. (1995) and Miglino et al. (1996). However, their critique mainly targeted naive simulation approaches that use idealized sensors and actuators in block-world-like environments. In fact, both Jakobi et al. (1995) and Brooks (1992) also note that simulations are indeed useful, provided that great care is taken to simulate, for example, actuators and sensors in a realistic manner and to validate the simulation results using real robots. Furthermore, even though the discrepancy between simulations and reality is likely to increase as ever more complex robotic tasks are considered (as argued by Lipson (2001)), it is also likely that carrying out proper tests of such robots will, in fact, *require* simulations to be used, particularly if the simulations can be executed faster than real time. In addition, careful simulations allow researchers to test various hardware configurations before embarking on the often costly and time-consuming task of constructing the real counterpart.

In our view, simulations are an essential tool in mobile robot research, but should only be applied in cases where one can argue convincingly that (i) the various hardware components (e.g. sensors and actuators) can be simulated in a reliable way, and (ii) the actions taken do not require extreme accuracy regarding time or space. As an example regarding simulation of hardware components, consider the sensors used on our robot; in the application considered here almost all relevant sensing is carried out using the LRF, a sensor that is, in fact, very simple to simulate, in view of its high accuracy: Our LRF has a typical error of $\pm$ 1 mm out to distances of around 4 m. On the other hand, if a *camera* had been used on the real robot, it is unlikely that a reliable simulation could have been carried out. Even though GPRSim allows an almost photo-realistic representation of the environment, the readings from a simulated camera would be unlikely to capture all aspects (such as shadows, variations in light levels etc.) of vision. Regarding time and space, the implementation of the navigation method (described above) introduces, by construction, a margin between the robot and the arena objects. Thus, while the actions of the simulated robot may differ somewhat from those of a real robot (and, indeed, also differ between different runs in the simulator, depending on noise in actuators and sensors), there is little risk of *qualitative* differences in the results obtained.

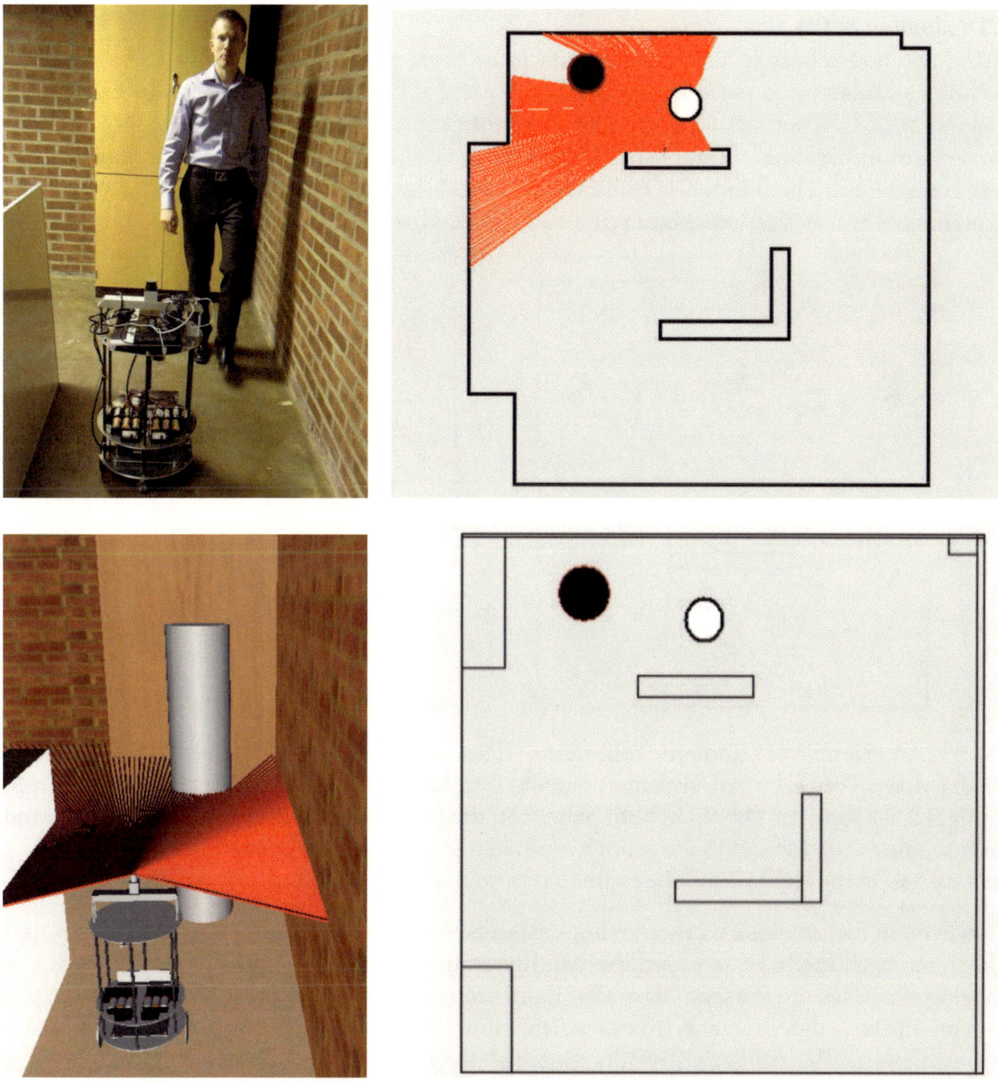

Fig. 10. Upper panels: Detection of a moving obstacle by the real robot. Left: The person approaching the robot. Right: a screenshot showing the LRF rays (for the real robot). The robot is shown as a white disc at its estimated pose, and the detected moving obstacle as a black disc. Lower panels: The same situation, shown for the simulator. Here, the laser rays (which obviously are not visible in the real robot) have been plotted in the screenshot rather than in the map view shown in the right panel.

Nevertheless, before using any simulator for realistic robotics experiments, careful validation and system identification must be carried out, a process that will now be described briefly, in the case of our simulator.

5.3 Validation of the simulator

Using the real robot, several experiments have been carried out in order to validate the GPRSim simulator and to set appropriate parameter values for the simulator. In fact, careful validation of GPRSim (as well as system identification of sensors and actuators) has been carried out in previous (unpublished) work, in which navigation over distances of 10-20 m was considered. The odometric calibration method has been carefully tested both in those experiments and as a separate brain process; see Sandberg et al. (2009).

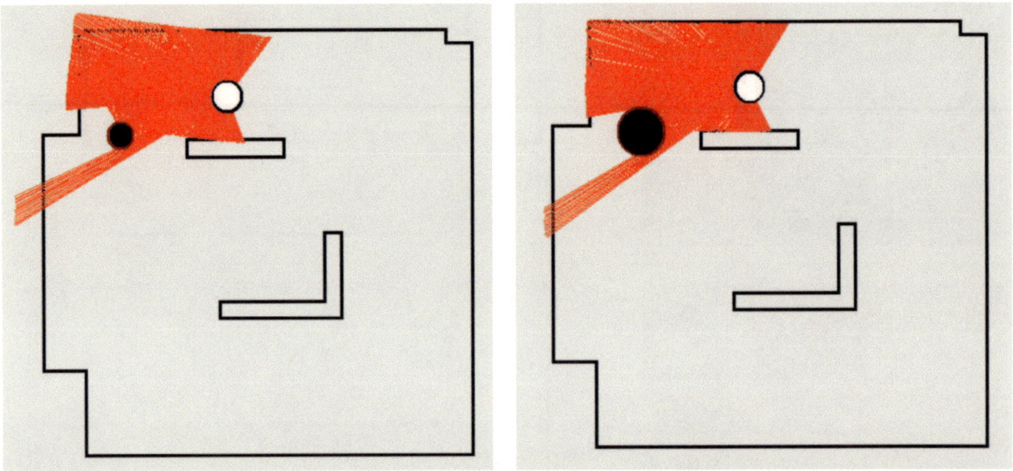

Fig. 11. An example of odometric calibration in the presence of a moving obstacle (shown as a black disc). The real robot, shown as a white disc, was used. The same laser readings (from a single laser scan) are shown in both panels. In the left panel, the origin and direction of the laser readings correspond to the robot's estimated (and incorrect) pose before odometric calibration. In the right panel, the estimated pose has been corrected.

However, in the previous work, moving obstacles were not included. Thus, here, a thorough effort has been made to ascertain the validity of the simulation results in situations where moving obstacles are present. Basically, there are two things that must be tested namely (i) moving obstacle detection and (ii) odometric calibration in the presence of moving obstacles. An example of the validation results obtained for moving obstacle detection is shown in Fig. 10. The upper panels show the results obtained in the real robot, and the lower panel show the results from GPRSim, for a given robot pose and obstacle configuration. As can be seen, the robot (shown as a white disc) detected the moving obstacle (shown as a black disc) in both cases.

Fig. 11 shows the results (obtained in the real robot) of a validation test regarding odometric calibration. In this particular test, the actual position of the robot was $(x, y) = (2.45, 4.23)$ m. The heading angle was equal to π radians (so that the robot was directed towards the negative x-direction). For the purposes of testing the odometric calibration process, an error was introduced in the *estimated* pose; the estimated position was set to $(x, y) = (2.25, 4.20)$ m, and the estimated heading to 3.0 radians. The left panel shows the robot and the moving obstacle (black disc), as well as the actual LRF readings, plotted with the origin and direction provided by the robot's *estimated* pose. As can be seen, the map is not very well aligned with the readings. The right panel shows the same laser readings, again with origin and direction given by the robot's estimated pose, but this time using the estimate obtained *after* odometric calibration. At this point, the alignment is almost perfect, despite the presence of the moving obstacle. Repeating the experiment 10 times, the estimated position (after odometric calibration) was found to be $(x, y) = (2.40 \pm 0.02, 4.22 \pm 0.01)$ m and the estimated heading was 3.11 ± 0.02 radians, where the errors correspond to one standard deviation. The corresponding values found in the simulator were $(x, y) = (2.38 \pm 0.03, 4.23 \pm 0.01)$ m, whereas the estimated heading was found to be 3.13 ± 0.02 radians.Several experiments of this kind were carried out. In all cases, both the simulated robot and the real robot were able to correct their estimated pose using odometric calibration, with essentially identical level of precision.

6. Results

The (simulated) arena used for studying long-term robustness is shown from above in the upper left panel of Fig. 12. The size of the arena, which represents a warehouse environment, is 10×10 m. The large object near the lower right corner represents an elevator shaft (inaccessible to the robot).
Regarding the decision-making system, a single state variable was used, namely the obstacle danger level described in Subsect. 4.4 above. The four cognitive brain processes (*Odometry, Odometric calibration, Moving obstacle detection,* and *Long-term memory*) were allowed to run continuously. Hence, in their utility functions, the parameters a_{ij} were equal to zero, and b_i was positive. The main task of the decision-making system was thus to select between the two motor behaviors, namely *Grid navigation* (B_1) and *Moving obstacle avoidance* (B_5), using the obstacle danger level as the state variable (z_1). Since only the relative utility values matter (see Sect. 3), the parameters a_{1j} were set to zero, b_{1j} was set to a positive value (0.25), and the time constant τ_1 was set to 0.10 s, thus making u_1 rapidly approach a constant value of around 0.245. After extensive testing, suitable parameters for B_5 were found to be: $a_{51} = 1.00$, $b_5 = -0.45$, and $\tau_1 = 0.10$ s. The gamma parameter Γ_5 (used as described in Subsect. 4.5 above) was set to 3.0 in State 1 of B_5. The corresponding time constant (for exponential decay of the gamma parameter) was set to 1.0 s. The sigmoid parameters c_i were set to 1 for all brain processes.
Several runs were carried out with GPRSim, using the robot configuration described above. Here, the results of two runs will be presented.

6.1 Run 1: Stationary arena

As a first step, a long run (*Run 1*) was carried out, without any moving obstacle present, with the intention of testing the interplay between navigation, odometry, and odometric calibration. The robot was released in the long corridor near the left edge of the arena, and was then required to reach a set of target points in a given order. A total of 11 different target points were used. When the final point was reached, the robot's next target was set as the first target point, thus repeating the cycle. The right panel of Fig. 12 shows the locations of the 11 target points. The robot's trajectory during the first 100 meters of movement (256 s) is shown in the lower left panel of the figure. Both the actual trajectory (thick green line) and the odometric trajectory (thin red line) are shown. As can be seen, the robot successfully planned and followed the paths between consecutive target points. Over the first 100 m of the robot's

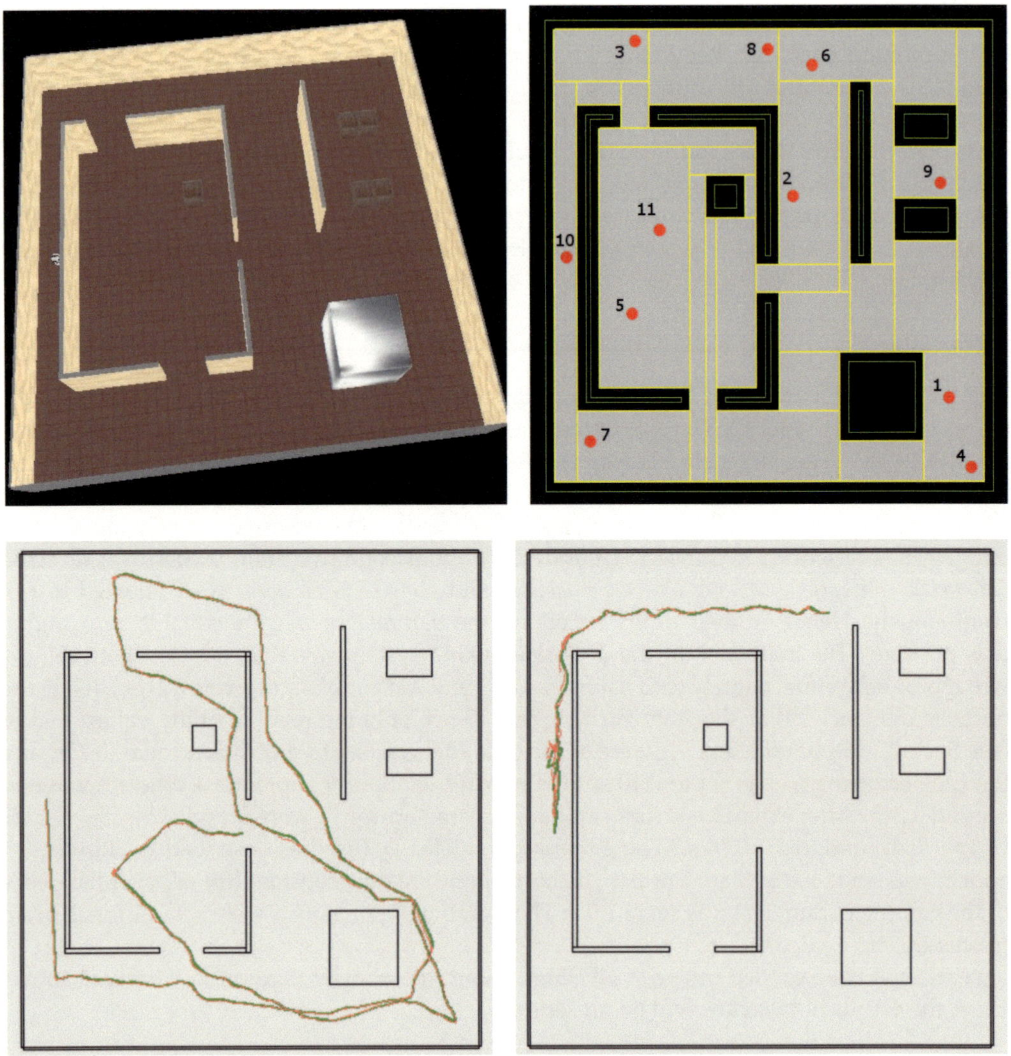

Fig. 12. Upper left panel: The arena used for the simulations. Upper right panel: The 11 target points, with enumeration. Note that, after reaching the 11th target point, the robot's next target becomes the first target point etc. The robot is initially located near target point 10, as can be seen in the upper left panel. Lower left panel: The trajectory for the first 100 m of navigation. The actual trajectory is shown as a thick green line, and the odometric trajectory as a thin red line. Lower right panel: A later part of the trajectory showing the robot approaching 200 m of distance covered. As can be seen, in the long corridor, the robot suffered a rather large pose error at one point. The error was swiftly corrected, however, so that the robot could continue its task.

movement, the average position error[8] was around 0.063 m, and the average heading error

[8] The average position error is computed by averaging the *modulus* of the position error over time. Similarly the average heading error is obtained by averaging the *modulus* of the heading error (modulo 2π) over time.

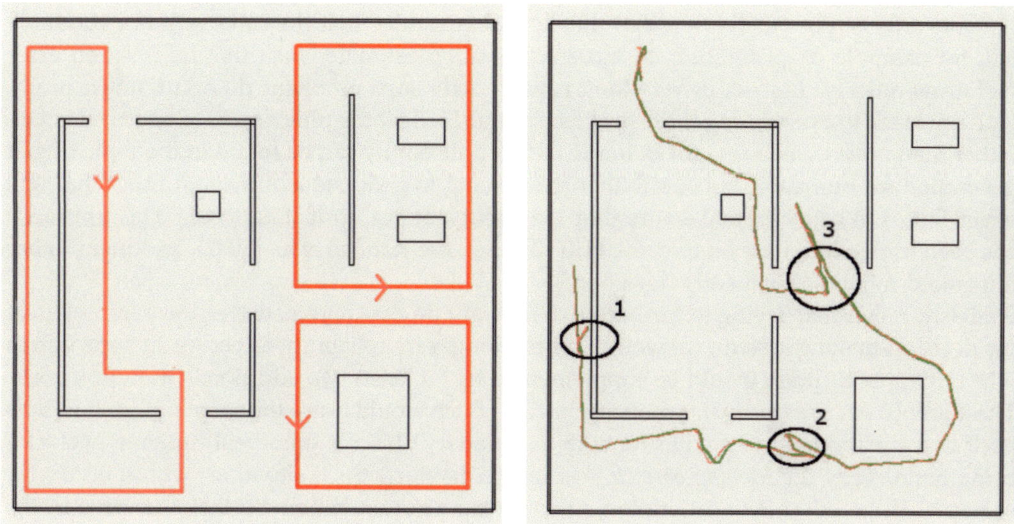

Fig. 13. Left panel: The trajectories of the moving obstacles in Run 2. Right panel: The robot's trajectory during the first 50 m of movement. The ellipses highlight the robot's three evasive maneuvers.

0.031 radians (1.8 degrees). Note, however, that most of the contribution to the error comes from a few, rather rare occasions, where the robot's pose accuracy was temporarily lost. In all cases, the robot quickly recovered (through continuous odometric calibration) to find a better pose estimate. An example of such an error correction is shown in the lower right panel of Fig. 12.

6.2 Run 2: Arena with moving obstacles

In the second run (*Run 2*) presented here, the robot moved in the same arena as in Run 1, and with the same sequence of target points. However, in Run 2, there were three moving obstacles present in the arena. The approximate trajectories of the moving obstacles are illustrated in the left panel of Fig. 13. The right panel shows the first 50 m of the robot's movement. Here, the average position error was around 0.068 m and the average heading error around 0.025 radians (1.4 degrees). The ellipses highlight the three cases where the robot was forced to make an evasive maneuver (during the first 50 m of movement) in order to avoid a collision with a moving obstacle. Note that the odometric calibration process managed to maintain a low pose error even during the evasive maneuver. The run was extended beyond 100 m, with only a small increase in pose error; after 100 m, the average position error was 0.096 m and the average heading error 0.038 radians (2.2 degrees). In this run, the first 100 m took 352 s to traverse (compared to the 256 s in Run 1), due to the evasive maneuvers. Note that, after the third evasive maneuver, the robot modified the original path towards target point 3 (cf. the lower left panel of Fig. 12), having realized that a shorter path was available after evading the moving obstacle.

7. Discussion and conclusion

Even though the results presented in Sect. 6 show that the method used here is capable of achieving robust navigation over long distances in the presence of moving obstacles, a possible criticism of our work is that there is no guarantee that the robot will not suddenly fail, for example by generating an incorrect heading estimate, such that its position error will grow quickly. Indeed, in very long runs, exactly such problems do occur, albeit rarely. Our approach to overcoming those problems would follow the philosophy of error correction rather than error avoidance. An example of this philosophy can be found in the topic of gait generation for humanoid robots: Rather than trying to generate a humanoid robot that will never fall, it is better to make sure that the robot can get up if it *does* fall. This approach has been applied in, for example, the HRP-2 robot, see Kaneko et al. (2002), and many other humanoid robots subsequently developed.

Similarly, rather than trying to further tune the brain process parameters or the parameters of the decision-making system, we would, for full long-term reliability, advocate an approach in which the robotic brain would be supplemented by (at least) two additional brain processes. These are: (i) An *Emergency stop* motor behavior which would cause the robot to stop if it finds itself in a situation where a collision with a stationary object is imminent (moving obstacles being handled by the *Moving obstacle avoidance* behavior). Such situations would occur, for example, if the odometric calibration suffers a catastrophic failure so that the error in the robot's estimated pose grows without bound. Once the robot has stopped it should turn towards the most favorable direction for calibration (which would hardly be the direction towards the stationary object and neither any direction in which *no* part of the map would be visible). Once a suitable direction has been found, the robot should activate the second added brain process, namely *Wake-up*. This cognitive process would attempt a global search (again matching the current laser scan to virtual scans taken in the map) for the robot's current pose. With these two brain processes, which are to be added in future work, the robot would be able to correct even catastrophic errors.

As for the currently available brain processes, the *Moving obstacle avoidance* brain process could be improved to handle also situations in which *several* moving obstacles are approaching the robot simultaneously. However, to some extent, the brain process already does this since, before letting the robot move, it generally checks whether or not there is a clear path in the intended direction of movement; see also the right panel of Fig. 9. In cases where multiple moving obstacles are approaching, typically the best course of action (at least for a rather slow robot) is to do nothing.

Another obvious part of our future work is to carry out long-term field tests with the real robot. Such field tests could, for example, involve a delivery task in an office environment, a hospital, or a factory.

To conclude, it has been demonstrated that the approach of using modularized brain processes in a given, unified format coupled with decision-making based on utility functions, is capable of robustly solving robot motor tasks such as navigation and delivery, allowing the robot to operate autonomously over great distances. Even so, the robot may occasionally suffer catastrophic errors which, on the other hand, we suggest can be dealt with by the inclusion of only a few additional brain processes.

8. References

Arkin, R. C. (1987). *Towards Cosmopolitan Robots: Intelligent Navigation in Extended Man-Made Environments*, PhD thesis, University of Massachusetts.

Arkin, R. C. (1998). *Behavior-Based Robotics*, MIT Press.

Brooks, R. A. (1986). A robust layered control system for a mobile robot, *IEEE J. of Robotics and Automation* RA-2(1): 14–23.

Brooks, R. A. (1992). Artificial life and real robots, *Proceedings of the First European Conference on Artificial Life*, pp. 3–10.

Bryson, J. J. (2007). Mechanisms of action selection: Introduction to the special issue, *Adaptive Behavior* 15: 5–8.

Dijkstra, E. W. (1959). A note on two problems in connexion with graphs, *Numerische Mathematik* 1: 269–271.

Gat, E. (1991). *Reliable Goal-directed Reactive Control for Real-world Autonomous Mobile Robots*, PhD thesis, Virginia Polytechnic Institute and State University.

Jakobi, N., Husbands, P. & Harvey, I. (1995). Noise and the reality gap: The use of simulation in evolutionary robotics, *in* F. Morán, A. Moreno, J. J. Merelo & P. Chacón (eds), *Proceedings of the Third European Conference On Artifical Life, LNCS 929*, Springer, pp. 704–720.

Kaneko, K., Kanehiro, F., Kajita, S. et al. (2002). Design of prototype humanoid robotics platform for HRP, *Proceedings of the IEEE/RSJ International Conference on Intelligent Robots and Systems (IROS2002)*, pp. 2431–2436.

Lipson, H. (2001). Uncontrolled engineering: A review of nolfi and floreano's evolutionary robotics (book review), *Artificial Life* 7: 419–424.

McFarland, D. (1998). *Animal Behaviour: Psychobiology, Ethology and Evolution, 3rd Ed.*, Prentice Hall.

Miglino, O., Lund, H. H. & Nolfi, S. (1996). Evolving mobile robots in simulated and real environments, *Artificial Life* 2: 417–434.

Okabe, A., Boots, B., Sugihara, K. & Chiu, S. N. (2000). *Spatial Tessellations: Concepts and Applications of Voronoi Diagrams*, 2nd edn, John Wiley and Sons, Ltd.

Pirjanian, P. (1999). Behavior coordination mechanisms - state-of-the-art, *Technical report*, Institute for Robotics and Intelligent Systems, University of Southern California.

Prescott, T. J., Bryson, J. J. & Seth, A. K. (2007). Introduction. modelling natual action selection, *Philos. Trans. R. Soc. Lond. B* 362: 1521–1529.

Sakagami, Y., Watanabe, R., Aoyama, C., Matsunaga, S., Higaki, N. & Fujimura, K. (2002). The intelligent ASIMO: System overview and integration, *Proceedings of the 2002 IEEE/RSJ International Conference on Intelligent Robots and Systems*, pp. 2478–2483.

Sandberg, D., Wolff, K. & Wahde, M. (2009). A robot localization method based on laser scan matching, *in* J.-H. Kim, S. S. Ge, P. Vadakkepat, N. Jesse et al. (eds), *Proocedings of FIRA2009, LNCS 5744*, Springer, pp. 179–186.

Singh, J. S. & Wagh, M. D. (1987). Robot path planning using intersecting convex shapes: Analysis and simulation, *IEEE Journal of Robotics and Automation* RA-3: 101–108.

Singh, S. & Agarwal, G. (2010). Complete graph technique based optimization in meadow method of robotic path planning, *International Journal of Engineering Science and Technology* 2: 4951–4958.

Thorpe, C. E. (1984). Path relaxation: Path planning for a mobile robot, *Proceedings of AAAI-84*, pp. 318–321.

von Neumann, J. & Morgenstern, O. (1944). *Theory of Games and Economic Behavior*, Princeton University Press.

Wahde, M. (2003). A method for behavioural organization for autonomous robots based on evolutionary optimization of utility functions, *Journal of Systems and Control Engineering (IMechI)* 217: 249–258.

Wahde, M. (2009). A general-purpose method for decision-making in autonomous robots, *in* B.-C. Chien, T.-P. Hong, S.-M. Chen & M. Ali (eds), *Next-Generation Applied Intelligence, LNAI 5579*, Springer, pp. 1–10.

14

Fuzzy Logic Based Navigation of Mobile Robots

Amur S. Al Yahmedi and Muhammed A. Fatmi
Sultan Qaboos University,
Oman

1. Introduction

Robots are no longer confined to engineered, well protected sealed corners, but they are currently "employed" in places closer and closer to "us". Robots are getting out of factories and are finding their way into our homes and to populated places such as, museum halls, office buildings, schools, airports, shopping malls and hospitals.

The gained benefit of the potential service and personal robots comes along with the necessity to design the robot in a way that makes it safe for it to interact with humans and in a way that makes it able to respond to a list of complex situations. This includes at least the possibility to have the robot situated in an unknown, unstructured and dynamic environment and to navigate its way in such an environment. One of the fundamental issues to be addressed in autonomous robotic system is the ability to move without collision. An "intelligent" robot should avoid undesirable and potentially dangerous impact with objects in its environment. This simple capability has been the subject of interest in robotic research.

Behavior based navigation systems (Arkin, 1987, 1989; Arkin & Balch, 1997; AlYahmedi et al., 2009; Brooks, 1986, 1989; Fatmi et al. 2006 and Ching-Chih et al. 2010) have been developed as an alternative to the more traditional strategy of constructing representation of the world and then reasoning prior to acting. The main idea of behavior based navigation is to identify different responses (behaviors) to sensory inputs. For example, a behavior could be "avoiding obstacles" in which sonar information about a close obstacle should result in a movement away from the obstacle. A given set of behaviors is then blended in a certain way to produce either a trade off behavior or a more complex behavior. However, a number of issues with regard to behavior based navigation are still under investigation. These issues range from questions concerning the design of individual behaviors to behavior coordination issues, to intelligently improve "behaviors" through learning.

An important problem in autonomous navigation is the need to deal with the large amount of uncertainties of the sensory information received by the robot which is incomplete and approximate as well as with the fact that the environment in which such robots operate contains dynamics and variability elements.

A fuzzy logic behavior based navigation approach is introduced in this chapter in order to deal with the uncertainty and ambiguity of the information the system receives. Issues of individual behavior design and action coordination of the behaviors will be addressed using fuzzy logic.

The approach described herein, consists of the following four tasks,

- The use of fuzzy sets to represent the approximate positions and possibly shapes of objects in the environment.
- The design of simple fuzzy behaviors (avoiding obstacles, goal reaching, wall following…etc.).
- The blending of the different fuzzy behaviors.

2. Behavior based navigation

One of the long standing challenging aspect in mobile robotics is the ability to navigate autonomously, avoiding modeled and unmodeled obstacles especially in crowded and unpredictably changing environment. A successful way of structuring the navigation task in order to deal with the problem is within behavior based navigation approaches (Arkin, 1987, 1989; Arkin & Balch, 1997; AlYahmedi et al., 2009; Brooks, 1986, 1989; Fatmi et al. 2006; Ching-Chih et al. 2010; Maes, 1990; Mataric, 1997; Rosenblatt et al. 1989, 1994, 1995; Saffiotti, 1997 and Seraji & Howard, 2002).

2.1 Introduction

The basic idea in behavior based navigation is to subdivide the navigation task into small easy to manage, program and debug behaviors (simpler well defined actions) that focus on execution of specific subtasks. For example, basic behaviors could be "avoid obstacles" or "moving to a predefined position". This divide-and-conquer approach has turned out to be a successful approach, for it makes the system modular, which both simplifies the navigation solution as well as offers a possibility to add new behaviors to the system without causing any major increase in complexity. The suggested outputs from each concurrently active behaviors are then "blended" together according to some action coordination rule. The task then reduces to that of coupling actuators to sensory inputs, with desired robot behaviors. Each behavior can take inputs from the robot's sensors (e.g., camera, ultrasound, infrared, tactile) and/or from other behaviors in the system, and send outputs to the robot's actuators(effectors) (e.g., wheels, grippers, arm, and speech) and/or to other behaviors.

A variety of behavior-based control schemes have been inspired by the success of (Brooks, 1986, 1989), with his architecture which is known by the subsumption architecture. In this architecture behaviors are arranged in levels of priority where triggering a higher level behavior suppresses all lower level behaviors. (Arkin, 1987, 1989; Arkin & Balch, 1997), has described the use of reactive behaviors called motor schemas. In this method, potential field is used to define the output of each schema. Then, all the outputs are combined by weighted summation. Rosenblatt et al. (Rosenblatt et al. 1989, 1994, 1995), presented DAMN architecture in which a centralized arbitration of votes provided by independent behaviors combines into a "voted" output. Others (Saffiotti, 1997), (Seraji et al., 2001, 2002), (Yang et al. 2004, 2005; Selekwa et al., 2005 and Aguirre & Gonzales, 2006) used fuzzy logic system to represent and coordinate behaviors.

2.2 Fuzzy behavior based navigation

An important problem in autonomous navigation is the need to deal with the large amount of *uncertainties* that has to do with the sensory information received by the robot as well as with the fact that the environment in which such robots operate contains elements of *dynamics* and *variability* that limit the utility of prior knowledge. Fuzzy theory has the features that enable it to cope with uncertain, incomplete and approximate information. Thus, fuzzy logic stirs more and more interest amongst researchers in the field of robot navigation. Further, in the majority of fuzzy logic applications in navigation, a mathematical model of the dynamics of the robot nor the environment is needed in the design process of the motion controller.

The theory of fuzzy logic systems is inspired by the remarkable human capacity to reason with perception-based information. Rule based fuzzy logic provides a formal methodology for linguistic rules resulting from reasoning and decision making with uncertain and imprecise information.

In the fuzzy logic control inputs are processed in three steps (Fuzzification, Inference and Deffuzification) as seen in Fig. 1.

Fig. 1. Fuzzy logic control steps.

In the fuzzification block one defines for example fuzzy set A in a universe of discourse X defined by its membership function $\mu_A(x)$ for each x representing the degree of membership of x in X. In fuzzy logic control, membership functions assigned with linguistic variables are used to fuzzify physical quantities. Next, in the inference block, fuzzified inputs are inferred to a fuzzy rules base. This rules base is used to characterize the relationship between fuzzy inputs and fuzzy outputs. For example, a simple fuzzy control rule relating input v to output u might be expressed in the condition-action form as follows,

$$IF\ v\ is\ W\ then\ u\ is\ Y \tag{1}$$

Where W and Y are fuzzy values defined on the universes of v and u, respectively.

The response of each fuzzy rule is weighted according to the degree of membership of its input conditions. The inference engine provides a set of control actions according to fuzzified inputs. Since the control actions are in fuzzy sense. Hence, a deffuzification method is required to transform fuzzy control actions into a crisp value of the fuzzy logic controller.

In behavior based navigation the problem is decomposed into simpler tasks(independent behaviors). In fuzzy logic behavior based navigation systems each behavior is composed of a set of fuzzy logic rule statements aimed at achieving a well defined set of objectives, for example a rule could be:

If goal is near and to the left then turn left and move forward with a low speed

In general the actions recommended by different behaviors are compiled to yield the most appropriate action according to certain criteria.

2.3 Behavior coordination

The main problem in robot behavior based navigation is how to coordinate the activity of several behaviors, which may be active concurrently with the possibility of having behavior conflict. For example, one may have "goal reaching" behavior and " obstacle avoidance" behavior active at the same time as seen in Fig 2.

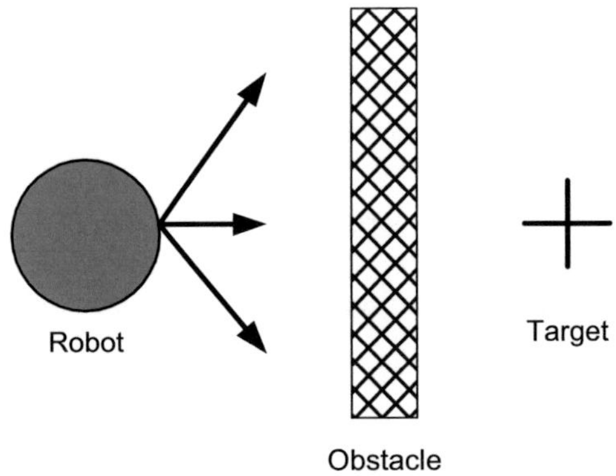

Fig. 2. Conflicts in bahviors.

The coordination task will be to reach a trade-off conclusion that provides the suitable command to the robot actuators which can result in choosing one behavior or a combination of all activated behaviors as shown in Fig.3. Behavior coordination is the point at which most strategies differ. Some of the earlier strategies are based on Brooks subsumption architecture (Brooks, 1986, 1989) uses a switching type of behavior coordination. In the Subsumption approach a prioritization scheme is used in which recommendation of only one behavior with the highest priority is selected, while recommendations of the remaining competing behaviors are ignored. This approach however, leads to inefficient results or poor performance in certain situations. For example if a robot is to encounter an obstacle right in front of it the action that will be selected is "avoid obstacle", the robot then decides to turn left to avoid the obstacle while the goal is to the right of the robot, so the "seek goal" behavior is affected in a negative way.

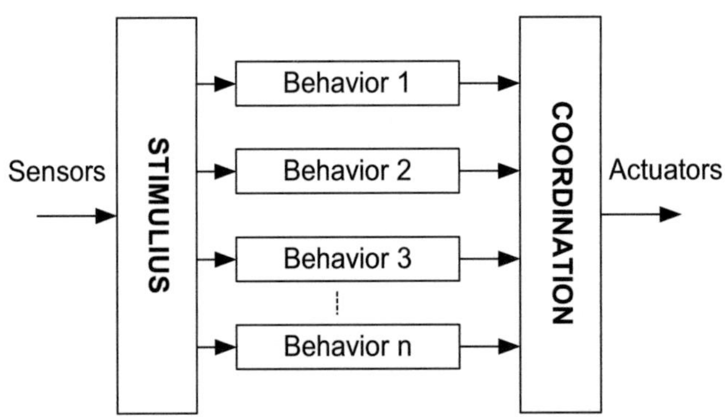

Fig. 3. Structure of behavior based navigation.

Other techniques combine the output of each behavior based on predetermined weighting factors, for example Arkin's motor schema approach(Arkin, 1987, 1989; Arkin & Balch, 1997), or Philipp A. and H.I. Christensen (ALthaus & Christensen, 2002) and the work of Rosenblatt(Rosenblatt et al. 1989, 1994, 1995), who developed the distributed architecture for mobile robot navigation, in which a centralized arbitration of votes provided by independent behaviors. In this method each behavior is allowed to vote for or against certain vehicle actions. The action that win the vote is carried out. These techniques may as well lead to poor performance in certain situations, for example if the robot is to encounter an obstacle right in front of it the "avoid obstacle" behavior may recommend the robot to turn left, while the "seek goal" behavior may request the robot to turn right since the goal is to the right of the robot, this may lead to trade off command that directs the robot forward resulting in a collision with the obstacle.

To deal with these limitations other schemes were recommended that achieve the coordination via considering the situation in which the robot is found, i.e each behavior is allowed to affect the robot motion based on the situational context. (Saffiott, 1997) uses the process of context-dependent-blending in which the current situation is used to decide the action taken using fuzzy logic. Independently (Tunstel et al., 1997) developed an approach similar to context-dependent-blending, in which adaptive hierarchy of multiple fuzzy behaviors are combined using the concept of degree of applicability. In this case certain behaviors are allowed to affect the overall behavior as required by the current situation and goal. The behavior fusion methodology in this chapter is motivated by the approaches used by Saffiot and Tunstel et al.

2.4 Context-dependent behavior coordination

The robot navigation tasks are divided into small independent behaviors that focus on execution of a specific subtask. For example, a behavior focuses on reaching the global goal, while another focuses on avoiding obstacles. Each behavior is composed of a set of fuzzy logic rules aimed at achieving a given desired objective. The navigation rules consist of a set of fuzzy logic rules for robot *velocity* (linear velocity m/s) and *steering* (angular velocity rad/s) of the form

$$IF\ C\ then\ A \tag{2}$$

Where the condition *C* is composed of fuzzy input variables and fuzzy connectives (And) and the action *A* is a fuzzy output variable. Equation (2) represents the typical form of natural *linguistic* rules .This rules reflect the human expert and reason to ensure logic, reliable and safe navigation. For example, obstacle avoidance behavior has inputs sensory data which can be represented by fuzzy sets with linguistic labels, such as {Near, Medium, Far}, corresponding to distance between robot and obstacle. Typical examples of fuzzy rules are as follow,

If *Front left* is Near And *Front right* is Far, Then *Steering* is Right

If *Front left* is Far And *Front right* is Near, Then *Velocity* is Zero

Where *Front left* and *Front right* are the distances acquired from sensors located in different locations on the robot.

Many behaviors can be active simultaneously in a specific situation or *context*. Therefore, a coordination technique, solving the problem of activation of several behaviors is needed. We call the method *context dependent behavior coordination*. The coordination technique employed herein is motivated by the approaches used by Saffiotti (Saffiotti, 1997). The supervision layer based on the context makes a decision as to which behavior(s) to process (activate) rather than processing all behavior(s) and then blending the appropriate ones, as a result time and computational resources are saved. Fig.4 and Fig.5 represent the architecture of the compared method and our approach, respectively.

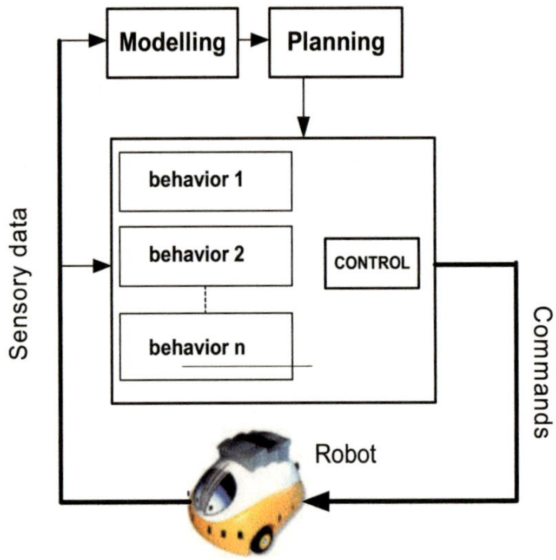

Fig. 4. Architecture of compared method.

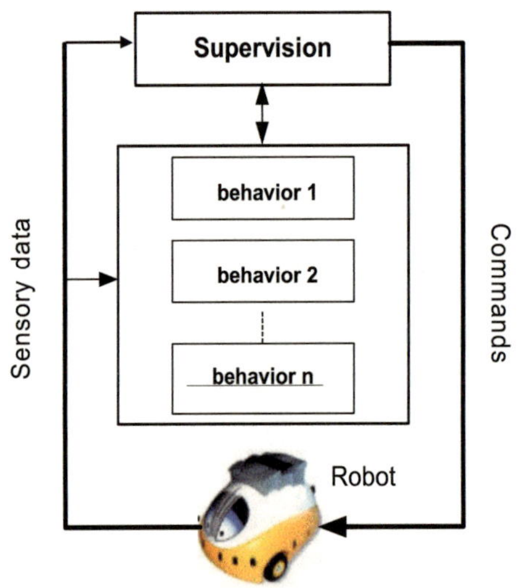

Fig. 5. Architecture of Context Dependent Behavior Coordination.

Our approach consists of the following characteristics.

- The robot navigation is comprised of four behaviors as follows,
  1. Goal reaching behavior
  2. Wall following behavior
  3. Emergency situation behavior
  4. Obstacle avoidance behavior.
- Each behavior is composed of a set of fuzzy logic rules achieving a precise goal.
- The output of each behavior represents the Steering angle and the Velocity.
- The supervision layer defines the priority of each behavior. It selects or activates and blends behaviors depending on situation or context.

3. Simulation & experimental results

3.1 Introduction

The navigation task can be broken down to a set of simple behaviors. The behaviors can be represented using fuzzy *if-then* rules. A context dependent coordination method can be used to blend behaviors.

To validate the applicability of the method simulation and experimental studies were performed.

3.2 Design of individual behaviors

To provide the robot with the ability to navigate autonomously avoiding modeled and unmodeled obstacles especially in crowded and unpredictably dynamic environment the following behaviors were designed: Goal reaching , Emergency situation, Obstacle avoidance, Wall following. Each behavior was represented using a fuzzy *if- then* rule base. The fuzzy rule base comprises the following *if-then* rules:

$$R^{(l)}: IF\ x_1 \text{ is } A_1^l \text{ and ... and } x_n \text{ is } A_n^l, THEN\ y \text{ is } B^l \tag{3}$$

Where $l = 1 \ldots m$, and m is the number of rules in a given fuzzy rule base, $x_1 \ldots x_n$ are the input variables which are the sensor data of the mobile robot, $A^l_1 \ldots A^l_2$ are the input fuzzy sets, B^l is the output fuzzy set and y is the output variable.

3.2.1 Goal reaching behavior

The goal reaching behavior tends to drive the robot from a given initial position to a stationary or moving target position. This behavior drives the robot to the left to the right or forward depending on θ_{error}, the difference between the desired heading (the heading required to reach the goal) and the actual current heading.

Fig.6 gives a schematic block diagram of the goal reaching architecture. From this figure we can notice that the inputs of the goal reaching controller are the distance robot to goal (D_{rg}) and θ_{error} which are given by Equation (4) and Equation (5), respectively.

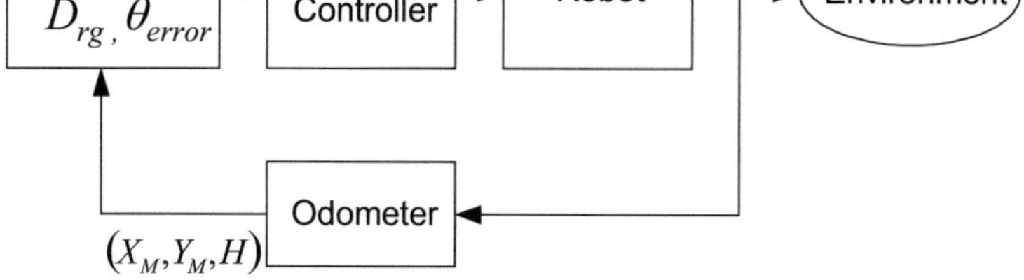

Fig. 6. Control architecture for goal reaching Behavior.

$$D_{rg} = \sqrt{(X_{Goal} - X_M)^2 + (Y_{Goal} - Y_M)^2} \tag{4}$$

$$\theta_{error} = \tan^{-1}\left(\frac{Y_{Goal} - Y_M}{X_{Goal} - X_M}\right) - H \tag{5}$$

Where, (X_M, Y_M, H) are the robot position and the heading measured by the robot odometer. (X_{Gaol}, Y_{Gaol}) is the target position

Although, there is no restriction on the form of membership functions, the appropriate membership functions for D_{rg} (in mm) and θ_{error} (in degrees) shown in Fig.7 were chosen.

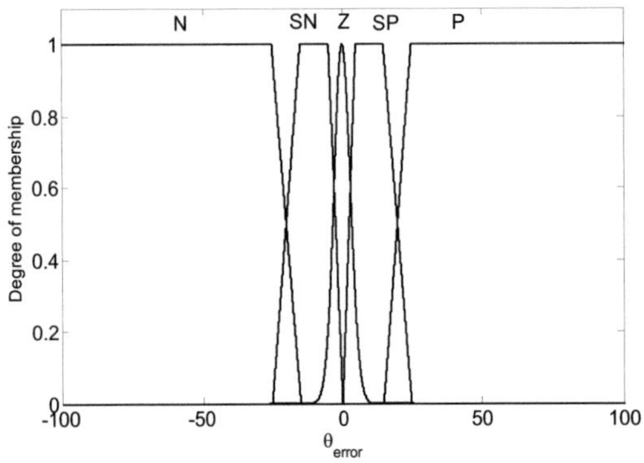

N: Negative, SN: Small Negative, Z: Zero, SP: Small Positive, P: Positive.

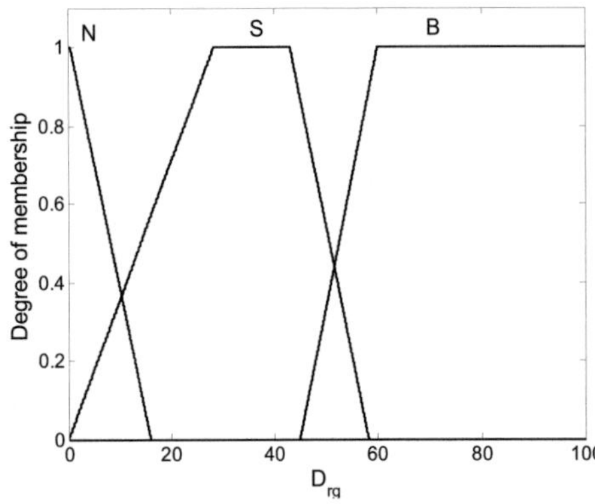

N for Near, S for Small, B for Big.

Fig. 7. Input membership functions for D_{rg} and θ_{error}.

The Goal reaching is expected to align the robot's heading with the direction of the goal so when θ_{error} is positive, the Steering is left or when θ_{error} is negative, the Steering is Right in a way that minimizes θ_{error}. For The Velocity it is proportional to the distance to goal D_{rg}. Example of goal reaching fuzzy rules.

If θ_{error} is *Positive* And D_{rg} *is Big then Velocity is Small Positive*
If θ_{error} is *Positive* And D_{rg} *is Big then Steering is Left*
If θ_{error} is *Negative* And D_{rg} *is Small then Velocity is Small Positive*
If θ_{error} is *Negative* And D_{rg} *is Small then Steering is Right*
If θ_{error} is *Small Negative* And D_{rg} *is Big then Velocity is Positive*
If θ_{error} is *Small Negative* And D_{rg} *is Big then Steering is Right Front*

Velocity

D_{rg} \ θ_{error}	*Z*	*SN*	*N*	*SP*	*P*
Near	*Z*	*Z*	*Z*	*Z*	*Z*
Small	*P*	*P*	*SP*	*SP*	*SP*
Big	*P*	*P*	*SP*	*P*	*SP*

Steering

D_{rg} \ θ_{error}	*Z*	*SN*	*N*	*SP*	*P*
Near	*F*	*RF*	*R*	*LF*	*L*
Small	*F*	*RF*	*R*	*LF*	*L*
Big	*F*	*RF*	*R*	*LF*	*L*

Table 1. Fuzzy table rules for goal reaching behavior.

For other behaviors the robots needs to acquire information about the environment. The Pekee Robot (Pekee is the robotic platform that will be used to validate the functionality of the proposed scheme experimentally) is endowed with 14 infrared sensors (See Fig.8). These sensors are used to detect obstacles in short-range distances and in a cone of 10 degrees. These sensors are clustered into 6 groups in the simulated robot in such a way as to be similar to Pekee(the simulated robot is very similar kinematically to Pekee).

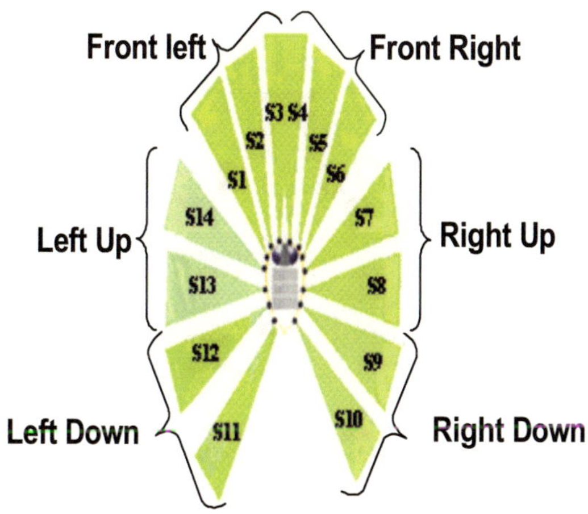

Fig. 8. Clustered sensors arrangement.

For each group, we refer to the minimum of distance measurement by group of sensors as:

Right Down= min (D_9, D_{10}) *Right Up= min* (D_7, D_8) *Front Right= min* (D_4, D_5, D_6)

Left Down=min (D_{11}, D_{12}) *Left Up=min* (D_{13}, D_{14}). *Front Left=min* (D_1, D_2, D_3),

D_i is the distance acquired by the sensor S_i, i=1….14.

These distances represent the inputs of the fuzzy controller for behaviors like Emergency situation, Obstacle avoidance, Wall following behaviors. Each distance is fuzzified using the following membership function described in Fig.9.

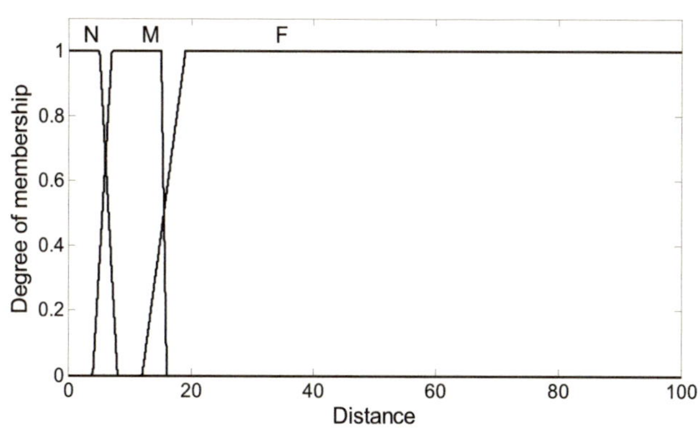

Fig. 9. Input membership function for distance to obstacle, N:Near, M:Medium, F:Far

The output variable are the Steering and the Velocity. Fig.10 illustrates the membership functions for these outputs.

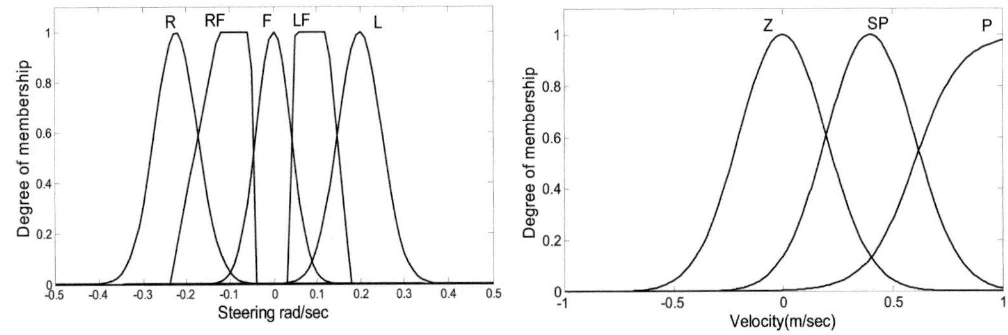

R: Right, RF: Right Front, FL: Front Left, F: Front, L: Left

Z: Zero, SP: Small Positive, P: Positive

Fig. 10. Output membership functions for steering & velocity.

3.2.1 Obstacle avoidance behavior

The obstacle avoidance behavior tends to avoid collisions with obstacles that are in the vicinity of robot. We take into account different cases as shown in table2.

		Inputs						Outputs	
Cases		Right Down	Right Up	Front Right	Front Left	Left Up	left Down	Steering	Velocity
1	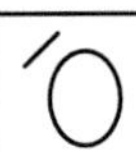			F	N			R	Z
				F	M			R	SP
2				N	F			L	Z
				M	F			L	SP
3			N					L	Z
			M					L	SP
4						N		R	Z
						M		FR	P
5							N	R	Z
6		N						L	Z
7	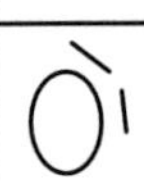		N	N	F			L	Z
			M	M	F			L	SP

Cases		Inputs						Outputs	
		Right Down	Right Up	Front Right	Front Left	Left Up	left Down	Steering	Velocity
8				F	N	N		R	Z
				F	M	M		R	SP
9				N	N	F		L	Z
				M	M	F		L	SP
				F	F			F	P
10			N	F	F	N		F	SP
			N		M	N		F	SP
			N	M		N		F	SP

Table 2. Fuzzy table rules for obstacle avoidance behavior.

Example of obstacle avoidance rules
If Front left is Near And Front right is Far, Then Steering is Right
If Front left is Near And Front right is Far, Then Velocity is Zero
If Front left is Far And Front right is Near, Then Steering is left
If Front left is Far And Front right is Near, Then Velocity is Zero

3.2.2 Wall following behavior

The objective of the control of the wall following behavior is to keep the robot at a safe close distance to the wall and to keep it in line with it. Example of wall following fuzzy rule:
If Right down is Medium and Front Right is Medium then Steering is Front
If Right down is Medium and Front Right is Medium then Velocity is Positive

Cases	Inputs						Outputs	
	Right Down	Right Up	Front Right	Front Left	Left Up	left Down	Steering	Velocity
	M	M	F	F			F	P
	M	M	M	F			F	P
			F	F	M	M	F	P
			F	M	M	M	F	P

Table 3. Fuzzy table rules for wall following behavior.

3.2.3 Emergency situation behavior

The emergency situation behavior drives the robot to the left or to the right when it is surrounded by obstacles in away depicted in Table 4.

	Inputs				Outputs	
Cases	Right Up	Front Right	Front Left	Left Up	Steering	Velocity
	Far	M	M	M	R	SP
	M	M	M	F	L	SP
	M		F	M	L	Z
	M	F		M	L	Z
	M	M	M	M	R	Z

Table 4. Fuzzy table rules for "emergency" behavior.

3.3 Blending of behaviors

The question to answer once the behaviors are designed is how best decide what the actuators shall receive(in terms of steering and velocity)taking into account the context in which robot happens to be in and relative importance of each behavior. To achieve that the work herein proposes the following architecture with details.

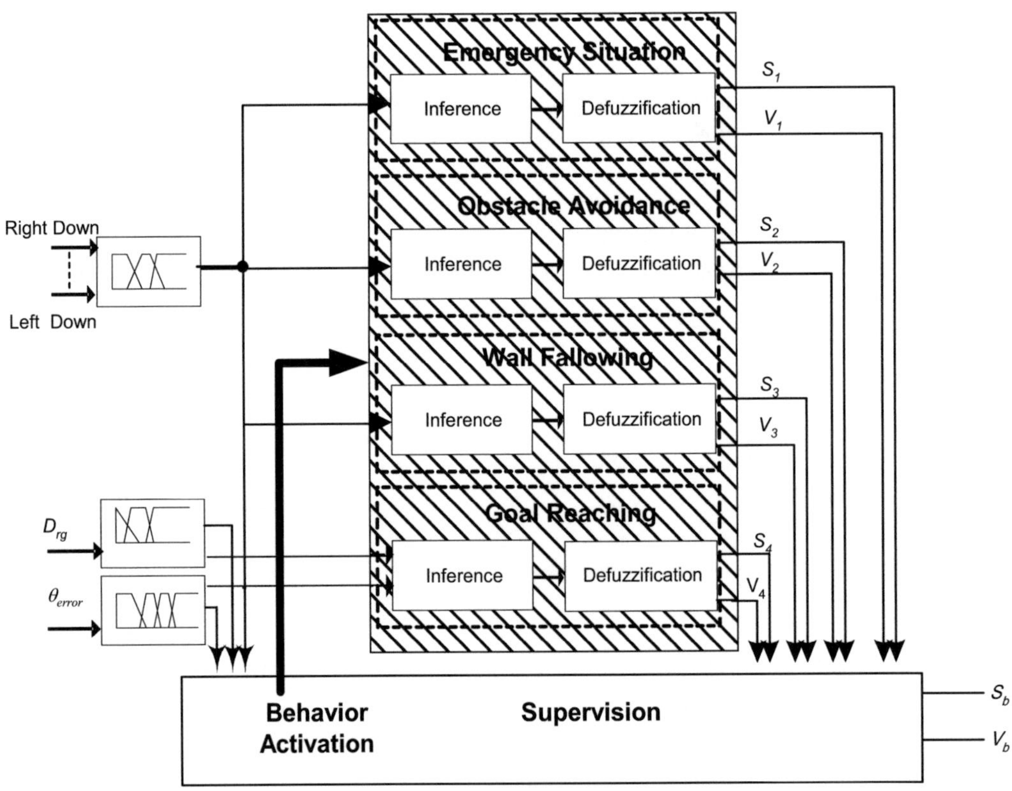

Fig. 11. Supervision Architecture.

S_i and V_i are the output Steering and Velocity of each behavior, i=1...4
S_b and V_b are the output steering and velocity to motor.
The inputs of supervision layer are the degree of membership of each group of sensors in all membership function, D_{rg} and θ_{error},in addition to the Steering and Velocity outputs of all behaviors (S_1, V_1, S_2, V_2, S_3, V_3, S_4 and V_4).
The supervision layer program is based on fuzzy rules such as,

$$IF\ context\ then\ behavior \tag{6}$$

Cases	RD	RU	FR	FL	LU	LD	θ_{error}	D_{rg}	Behavior
		F	F	F	F				Goal Reaching
								N	
				F	F	F	SP		Goal Reaching
				F	F	F	P		
				F	F	F	Z		
			N	N			SN		Fusion
			N	N			N		
			M	M			SN		
			M	M			N		
	F	F	F				SP		Goal Reaching
	F	F	F				P		Goal Reaching
	M	M		F					Wall Following
			F		M	M			Wall Following

Table 5. Context priority.

S_b and V_b are the appropriate velocity and steering control commands sent to the motors in a given robot situation as a result of the decision to activate a certain behavior (avoid obstacle, wall following...). However, S_b and V_b can be the fusion result of many behaviors. As seen at Table 5 the direction of robot to avoid obstacles is taken as default to the left, but, the supervision layer takes into account the position of goal by θ_{error} in such away as to minimize the distance traveled towards the goal as shown in Fig.12.
The advantage of this method is the fact that the number of rules are reduced by reducing the number of input variables to the fuzzy rules of each behavior, for example in the obstacle avoidance behavior there was no need to add θ_{error} as an input.

3.4 Simulation results

To verify the validity of the proposed scheme, some typical cases are simulated in which a robot is to move from a given current position to a desired goal position in various unknown environment. In all cases the robot is able to navigate its way toward the goal while avoiding obstacles successfully.

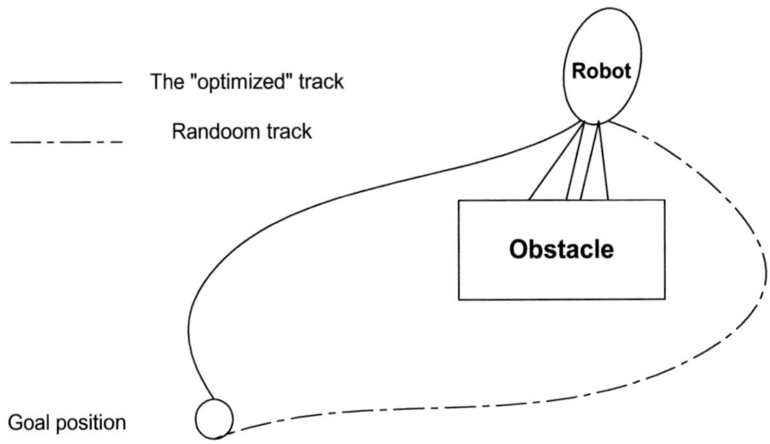

Fig. 12. Fusion of obstacle avoidance and goal reaching behaviors.

In the experimental work, the functionality of the proposed scheme was tested in an environment mimicking a crowded dynamic environment. The environment was designed taking into account several situations such as: simple corridor, narrow space and an area with many different shapes as obstacles (round, rectangle, trapezoidal)in away that mimics an image of office or equipped room (indoor environment).

In experiment (1) Fig.13 the robot has to reach Goal 1 for the start point placed between two walls. The robot begins to execute the behavior according to the rule base of the supervision layer depending on the current context. First, the robot follows wall 1 with maximum velocity until it senses obstacle 1, then it changes its behavior to obstacle avoidance at point A up to point B during which the robot crosses a narrow space between obstacle 1 and wall 2. The goal reaching behavior and the obstacle avoidance behavior are activated separately or fused to leave the narrow space, until point C. The robot then was encountered by obstacle 2, wall 2 and wall 3, so the emergency situation behavior was active. Next, the presence of obstacle 2 in front of the robot makes the obstacle avoidance behavior active until point D. For the route between of point D to E the robot just follows wall 2. From point E, three behaviors are activated (wall following, obstacle avoidance and goal reaching). The wall follow behavior is activated when a medium distance between the robot and obstacle 4 and 5 is established(corresponding to the last two cases in Table 5). The goal reaching behavior is activated to guide the robot to the goal. Between point F and G the robot is situated far from obstacles which in turn makes the goal reaching behavior active up to the presence of obstacle 6 at which point both goal reaching and obstacle avoidance are active. Finally, the goal reaching behavior is activated to reach the goal when it is near to the robot.

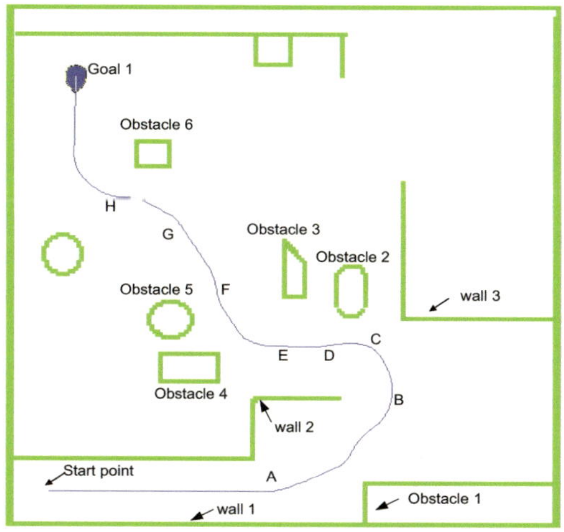

Fig. 13. Experiment 1.

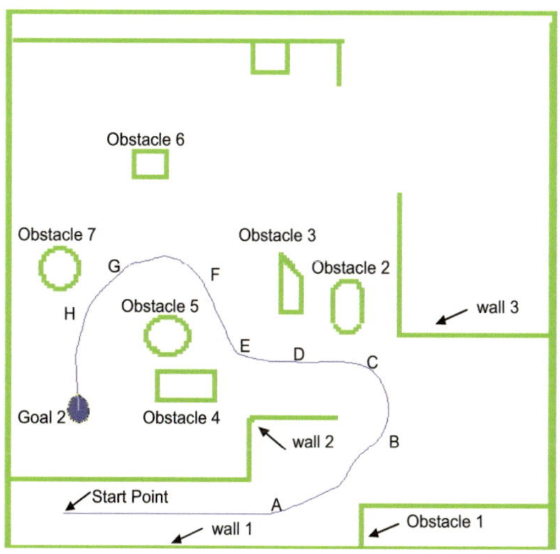

Fig. 14. Experiment 2.

For the experiment (2) Fig.14 the robot had to reach goal 2. It uses the same path as experiment 1 up to the point C. Then, avoiding obstacle 2 and goal reaching behavior are activated up to point D. From this point the robot follows obstacle 3 as a wall up to point E. The presence of obstacle 4 changes the behavior of the robot to obstacle avoidance. The robot avoids obstacle 4 then it follows obstacle 5 as wall by keeping medium distance to it. After that, the robot is situated in a free space and goal reaching behavior is activated up to point G. The obstacle avoidance behavior is activated when the robot senses obstacle 7 the robot avoids obstacle 7 and the behavior is changing to wall following when a medium distance to obstacle is measured by the robot. Finally the robot reaches the goal.

Figs.15-16 illustrate the levels of activation of each behavior during experiment 1 and experiment 2, respectively.

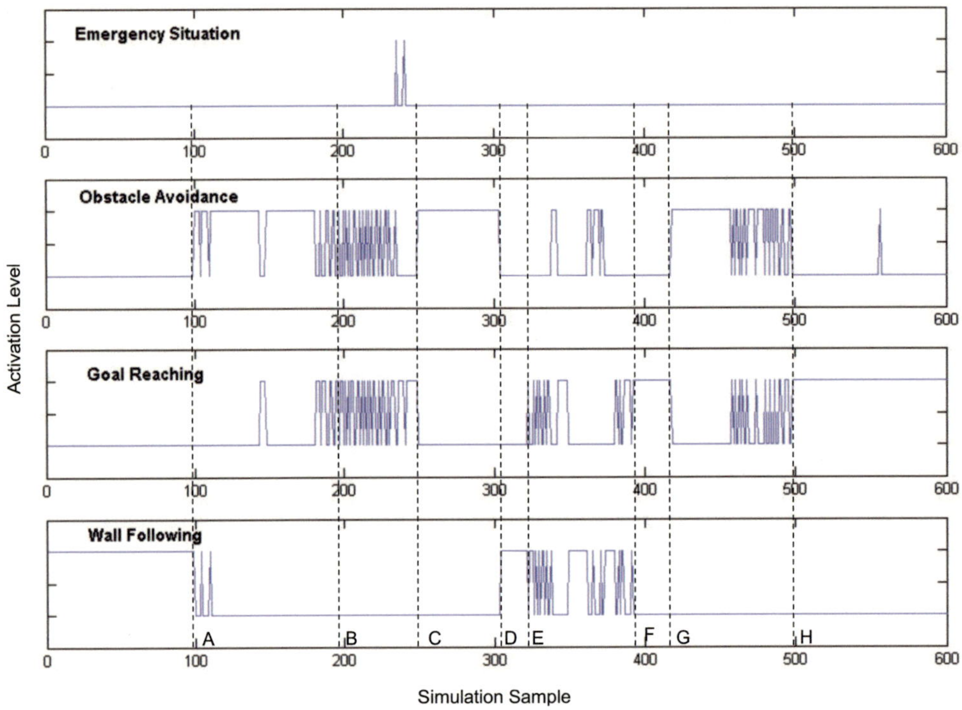

Fig. 15. Levels of activation of each behavior during Experiment 1.

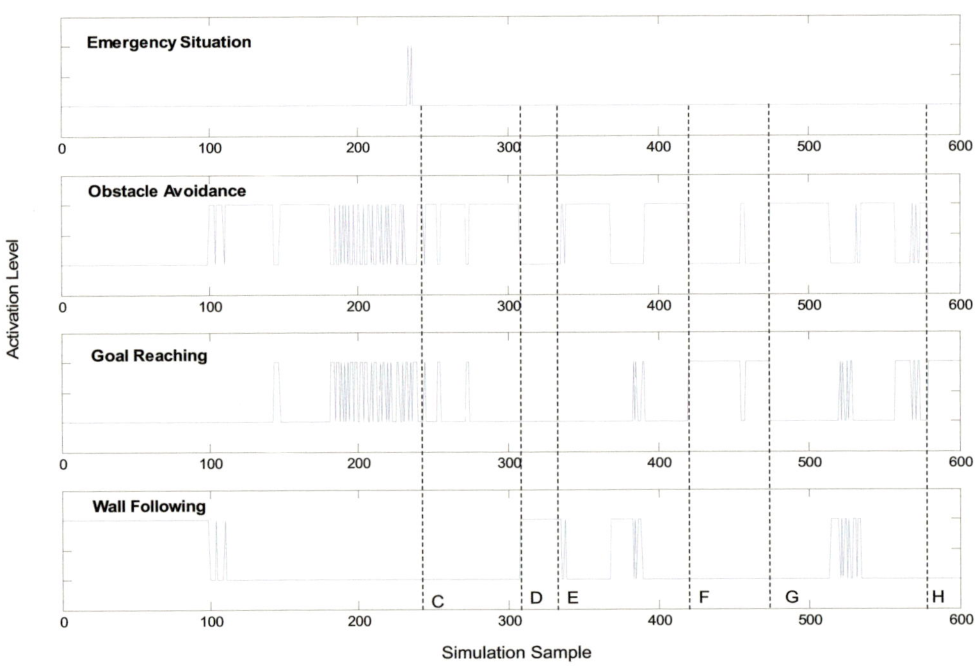

Fig. 16. Levels of activation of each behavior during Expriment 2.

The aim of experiment (3) (see Fig.17) is to show the ability of the robot to manage to escape from U shape obstacle and reaching a goal between two walls. From the start point the robot is situated in a U shape. In this context three behaviors are activated (the goal reaching behavior, obstacle avoidance and wall following). The avoid obstacle behavior always guided by the activation of goal reaching behavior especially at point A and B as shown in Fig.18.

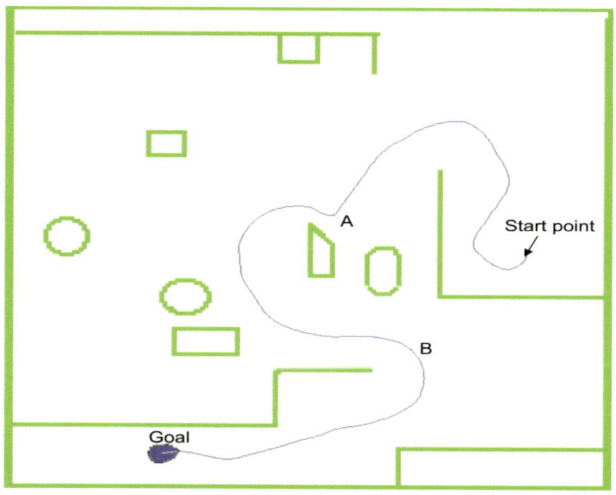

Fig. 17. Expriment 3.

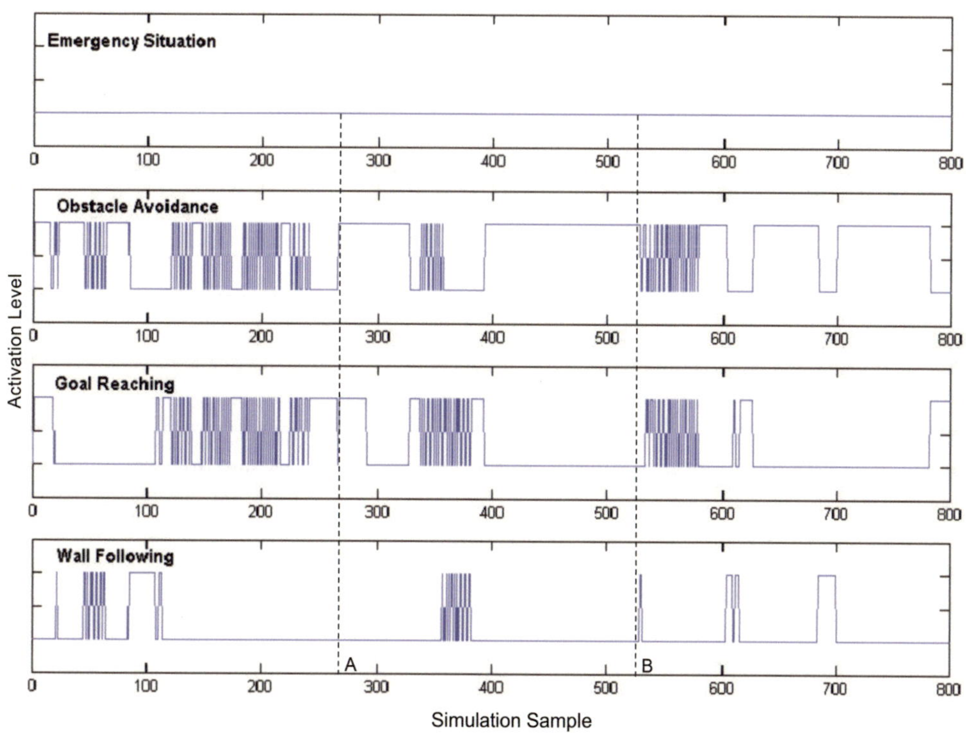

Fig. 18. Levels of activation of each behavior during Experiment 3.

Experiment (4) (Fig.19) shows the ability of the robot to escape from a trap situation and searching another path for reaching the goal.

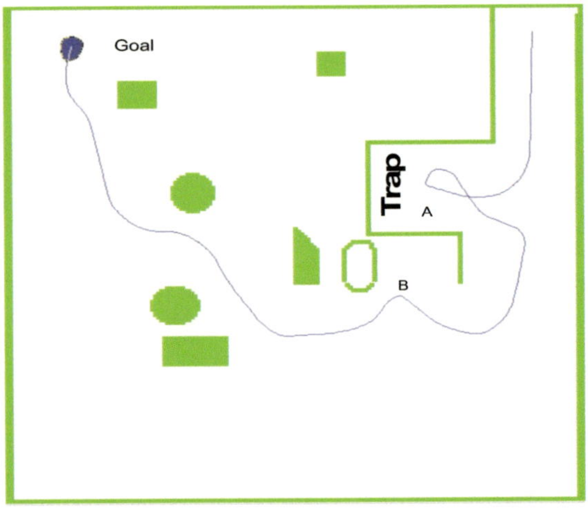

Fig. 19. Expriment 4.

After crossing the corridor, no obstacles are in front of the robot. Thus, the robot goes to the direction of the goal. At point A the robot senses the existence of obstacles around it, in the front, left and right, and then it made a round to right to escape from this trap and continues its navigation looking for another path.
In point B, the activation of goal reaching behavior and obstacle avoidance behavior is observed to take place concurrently. Here, the supervision layer fused these 2 behaviors to get the appropriate action as shown in section 3.3 Fig.12. The orientation of the robot depends on the goal position.
The Fig.20 shows the level of activation of emergency situation in the point A and the activation of goal reaching behavior and obstacle avoidance at the point B.

3.5 Experimental work

The effectiveness of the suggested navigation approach was experimentally demonstrated on a robotic platform named Pekee (Pekee™ robot is an open robotic development toolkit of Wany Robotics).

3.5.1 Pekee mobile robot

Pekee is equipped with two driving wheels with an additional supporting wheel. Its length is 40 cm and width is 25.5 cm, max speed 1 meter/second rotation 360 degree in a circle of 70 cm. The velocities of driven wheels are independently controlled by a motor drive unit. In addition the robot is endowed by,

- 2 odometers (180 impulses/wheel-turn).
- 2 gyro meters (pan and tilt)
- 2 temperature sensors.
- 1 variable frequency buzzer.
- Infrared link for communication between robots and peripherals.
- Serial infrared link for data transfer between Pekee and docking station or PC.

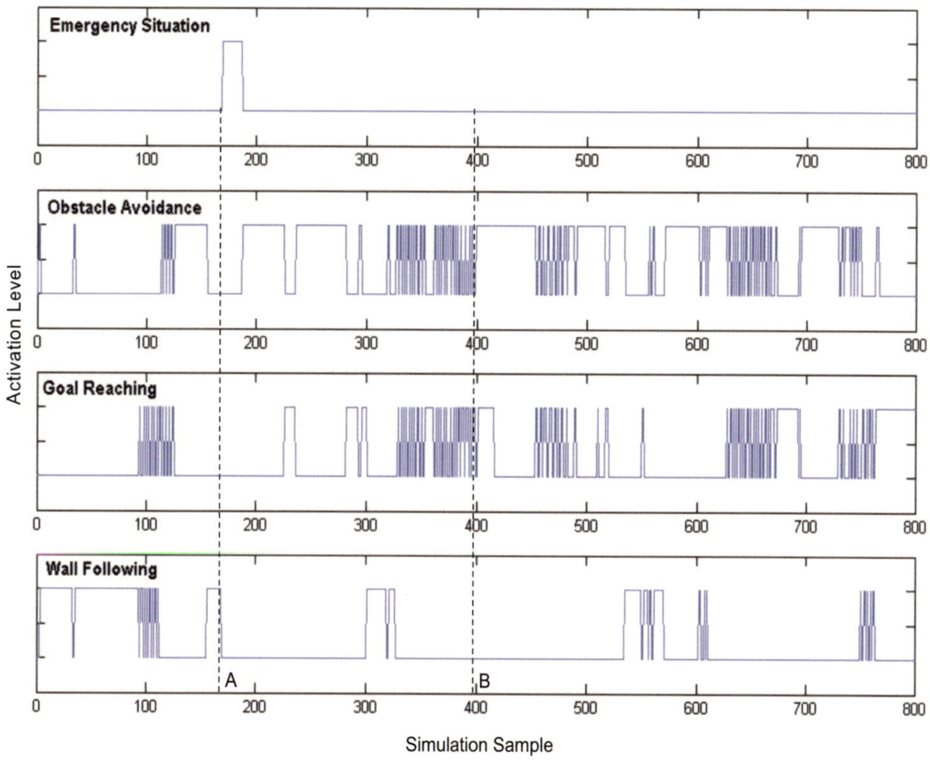

Fig. 20. Levels of activation of each behavior during expriment 4.

- Embedded Pc X86
- Camera
- 16-Mhz Mitsubishi micro-controller (16-bit), with 256 KB Flash-ROM and 20 KB RAM.
- 15 infrared sensors infrared telemeters (up to 1 measurement) arranged as shown in Fig21

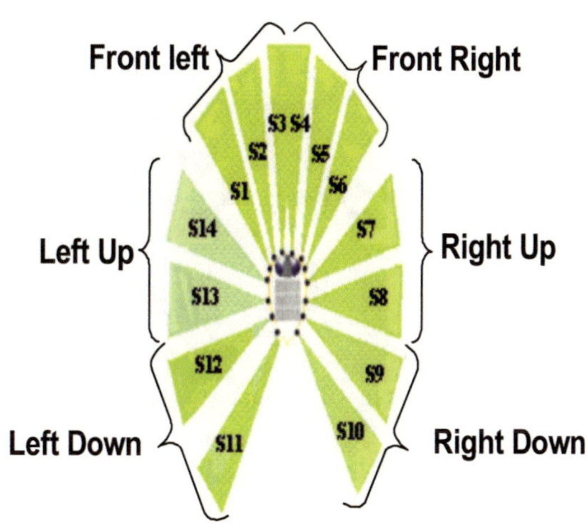

Fig. 21. IR sensors arrangement.

3.5.2 Pekee to PC communication

In the experimental work, the robot and the PC are connected via TCP/IP protocol using 2 network wireless cards one on the robot and the other on the computer or using a RJ45 cable.

The PC is equipped by a 54G network card compiles with the IEEE 802.11b standard in order to communicate with other 802.11b compliant wireless devices at 11 Mbps (the robot wireless network card). The card runs at speed of up to 54Mbps and operates on the same 2.4 GHz frequency band as 802.11b WI-FI products. This frequency band is suitable in industrial, science, and medical band operation. The work space of this card is as maximum 300 m.

The PC compiles, links the source code and executes the program. Then, it transmits the frames to the robot embedded PC via TCP\IP. These frames will be transmitted to the micro controller via the OPP bus. The micro controller is responsible to execute frames and transmits order to actuators (motors, buzzer…) also it sends data about the robot sensors status, robot position, measured distances ….

3.5.3 Experimental results

In order to validate the results of simulated experiments, a navigation task has been tested in a real world as an environment similar to a small equipped room. Fig. 22-26 shows that Pekee was able to navigate from a given starting point to a target point while avoiding obstacles.

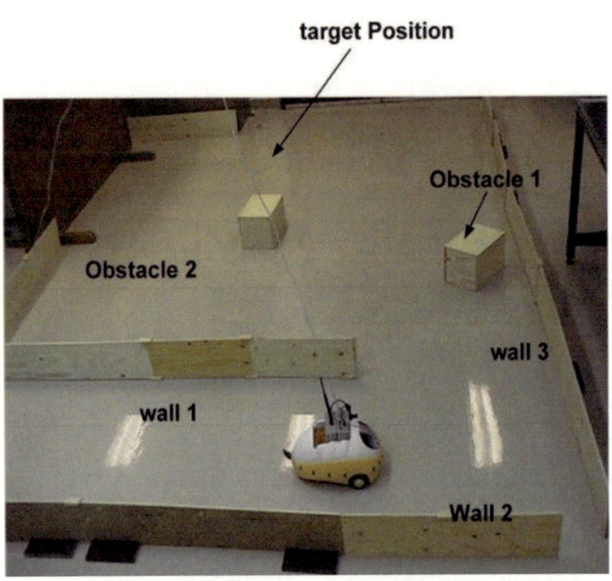

Fig. 22. Pekee Navigates in a crowded environment.

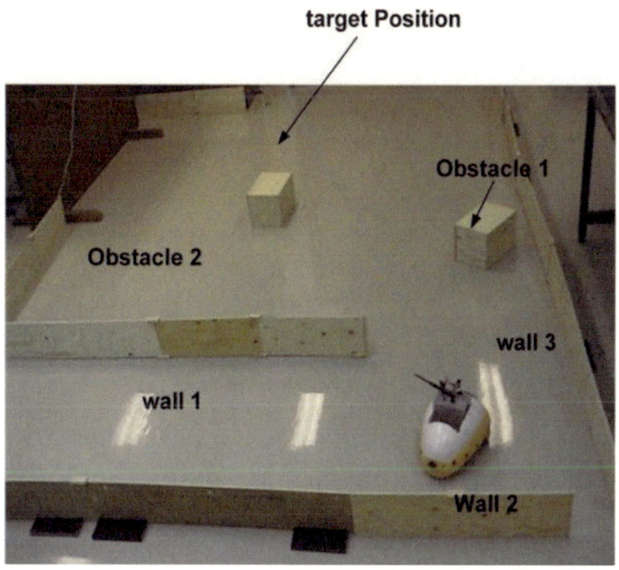

Fig. 23. Pekee Navigates in a crowded environment.

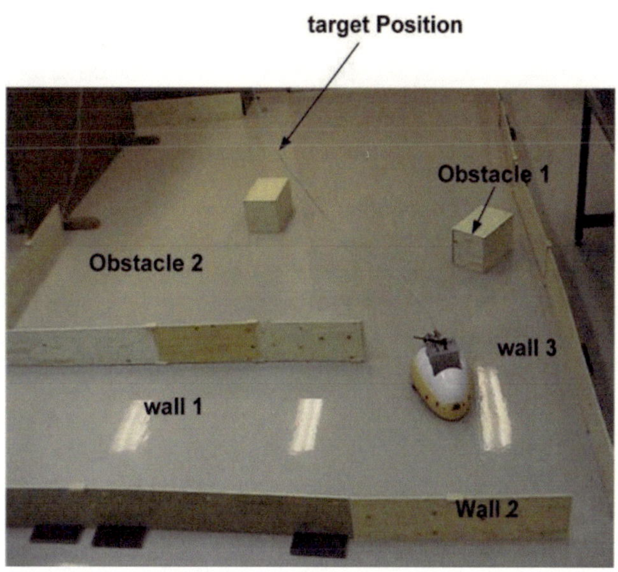

Fig. 24. Pekee Navigates in a crowded environment.

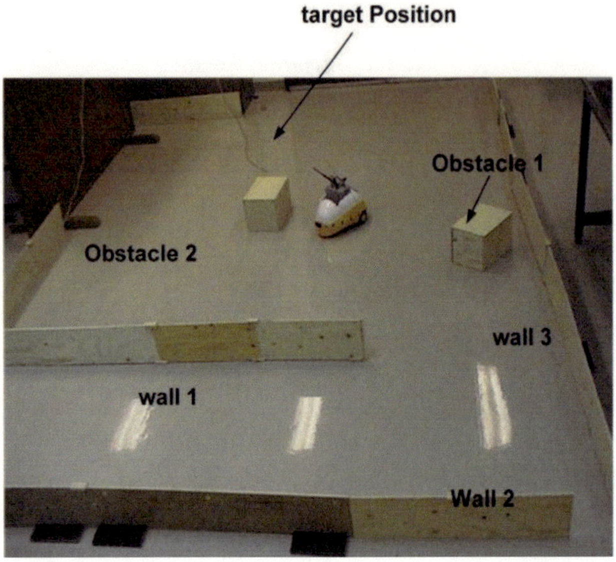

Fig. 25. Pekee Navigates in a crowded environment.

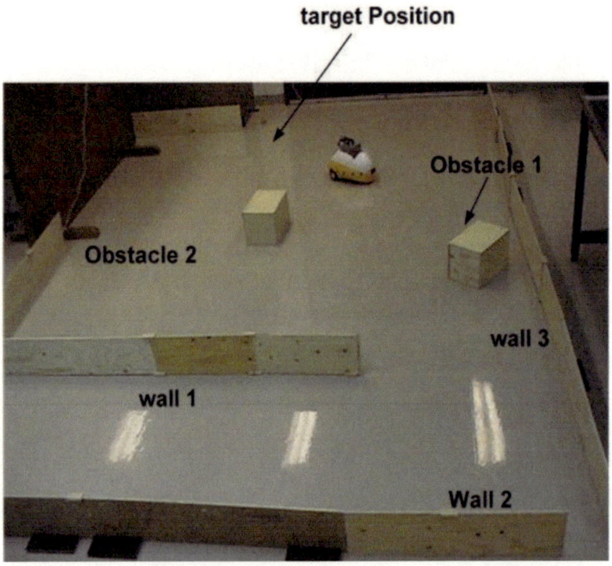

Fig. 26. Pekee Navigates in a crowded environment.

3.6 Conclusions

A successful way of structuring the navigation task in order to deal with the problem of mobile robot navigation is demonstrated. Issues of individual behavior design and action coordination of the behaviors were addressed using fuzzy logic. The coordination technique employed in this work consists of two layers. A Layer of primitive basic behaviors, and the supervision layer which based on the context makes a decision as to which behavior(s) to process (activate) rather than processing all behavior(s) and then blending the appropriate ones, as a result time and computational resources are saved. Simulation and experimental studies were done to validate the applicability of the proposed strategy.

As an improvement to our implemented system we are planning to incorporate learning to improve the effectiveness of the navigation approach.

As a further work the algorithm is to be tested on a "robotic cane", which is a device to help the blind or visually impaired users navigate safely and quickly among obstacles and other hazards. During operation, the user will be able to push the lightweight "Robotic Cane" forward. When the "Robotic Cane's ultrasonic sensors detect an obstacle, the behavior based navigation that will be developed determines a suitable direction of motion that steers the user around it. The steering action results in a very noticeable force felt in the handle, which easily guides the user without any conscious effort on his/her part. The navigation methodology that will be followed in the robotic cane will be based on the behavior based navigation developed herein, in particular the work will use fuzzy logic based navigation scheme to steer the "Robotic Cane". Further the user is continually interacting with the robotic cane, this calls for the need to address the issue of cooperation and/or conflict resolution between the user and the robotic cane.

4. References

Aguirre E. & Gonzales A. (2000). Fuzzy behaviors for mobile robot navigation:design, coordination and fusion, *Int. J. of Approximate Reasoning*, Vol. 25, pp. 255-289.

Althaus P. & Christensen H. I.(2002). Behavior coordination for navigation in office environment, *Proceedings of 2002 IEEE/RSJ Int. Conference on Intelligent Robots and Systems*, pp. 2298-2304, Switzerland, 2002

AlYahmedi, A. S., El-Tahir, E., Pervez, T. (2009). Behavior based control of a robotic based navigation aid for the blind, *Control & Applications Conference, CA2009*, July 13-July 15, 2009, Cambridge, UK.

Arkin, R. C. (1987). Towards Cosmopolitan Robots: Intelligent Navigation in Extended Man-made Environments, *PhD Thesis*, University of Massachusetts, Department of Computer and Information Science.

Arkin, R. C. (1989). Motor schema-based mobile robot navigation, *Int. J. of Robotic Research*, Vol 8, pp. 92-112.

Arkin, R. C. & Balch, T.(1997) AuRA: Principles and Practice in Review, *Journal of Experimental and Theoretical Artificial Intelligence(JETAI)*, Vol. 9, No. 2/3, pp. 175-188.

Brooks R. A.(1986). A Robust Layered Control System for a Mobile Robot, *IEEE Journal of Robotics and Automation*, Vol. 2, No. 1, (March 1986), pp. 14–23.

Brooks R. A.(1989). A Robot that Walks; Emergent Behavior from a Carefully Evolved Network, *IEEE International Conference on Robotics and Automation*, Scottsdale, AZ, pp. 292–296, May 1989

Ching-Chih, T., Chin-Cheng, C., Cheng-Kain, C, Yi Yu, L. (2010). Behavior-based navigation using heuristic fuzzy kohonen clustering network for mobile service robots, *International Journal of Fuzzy Systems, Vol. 12, No. 1, March, 2010*, pp. 25-32.

Fatmi, A., ALYahmedi, A. S., Khriji, L., Masmoudi, N(2006). A fuzzy logic based navigation of a mobile robot, *World academy of science, Engineering and Technology*, issue 22, 2006, pp. 169-174.

Huq, R., Mann, G. K. I., Gosine, R. G.(2008). Mobile robot navigation using motor schema and fuzzy context dependent behavior modulation, *Applied soft computing,* 8, 2008, pp. 422-436.

Langer D., Rosenblatt J.K. & Hebert M. (1994). A Behavior-Based System For Off-Road Navigation, *IEEE Journal of Robotics and Automation* , Vol. 10, No. 6, pp. 776-782.

Maes P. (1990). How to do the Right Thing, *Connection Science Journal, Special Issue on Hybrid Systems,* Vol. 1.

Mataric´ M. J.(1997). Behavior-Based Control: Examples from Navigation, Learning, and Group Behavior, *Journal of Experimental and Theoretical Artificial Intelligence,* special issue on Software Architectures for Physical Agents, Vol. 9, No.2/3, pp. 323-336

Rosenblatt J. & Payton D. W.(1989). A Fine-Grained Alternative to the Subsumption Architecture for Mobile Robot Control, *Proceedings of the IEEE/INNS International Joint Conference on Neural Networks*, Washington DC, June 1989, vol. 2, pp. 317-324.

Rosenblatt J.(1995). DAMN: A Distributed Architecture for Mobile Navigation, *Ph.D. dissertation, Carnegie Mellon University Robotics Institute Technical Report CMU-RI-TR-97-01*, Pittsburgh, PA, 1995.

Saffiotti A.(1997). The uses of fuzzy logic for autonomous robot navigation: a catalogue raisonn'e, *Soft Computing Research journal,* Vol. 1, No. 4, pp. 180-197.

Selekwa M. F., Damion D., & Collins, Jr. E. G. (2005). Implementation of Multi-valued Fuzzy Behavior Control for Robot Navigation in Cluttered Environments, *Proceedings of the 2005 IEEE International Conference on Robotics and Automation*, Barcelona, Spain, pp., 3699-3706, April 2005

Seraji H. & Howard A.(2002). Behavior-based robot navigation on challenging terrain: A fuzzy logic approach, *IEEE Trans. Rob. Autom.* Vol. 18, No. 3, pp. 308-321

Seraji H., Howard A. & Tunstell E.(2001). Terrain-Based Navigation of Planetary Rovers: A Fuzzy Logic Approach, *Proceeding of the 6th International Symposium on Artificial Intelligence and Robotics & Automation in Space*, Canada, June 18-22, 2001.

Tunstel E., Lippincott T. & Jamshidi M.(1997). Behavior Hierarchy for Autonomous Mobile Robots: Fuzzy-behavior modulation and evolution, *International Journal of Intelligent Automation and Soft Computing, Special Issue: Autonomous Control Engineering at NASA ACE Center*, Vol. 3,No. 1,pp. 37--49.

Yang S. X., Li H., Meng M. Q.-H , & Liu P. X. (2004). An Embedded Fuzzy Controller for a Behavior-Based Mobile Robot with Guaranteed Performance, *IEEE Transactions on Fuzzy Systems*, Vol. 12, No. 4, pp.436-446.

Yang S. X., Moallem M., & Patel R. V. (2005). A Layered Goal-Oriented Fuzzy Motion Planning Strategy for Mobile Robot Navigation, *IEEE transactions on systems, man, and cybernetics – part b: cybernetics*, Vol. 35, no. 6, 1214-1224.

15

Navigation of Quantum-Controlled Mobile Robots

Eugene Kagan and Irad Ben-Gal
Tel-Aviv University
Israel

1. Introduction

The actions of autonomous mobile robots in stochastic medium imply certain intellectual behavior, which allows fulfilling the mission in spite of the environmental uncertainty and the robot's influence on the characteristics of the medium. To provide such a behavior, the controllers of the robots are considered as probabilistic automata with decision-making and, in some cases, learning abilities. General studies in this direction began in the 1960s (Fu & Li, 1969; Tsetlin, 1973) and resulted in practical methods of on-line decision-making and navigation of mobile robots (Unsal, 1998; Kagan & Ben-Gal, 2008).

Along with the indicated studies, in recent years, the methods of mobile robot's navigation and control are considered in the framework of quantum computation (Nielsen & Chuang, 2000), which gave rise to the concept of quantum mobile robot (Benioff, 1998; Dong, et al., 2006). Such approach allowed including both an environmental influence on the robot's actions and the changes of the environment by the robot by the use of the same model, and the ability to apply the methods of quantum communication and decision-making (Levitin, 1969; Helstrom, 1976; Davies, 1978; Holevo, 2001) to the mobile robot's control.

Following Benioff, *quantum robots* are "mobile systems that have a quantum computer and any other needed ancillary systems on board... Quantum robots move in and interact (locally) with environments of quantum systems" (Benioff, 1998). If, in contrast, the quantum robots interact with a non-quantum environment, then they are considered as *quantum-controlled mobile robots*. According to Perkowski, these robots are such that "their controls are quantum but sensors and effectors are classical" (Raghuvanshi, et al., 2007). In the other words, in the quantum-controlled mobile robot, the input data obtained by classical (non-quantum) sensors are processed by the use of quantum-mechanical methods, and the results are output to classical (non-quantum) effectors.

In this chapter, we present a brief practical introduction into quantum computation and information theory and consider the methods of path planning and navigation of quantum-controlled mobile robots based on quantum decision-making.

2. Quantum information theory and models of quantum computations

We begin with a brief introduction into quantum information theory and stress the facts, which are required for the tasks of mobile robots' navigation. An application of quantum-mechanical methods for the mobile robot's control is based on the statistical interpretation of quantum mechanics (Ballentine, 2006); for a complete review of quantum computation and information theory, see, e.g., (Nielsen & Chuang, 2000).

2.1 Basic notation and properties of quantum-mechanical systems

In general, in the considerations of the finite quantum-mechanical systems, it is postulated (Ballentine, 2006) that the state of the quantum-mechanical system is represented by a Hermitian complex matrix $\sigma = \|\sigma_{jk}\|$, $\sigma_{jk} = \sigma_{kj}^*$, where $(\cdot)^*$ stands for complex conjugate, with the unit sum of the diagonal elements, $\mathrm{tr}(\sigma) = 1$; and that the observed value of the quantum-mechanical system is specified by the eigenvalues of the matrix σ. Since matrix σ is Hermitial, its eigenvalues are real numbers. If matrix σ is diagonal, then such a representation of the state is equivalent to the representation of the state of stochastic classical system, in which diagonal elements σ_{jj} form a probability vector (Holevo, 2001).

Let $\bar{s} = (s_1, s_2, \ldots)$ be a vector of complex variables $s_j = a_j + ib_j$, $j = 1, 2, \ldots$, and let $s_j^* = a_j - ib_j$ be its complex conjugate. According to the Dirac's "bra[c]ket" notation, row-vector $\langle s| = (s_1, s_2, \ldots)$ is called *bra-vector* and a column-vector $|s\rangle = (s_1^*, s_2^*, \ldots)^T$, where $(\cdot)^T$ stands for the transposition of the vector, is called *ket-vector*. Vector $\bar{s}$ or, equivalently, a Hermitian matrix $\rho = |s\rangle\langle s|$, are interpreted as a state of quantum object; vector $\bar{s}$ is called *state vector* and matrix ρ is called *state matrix* or *density matrix*.

Let $\bar{s}_1 = (s_{11}, s_{12}, \ldots)$ and $\bar{s}_2 = (s_{21}, s_{22}, \ldots)$ be two state vectors. The inner product $\langle s_2|s_1\rangle = (s_{21}, s_{22}, \ldots) \cdot (s_{11}^*, s_{12}^*, \ldots)^T = s_{21} \cdot s_{11}^* + s_{22} \cdot s_{12}^* + \ldots$ is called *amplitude* of the event that the object moves from the state $\bar{s}_1$ to the state $\bar{s}_2$. The *probability* of this event is defined by a square of the absolute value of the amplitude $\langle s_2|s_1\rangle$, i.e. $P(s_2 \mid s_1) = |\langle s_2|s_1\rangle|^2$, where $|\langle s_2|s_1\rangle| = |\alpha + i\beta| = \sqrt{\alpha^2 + \beta^2}$. By definition, regarding the probability of any state $\bar{s}$, it is assumed that $0 \le P(s \mid s) \le 1$. For example, let $\langle s_1| = (1, 0)$ and $\langle s_2| = (1/\sqrt{2}, 1/\sqrt{2})$, where both $P(s_1 \mid s_1) = 1$ and $P(s_2 \mid s_2) = 1$. However, the probability of transition from the state $\bar{s}_1$ to the state $\bar{s}_2$ is $P(s_2 \mid s_1) = |\langle s_2|s_1\rangle|^2 = 1/2$.

The *measurement* of the state of the object is described by an *observation operator* O, which in the finite case is determined by a squire matrix $\mathrm{O} = \|o_{jk}\|$. The result of the observation of the state $\bar{s}$ is defined by a multiplication $\langle s|\mathrm{O}|s\rangle = (s_1, s_2, \ldots) \cdot \|o_{jk}\| \cdot (s_1^*, s_2^*, \ldots)^T$, and the state, which is obtained after an observation is defined as $|s'\rangle = \mathrm{O}|s\rangle$. For example, assume that the measurement is conducted by the use of the operator $\mathrm{O} = \begin{pmatrix} 1 & 0 \\ 0 & -1 \end{pmatrix}$. Then, the observation of the states $\langle s_1| = (1, 0)$ and $\langle s_2| = (1/\sqrt{2}, 1/\sqrt{2})$ results in $\langle s_1|\mathrm{O}|s_1\rangle = 1$ and $\langle s_2|\mathrm{O}|s_2\rangle = 0$, i.e. operator O unambiguously detects the states $\bar{s}_1$ and $\bar{s}_2$. Moreover, an observation of the state $\bar{s}_1$ does not change this state, i.e. $\mathrm{O}|s_1\rangle = (1, 0)^T$, while the observation of the state $\bar{s}_2$ results in the new state $\mathrm{O}|s_2\rangle = (1/\sqrt{2}, -1/\sqrt{2})^T$. From such a property of observation, it follows that in contrast to the classical systems, the actual state of the quantum-mechanical system obtains a value, which was measured by the observer, and further evolution of the system starts from this value. In the other words, the evolution the quantum-mechanical system depends on the fact of its observation.

An actual evolution of the quantum-mechanical system is governed by the evolution operators, which are applied to the state matrix σ or state vector $\vec{s}$. Below, we consider the states and operators, which are used in quantum information theory.

2.2 Concepts of the quantum information theory

The elementary state, which is considered in quantum information theory (Nielsen & Chuang, 2000), is called *qubit* (quantum bit) and is represented by a two-element complex vector $|s\rangle = (s_1, s_2)^T$, $0 \le |s_1|^2 + |s_2|^2 \le 1$. Among such vectors, two vectors are specified $|0\rangle = (1,0)^T$ and $|1\rangle = (0,1)^T$, which correspond to the bit values "0" and "1" known in classical information theory (Cover & Thomas, 1991). In general, vectors $|0\rangle$ and $|1\rangle$ determine the states "spin up" and "spin down" of electron, i.e. $|0\rangle \equiv "\uparrow"$ and $|1\rangle \equiv "\downarrow"$.

Given vectors $|0\rangle = (1,0)^T$ and $|1\rangle = (0,1)^T$, any qubit $|s\rangle = (s_1, s_2)^T$ is represented as $|s\rangle = (s_1, s_2)^T = s_1|0\rangle + s_2|1\rangle$. In particular, if there are defined two states $|"\rightarrow"\rangle = \left(1/\sqrt{2}, 1/\sqrt{2}\right)^T$ and $|"\leftarrow"\rangle = \left(1/\sqrt{2}, -1/\sqrt{2}\right)^T$, which represent the electron states "spin right" and "spin left", then $|"\rightarrow"\rangle = \left(1/\sqrt{2}\right)\cdot|1\rangle + \left(1/\sqrt{2}\right)\cdot|0\rangle$ and $|"\leftarrow"\rangle = \left(1/\sqrt{2}\right)\cdot|1\rangle - \left(1/\sqrt{2}\right)\cdot|0\rangle$. Moreover, since it holds true that $|0\rangle = \left(1/\sqrt{2}\right)\cdot|"\rightarrow"\rangle + \left(1/\sqrt{2}\right)\cdot|"\leftarrow"\rangle$ and $|1\rangle = \left(1/\sqrt{2}\right)\cdot|"\rightarrow"\rangle - \left(1/\sqrt{2}\right)\cdot|"\leftarrow"\rangle$, the pairs of the vectors $|0\rangle$ and $|1\rangle$, and $|"\rightarrow"\rangle$ and $|"\leftarrow"\rangle$ can be used interchangeably.

In general, the evolution of the qubits is governed by the use of the following operators:

Pauli operators: $\mathrm{I} = \begin{pmatrix} 1 & 0 \\ 0 & 1 \end{pmatrix}$, $\mathrm{X} = \begin{pmatrix} 0 & 1 \\ 1 & 0 \end{pmatrix}$, $\mathrm{Y} = \begin{pmatrix} 0 & -i \\ i & 0 \end{pmatrix}$, and $\mathrm{Z} = \begin{pmatrix} 1 & 0 \\ 0 & -1 \end{pmatrix}$;

- Hadamard operator: $\mathrm{H} = \frac{1}{\sqrt{2}}\begin{pmatrix} 1 & 1 \\ 1 & -1 \end{pmatrix}$, Phase shift operator: $\mathrm{S} = \begin{pmatrix} 1 & 0 \\ 0 & e^{i\pi/8} \end{pmatrix}$;
- Controlled NOT (CNOT) operator: $\mathrm{CNOT} = \begin{pmatrix} 1 & 0 & 0 & 0 \\ 0 & 1 & 0 & 0 \\ 0 & 0 & 0 & 1 \\ 0 & 0 & 1 & 0 \end{pmatrix} = \begin{pmatrix} \mathrm{I} & 0 \\ 0 & \mathrm{X} \end{pmatrix}$.

The Pauli operators are the most known qubits operators that are in use in general quantum mechanics, while the other three operators are more specific for quantum information theory. According to the Kitaev-Solovey theorem (Nielsen & Chuang, 2000), an algebra $\mathsf{U} = \{\{|0\rangle, |1\rangle\}, \mathrm{CNOT}, \mathrm{H}, \mathrm{S}\}$, which consists of the qubits $|0\rangle$ and $|1\rangle$, and CNOT, Hadamard and phase shift operators, forms a universal algebra that models any operation of the Boolean algebra $\mathfrak{B} = \{\{0,1\}, \neg, \&, \vee\}$. Notice that the qubit operators are reversible. In fact, direct calculations show that $\mathrm{NOT}(\mathrm{NOT}(\vec{s})) = \mathrm{X}\cdot\mathrm{X}\cdot|s\rangle = |s\rangle$, $\mathrm{H}\cdot\mathrm{H}\cdot|s\rangle = |s\rangle$ and so on.

To illustrate the actions of the simplest qubit operators and their relation with classical logic operations, let us present the corresponding quantum gates and their pseudocode.

NOT ($x' := \overline{x}$): **if** $x=1$ **then** $x'=0$ **else** $x'=1$		CNOT: $x' = x$ **if** $x=1$ **then** $y' = \overline{y}$ **else** $y' = y$	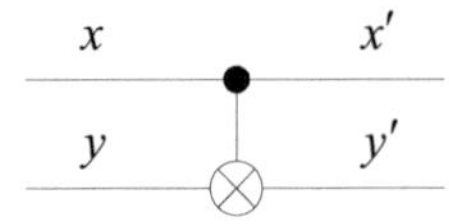
CCNOT (TOFFOLI GATE): $x' = x$, $y' = y$ **if** $x=1$ **and** $y=1$ **then** $z' = \overline{z}$ **else** $z' = z$	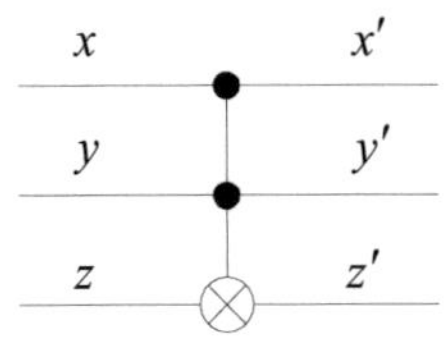	FAN OUT: CNOT with constant input $y := 0$	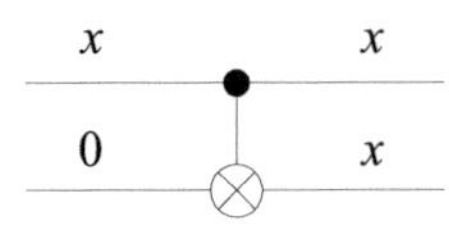
EXCHANGE: $x' := y$, $y' := x$	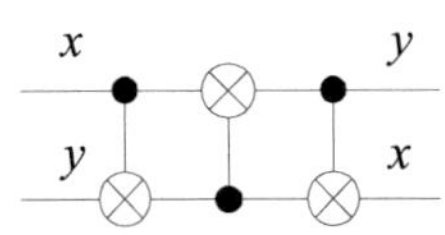	C-EXCHANGE (FREDKIN GATE):	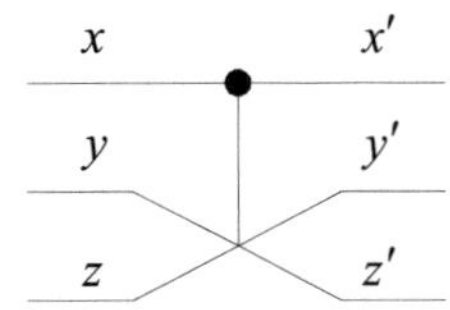

The other types of the qubit gates, e.g. phase shift operator S and its derivatives, cannot be represented by the classical operators and require quantum computing devices. In such computing, it is assumed that each matrix operation is conducted in one computation step providing a power of quantum computing. The indicated dependence of quantum states on the observation process allows an implementation of such operations by the use of adaptive computation schemes. Below, we demonstrate a relation between quantum operations and evolving algebras and consider appropriate probabilistic decision-making.

3. Probabilistic automata and mobile robots control

In this section we describe a simulation example, which illustrates an implementation of qubit operators for the mobile robot's navigation. Then, based on this example, we consider the model of probabilistic control and algorithmic learning methods based on evolving algebras.

3.1 Control of the mobile robot by qubit operators

Let us consider a mobile robot, which moves on the plain, and assume that the inner states of the robots correspond to its direction on a plane. Let $S = \{ "\uparrow", "\downarrow", "\rightarrow", "\leftarrow" \}$ be a set of pure or errorless states such that each state corresponds to a certain direction of the robot. The set $V = \{ "\textit{step forward}", "\textit{step backward}", "\textit{turn left}", "\textit{turn right}", "\textit{stay still}" \}$ includes the actions that are available to the robot, where the steps *"forward"* and *"backward"* are restricted by a certain fixed distance. Being at time moment t in the state $s^t \in S$ and choosing the action $v^t \in V$, the robot receives to its input a finite scalar value $\varepsilon^t = \varepsilon\left(s^t, v^t\right)$, which depends on the robot's position in the environment. Notice that the elements of the set V form a group with multiplication acting on the set of states S. The steps *"forward"*, *"backward"* and *"stay still"* action do not change the state of the robot but change its position relatively to the environment that changes the obtained value ε.

Assume that the states of the robot are described by four qubits: $"\uparrow"=|s_1\rangle=(1,0)^T$, $"\downarrow"=|s_2\rangle=(0,1)^T$, $"\rightarrow"=|s_3\rangle=\left(1/\sqrt{2},1/\sqrt{2}\right)^T$ and $"\leftarrow"=|s_4\rangle=\left(1/\sqrt{2},-1/\sqrt{2}\right)^T$, and the actions "*turn right*" and "*turn left*" are, respectively, conducted by the use of the Hadamard $\mathrm{H}=\frac{1}{\sqrt{2}}\begin{pmatrix}1 & 1\\ 1 & -1\end{pmatrix}$ and reverse Hadamard $\mathrm{H}^R=\frac{1}{\sqrt{2}}\begin{pmatrix}-1 & 1\\ 1 & 1\end{pmatrix}$ operators. Then for the Hadamard operator it holds true that $\mathrm{H}|s_1\rangle=|s_3\rangle$, $\mathrm{H}|s_3\rangle=|s_1\rangle$, $\mathrm{H}|s_2\rangle=|s_4\rangle$, and $\mathrm{H}|s_4\rangle=|s_2\rangle$; and similarly, for the reverse Hadamard it holds true that $\mathrm{H}^R|s_1\rangle=|s_4\rangle$, $\mathrm{H}^R|s_4\rangle=|s_1\rangle$, $\mathrm{H}^R|s_2\rangle=|s_3\rangle$ and $\mathrm{H}^R|s_3\rangle=|s_2\rangle$, where the states $|s\rangle=a|0\rangle+b|1\rangle$ and $-|s\rangle=(-1)(a,b)^T$ conventionally are not distinguished.

Let us consider the states and the actions of the robot in particular. If there is no influence on the environment, then the relations between the states can be represented as follows. Assume that the robot is in the state $|s_1\rangle$. Then according to the above-indicated equality $|s_1\rangle=\left(1/\sqrt{2}\right)\cdot|s_3\rangle+\left(1/\sqrt{2}\right)\cdot|s_4\rangle$, and the probability that at the next time moment the robot will be in the state $|s_3\rangle$ is $p(s_3\,|\,s_1)=\Pr\left\{s^{t+1}=|s_3\rangle\,|\,s^t=|s_1\rangle\right\}=\left|1/\sqrt{2}\right|^2=1/2$ and the probability that it will be at the state $|s_4\rangle$ is also $p(s_4\,|\,s_1)=\Pr\left\{s^{t+1}=|s_4\rangle\,|\,s^t=|s_1\rangle\right\}=\left|1/\sqrt{2}\right|^2=1/2$. The similar equalities hold true for the other remaining states $|s_2\rangle$, $|s_3\rangle$ and $|s_4\rangle$. In general, for complex amplitudes s_{jk}, it follows that $|s_1\rangle=s_{13}|s_3\rangle+s_{14}|s_4\rangle$, $|s_2\rangle=s_{23}|s_3\rangle-s_{24}|s_4\rangle$, $|s_3\rangle=s_{31}|s_1\rangle+s_{32}|s_2\rangle$, and $|s_4\rangle=s_{41}|s_1\rangle-s_{42}|s_2\rangle$; thus $p(s_k\,|\,s_j)=\Pr\left\{s^{t+1}=|s_k\rangle\,|\,s^t=|s_j\rangle\right\}=\left|s_{jk}\right|^2$.

Now let us take into account an influence of the environment. Recall that the quantum state of the qubit is equivalently defined both by the state-vector $|s_i\rangle$ and by the density matrix. Let $|s_1\rangle=(1,0)^T$ and $|s_2\rangle=(0,1)^T$, and let $|s_3\rangle=(s_{31},s_{32})^T$, $|s_4\rangle=(s_{41},s_{42})^T$. Then, by definition, the states are defined by the density matrices $\sigma_1=|s_1\rangle\langle s_1|=\begin{pmatrix}1 & 0\\ 0 & 0\end{pmatrix}$, $\sigma_2=|s_2\rangle\langle s_2|=\begin{pmatrix}0 & 0\\ 0 & 1\end{pmatrix}$, $\sigma_3=|s_3\rangle\langle s_3|=\begin{pmatrix}|s_{31}|^2 & s_{31}s_{32}^*\\ s_{31}^*s_{32} & |s_{32}|^2\end{pmatrix}$, and $\sigma_4=|s_4\rangle\langle s_4|=\begin{pmatrix}|s_{41}|^2 & s_{41}s_{42}^*\\ s_{41}^*s_{42} & |s_{42}|^2\end{pmatrix}$, while $\mathrm{tr}(\sigma_i)=1$. Since the non-diagonal elements of the matrices can obtain arbitrary values, let us use these elements for specifying the relation with the environmental variable ε. In particular, assume that $\sigma_3(\varepsilon)=\sigma_3+\begin{pmatrix}0 & i\cdot\varepsilon\\ -i\cdot\varepsilon & 0\end{pmatrix}$. Then, the application of the Hadamard operators H and H^R to the state $\sigma_3(\varepsilon)$ after normalization results in the states $|s'\rangle=\left(1/\sqrt{2}-i\left(1/\sqrt{2}\right)\varepsilon,\,i\left(1/\sqrt{2}\right)\varepsilon\right)^T$ and $|s''\rangle=\left(-i\left(1/\sqrt{2}\right)\varepsilon,\,1/\sqrt{2}+i\left(1/\sqrt{2}\right)\varepsilon\right)^T$. If, e.g., $\varepsilon=1/3$, then $|s'\rangle=(0.72-0.24i,\,0.24i)^T$ and

$|s''\rangle = (-0.24i, 0.72 + 0.24i)^T$ instead of the states $|s_1\rangle = (1,0)^T$ and $|s_2\rangle = (0,1)^T$, which are obtained in the errorless case. The transition probabilities are $\rho(s_1 \mid s') = \rho(s_2 \mid s'') = 0.76$ and $\rho(s_2 \mid s') = \rho(s_1 \mid s'') = 0.24$. The presented method of control is illustrated by the example, shown in Fig. 1. In the simulation (Rybalov et al., 2010), the robot was programmed to follow the trajectory by the use of a compass sensor, and its turns were controlled by Hadamard and reverse Hadamard operators.

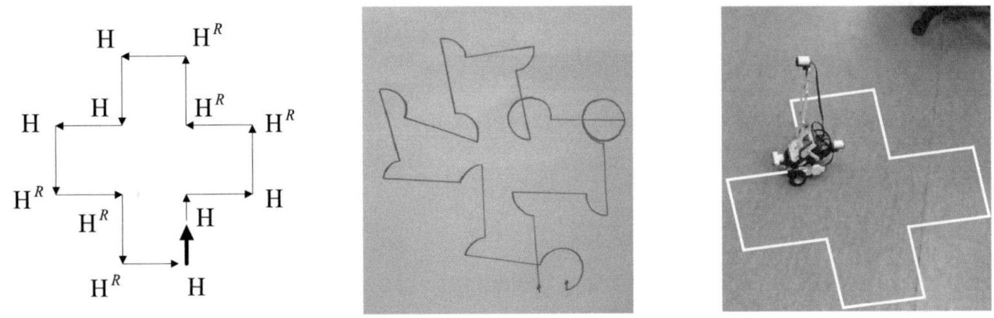

Fig. 1. From left to right: the planned trajectory with corresponding Hadamard and reverse Hadamard operator; the resulting trajectory with the directions of turns; and the mobile robot following the trajectory by the use of the compass sensor.

The presented approach to control the robot is a particular case of the dynamics of open quantum dynamical system (Holevo, 2001). In general, the dynamics of such systems is determined by transitions $|s\rangle \to \mathrm{U}|s\rangle$, where U is an appropriate Hermitian operator.

3.2 Probabilistic model and algorithmic learning

Let us take into account the observation process and present a brief description of the algorithmic model of the robot's control system with variable evolution structure. As it is usual for automata models, let X be a set of input values, Y be a set of output values, and S be a set of inner states. The automaton is defined by two characteristic functions $f: X \times S \to Y$ and $g: X \times S \to S$, such that discrete time dynamics is defined as $y^t = f(x^t, s^t)$ and $s^{t+1} = g(x^t, s^t)$, where $x^t \in X$, $y^t \in Y$, $s^t, s^{t+1} \in S$, $t = 0, 1, 2, \ldots$ In the case of *learning automata*, the probability distributions are defined over the sets X, Y and S, and functions f and g act on such distributions (Fu & Li, 1969; Tsetlin, 1973), while the metrical or topological structures of the sets are constant. By the other approaches, the *program structure learning* is specified by a convergence to the appropriate transition function g, or by the choice of a metric or topology over the set S, called *data structure learning*.
Algorithmically, the variable data structure is defined as follows (Gurevich, 1991). Let G be a global namespace with three distinguished elements $1 = \text{"}true\text{"}$, $0 = \text{"}false\text{"}$ and $\Diamond = \text{"}undef\text{"}$. A map $\varphi: G^r \to G$ is called basic function of arity r, while a basic function of arity $r = 0$ is considered as a distinguished element, and basic functions $\psi: G^r \to \{1, 0, \Diamond\}$ are considered as terms. The evolving algebra $\mathfrak{A} = \{G, \varphi_0, \varphi_1, \ldots\}$ is defined by the following updates:

- local update: $\varphi(a_0, a_1, \ldots, a_r) \leftarrow a$, which specifies the value to the basic function,
- guarded update: if $\psi = 1$ then $\varphi(a_0, a_1, \ldots, a_r) \leftarrow a$,

where $a_0, a_1, \ldots, a_r, a \in G$ and ψ is a term. Thus, evolving algebra $\mathfrak{A}$ permits its functions to change their domains according to the data flow, as it is required for the learning property. Notice that the implementation of the updates of the algebra $\mathfrak{A}$ by the use of quantum gates results in the universal algebra $\mathfrak{U}$, which defines quantum computations. Nevertheless, similar to the algebra $\mathfrak{U}$, the direct implementation of the evolving algebra $\mathfrak{A}$ is possible only in particular cases of computable operators.

Let us consider an implementation of the evolving algebra for probabilistic control of the mobile robots control (Kagan & Ben-Gal, 2008). Since during its mission the robot acts in a stochastic environment, the input variable $x \in X$ is random; thus, given characteristic f and g, both inner states $s \in S$ and outputs $y \in Y$ are also random variables. Assume that the namespace G, in addition to elements 1, 0 and $\Diamond$, includes all possible realizations of the inputs, inner states and outputs, i.e. $G = \{1, 0, \Diamond\} \cup X \cup Y \cup S$. For the realizations of inputs, we define the terms $\psi_1(x) \in \{1, 0, \Diamond\}$, and for the inner states we define the terms $\psi_2(s) \in \{1, 0, \Diamond\}$ and the local updates $\varphi(s') \leftarrow s''$, $s', s'' \in G$.

Now we can formulate transition function g in the terms of evolving algebra. We say that the pairs $(x', s'), (x'', s'') \in X \times S$ are equal (in the sense of the function g) if $g : (x', s') \mapsto s$ and $g : (x'', s'') \mapsto \overline{s}$. Then, since there are at most m distinguished realizations of the inner states and there may exist the pairs from $X \times S$, such that the map g is not defined, the space $X \times S$ is partitioned into at most $m+1$ equivalence classes. Denote such equivalence classes by A^g. Then, the transition function g is defined as a guarded update if $(x, s) \in A_j^g$ then $\phi(s) \leftarrow s_j$, which is checked for each pair $(x, s) \in X \times S$, and if the appropriate class A^g is not found, then $\phi(s) \leftarrow \Diamond$ is specified.

The presented algorithmic model allows simulations of quantum-control of the mobile robot and its navigation on the basis of the qubits model of states. Below, we consider an example of such simulations with probabilistic decision-making.

4. Navigation of quantum-controlled mobile robot along predefined path

Let us start with a simulation example. The simulation follows an idea of the experiment of checking a spin of elementary particle by three Stern-Gerlach apparatus, which are defined by a sequence of certain quantum operators (Albert, 1994).

As above, let the inner states of the robot be specified by four qubits $"\uparrow" = |s_1\rangle = (1, 0)^T$, $"\downarrow" = |s_2\rangle = (0, 1)^T$, $"\rightarrow" = |s_3\rangle = (1/\sqrt{2}, 1/\sqrt{2})^T$ and $"\leftarrow" = |s_4\rangle = (1/\sqrt{2}, -1/\sqrt{2})^T$. In addition, assume that there are two types of detectors defined by the above-defined Pauli operators X and Z, so that $\langle s_3|\mathrm{X}|s_3\rangle = +1$, $\langle s_4|\mathrm{X}|s_4\rangle = -1$, $\langle s_1|\mathrm{Z}|s_1\rangle = +1$ and $\langle s_2|\mathrm{Z}|s_2\rangle = -1$.

The robot starts with a random initial state $|s^0\rangle$ and arrives to the first detector Z. Detector Z checks the state $|s^0\rangle$ and the robot obtains a new state $|s^1\rangle$. According to the maximum of probabilities $P(s_3 \mid s^1)$ and $P(s_4 \mid s^1)$, the robot chooses the left or right trajectory and arrives to the second detector X. Then, after the similar actions and obtaining the state $|s^2\rangle$, the robot continues to the third detector Z, which checks the robot's state and results in the state $|s^3\rangle$. The fragment of the experiment with the robot following its path is shown in Fig. 2.

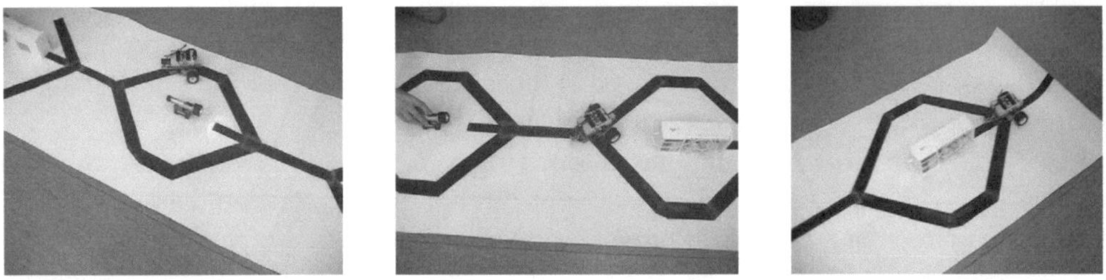

Fig. 2. The mobile robot follows a path (from right to left) with three simulated detectors using touch sensor ("Z detector") and light sensors ("X detector").

Let us consider the actions of the robot in particular. As indicated above, the robot starts with an initial state $|s^0\rangle = (s_1^0, s_2^0)^T$, where $s_1^0, s_2^0 \in [0,1]$ are random values such that $0 < (s_1^0)^2 + (s_2^0)^2 \le 1$. Then the first detector Z results in the state $|s^1\rangle = Z|s^0\rangle$ and the decision-making is conducted regarding the further left or right trajectory, which is based on the probabilities $P(s_3 \,|\, s^1) = |\langle s^1 | s_3 \rangle|^2$ and $P(s_4 \,|\, s^1) = |\langle s^1 | s_4 \rangle|^2$. If $P(s_3 \,|\, s^1) > P(s_4 \,|\, s^1)$, then the robot turns left, and if $P(s_3 \,|\, s^1) < P(s_4 \,|\, s^1)$, then the robot turns right (the ties are broken randomly). Following the chosen trajectory, the robot arrives to the second detector X. The check with this detector results in the state $|s^2\rangle = X|s^1\rangle$, and the decision regarding the further trajectory is obtained on the basis of the probabilities $P(s_1 \,|\, s^2) = |\langle s^2 | s_1 \rangle|^2$ and $P(s_2 \,|\, s^2) = |\langle s^2 | s_2 \rangle|^2$. Similar to the above, if $P(s_1 \,|\, s^2) > P(s_2 \,|\, s^2)$, then the robot turns left, and if $P(s_1 \,|\, s^2) < P(s_2 \,|\, s^2)$, then the robot turns right (the ties are broken randomly). The third check is again conducted by the Z detector, which results in the state $|s^3\rangle = Z|s^2\rangle$, and the decision-making is conducted by the same manner as for the state $|s^1\rangle$.

Now let us present a general description of the process (Kagan & Ben-Gal, 2008), that implements the above-indicated equivalence classes A^f and A^g. As indicated above, the evolution of the quantum-mechanical system with observations does not depend on the previous states and starts from the value of the state, which is obtained by the measurement. Thus, the outputs $y \in Y$ of the system are specified by a Markov process, which is controlled by input states $x \in X$ and inner states $s \in S$. Then the probability $p(y) = p(A^f(y))$ of the equivalence class $A^f(y) = f^{-1}(y)$ is defined as a sum

$p(y)=\sum_{(x,s)\in A^f(y)}p(x,s)$ of the probabilities of the pairs $(x,s)\in X\times S$. Similarly, the probability $p(s)$ of the inner state $s\in S$ of the system is defined by the use of equivalence class $A^g(s)=g^{-1}(s)$ as a sum $p(s)=\sum_{(x,s)\in A^g(s)}p(x,s)$.

Recall that, by definition, the equivalence classes $A^f(y)$, $y\in Y$, and A^g, $A^g(s)$, form partitions α^f and α^g of the space $X\times S$. Then, the relation between the dynamics of inner states is determined by the relation between the partitions α^f and α^g. For example, let $d_{Orn}\left(\alpha^g,\alpha^f\right)=\sum_{i=0}^{k-1}\left|p\left(A_i^g\right)-p\left(A_i^f\right)\right|$, where $k=\max\left\{\left|\alpha^g\right|,\left|\alpha^f\right|\right\}$, and if $\left|\alpha^g\right|>\left|\alpha^f\right|$, then α^f is completed by empty sets, while if $\left|\alpha^g\right|<\left|\alpha^f\right|$, then empty sets are added to α^g, being Ornstein distance between the partitions (Ornstein, 1974). Then, for the finite time case, $t=0,1,2,\ldots,T-1$, the Ornstein distance between partitions α_t^f and α_t^g, is defined as $d_{Orn}\left(\left\{\alpha_t^g\right\}_0^{T-1},\left\{\alpha_t^f\right\}_0^{T-1}\right)=\left(1/(T-1)\right)\sum_{t=0}^{T-1}d_{Orn}\left(\alpha_t^g,\alpha_t^f\right)$. Since the structure of the partitions α_t^f and α_t^g is constant, the distance represents the relation between probability measures defined by these partitions. Thus the goal of the robots navigation is to find a process for governing the inner states $s\in S$ such that the distance $d_{Orn}\left(\left\{\alpha_t^g\right\}_0^{T-1},\left\{\alpha_t^f\right\}_0^{T-1}\right)$ reaches its minimum over the considered time interval. Below we will consider general informational algorithms of local search, which can be implemented for such a task.

5. Information theoretic decision-making and path planning

The section presents information theoretic methods for quantum inspired decision-making and general path-planning algorithms. We start with a motivating example of informational decision-making, then we consider the logic of quantum mechanics and informational distance between the partitions of the events space. Finally, we present the navigation algorithms, which are based on the representation of the states' evolution by the use of partitions.

5.1 Decision-making by the use of quantum-mechanical information measure

An application of informational criteria for decision-making and path planning of the quantum-controlled mobile robots is motivated by the criteria of classical information theory (Cover & Thomas, 1991). Recall that in the classical case, an informational distance between the probability vectors $\vec{p}=(p_1,\ldots,p_n)$ and $\vec{q}=(q_1,\ldots,q_n)$, $\sum_{j=1}^{n}p_j=1$, $\sum_{j=1}^{n}q_j=1$, $p_j\geq 0$, $q_j>0$, $j=1,2,\ldots,n$, is specified by the *relative Shannon entropy* or *Kullback-Leibler distance* $KL(\vec{p}\,||\,\vec{q})=\sum_{j=1}^{n}p_j\log(p_j/q_j)=\sum_{j=1}^{n}p_j\log p_j-\sum_{j=1}^{n}p_j\log q_j$, where by convention it is assumed that $0\log 0=0$. The distance $KL(\vec{p}\,||\,\vec{q})$ satisfies both $KL(\vec{p}\,||\,\vec{q})\geq 0$ and $KL(\vec{p}\,||\,\vec{q})=0$ if and only if $p_j=q_j$ for all $j=1,2,\ldots,n$, and, in general

$KL(\vec{p}\,||\,\vec{q}) \neq KL(\vec{q}\,||\,\vec{p})$. Vectors $\vec{p}$ and $\vec{q}$ represent the states of the stochastic system, and distance $KL(\vec{p}\,||\,\vec{q})$ characterizes the information-theoretic difference between these states.
In contrast, in quantum-mechanical systems, the state is represented by the above-presented Hermitian density matrix σ with $\mathrm{tr}(\sigma)=1$. The informational measures for such states are defined on the basis of the von Neumann entropy (Nielsen & Chuang, 2000) $VN(\sigma) = -\mathrm{tr}(\sigma \log \sigma) = -\sum_{j=1}^{n} \lambda_j \log \lambda_j$, where λ_j are the eigenvalues of the matrix σ. Then the *relative von Neumann entropy* of the state σ' relative to the state σ'' is defined as $VN(\sigma'\,||\,\sigma'') = \mathrm{tr}(\sigma' \log \sigma') - \mathrm{tr}(\sigma' \log \sigma'') = -\sum_{j=1}^{n} \lambda'_j \log \lambda'_j - \sum_{j=1}^{n} \lambda'_j \log \lambda''_j$, where λ' and λ'' are eigenvalues of the matrices σ' and σ'', correspondingly.
Let σ^t be a state of the system at time moment t, and consider its representation $\sigma^t = \sum_{j=1}^{n} \lambda_j^t E_j$, where $\lambda_1^t < \lambda_2^t < \ldots < \lambda_n^t$ are eigenvalues of the matrix σ^t and $E = \{E_1, \ldots, E_n\}$ is a set of matrices such that $E_j E_k = \delta_{jk} E_k$ and $\sum_{j=1}^{n} E_j = I$, where $\delta_{jk} = 1$ if $j = k$, and $\delta_{jk} = 0$ otherwise. According to the dynamics of the quantum system, if the system is in the state σ^t, then its next state σ^{t+1} is specified by the use of the selected operator E_j according to the projection postulate $\sigma^{t+1} = E_j \sigma^t E_j \big/ \mathrm{tr}\left(\sigma^t E_j\right)$. This postulate represents the above indicated influence of the measurement, i.e. of the application of the operator E_j, to the state of the system, and the generalized the Bayesian rule to the evolution of quantum-mechanical systems. The decision-making problem required a definition of such projection that given a state σ^t, the next state σ^{t+1} is optimal in a certain sense. Since for a state σ^t, there exist several sets E of matrices with the above indicated properties, the decision-making includes two stages (Davies, 1978; Holevo, 2001) and requires finding the set $E = \{E_1, \ldots, E_n\}$, and then selecting an operator E_j from the set E according to optimality criteria.
One of the methods, which is widely used in classical information theory, implies a choice of such probability vector $\vec{p}^{t+1}$, that given a vector $\vec{p}^t$, the Kullback-Leibler distance $KL\left(\vec{p}^t\,||\,\vec{p}^{t+1}\right)$ reaches its maximum. Likewise, in the simulations, we implemented the choice of the set E and operator $E_j \in E$ such that it maximizes the relative von Neumann entropy $VN\left(\sigma^t\,||\,\sigma^{t+1}\right)$. In the simulations (Kagan et al., 2008), the mission of the robot was to navigate in the environment and to find the objects, which randomly change their location. The amplitudes, which defined the states, were derived from the distances between the objects and as they were measured by the ultra-sonic sensor. The scheme of the simulation and the fragment of the robot's movement are shown in Fig. 3.
The robot scans the environment, chooses such an object that maximizes the von Neumann relative entropy, and moves to this object. After the collision, the object was moved to a new random location. The comparison between quantum and classical decision-making demonstrated the difference in nearly 50% of the decisions, and in the search for variable number of objects, quantum decision-making demonstrated nearly 10% fewer decision errors than the classical one. Such results motivated an application of information theoretic methods for navigation of quantum-controlled mobile robots. In the next section we consider the algorithms which follow this direction.

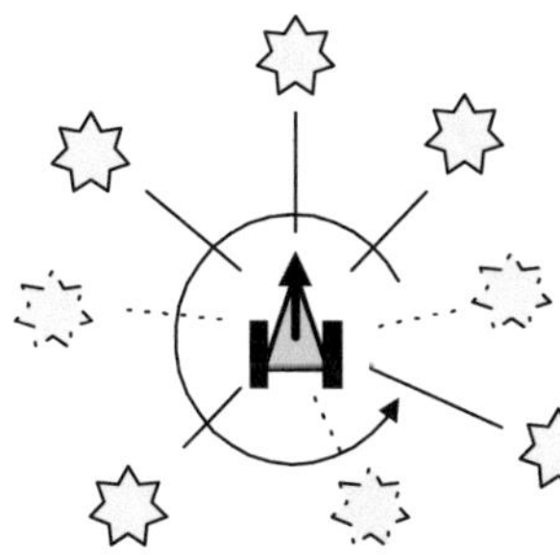

Fig. 3. The mobile robot with ultra-sonic sensor acting in the stochastic environment: after collision, the obstacle randomly changes its location in the environment.

5.2 Logic of quantum mechanics and informational distance

Now let us consider the logic of quantum mechanics and the structure of quantum events over which the informational algorithms act. Such a formal scheme of quantum events, which is called *quantum logic* (Cohen, 1989), was introduced by Birkhoff and von Neumann (Birkhoff & Neumann, 1936) as an attempt of to find an axiomatic description of quantum mechanics.

A quantum logic is a lattice Λ of events A, which contains the smallest element $\varnothing$, the greatest element I, relation $\subset$, unary operation $'$, and binary operations $\cup$ and $\cap$. It is assumed that for the events $A \in \Lambda$ the following usual set properties hold true: a) For any element $A \in \Lambda$ there exist an event $A' \in \Lambda$ such that $(A')' = A$; b) $A \cap A' = \varnothing$ and $A \cup A' = I$; c) For any pair of events $A, B \in \Lambda$, $A \subset B$ implies $B' \subset A'$, and $A \subset B$ implies $B = A \cup (B \cap A')$; d) For any countable sequence $A_1, A_2, \ldots$ of events from Λ their union is in Λ, i.e., $\left(\cup_j A_j\right) \in \Lambda$. Notice that is not required that the events $A \in \Lambda$ are subsets of the same set. That leads to the fact that under the requirements of quantum logic the distributive rule may not hold and, in general, $B \cap (A \cup A') \geq (B \cap A) \cup (B \cap A')$. In contrast, in the probabilistic scheme, it is postulated that the events are the subsets of the set of elementary outcomes, so the distribution rule is satisfied.

In quantum logic Λ, events $A \in \Lambda$ and $B \in \Lambda$ are called *orthogonal*, and denoted by $A \perp B$, if $A \subset B'$, and a finite set $\alpha = \{A_1, \ldots, A_n\}$, of events $A_j \in \Lambda$, is called $\vee$*-orthogonal system* if $\left(\cup_{j=1}^{k} A_j\right) \perp A_{k+1}$, $k = 1, 2, \ldots, n-1$. The state μ over the quantum logic Λ is defined as a map $\mu : \Lambda \to [0,1]$ such that $\mu(I) = 1$ and for any $\vee$-orthogonal system $\alpha = \{A_1, \ldots, A_n\}$ it holds true that $\mu\left(\cup_{j=1}^{n} A_j\right) = \sum_{j=1}^{n} \mu\left(A_j\right)$. Given a state $\mu : \Lambda \to [0,1]$, an $\vee$-orthogonal system α is a *partition* of the logic Λ with respect to the state μ if $\mu\left(\cup_{j=1}^{n} A_j\right) = 1$, $A_j \in \alpha$ (Yuan, 2005).

Following classical ergodic theory (Rokhlin, 1967), *entropy* of the partition α is defined as $H_\mu(\alpha) = -\sum_{A \in \alpha} \mu(A) \log \mu(A)$, and *conditional entropy* of partition α relatively to partition β is $H_\mu(\alpha \mid \beta) = -\sum_{B \in \beta} \sum_{A \in \alpha} \mu(A, B) \log \mu(A \mid B)$, where α and β are partitions of quantum logic Λ (Yuan, 2005; Zhao & Ma, 2007). In addition, similarly to the ergodic theory (Rokhlin,

1967), a Rokhlin distance is defined as $d_\mu(\alpha,\beta)=H_\mu(\alpha|\beta)+H_\mu(\beta|\alpha)$ between the partitions α and β of the quantum logic Λ, which preserves the metric properties (Khare & Roy, 2008), as it holds for the probabilistic scheme.

5.3 Actions of the robot over the quantum events

The indicated properties of the partitions and states of quantum logic allow an application of informational A*-type search algorithms acting over partitions space (Kagan & Ben-Gal, 2006) for navigation of quantum-controlled mobile robots. Let χ be a set of all possible partitions of the logic Λ given a state μ. By $d_\mu(\alpha,\beta)$, $\alpha,\beta\in\chi$, we denote a Rokhlin distance between the partitions α and β, and by $\tilde{d}_\mu(\alpha,\beta)$ an estimated distance such that $\tilde{d}_\mu(\alpha,\beta)\le d_\mu(\alpha,\beta)$; as a distance $\tilde{d}_\mu$ the above-defined Ornstein distance $d_{Orn}(\alpha,\beta)$ can be applied. Let $r\in R^+$ be a constant value. The non-empty neighborhood $N(\alpha,r)\subset\chi$ of partition $\alpha\in\chi$ is a set of partitions such that $\alpha\notin N(\alpha,r)$ and for every partition $\beta\in N(\alpha,r)$ it holds true that $\tilde{d}_\mu(\alpha,\beta)\le r\le d_\mu(\alpha,\beta)$. In the algorithm, α stands for a partition specified by the robot, and τ stands for a partition specified by the environment. The robot is located in the environment, which is specified by a partition $\tau_{cur}=\vartheta$ and starts with the initial partition $\alpha_{cur}=\theta$. Then, the actions of the robot are defined by the following sequence:

The robots actions given the environment τ_{cur}:

- Choose the next partition $\alpha_{next}\leftarrow \arg\min_{\alpha\in N(\alpha_{cur},r)}\left\{\tilde{d}_\mu(\alpha,\tau_{cur})\right\}$;
- Update estimation $\tilde{d}_\mu(\alpha_{cur},\tau_{cur})\leftarrow \max\left\{\tilde{d}_\mu(\alpha_{cur},\tau_{cur}),\, r+\min_{\alpha\in N(\alpha_{cur},r)}\tilde{d}_\mu(\alpha,\tau_{cur})\right\}$;
- Set current partition $\alpha_{cur}\leftarrow\alpha_{next}$.

The robots actions while the environment changes ($\tau_{cur}\leftarrow\tau$):

- Update estimation $\tilde{d}_\mu(\alpha_{cur},\tau)\leftarrow\max\left\{\tilde{d}_\mu(\alpha_{cur},\tau),\, \tilde{d}_\mu(\alpha_{cur},\tau_{cur})-r\right\}$.

Let us clarify the application of the algorithms to the quantum logic Λ by half steps, $t=0,1/2,1,3/2,\ldots$, which correspond to the robot's and the environment's actions. Let the initial map by $\mu^0:\Lambda\to[0,1]$. Regarding the robot's states, it means a choice of the basis for the states; in the above-given examples, such a choice corresponds to the definition of the qubits for the states $"\uparrow"=|s_1\rangle$ and $"\downarrow"=|s_2\rangle$, or that is equivalent to the states $"\rightarrow"=|s_3\rangle$ and $"\leftarrow"=|s_4\rangle$. By the use of the map μ^0 the robot chooses the state $|s^0\rangle$ and a certain partition which specifies the next map $\mu^{1/2}$. According to the obtained map $\mu^{1/2}$, the environment changes and the map μ^1 is specified. Now the robot chooses the $|s^1\rangle$ and according to the chosen partition specifies the map $\mu^{3/2}$. The environment, in its turn, changes according to the map $\mu^{3/2}$, and so on.

The presented actions over the partitions of quantum logic provide a representation, which differs from the one use in the quantum learning algorithms (Chen & Dong, 2008; Dong, et al., 2010) and implements information theoretic decision-making using the Rokhlin distance. However, the meaning of the actions is similar, and the goal of the robot is to determine the actions such that the inner states of the robot correspond to the environmental states, which, in their turn, are changed by the robot.

6. Notes on fuzzy logic representation of quantum control

Let us return to the representation of the robot's states by qubits $|s\rangle = a|0\rangle + b|1\rangle = a(1,0)^T + b(0,1)^T$, where both amplitudes a and b are real numbers. In such a case, the state-vector $|s\rangle$ can be represented by two membership functions $\mu_a : X \to [0,1]$ and $\mu_b : X \to [0,1]$ for some universal set X, which are defined as (Hannachi, et al., 2007) $\mu_a = (2/\pi)\arcsin\sqrt{(sign(a)a^2 - sign(b)b^2 + 1)/2}$ and $\mu_b = (2/\pi)\arcsin\sqrt{(sign(a)a^2 + sign(b)b^2 + 1)/2}$, with backward transformations $a = sign(\mu_a + \mu_b - 1)\sqrt{|\sin^2\mu_a\pi/2 + \sin^2\mu_b\pi/2 - 1|}$ and $b = sign(\mu_b - \mu_a)\sqrt{|\sin^2\mu_b\pi/2 - \sin^2\mu_a\pi/2|}$. Over the pairs of membership functions μ_a and μ_b, fuzzy analogs of quantum mechanical operators are defined (Hannachi, et al., 2007). Let us consider the Hadamard operators, which represent the turns of the robot.

Fuzzy analog $\tilde{\mathrm{H}}$ of the Hadamard operator H is the following (Hannachi, et al., 2007): $\tilde{\mathrm{H}}(\mu_a, \mu_b) = (\max(0, \min(1, 1/2 - \mu_a + \mu_b)), \min(1, \max(0, \mu_a + \mu_b - 1/2)))$. Similarly, reverse fuzzy Hadamard operator is defined by the following transformation (Rybalov et al., 2010): $\tilde{\mathrm{H}}^R(\mu_a, \mu_b) = (\min(1, \max(0, 3/2 - \mu_a - \mu_b)), \max(0, \min(1, 1/2 - \mu_a + \mu_b)))$.

Straightforward calculations show that the direct and reverse fuzzy Hadamard operators $\tilde{\mathrm{H}}$ and $\tilde{\mathrm{H}}^R$ are reversible and preserve the properties of corresponding quantum Hadamard operators. Trajectories of the robot that acts according to the fuzzy Hadamard operators are illusrated by Fig. 4; the simulattions have been conducted likewise in the example shown in Fig. 1.

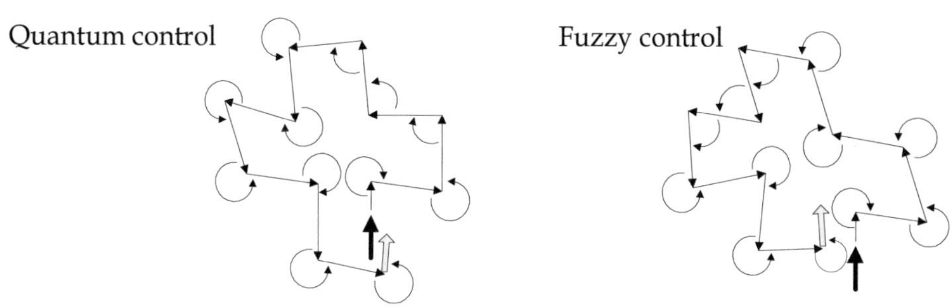

Fig. 4. Trajectories and turns of the mobile robot according to quantum and fuzzy control.

According to simulations, the turns of quantum and fuzzy controlled robot are different; however, the states of the robot and the results of its actions are statistically equivalent. Such preliminary results show that in case of real amplitudes, fuzzy logic models of quantum control may be applied.

7. Conclusion

In the chapter we presented a brief introduction into the methods of navigation of quantum-controlled mobile robots and considered the ideas of its implementation by the use of probabilistic and information theoretic techniques. The described methods represent such a property of the quantum-controlled mobile robots that the state of quantum-mechanical system includes the state of the environment as a part of its inner state.

In particular, the state of the mobile robot in the environment was defined by the use of the density matrix, which, in addition to the inner state of the robot, included the state of the environment. Such a specification of the state allowed calculations of both the robot's influence on the environment and the environmental influence on the robot by the use of the unified techniques.

The decision-making methods, which define the robot's behavior, implemented the indicated representation of the state and were based on the probabilistic and informational schemes. These schemes generalize the known maximum probability and maximum information criteria while taking into account additional information regarding the robot's influence on the environment, and correspond to the statistical considerations of quantum-mechanical methods (Malley & Hornstein, 1993; Barndorff-Nielsen & Gill, 2003).

In general, the actions of quantum-controlled mobile robot were specified by the choices of certain partitions of quantum logic. The choices were based on informational distances following the line of informational search algorithms (Kagan & Ben-Gal, 2006). As indicated above, such a method gives an alternative view to quantum learning and path-planning algorithms (Chen & Dong, 2008; Dong, et al., 2010).

The presented methods were simulated by the use of small mobile robots, while the complete realization of quantum control requires quantum-mechanical on-board computers. However, as it follows from preliminary considerations (Rybalov, et al., 2010), fuzzy control of the mobile robot demonstrates similar results as probabilistic and informational schemes of quantum control; thus in some cases fuzzy logic models of quantum control may be applied.

8. References

Albert, D. Z. (1994). *Quantum Mechanics and Experience*. Harvard University Press, Cambridge, Massachusetts and London, England.

Ballentine, L. E (2006). *Quantum Mechanics. A Modern Development*. Word Scientific, Singapore.

Barndorff-Nielsen, O. E. & Gill, R. D. (2003). On Quantum Statistical Inference. *J. Royal Statistical Society B*, Vol. 65, No. 4, pp. 775-816.

Benioff, P. (1998). Quantum Robots and Environments. *Phys. Rev. A*, Vol. 58, pp. 893-904.

Birkhoff, G. & Neumann, J. von. (1936). The Logic of Quantum Mechanics. *Annals Math.*, Vol. 37, No. 4, pp. 823-843.

Chen, C.-L. & Dong, D.-Y. (2008). Superposition-Inspired Reinforcement Learning and Quantum Reinforcement Learning. In *Reinforcement Learning: Theory and Applications*, C. Weber, M. Elshaw, N. M. Mayer, (Eds.), InTech Education and Publishing, Vienna, Austria, pp. 59-84.

Cohen, D. W. (1989). *An Introduction to Hilbert Space and Quantum Logic*. Springer-Verlag, New York.

Cover, T. M. & Thomas, J. A. (1991). *Elements of Information Theory*. John Wiley & Sons, New York.

Davies, E. B. (1978). Information and Quantum Measurement. *IEEE Trans. Inform. Theory*, Vol. 24, No. 5, pp. 596-599.

Dong, D.-Y.; Chen, C.-L.; Zhang, C.-B. & Chen, Z.-H. (2006). Quantum Robot: Structure, Algorithms and Applications. *Robotica*, Vol. 24, No. 4, pp. 513-521.

Dong, D.; Chen, C.; Chu, J. & Tarn, T.-J. (2010). Robust Quantum-Inspired Reinforcement Learning for Robot Navigation. *IEEE/ASME Trans. Mechatronics*, To appear.

Fu, K. S. & Li, T. J. (1969). Formulation of Learning Automata and Automata Games. *Information Science*, Vol. 1, No. 3, pp. 237-256.

Gurevich, Y. (1991). Evolving Algebras: An Attempt to Discover Semantics. *Bull. European Assoc. Theor. Comp. Science*, Vol. 43, pp. 264-284.

Hannachi, M. S.; Hatakeyama, Y. & Hirota, K. (2007). Emulating Qubits with Fuzzy Logic. *J. Advanced Computational Intelligence and Intelligent Informatics*, Vol. 11, No. 2, pp. 242-249.

Helstrom, C. W. (1976). *Quantum Detection and Estimation Theory*. Academic Press, New York.

Holevo, A. S. (2001). *Statistical Structure of Quantum Theory*. Springer-Verlag, Berlin.

Kagan, E. & Ben-Gal, I. (2008). Application of Probabilistic Self-Stabilization Algorithms to Robot's Control. *Proc. 15-th Israeli Conf. IE&M'08*, Tel-Aviv, Israel.

Kagan, E. ; Salmona, E. & Ben-Gal, I. (2008). Probabilistic Mobile Robot with Quantum Decision-Making. *Proc. IEEE 25-th Conv. EEEI*. Eilat, Israel.

Kagan, E. & Ben-Gal, I. (2006). An Informational Search for a Moving Target. *Proc. IEEE 24-th Convention of EEEI*. Eilat, Israel.

Khare, M. & Roy, S. (2008). Conditional Entropy and the Rokhlin Metric on an Orthomodular Lattice with Bayesian State. *Int. J. Theor. Phys.*, Vol. 47, pp. 1386-1396.

Levitin, L. B. (1969). On a Quantum Measure of the Quantity of Information. *Proc. Fourth All-Union Conf. Problems of Information Transmission and Coding*. IPPI AN USSR, Moscow, pp. 111-115 (In Russian). *English translation*: A. Blaquieve, et al., eds. (1987). *Information Complexity and Control in Quantum Physics*. Springer-Verlag, New York, pp. 15-47.

Malley, J. D. & Hornstein, J. (1993). Quantum Statistical Inference. *Statistical Sciences*, Vol. 8, No. 4, pp. 433-457.

Nielsen, M. A. & Chuang, I. L. (2000). *Quantum Computation and Quantum Information*. Cambridge University Press, Cambridge, England.

Ornstein, D. S. (1974). *Ergodic Theory, Randomness, and Dynamical Systems*. Yale University Press, New Haven, 1974.

Raghuvanshi, A.; Fan, Y.; Woyke, M. & Perkowski, M. (2007). Quantum Robots for Teenagers, *Proc. IEEE Conf. ISMV'07*, Helsinki, Finland.

Rokhlin, V. A. (1967). Lectures on the Entropy Theory of Measure-Preserving Transformations. *Rus. Math. Surveys*, Vol. 22, pp. 1-52.

Rybalov, A.; Kagan, E.; Manor, Y. & Ben-Gal, I. (2010). Fuzzy Model of Control for Quantum-Controlled Mobile Robots. *Proc. IEEE 26-th Conv. EEEI*. Eilat, Israel.

Tsetlin, M. L. (1973). *Automaton Theory and Modeling of Biological Systems*. Academic Press, New York.

Unsal, C. (1998). *Intelligent Navigation of Autonomous Vehicles in an Automated Highway System: Learning Methods and Interacting Vehicles Approach*. PhD Thesis, Virginia Polytechnic Institute and State University, VI, USA.

Yuan, H.-J. (2005). Entropy of Partitions on Quantum Logic. *Commun. Theor. Phys.*, Vol. 43, No. 3, pp. 437-439.

Zhao, Y.-X. & Ma, Z.-H. (2007). Conditional Entropy of Partitions on Quantum Logic. *Commun. Theor. Phys.*, Vol. 48, No. 1, pp. 11-13.

16

Tracking Control for Reliable Outdoor Navigation Using Curb Detection

Seung-Hun Kim
Korea Electronics Technology Institute,
Korea

1. Introduction

In the past several years there has been increasing interest in applications of a mobile robot. Personal service robots perform the missions of guiding tourists in museum, cleaning room and nursing the elderly [1]. Mobile robots have been used for the purpose of patrol, reconnaissance, surveillance and exploring planets, etc [2].
The indoor environment has a variety of features such as walls, doors and furniture that can be used for mapping and navigation of a mobile robot. In contrast to the indoor cases, it is hard to find any specific features in outdoor environment without some artificial landmarks [3]. Fortunately, the existence of curbs on roadways is very useful to build a map and localization in outdoor environment. The detected curb information could be used for not only map building and localization but also navigating safely [4]. And also the mobile robot decides to go or stop using the width of the road calculated from the detected curb.
The present paper deals with the development of a robot which can patrol areas such as industrial complexes and research centers. We have already reported about outdoor navigation of a mobile robot [5]. The extended Kalman filter is applied for the fusion between odometry and Differential Global Positioning System(DGPS) measurement data. It is insufficient for reliable navigation since the error of DGPS measurement data increased near high buildings and trees [6]. Hence, it is necessary to correct the pose of a mobile robot when the position data from the odometry and DGPS are inaccurate. This chapter proposes the curb detection algorithm and calculate the pose error from the curb edge data and then use it to correct the pose of the mobile robot.

2. Outdoor mobile robot system

The outdoor mobile robot system on an ATRV-mini from RWI Inc. is constructed as shown in Fig. 1. It is a four-wheeled robot designed for outdoor use and equipped with odometry. This robot is chosen because it can climb up and down over speed bumps and slopes. It is added a laser range finder to detect obstacles and build a map of the environment. For global localization, a DGPS receiver is added. A camera that can zoom-in and zoom-out, was installed to send the image data to the control station.
Fig. 2 shows the software architecture of the mobile robot system. The system consists of three parts which are the mobile server, the navigation server and the remote server. The mobile sever controls the motor in the ATRV according to the velocity and angular velocity

Fig. 1. Mobile robot system configuration.
(a) DGPS antenna (b) PC (c) Image camera (d) LMS200 laser range finder
(e) Attitude/heading sensor (f) Wireless LAN bridges (g) Joypad (h) Control station

from the navigation server and transfers the odometry to the navigation server. The navigation server makes the mobile robot localize using latitude and longitude data from the DGPS and attitude from 3-axis angle sensor. It detects curbs using distance from the tilted laser range finder and sends the distance, image and navigation data to the remote server through the wireless LAN. The remote server operates in two modes, autonomous mode or teleoperated mode, display the image and the navigation information.

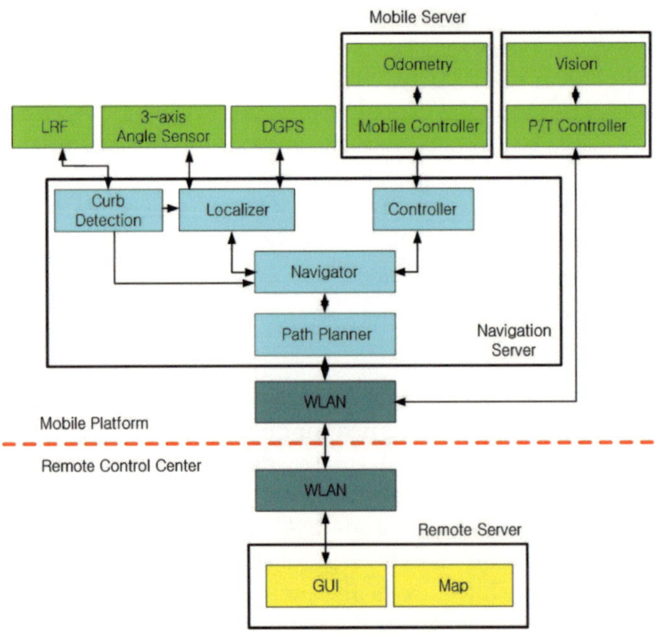

Fig. 2. Software diagram of the mobile robot system.

3. Curb detection

The outdoor mobile robot requires ability to navigate in a dynamic and potentially dangerous environment due to curbs and obstacles. Using the laser range finder, it can measure the distance from the obstacles and calculate the distance between obstacles. These data obtained from the laser range finder can be used in environment mapping and obstacle avoidance.

An important feature of road environment is the existence of curbs as shown in Fig. 3. The curbs could be used in mobile robot navigation.

Fig. 3. Road and curb on the road.

The proposed algorithm is to detect curbs through the laser range finder, which was positioned on the mobile robot and looking down the road with a small tilt angle as shown in Fig. 4. The tilted laser range finder projects three-dimensional shape of curbs on two-dimension space. And then it extracts curb edge points.

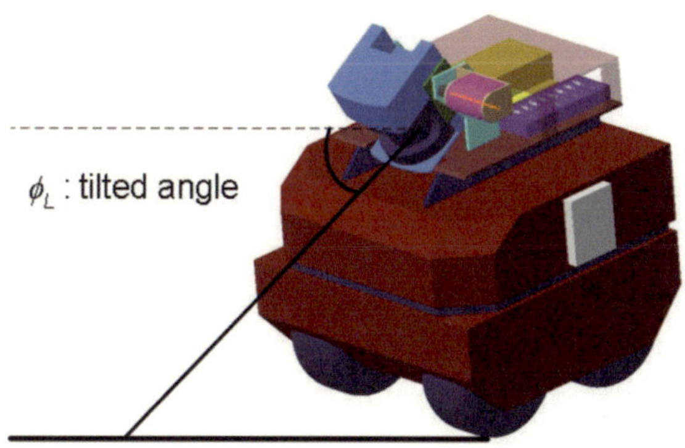

Fig. 4. Tilted laser range finder on the mobile robot.

The mobile robot builds the map of the curbs of roads and the map is used for tracking and localization. (n, x_L, y_L) is obtained by the laser range finder, where n represents n-th data and x_L, y_L are the distance data in the laser range finder's coordinate as shown in Fig. 5. The curb edges on the both sides of the mobile robot is used for mapping and tracking.

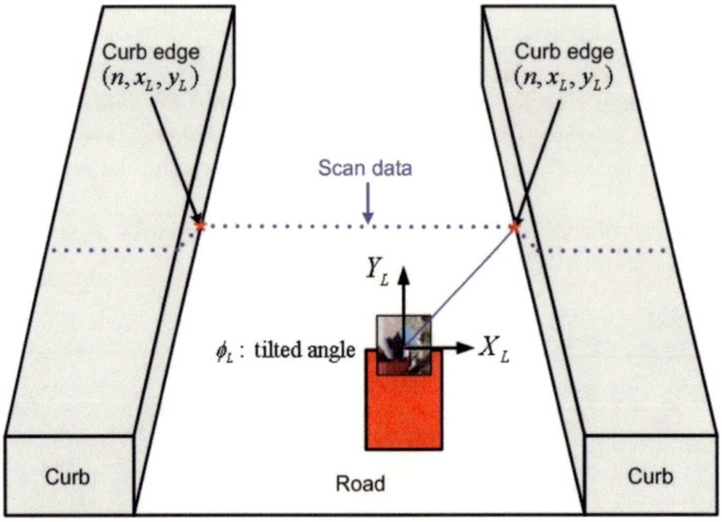

Fig. 5. Curb detection on the road.

Even though the angle between the robot and the curb changes, the feature points of the curb do not change as shown in Fig. 6. The points around 1600mm of the axis Y represent the road surface and the curved points between 0mm and 2000mm of the axis X are the curb of the road.

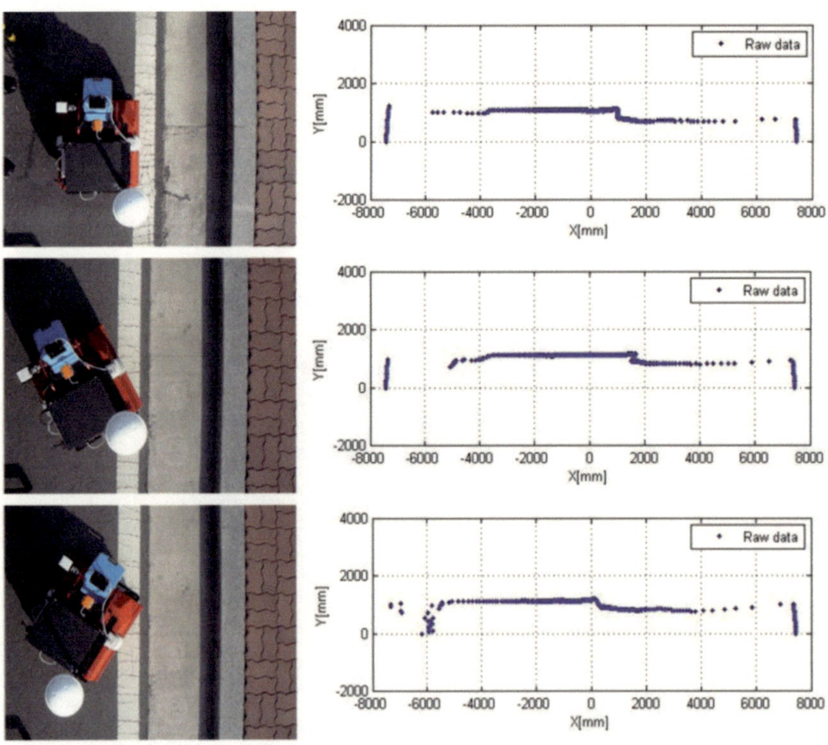

Fig. 6. Angles (0°, 30°, -30°) between the mobile robot and the curb and raw data.

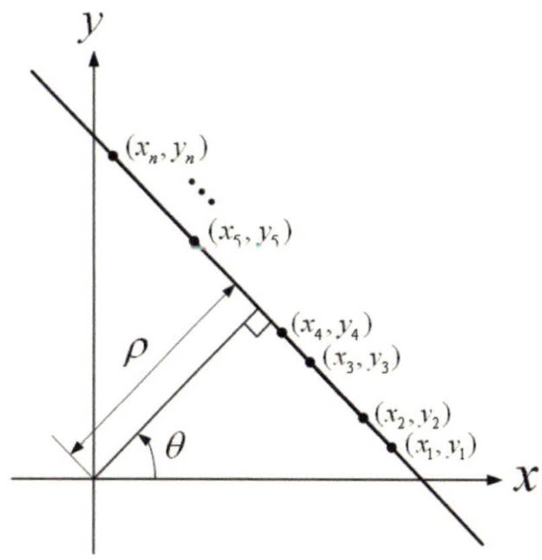

Fig. 7. Projection of collinear points onto a line.

3.1 Extracting the longest straight line

The laser range finder senses the surface of roads due to the small angle between the laser range finder and the mobile robot. Therefore, the longest straight line of the raw data from laser range finder is the surface of roads and the curbs locates the edges of the road. The longest straight line is extracted using Hough Transform [7], [8], [9]. In Hough transform, each line is represented by two parameters, commonly called ρ and θ, which represent the length and angle from the origin of a normal to the line. Using this parameterization, an equation of the line can be written as:

$$\rho = x_n \cos\theta + y_n \sin\theta \tag{1}$$

(x_n, y_n) is the n-th measurement data, where n is 1 to 361 and $\theta \in [0, 2\pi)$. The dominant ρ and θ are obtained by Hough Transform, called ρ_d and θ_d as shown in Fig. 8.

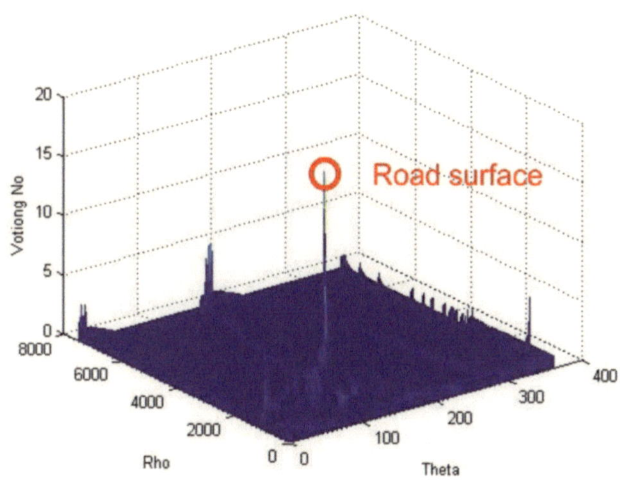

Fig. 8. Dominant ρ_d and θ_d found by Hough Transform.

The straight line in Fig. 9. shows the longest straight line represented by ρ_d and θ_d.

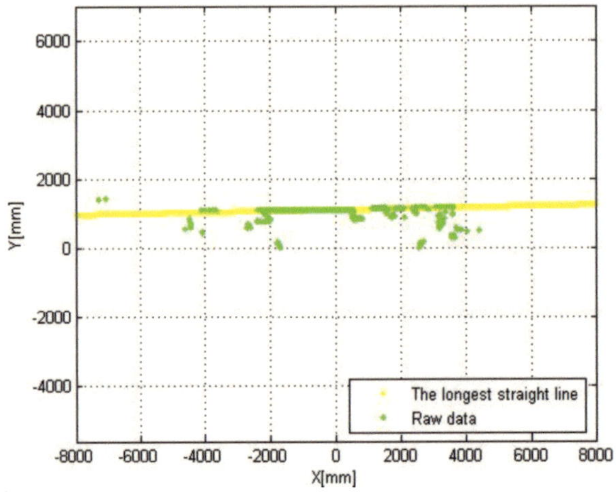

Fig. 9. Raw data and the longest straight line.

3.2 Fitting the points around the longest straight line

It is needed to fit the raw data points corresponding to the straight line for identifying the road surface and the curbs. The road surface and curbs exist on the boundary of the straight line from Fig. 10. To obtain the boundary points of the line, ρ is first calculated by Eq. (1) for each raw scan data with θ_d. And then, choose the points satisfying the following Eq. (2). The points of Fig. 6 show the fitting points with $\Delta\rho$ of 100mm.

$$\rho_d - \Delta\rho \le \rho \le \rho_d + \Delta\rho \tag{2}$$

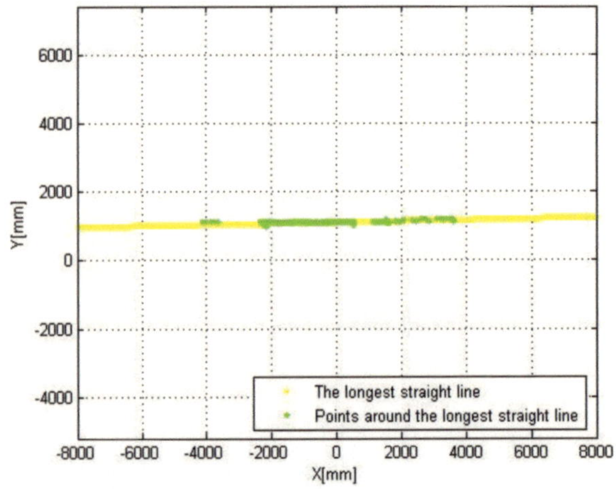

Fig. 10. Points around the longest straight line.

3.3 Grouping the points and finding the road surface

The points around the longest straight line include the road surface and obstacles. To extract the road surface, we classify into six groups (ellipses in Fig. 11). Points of each group are continuous. The biggest group among the six groups indicates the sensed road surface.

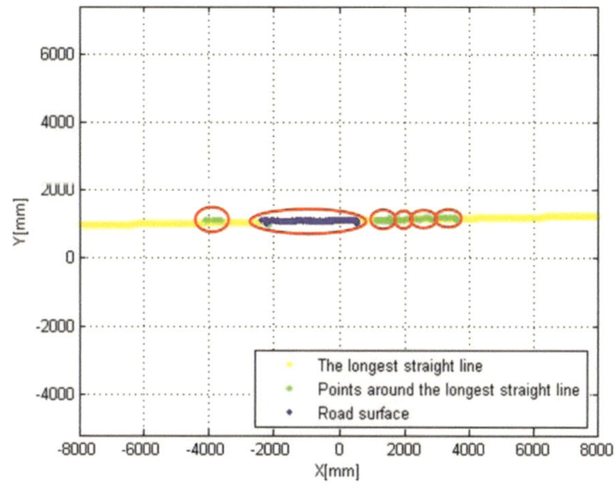

Fig. 11. Grouped points and found road surface.

3.4 Extracting the curb edge points

The scanned data of a flat road surface has two specific characteristics. One is that the scanned data of road surface is on the straight line and the interval between adjacent two points is small. The other is that the slope of the straight line is parallel to the direction X_L of a laser range finder.

When the distance between adjacent two points is continuously within d and the slope between them is less than σ, the two points are considered as being on the straight line (Eq. (3)). The end points (stars in Fig. 12) of the straight line are the road edges and also the start points of the curbs. Therefore the curb edge points could be obtained by finding the end points of the straight line.

$$\sqrt{(x_i - x_{i+1})^2 + (y_i - y_{i+1})^2} \le d \text{ and } \left| \tan^{-1}\left(\frac{y_i - y_{i+1}}{x_i - x_{i+1}} \right) \right| \ge \sigma \tag{3}$$

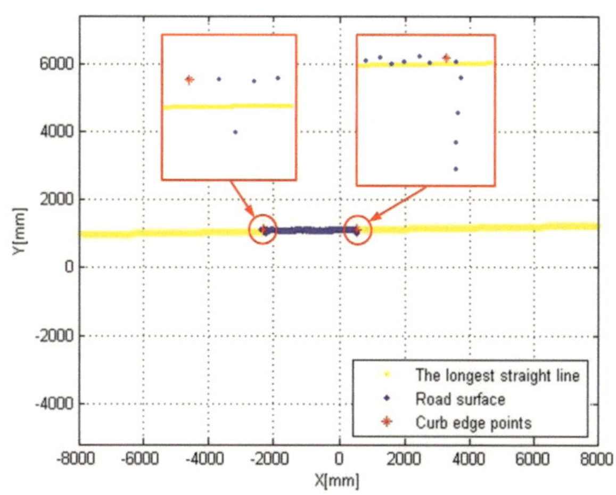

Fig. 12. Extracted curb edge points.

3.5 Parameter estimation of curb

The position of the mobile robot and the extracted curb point can be simultaneously represented in global coordinates. If we can calculate the relative pose of the mobile robot to the curb position, we can correct the mobile robot's pose against its colliding with the curb. Fig. 13 represents the both position in global coordinates.

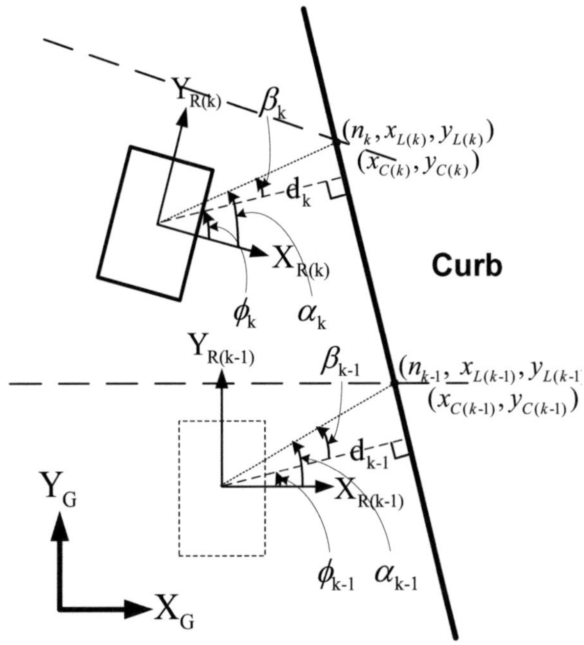

x_k : x position of the robot in global coordinate
y_k : y position of the robot in global coordinate
θ_k : orientation of the robot in global coordinate
$x_{L(k)}$: x position of the curb in robot coordinate
$y_{L(k)}$: y position of the curb in robot coordinate
$x_{C(k)}$: x position of the curb in global coordinate
$y_{C(k)}$: y position of the curb in global coordinate

Fig. 13. Positions of the mobile robot and curbs in global coordinate.

We have obtained $(n_k, x_{L(k)}, y_{L(k)})$ from curb detection method as explained in Section III. Curb position $(x_{C(k)}, y_{C(k)})$ in global coordinate can be obtained from the curb position $(x_{L(k)}, y_{L(k)})$ as shown in Eq. (4).

$$\begin{pmatrix} x_{C(k)} \\ y_{C(k)} \\ 1 \end{pmatrix} = \begin{pmatrix} -\sin\theta_k & -\cos\theta_k & x_k \\ \cos\theta_k & -\sin\theta_k & y_k \\ 0 & 0 & 1 \end{pmatrix} \begin{pmatrix} x_{L(k)} \\ y_{L(k)} \\ 1 \end{pmatrix} \tag{4}$$

3.6 Curb line equation

If we assume that the curb is straight, we can express the curb edges to a line equation such as Eq. (5).

$$ax + by + c = 0 \tag{5}$$

From the subsequent two curb points $(x_{C(k)}, y_{C(k)})$ and $(x_{C(k-1)}, y_{C(k-1)})$, we can find the line equation parameters (a,b,c) as shown in Eq. (6).

$$\begin{aligned}
&y - y_{C(k)} = \frac{y_{C(k)} - y_{C(k-1)}}{x_{C(k)} - x_{C(k-1)}}(x - x_{C(k)}) \\
&(y_{C(k)} - y_{C(k-1)})x + (x_{C(k-1)} - x_{C(k)})y \\
&\quad + (x_{C(k)}y_{C(k-1)} - x_{C(k-1)}y_{C(k)}) = 0 \\
&a = y_{C(k)} - y_{C(k-1)} \\
&b = x_{C(k-1)} - x_{C(k)} \\
&c = x_{C(k)}y_{C(k-1)} - x_{C(k-1)}y_{C(k)}
\end{aligned} \tag{6}$$

3.7 Distance and angle between robot and curb

We can also obtain the distance d_k between the robot and the curb from the line equation and angle ϕ_k between robot and curb by geometric relation as shown in Fig. 11.
Eq. (7) shows the distance d_k.

$$d_k = \frac{|ax_k + by_k + c|}{\sqrt{a^2 + b^2}} \tag{7}$$

The angle ϕ_k can be obtained such as Eq. (8).

$$\begin{aligned}
&\phi_k = \alpha_k - \beta_k \\
&\alpha_k = \text{atan2}(y_{L(k)}, x_{L(k)}) \\
&\beta_k = \text{acos}\left(\frac{d_k}{\sqrt{x^2_{L(k)} + y^2_{L(k)}}}\right) \\
&\phi_k = \text{atan2}(y_{L(k)}, x_{L(k)}) - \text{acos}\left(\frac{d_k}{\sqrt{x^2_{L(k)} + y^2_{L(k)}}}\right)
\end{aligned} \tag{8}$$

The proposed curb detection algorithm is experimented to identify the performance of it. The mobile robot was driven along a road with the distance (1m) between the mobile robot and curbs. The tilted angle (ϕ_L in Fig. 4) between the mobile robot and the laser range finder is 25°. The values d and σ as 50mm and $\pi/4$ are selected in Eq. (4.3). The navigation server sends the data of curb edge points (position data of stars in Fig. 14) to the remote server. The remote server receives the image and curbs edges data from the navigation server and displays the map of curbs and navigation information as shown in Fig. 16. The rectangular and the dots in the map represent the mobile robot and curb edge points, respectively.

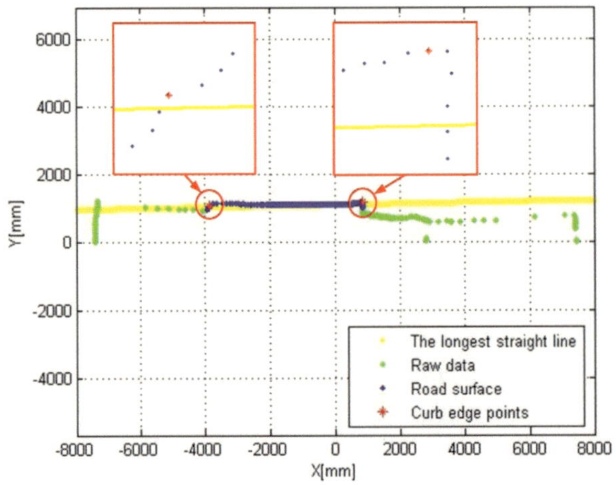

Fig. 14. Extracted curb edge points on the road.

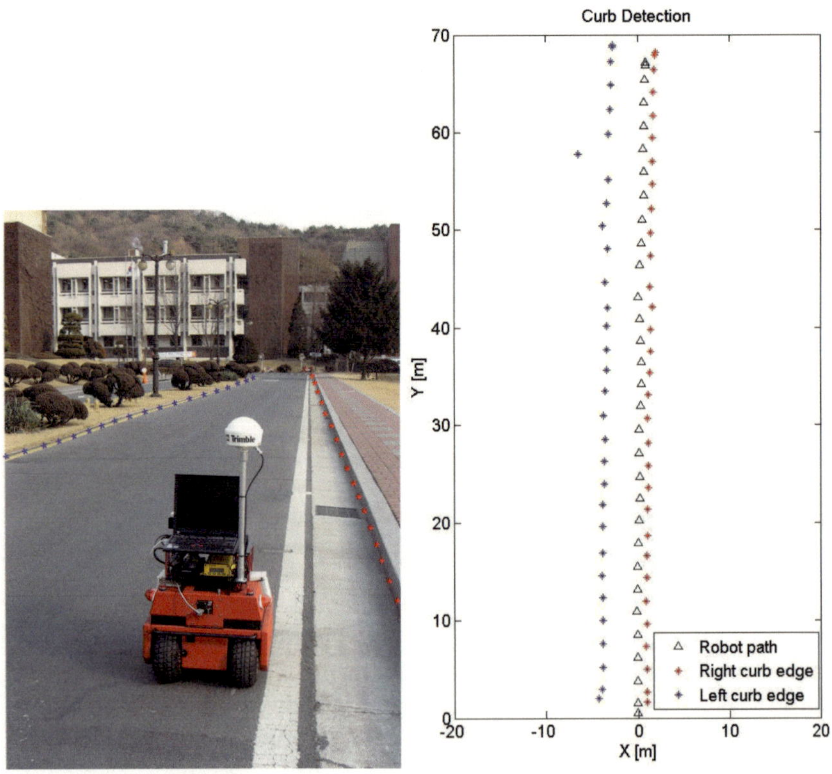

(a) (b)

Fig. 15. Curb detection experiment.
(a) The mobile robot tracking the curbs (b) Robot path and detected curbs.

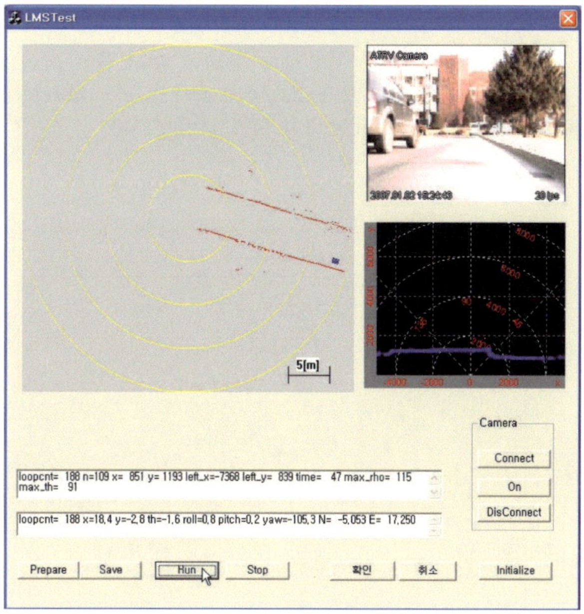

Fig. 16. Screen of the remote server.

The left and right-side stars and the black triangles in Fig. 15(b), 17 show the detected curbs and the robot path, respectively. The left-side stars in Figures are irregular because the right-side curb has regular shape (height: 13cm), but the left-side of the road is not the curb. It is just the structure with low height (2cm). Even though the mobile robot moves zigzag, the detected right-side curb path is smooth. The average error of the curb points is 2cm.

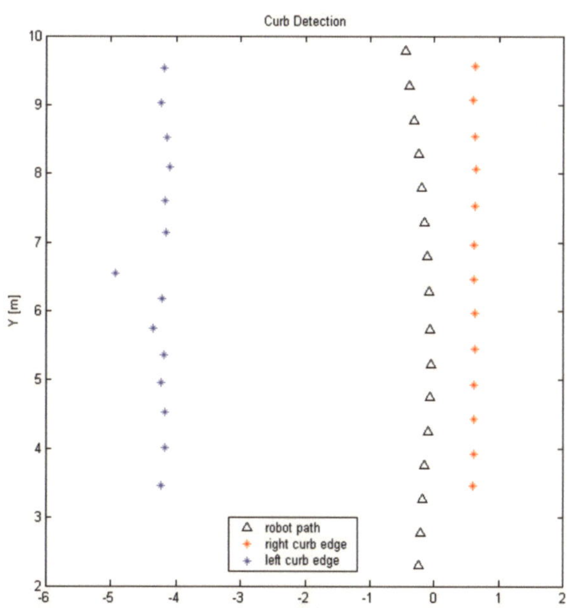

Fig. 17. Robot path and extracted curb positions.

Fig. 18 shows the distance between robot and curb, the curb edge angle α and angle ϕ between robot and curb, respectively. We can know that the proposed curb detection algorithm produces proper results and the distance and angle information calculated from the curb position could be used for reliable and safe navigation.

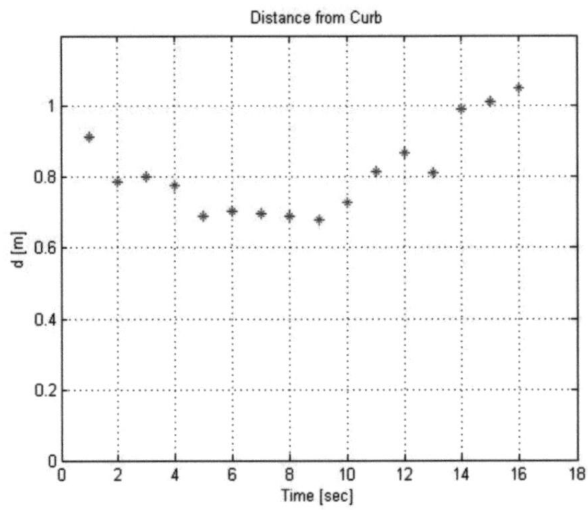

(a)

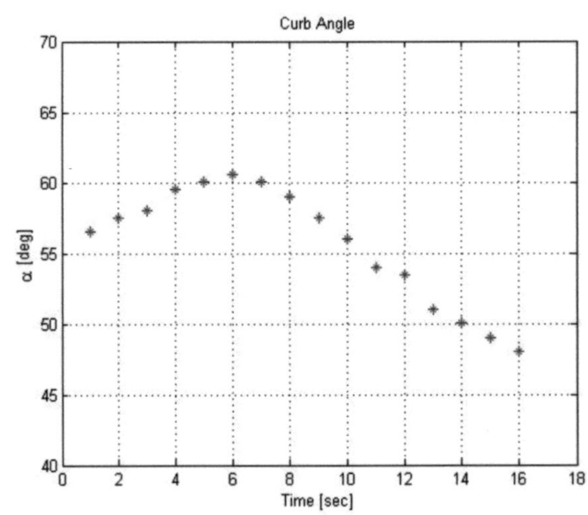

(b)

Fig 18. Results for curb detection. (a) Distance d between robot and curb (b) Curb edge angle α

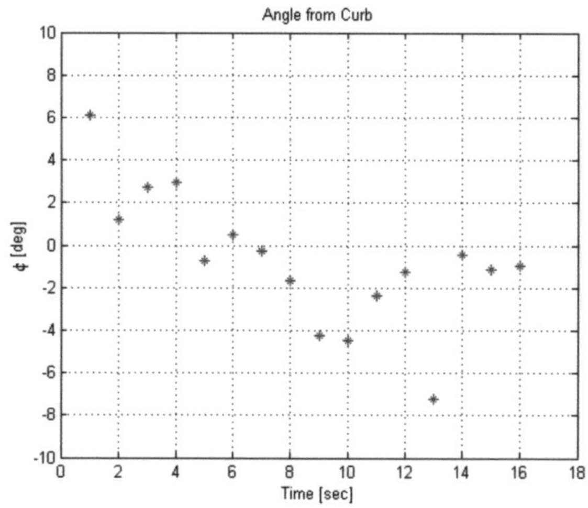

(c)

Fig. 18. Results for curb detection. (c) Angle ϕ between robot and curb.

4. Tracking control using curb position

An ATRV-mini is a nonholonomic mobile robot, equivalent to a four-wheeled differential drive. Kanayama proposed a kinematic control law in Cartesian coordinates for solving the trajectory tracking problems [10]. We add a term related to the distance between the robot and the right-side curb of roads to Kanayama controller.

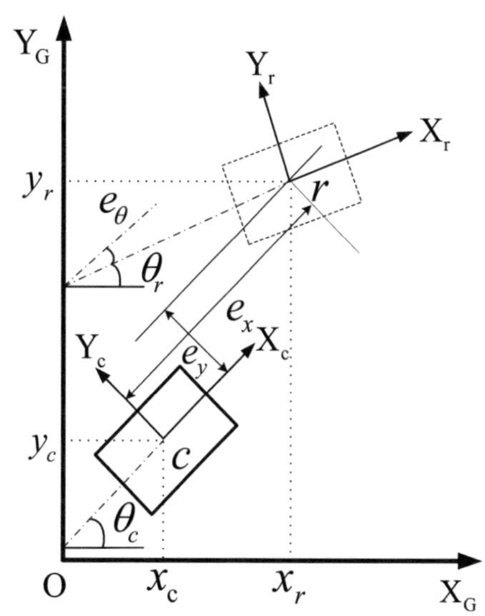

Fig. 19. Mobile robot and reference robot.

Considering the Fig. 19, in which the mobile robot c is represented by the coordinates $\mathbf{p} = [x_c\ y_c\ \theta]^T$ of the center of mass, moving over plane with linear velocity υ and angular velocity ω. The matrix equation that describes the kinematic model of a mobile robot is given by

$$\begin{bmatrix} \dot{x}_c \\ \dot{y}_c \\ \dot{\theta} \end{bmatrix} = \begin{bmatrix} \cos\theta & -\sin\theta \\ \sin\theta & \cos\theta \\ 0 & 1 \end{bmatrix} \begin{bmatrix} \upsilon \\ \omega \end{bmatrix} \tag{9}$$

The posture tracking error vector e_p is calculated in the reference robot frame X_r, Y_r.

$$e_p = R(\theta)(p_r - p)$$
$$e_p = \begin{bmatrix} e_x \\ e_y \\ e_\theta \end{bmatrix} = \begin{bmatrix} \cos\theta & \sin\theta & 0 \\ -\sin\theta & \cos\theta & 0 \\ 0 & 0 & 1 \end{bmatrix} \begin{bmatrix} x_r - x_c \\ y_r - y_c \\ \theta_r - \theta_c \end{bmatrix}, \tag{10}$$

where $R(\theta)$ is the rotation matrix.
Differentiating the Eq. (10), yields

$$\begin{bmatrix} \dot{e}_x \\ \dot{e}_y \\ \dot{e}_\theta \end{bmatrix} = \begin{bmatrix} \omega e_y - \upsilon + \upsilon_r \cos(e_\theta) \\ -\omega e_x + \upsilon_r \sin(e_\theta) \\ \omega_r - \omega \end{bmatrix} \tag{11}$$

The asymptotic convergence of posture error can be obtained

$$V_c = \begin{bmatrix} \upsilon_c \\ \omega_c \end{bmatrix} = \begin{bmatrix} k_x e_x + \upsilon_r \cos(e_\theta) \\ \omega_r + k_y \upsilon_r e_y + k_\theta \upsilon_r \sin(e_\theta) \end{bmatrix}, \tag{12}$$

where k_x, k_y and k_θ are positive constants, defined as controller gains.

For reliable and safe navigation, we design a new controller V_d by adding the extracted curb distance and angle information to the above Kanayama controller.

$$e_d = d_{ref} - d$$
$$V_d = \begin{bmatrix} \upsilon_d \\ \omega_d \end{bmatrix} = \begin{bmatrix} k_x e_x + \upsilon_r \cos(e_\theta) \\ \omega_r + \upsilon_r (k_y e_y + k_\theta \sin(e_\theta) + k_d e_d + k_\phi \sin\phi) \end{bmatrix}, \tag{13}$$

where d_{ref} is the reference distance from the curb for safety and k_d, k_ϕ are positive constants as controller gains.
If the robot is too close to the curb, the added terms of the proposed controller try to prohibit the robot colliding with curbs, and vice versa.
We have performed two kinds of experiments to evaluate the performance of the proposed path tracking controller in outdoor environment. One is that the mobile robot has to follow a straight path. The other is that the mobile robot has to follow a curved path.

4.1 Straight path tracking

We have performed a straight path tracking experiment, which target point is (78m, -22m) and linear velocity is 0.5m/sec. The gains of controller were determined by some simulations and experiments. Fig. 20 shows the reference path and the error of path, position and angle of the robot. Fig. 20(a) shows the reference path and tracking path and the position error of 20cm at the final point, respectively. Even though DGPS position error of 1m has occurred at the point of 12m from X-axis, the mobile robot returned to its reference path by the stable tracking controller.

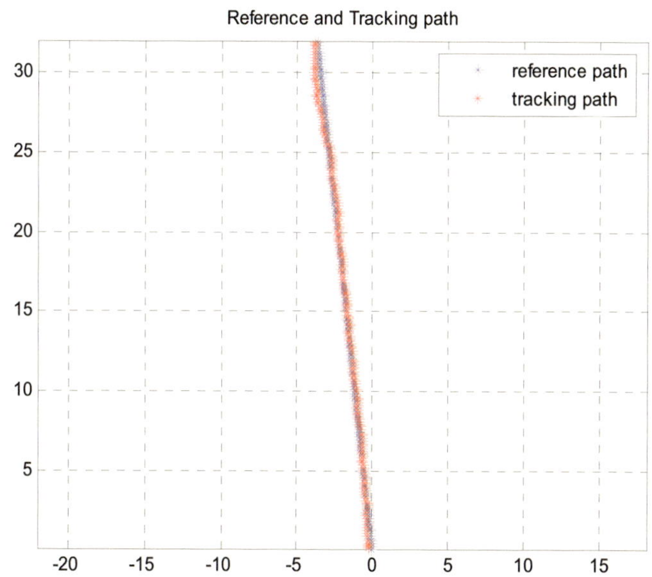

(a)

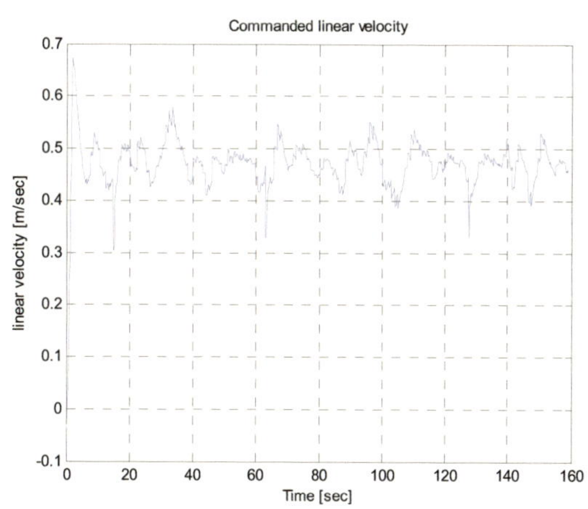

(b)

Fig. 20. (Continued)

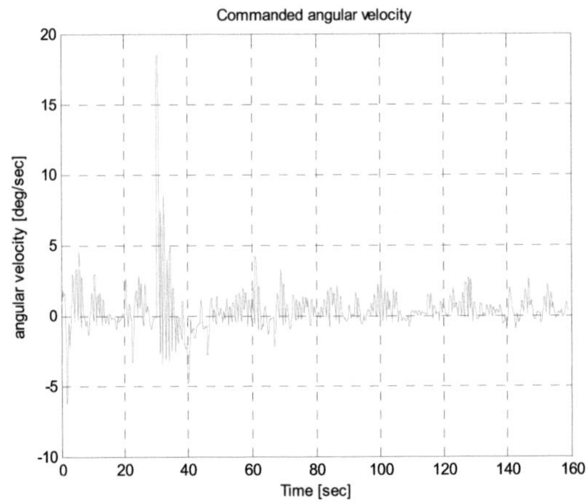

(c)

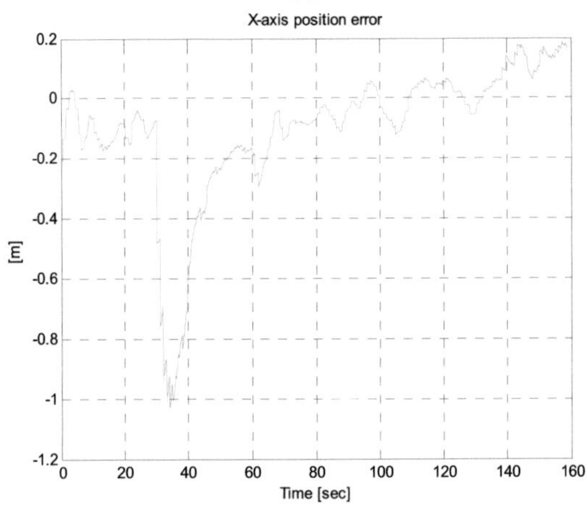

(d)

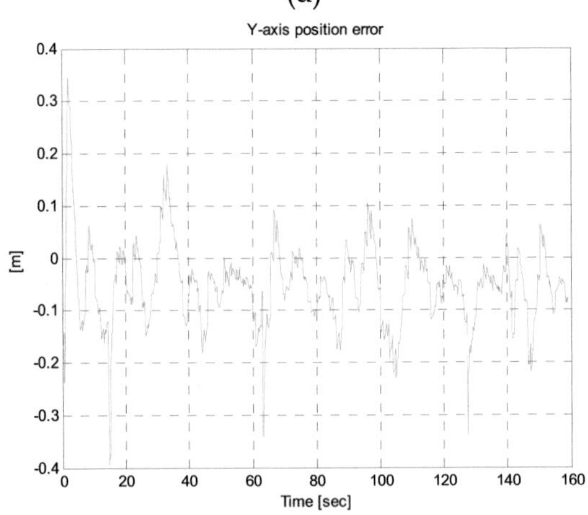

(e)

Fig. 20. (Continued)

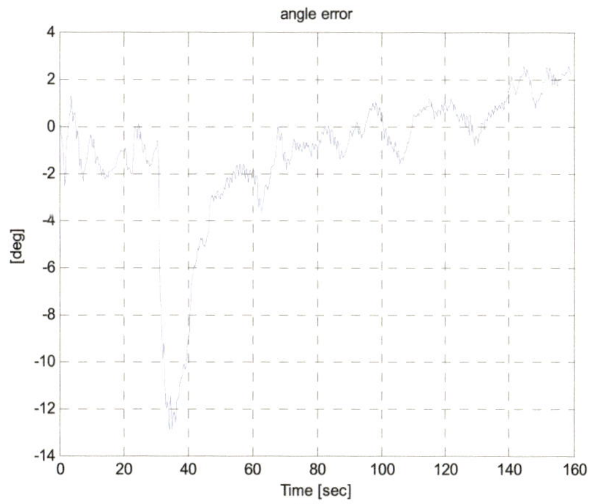

(f)

Fig. 20. Results for straight path tracking. (a) Reference path and tracking path (b) Controller linear velocity (c) Controller angular velocity (d) X-axis position error for robot coordinate (e) Y-axis position error for robot coordinate (f) Angle error for robot coordinate.

4.2 Curved path tracking

We have performed a curved path tracking experiment, which linear velocity is 0.3m/sec and angular velocity is 0.06rad/s. Fig. 21 shows the reference path and the error of path, position and angle of the robot. Fig. 21(a) shows the reference path and tracking path. The X-axis position error and the angular error are 40cm and 7° at the final point, respectively. The reason that the errors are bigger than the straight path tracking experiment is related to the characteristic of skid steering system of our mobile robot platform. It is necessary to optimize the gains of controller to reduce the errors.

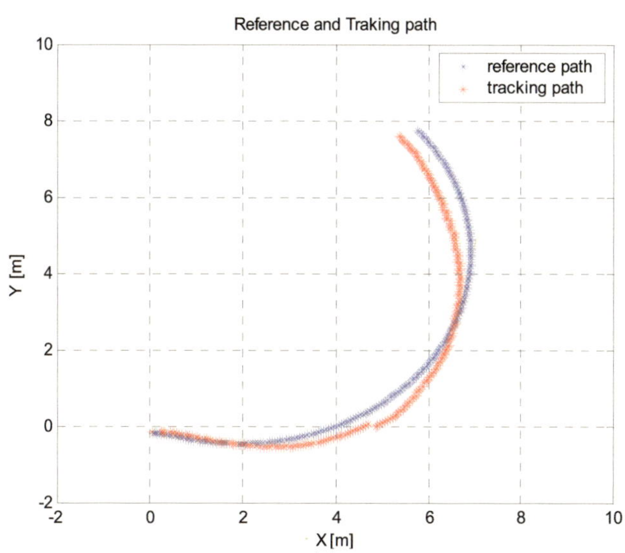

(a)

Fig. 21. (Continued)

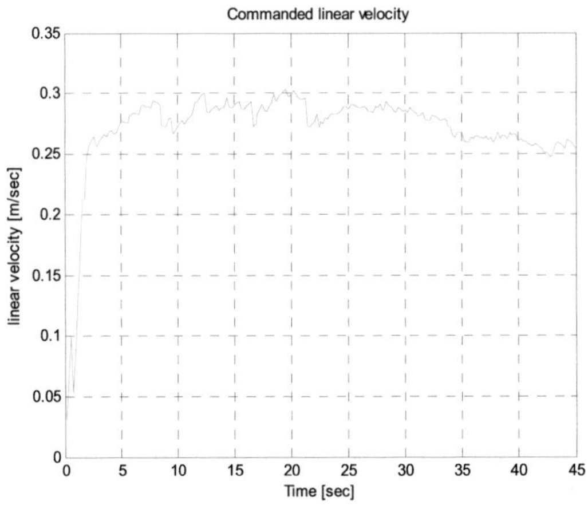

(b)

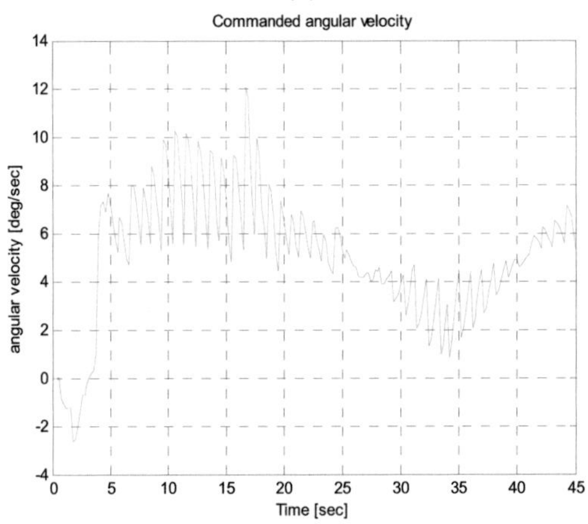

(c)

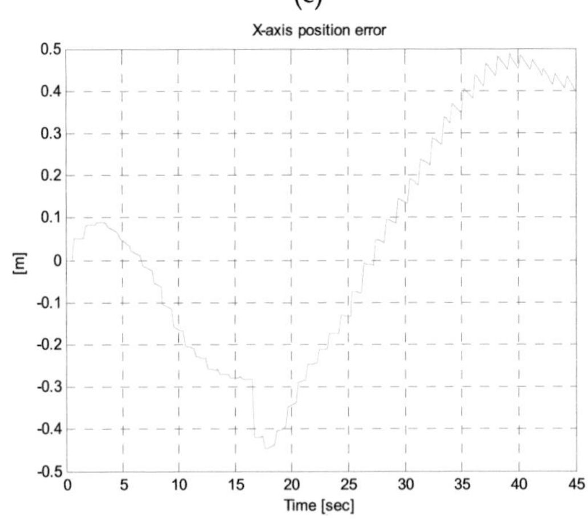

(d)

Fig. 21. (Continued)

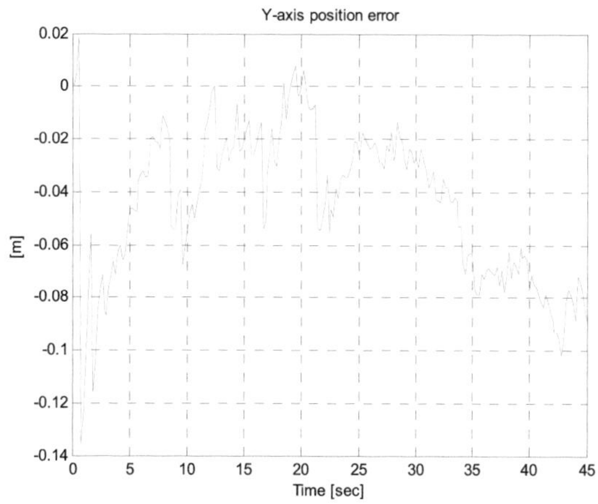

(e)

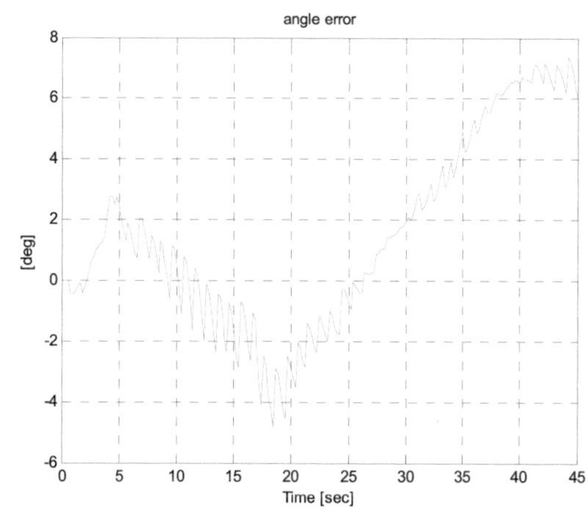

(f)

Fig. 21. Results for curved path tracking.
(a) Reference path and tracking path (b) Controller linear velocity (c) Controller angular velocity (d) X-axis position error for robot coordinate (e) Y-axis position error for robot coordinate (f) Angle error for robot coordinate

5. Conclusion

In this chapter, an algorithm is described to detect road curb position through a laser range finder. The curbs map was built and the width of the road was calculated by the detected curbs. The proposed algorithm is to calculate the distance and the angle between the mobile robot and the curb. A new controller is designed by adding the extracted curb distance and angle information to the tracking controller. It helps the mobile robot to navigate safely and reliably. To identify the availability of the proposed algorithm, navigation experiments was performed in outdoor environment and proper results was obtained.

6. References

[1] S. Thrun at el. (2000). Probabilistic Algorithms and the Interactive Museum Tour-Guide Robot Minerva, *International Journal of Robotics Research*, Vol.19, No.11, pp. 972-999

[2] Richard Thrapp at el. (2001). Robust localization algorithms for an autonomous campus tour guide, *Proc. of IEEE Int'l Conf. on Robotics and Automation*, pp. 2065-2071

[3] C.C. Wang and C. Thorpe, "Simultaneous Localization and Mapping with Detection and Tracking of Moving Objects", Proc. of IEEE Int'l Conf. on Robotics and Automation, 2002, pp. 2918-2924

[4] W. S. Wijesoma, K. R. S. Kodagoda, and Arjuna P. Balasuriya, "Road-Boundary Detection and Tracking Using Ladar Sensing", IEEE Trans. On Robotics and Automation, Vol.20, No.3, 2004

[5] Seung-Hun Kim at el. (2006). A hybrid autonomous/teleoperated strategy for reliable mobile robot outdoor navigation, *SICE-ICASE Int. Joint Conf.*, pp. 3120-3125

[6] K. Ohno at el. (2003). Outdoor Navigation of a Mobile Robot between Buildings based on DGPS and Odometry Data Fusion, *Proc. of IEEE Int'l Conf. on Robotics and Automation*, pp. 1978-1984

[7] J. Forsberg at el. (March 1995). Mobile robot navigation using the rangeweighted Hough transform, *IEEE Robotics & Automation Magazine*, vol. 2, pp. 18-26

[8] R. O. Duda. & P. E. Hart. (1972). Use of the Hough transformation to detect lines and curves in pictures, *Communications of the ACM*, vol. 15, pp. 11-15

[9] P. V. C. Hough. (December 1962). Methods and means for recognizing complex patterns, in U.S. Patent 3 069 654

[10] Y. Kanayama at el. (1990). A stable tracking control method for an autonomous mobile robot, *IEEE Int. Conf. on Robotics and Automation*, vol. 1, pp. 384-389

Part 4

Methods for Control

17

Statistical Video Based Control of Mobile Robots

Krzysztof Okarma and Piotr Lech
West Pomeranian University of Technology, Szczecin
Poland

1. Introduction

A typical approach to the control of mobile robots is based on the analysis of signals from various kinds of sensors (e.g. infra-red or ultrasound). Another type of input data used for motion control of robots can be video signals acquired by the cameras. In such case many algorithms can be utilised e.g. SLAM (Simultaneous Localization and Mapping) and its modifications, being still an active field of research (Kalyan et al., 2010), which require usually the full analysis of the images (or video frames) being the input data (Se et al., 2001; 2002), in some applications using also additional operations such as Principal Component Analysis (Tamimi & Zell, 2004). Similar problems may also occur in the robot localisation and map building processes (Hähnel et al., 2003).
An interesting alternative can be the application of the fast image analysis techniques based on the statistical experiments. Presented statistical approach is related to the first step of such control systems, which can be considered as the pre-processing technique, reducing the amount of data for further analysis with similar control accuracy. Discussed methods can be utilised for many control algorithms based on the image and video data e.g. the proportional steering of line following robots (Cowan et al., 2003) considered as a representative example in further discussions. Nevertheless, some other applications of the presented approach are also possible e.g. fast object classification based on the shape analysis (Okarma & Lech, 2009) or motion detection, as well as fast reduced-reference image quality estimation (Okarma & Lech, 2008b).

2. Reduction of power consumption by mobile robots

Considering the autonomous mobile robots, one of their most important parameters, corresponding directly to their working properties, is the power consumption related to the maximum possible working time. Limited energy resources are often the main element reducing the practical applicability of many such constructions regardless of many modern energy sources which can be used such as e.g. solar based solutions. The limited capacity of the installed batteries reduces the range and capabilities of these devices and each of the possible attempts to optimise energy consumption usually affects the mobility and efficiency of the robots.
It is possible to reduce the consumed energy by finding the shortest path to the target (robot's drive) or conducting the optimisation of the amount of processed data by the control algorithm. Since the computational cost of mapping algorithms, searching for the optimal path, finding robot's own location, visual SLAM algorithms etc., based on the image analysis is strongly dependent on the resolution and representation of the analysed image, a significant "smart" decrease of the amount of processed data seems to be an interesting direction of

research in this area. Reducing the computational load of the processor the overall demand for the electrical energy decreases, so it can be assumed that limiting the amount of processed data in the system, energy savings can be obtained, despite using a non-optimal (longer) motion path. For the video based control algorithms such restriction can be achieved simply by reducing the image resolution or reducing the frame rate of the acquired image sequence used for he robot's motion control. Nevertheless, such decrease of the amount of analysed data may lead to some steering errors, as illustrated in Fig. 1, where the optimal shortest path cannot be used due to the presence of some obstacles. Each of the indicated alternative paths has a different length related to the different energy necessary for the robot to reach the target point.

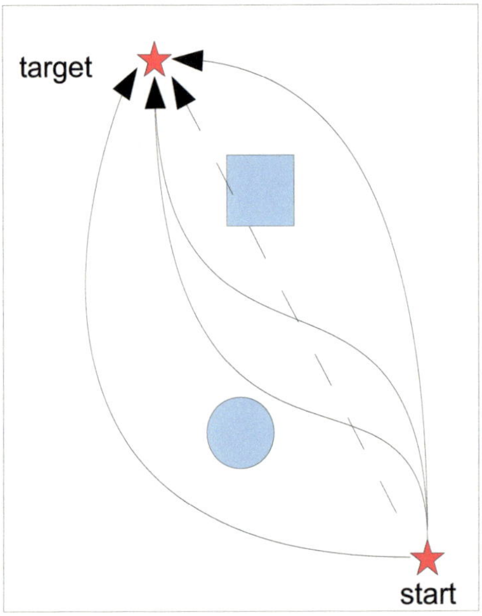

Fig. 1. Illustration of the problem of the optimal path searching in the presence of obstacles

The optimal robot's path in the known environment can be chosen using the machine vision approach with motion detection and robot's motion parameters estimation, using the general algorithm presented in Fig. 2 such that the total energy consumption, related both to the robot's drive and the control part dependent mainly on the processing of the vision data, is minimum. Conducting the experimental verification of the idea for the mobile robot based on a simple netbook platform using the joystick port for controlling the drives, for the testing area of 6 square meters illustrated in Fig. 1, the maximum power consumption is about 5.2 A (42.5% by the drive part and 57.5% by the control part of the system). The length of the optimum straight path, assuming no obstacles on the scene, is equal to 2 meters so all the alternative trajectories are longer.
The experiments conducted in the environment described above has been related to the decrease of the resolution of the image (obtained results are presented in Table 1) and the frame rate for the fixed resolution of 320×240 pixels (results shown in Table 2). For the reduction to 8 frames per second the vision based control algorithm could not find any appropriate path.

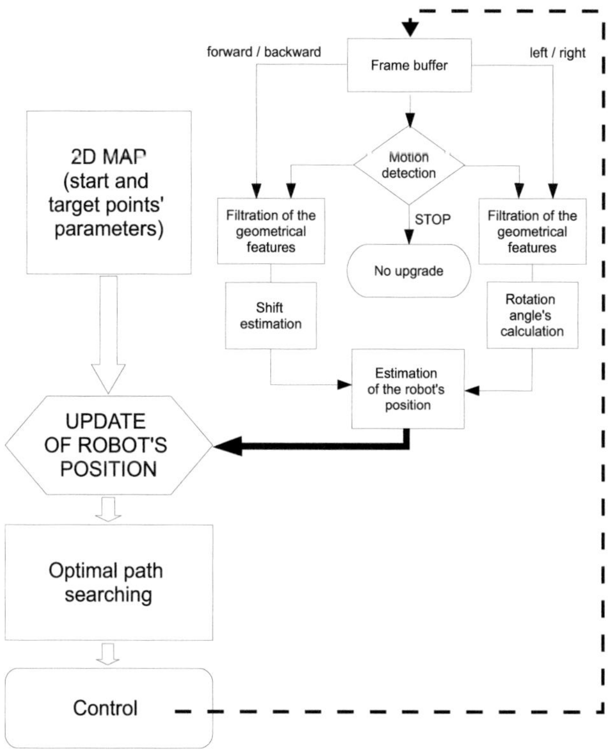

Fig. 2. Illustration of the general vision based control algorithm of the mobile robot

Resolution in pixels	Length of the path [m]	Total energy consumption [kJ]
640 × 480	2.15	5.022
352 × 288	2.39	4.519
320 × 240	2.69	4.268
176 × 144	2.80	4.168
160 × 120	2.95	4.369

Table 1. Path length and power consumption of the mobile robot used in the experiments dependent on the image resolution

Frame rate [fps]	Length of the path [m]	Total energy consumption [kJ]
25	2.39	4.519
21	2.44	4.432
16	2.66	4.332
12	3.23	4.668
10	3.33	4.798

Table 2. Path length and power consumption of the mobile robot used in the experiments dependent on the frame rate of the analysed video sequence

Analysing the results presented in Tables 1 and 2 a significant influence of the amount of the analysed video data on the obtained results and the power consumption can be noticed. The decrease of the resolution as well as the reduction of the video frame rate causes the increase of the path's length but due to the reduction of the processor's computational load the total energy consumption may decrease. Nevertheless, the decrease of the amount of the processed data should be balanced with the control correctness in order to prevent an excessive extension of the robot's path.

Since the simple decrease of the resolution may cause significant control errors as discussed above, a "smart" decrease of the number of analysed pixels, which does not cause large control errors should be used for such purpose much more efficiently. In the proposed statistical approach the main decrease of the amount of data is related to the random draw of a specified number of pixels for further analysis using the modified Monte Carlo method. In a typical version of this method (as often used for the fast object's area estimation) the pixels are chosen randomly by an independent drawing of two coordinates (horizontal and vertical), what complicates the error estimation. For the simplification, the additional mapping of pixels to the one-dimensional vector can be used, what is equivalent to the data transmission e.g. as an uncompressed stream. The additional reduction takes place during the binarization performed for the drawn pixels.

3. The Monte Carlo method for the fast estimation of the objects' geometrical properties

3.1 The area estimation

Conducting a statistical experiment based on the Monte Carlo method, an efficient method of estimation of the number of pixels in the image fulfilling a specified condition can be proposed, which can be used for the motion detection purposes as well as for the area estimation of objects together with their basic geometrical features. Instead of counting all the pixels from a high resolution image, the reduction of the number of pixels used for calculations can be achieved by using a pseudo-random generator of uniform distribution. Due to that the performance of the algorithm can be significantly increased. The logical condition which has to be fulfilled can be defined for a greyscale image as well as for the colour one, similarly to the chroma keying commonly used in television. leading to the binary image, which can be further processed. In order to prevent using two independent pseudo-random generators it is assumed that the numbered samples taken from the binary image are stored in the one-dimensional vector, where "1" denotes the black pixels and "0" stands for the white ones representing the background (or reversely). Then a single element from the vector is drawn, which is returned, so each random choice is conducted independently. For simplifying the further theoretical analysis the presence of a single moving dark object (Ob) in the scene (Sc) with a light background can be assumed.

In order to determine the object's geometrical features as well as some motion parameters, such as direction and velocity, the random choice of pixels cannot be performed using the whole image due to the integrating properties of the Monte Carlo method. For preventing possible errors caused by such integration the scene can be divided into smaller and smaller squares, so the K_N squares from N elements of the scene would represent the object. In such case the probability of choosing the point on the object's surface (assuming the generator's uniform distribution) is equal to:

$$p = \frac{K_N}{N} \tag{1}$$

For the infinite number of samples the reduction of the sampling distance takes place and the probability of choosing the point representing the object on the image can be expressed as:

$$p = \lim_{N \to \infty} \frac{K_N}{N} = \frac{A_{Ob}}{A_{Sc}} \tag{2}$$

so the estimated area of the object is:

$$A_{Ob} \approx K_N \cdot \frac{A_{Sc}}{N} \tag{3}$$

Since the total area of the scene is equal to $A_{Sc} = N \cdot k_x \cdot k_y$, the estimated object's area is equal to $A_{Ob} \approx K_N \cdot k_x \cdot k_y$, where k_x and k_y are the scale factors for horizontal and vertical coordinates respectively, equivalent to the number of samples per unit.
Considering the above analysis, the probability of choosing the point belonging to the object can be used in the proposed algorithm with a reduced number of analysed samples instead of the full image analysis using a statistical experiment based on the Monte Carlo method. This method originates directly from the law of large numbers, because the sequence of successive approximations of the estimated value is convergent to the sought solution and the distance of the actual value after performing the specified number of statistical tests to the solution can be determined using the central limit theorem.
After the binarization the luminance (or colour) samples represented as "ones" or "zeros" corresponding to the allowed values of the specified logical condition, which are stored in one-dimensional vector are chosen randomly. For a single draw a random variable X_i of the two-way distribution is obtained:

$$X_i = \begin{cases} 1 & \text{for black samples} \\ 0 & \text{for white samples,} \end{cases} \tag{4}$$

leading to the following probability expressions:

$$P(X_i = 1) = p \qquad P(X_i = 0) = q \tag{5}$$

where $p + q = 1$, $E(X_i) = p$, $V(X_i) = p \cdot q$.
An important element of the method is the proper choice of the logical condition for the binarization allowing a proper separation of the object from the background. Such choice depends on the specific application, for example for the light based vehicles' tracking purposes the chroma keying based on the CIELAB colour model can be appropriate (Mazurek & Okarma, 2006).
For n independent draws the variable Y_n is obtained:

$$Y_n = \frac{1}{n} \cdot \sum_{i=1}^{n} X_i \tag{6}$$

According to the Lindberg-Levy's theorem, the distribution of Y_n tends to the normal distribution $N(m_y, \sigma_y)$ if $n \rightarrow \infty$. Since the expected value and variance of Y_n are equal to $E(Y_n) = p$ and $V(Y_n) = \frac{p \cdot q}{n}$ respectively, the distrubution of the random value Y_n is normal with the following parameters:

$$m_y = p \tag{7}$$

$$\sigma_y = \sqrt{\frac{p \cdot q}{n}} \tag{8}$$

Considering the asymptotic normal distribution $N(p, \sqrt{p \cdot q/n})$ it can be stated that the central limit theorem is fulfilled for the variable Y_n.
Substituting:

$$U_n = \frac{Y_n - m_y}{\sigma_y} \tag{9}$$

obtained normal distribution can be standardized towards the standard normal distribution $N(0, 1)$.
In the interval estimation method the following formula is used:

$$p(|U_n| \leq \alpha) = 1 - \alpha \tag{10}$$

Assuming the interval:

$$|U_n| \leq u_\alpha \tag{11}$$

considering also the formulas (3), (7), (8) and (9), the following expression can be achieved:

$$\left| Y_n - \frac{K_N}{N} \right| \leq \varepsilon_\alpha \tag{12}$$

where

$$\varepsilon_\alpha = \frac{u_\alpha}{\sqrt{n}} \cdot \sqrt{\frac{K_N}{N} \cdot \left(1 - \frac{K_N}{N}\right)} \tag{13}$$

The probability estimator p (eq. 1), for k elements from n draws, fulfilling the specified logical condition used for the definition of X_i (eq. 4) and representing the number of the drawn pixels representing the object, can be expressed as:

$$\hat{p} = \frac{k}{n} = \frac{1}{n} \cdot \sum_{i=1}^{n} X_i = Y_n \tag{14}$$

and the object's area estimator as:

$$\hat{A}_{Ob} = \hat{p} \cdot A_{Sc} = \frac{k}{n} \cdot A_{Sc} \tag{15}$$

Using the equations (12) and (15) the obtained formula describing the interval estimation for the object's area is:

$$\left| \frac{\hat{A}_{Ob}}{A_{Sc}} - \frac{K_N}{N} \right| \leq \varepsilon_\alpha \tag{16}$$

where ε_α is specified by the equation (13).
It is worth to notice that all the above considerations are correct only for a random number generator with the uniform distribution, which should have as good statistical properties as possible. The discussed algorithm are identical to the method of area estimation of the 2-D object's (expressed in pixels). Nevertheless, the applicability of such approach is limited by the integrating character of the method so an additional modification based on the block approach is necessary, as mentioned earlier.

Such prepared array, considered as a reduced resolution "greyscale" image, can be directly utilised in some typical control algorithms. Further decrease of the computational cost can be obtained by the additional binarization of this image leading to the ultra-low resolution binary image considered in further discussions, where each of the blocks can be classified as a representative of an object or the background. For the line following robot control purposes the objects is equivalent to the path. The control accuracy of the mobile robots corresponds to the quality loss of the data present in this image. Such quality can be treated as the quality of the binary image assuming the knowledge of the reference image (without any distortions). Unfortunately, most of the image quality assessment methods, even the most recent ones, can be successfully applied only for greyscale or colour images and the specific character of the binary images is not respected by them. For this reason only the methods designed exclusively for the binary images can be used. Such binary image quality assessment methods can be used as the optimisation criteria for the proposed algorithm. In the result the proper choice of the block size, binarization threshold and the relative number of randomly drawn pixels can be chosen.

Depending on the logical condition used for the statistical analysis (the construction of the 1-D binary vector), the algorithm can be implemented using various colour spaces (or only some chosen channels), with the possibility of utilising independent logical conditions for each channel (chroma keying).

3.2 The Monte Carlo based motion detection

The real-time navigation of an autonomous mobile robot based on the machine vision algorithms requires a fast processing of images acquired by the camera (or cameras) mounted on the robot. In some more sophisticated systems the data fusion based on the additional informations acquired by some other sensors, such as PIR sensors, ultrasound detectors, a radar technique equipment or optical barriers, is also possible. Most of such sensors are responsible for the motion detection, especially if the presence of some moving obstacles in the robot's surrounding is assumed. The vision based sensors can be divided into three main groups: analogue (i.e. comparing the luminance with a given threshold value), digital (obtained by the sampling and quantization of acquired analogue signal) and the digital ones acquired directly from digital video cameras. There are also some restrictions of typical alternative solutions, e.g. passive infrared detectors are sensitive to temperature changes and they are useless for the motion detection of objects with the same temperature as the background. Some more robust solutions, such as ultrasound and radar detectors, are usually active (energy emitting) and require an additional power supply. The application of a vision based motion detection procedure utilising the Monte Carlo method for the mobile robots control purposes is caused mainly by the fact that no additional hardware is necessary, since the images used by the motion detector are acquired by the same camera as in the classification procedure.

Assuming the scene with the constant light conditions without any moving objects, it can be characterised by a constant value defined as the number of black (or white) pixels of the respective binary image, which usually represent the object visible on the image obtained from the camera. The algorithm of the motion detection utilises the rapid changes of the number of such pixels (treated as "ones") when an object moves respectively to the light sources. Such change is caused by the dependence of the image pixel's luminance (and colour) on the angle between normal vector to its surface and the direction of the light ray passing the pixel, as well as the influence of the surface's reflection coefficient, shape, roughness etc. Real objects usually have a heterogeneous structure of their surface so the number of analysed points changes dynamically (assuming constant threshold or chroma keying range). For slow

changes of the light conditions, especially for outside environment, the binarization threshold can be updated. It it worth to notice that in the proposed approach the necessity of the storage of the single value only (the number of pixels corresponding to the "ones") allows the reduction of the required system's operating memory size.
A typical well known video based algorithm of motion detection is based on the comparison of two neighbouring frames of the video signal but it requires the storage of the images used during the comparison so relatively large amount of memory is needed. Some more robust algorithms of background estimation, directly related to motion detection, usually based on moving average or median filter, are also more computationally demanding and require the analysis of several video frames (Cucchiara et al., 2003; Piccardi, 2004). Even a comparison of all the pixels from two video frames can be treated as a time consuming operation for an autonomous mobile robot with limited computational performance and relatively small amount of operating memory, so in the real-time applications some high performance processing units would be required. Instead of them the statistical Monte Carlo approach described above can be used for the reduction of the computational complexity, where only some randomly chosen pixels are taken into account. Therefore the comparison of full frames is reduced to the comparison of only two values representing the number of "ones" in each frame. Any rapid change of that value is interpreted as the presence of a moving object.
A slightly modified method based on the additional division of the image into smaller blocks can be applied for a more robust detection. In the presence of moving objects the changes can be detected only in some of obtained blocks so the integrating character of the Monte Carlo method is eliminated.
The fast motion detection based on the comparison of two values representing the estimated areas of the object in the two consecutive binary frames is based on the following formula:

$$\hat{A}_{Ob}(i+\delta) - \hat{A}_{Ob}(i) > threshold \tag{17}$$

where $threshold >> \varepsilon_\alpha$ and $\delta \geq 1$ denotes the shift between both analysed frames.
In the proposed block based approach the application of the formula (17) for each of the $r \times r$ pixels blocks allows proper motion detection, using the array of the Monte Carlo motion detectors, also in the situation when the motion of the mobile robot causes the relative motion of the objects observed by the integrated camera. For the perspectively looking camera, typical for such applications, even if the objects do not change their size on the image, the changes can be observed among the blocks. The objects moving towards the camera always cause the increase of their estimated area in the image used by the motion detector. The estimation of the geometrical parameters of the objects, often necessary for the proper navigation of the controlled robot, can be performed for the frame with the maximum value of the estimated area occupied by the object on the image with eliminated background, assuming that robot is not moving directly towards the object. If the size of the object increases in consecutive frames, such estimation should be performed immediately preventing possible collision with the object.

3.3 The estimation of geometrical parameters

The extraction of some geometrical parameters, such as perimeters or diameters, using the discussed approach is also possible. For this purpose the analysed binary image has to be divided into $T \times S$ blocks of $r \times r$ pixels each using a square grid. Then the area of each object's fragment in the elementary square elements (blocks) is calculated and the results of such estimation can be stored in the array P containing $T \times S$ elements.

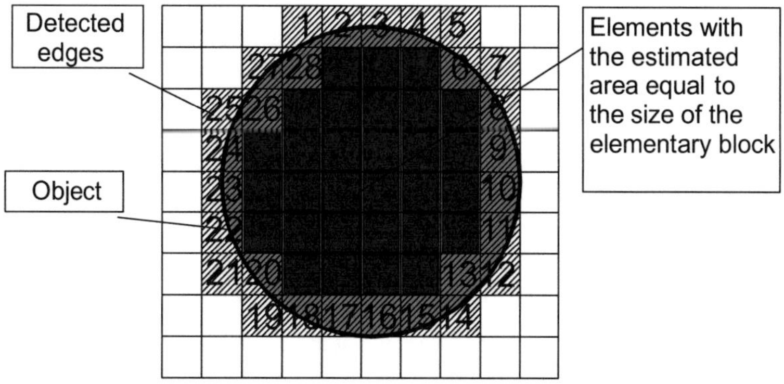

Fig. 3. The idea of the block based edge estimation

The simplest approach to the edge detection, which is necessary for further processing, is based on the array *P*. For this purpose the second array *K* with the same size is created and its elements have the following values:

- zero if the corresponding element's value in the array *P* is equal to zero (background),
- zero if the corresponding element's value in the array *P* is equal to the size of the elementary block (all pixels represent the inner part of the object) and none of its neighbouring blocks (using the 8-directional neighbourhood) has the zero value,
- one for the others (representing the edge).

The idea of the edge detection is illustrated in Fig. 3
The obtained array *K* is the projection of edges detected from the source image, so the number of its non-zero elements represents the estimated value of object's perimeter expressed in squares of $r \times r$ pixels. For a better estimation the number of square elements can be increased (smaller values of the parameter r) and then the previously discussed operation should be repeated. In such case the analysis of the blocks in the array *P* with the zero values (representing the inner parts of the object) is not necessary, so the further analysis can be conducted only for the significantly reduced number of indexed elements classified as the edge in the previous step. Such alternative edge detection method based on the two-step algorithm with the same initial step of the algorithm and another pass of the previous algorithm conducted as the second step with the edge detection calculated using the array *K* (the edge obtained in the first step is treated as the new object without fill) allows more accurate estimation of the edge with slight increase of the computational complexity. All the pixels with the zero values should be replaced from the array *K* and using such corrected image the two curves are obtained as the result (assuming the scale factor of the block greater than $t = 2$ - at least 3×3 elements). As the estimate of the edge the middle line between the two obtained curves can be used, providing slightly lower accuracy but also a reduced computational cost. The estimate of the total object's area can also be determined with similarly increased accuracy as the sum of the values in the array *P*. The limit accuracy of the algorithm is determined by the size of the elementary block equal to 1 pixel what is equivalent to the well-known convolution edge detection filters.
Geometrical parameters of the objects, which can be estimated using the modified Monte Carlo method, can be divided into the two major groups: local (such as mean diameter or the

average area) and global (e.g. the number of objects in the specified area). The estimation of the relative global parameters is not necessary if only a single object is present on the analysed fragment of the image. The most useful parameters used for image analysis, which can be also used for robot's control purposes, are insensitive to image deformations which can occur during the acquisition, as well as to some typical geometrical transformations. In such sense the usefulness of the simplest parameters such as area and perimeter is limited. Nevertheless, the majority of more advanced parameters (such as moments) can be usually determined using the previously described ones, similarly as some motion parameters e.g. direction or velocity (Okarma & Lech, 2008a).

4. Application for the line following robots

One of the most important tasks of the control systems designed for mobile robots is line tracking. Its typical optical implementation is based on a line of sensors receiving information about the position of the traced line, usually located underneath the robot. Such line sensor is built from a specified number of cells limiting the resolution of the optical system. Another important parameter of such systems is the distance from the sensors to the centre of steering, responsible for the maximum possible speed of the properly controlled robot. The smoothness of the robot's motion is also dependent on the spacing between the cells forming the line sensor.

A significant disadvantage of such optical systems is relatively low resolution of the tracking system, which can be increases using some other vision based systems with wide possibilities of analysing data acquired from the front Dupuis & Parizeau (2006); Rahman et al. (2005). Nevertheless the analysis of the full acquired image even with low resolution is usually computationally demanding and time consuming, especially in the presence of some obstacles, some line intersections etc., assuming also varying lighting conditions. Another serious drawback of using line of sensors is the short time for the reaction limited by the distance between the sensors and the steering centre (or wheels). The proposed approach, based on the fast image analysis using the Monte Carlo method, preserves the main advantages of the vision systems allowing the reduction of the amount of processed data. Considering its application for the control of the mobile robots, the camera moves relatively to the static scene, differently than in the primary version of the algorithm, but the working properties of the method are similar.

In order to filter the undesired data, usually related to some contaminations which should not be used for the robot control purposes, the binarization threshold should be properly set. In the conducted experiments the additional "cut-off" value has been set as minimum 20% black pixels possibly representing the object within the block in order to avoid the influence of the small artifacts, especially close to the region of interest. Finally, the simplified binary representation of the image is obtained where such artifacts (the elements which in fact do not represent the followed line) have been removed during the "cut-off" operation illustrated in Fig. 4 where the original binary image and the intermediate result obtained by the Monte Carlo method are also presented.

4.1 A simple proportional control algorithm for controlling a line following robot

The mobile robot control process can be based on the popular proportional control approach. It is typically used for controlling the robots with sensors based on the infra-red receivers grouped into the line containing a specified number of the infra-red cells. In the proposed method the differential steering signals for the motors are obtained using the first (the lowest)

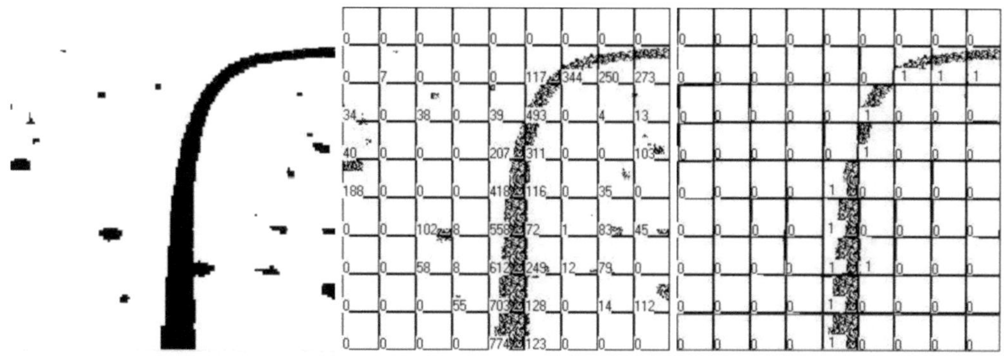

Fig. 4. The idea of the "cut-off" operation

row of the obtained simplified binary image which is equivalent to the typical line of the infra-red sensors.

Nevertheless, the machine vision algorithm can be disturbed by some artifacts caused by the errors during the preprocessing step (e.g. insufficient filtration of some undesired elements on the image). An optimal situation would take place for a single "one" in the lowest row of the image representing the tracked line.

In the simplified version of the algorithm no line crossings can be assumed. The reduction of the possible disturbances can be achieved using the following additional operations:

- limitation of the line thickness by the proper choice of the binarization threshold (lines containing more than a single "one" can be ignored assuming no line crossings),
- orphan blocks removal (applicable for all blocks with the value "one" without any neighbouring "ones"),
- ignoring the empty lines (without any "ones").

4.2 Extension of the robot control by vision based prediction

The control algorithm of the robot should point towards such a position that only a single block of the bottom image row will be filled (representing the line). In the case when two neighbouring blocks represent the line the choice of the block should be based on the simple analysis of the current trend, assuming that the turn angles of the line match the motion and control possibilities of the controlled robot.

The control algorithm for a single frame can be described as follows (Okarma & Lech, 2010):

- binarization of the current frame,
- orphan blocks removal,
- filling the gaps in the detected line using the approximation methods,
- control operation:
  - the detection of the crossing lines
    * if not turning: moving forward with the maximum speed for the symmetrical line-crossing or using the speed control
    * else: turning and using the speed control if the maximum value is not in the middle block
  - speed control: velocity should be proportional to the sum of the values in the middle blocks of each horizontal line (if zero, the minimum speed is set before the turning and the control flag is set to 0).

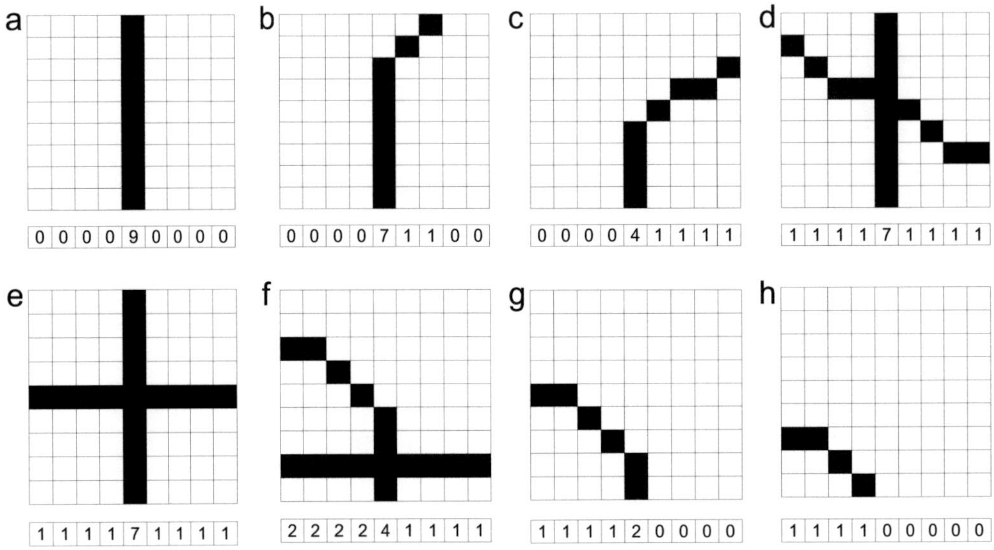

Fig. 5. The illustration of the various configurations of the followed line

The main element of the modified control system is the variable (flag), determined using the previous frames, informing the control system about the current state (e.g. 0 for moving forward or turning, 1 for turning with detected crossing lines).

In the case of detected crossing line the control depends on the symmetry. For the symmetrical line-crossing the robot should move forward with the maximum speed or the additional speed control. If the maximum value is not in the middle block the line-crossing is not symmetrical and the robot should turn left or right and decrease it speed. The speed of the robot should be proportional to the sum of the values in the middle blocks of each horizontal line. If that sum is equal to zero, the minimum speed should be set before the turning. In that case the mentioned above flag should be set to 1. The illustration of the various scenarios is presented in Fig. 5.

One of the typical features of the proportionally controlled systems is the oscillating character of the path for the robot reaching the specified line, causing also the characteristic oscillations of the images acquired from the camera. Nevertheless, the conducted experiments have verified the usefulness of the fast image analysis based on the Monte Carlo method for the efficient controlling of the line following robot in a simplified but very efficient way, especially useful for the real-time applications. The application of the proposed solution for the control of the mobile robot, despite of the simplicity of its control algorithm, is comparable with some typical ones based on the classical approach (proportional steering based on the infra-red sensors).

The main advantage of the proposed method is the possibility of using some prediction algorithms allowing the increase of the robot's dynamic properties and fluent changes of its speed depending on the shape of the line in front of the robot. Another relevant feature is its low computational complexity causing the relatively high processing speed, especially important in the real-time embedded systems.

4.3 Application of the Centre of Gravity

If the robot is located directly over the straight line its central point should be exactly between the two edges obtained using the two-step algorithm of edge detection discussed above. In such situation the sum of the pixel values on its left side should be identical as on its right side. Any significant disturbance can be interpreted as the beginning of a line curve with the assumption that no additional elements, except two line edges, are visible on the image, so the mobile robot should begin turning. An efficient and simple method for controlling the robot which can be used for such purpose is based on the image's Centre of Gravity (Centre of Mass). It can be determined for a grayscale image as:

$$x_c = \frac{\sum\limits_{i,j} i \cdot X_{i,j}}{\sum\limits_{i,j} X_{i,j}} \tag{18}$$

$$y_c = \frac{\sum\limits_{i,j} j \cdot X_{i,j}}{\sum\limits_{i,j} X_{i,j}} \tag{19}$$

where $X_{i,j}$ represents the luminance of the pixel located at position (i, j).
Calculating the coordinates of the Centre of Gravity (CoG), they can be used by the control algorithm as the current target point for the mobile robot. The horizontal coordinate x_c can be used for controlling the turning angle and the vertical y_c for the speed control, assuming the previous filtration of any artifacts from the image.
The approach discussed above can also be implemented using the block based Monte Carlo method using the "cut-off" operation for decrease of the influence of noise, leading to the hybrid control method. For each block the local area estimator $\hat{A}$ is used as the input for the "cut-off" operation leading to the simplified binary representation utilised for steering using the method discussed above. In comparison to the classical Center of Gravity method its implementation using the Monte Carlo approach causes about 8 times reduction of the computation time as has been verified in the conducted experiments.
One of the main modifications proposed in the hybrid method is the switch between the CoG and Monte Carlo based control in order to increase the robustness of the control. For this purpose the sum of each column of the simplified binary image is calculated and stored in an additional vector. If the vector containing the sums of the values, obtained by counted blocks with "ones", has more than one local minimum (e.g. 2 4 0 0 1 6 0 0 1), some additional objects are present on the image. In such situation the calculation of the Centre of Gravity coordinates would be disturbed by a neighbouring object located close to the line, so the motion control should be based on the previous value. Otherwise, the control values determined by the Monte Carlo method can be additionally verified by the CoG calculations.
A more complicated situation may take place for the presence of artifacts together with some discontinuities of the followed line. In this situation the vector containing the sum of the column in the simplified block representation of the image has more than one local maximum. In such case the additional elimination of the blocks which are not directly connected to the followed line (detected as represented by the maximum value in the vector) should be conducted e.g. using the morphological erosion, before the calculations of the Centre of Gravity used for determining the current trend (direction of the line).

5. Simulations in the Simbad environment

In order to verify the working properties of the proposed methods, regardless of some tests conducted using a netbook based prototype, a test application has been created using Java programming language in NetBeans environment allowing the integration of many libraries developed for Java e.g JARToolKit (Geiger et al., 2002) allowing to access the Augmented Reality Toolkit (ARToolKit) functionality via Java and the use of different rendering libraries allows high and low level access. The environment of the mobile robot as well as the robot itself have been modelled using the Simbad simulator (Hugues & Bredeche, 2006), which is available for this platform.
A significant advantage of the simulator is its 3D environment, due to the utilisation of Sun Java3D technology. Nevertheless, an important restrictions of Simbad, in opposite to some other similar products, are the limited working properties for the controlling the real robots die to its internal structure related to the modelling of physical phenomena. However, this results in a simplified and transparent programming source code, so the implementation of the presented algorithms does not cause any considerable problems. The only restriction in the initialisation part of the robot controller is the directing the camera towards the "ceiling", caused by the inability to put the followed line on the ground. Instead of it the line has been located over the robot and the lines in the front of the robot are shown in the top part of the image. The only negative consequence is the reverse transformation of the left and right directions between the robot and the camera. The proportional controlled differential drive robots used in the experiments are typically implemented in Simbad and conducted simulations have confirmed the working properties of the presented steering methods for all control modes with different simulation speed values. The test application used for the simulations is illustrated in Fig . 6.

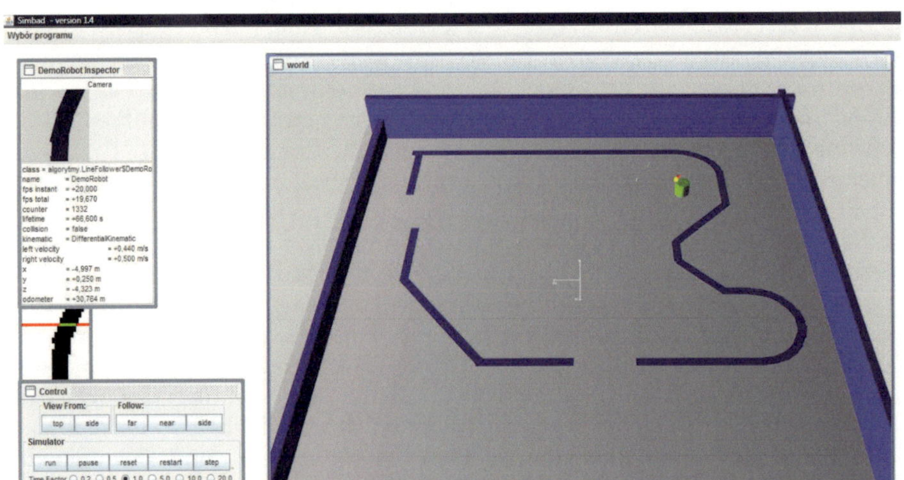

Fig. 6. Illustration of the test application

6. Conclusions and future work

Obtained results allowed the full control of the line following robot (speed, direction and "virtual" line reconstruction). Due to the ability of path prediction, in comparison to the typical line following robots based on the line of sensors mounted under the robot, the "virtual" line reconstruction can be treated as one of the most relevant advantages of the image based motion control. Such reconstruction can also be performed using the ultra-low resolution binary image as the input data. Another possibility, being analysed in our experiments, is the adaptation of the Centre of Gravity method for the real time steering based on the binary image analysis.

The additional robustness of the method can be achieved by the adaptive change of the block size used in the modified Monte Carlo method according to the quality assessment of the binary image and the additional analysis of the line continuity.

Positive results of the simulations performed in Java 3D based Simbad environment allow the proper implementation of the proposed soft computing approach in the experimental video controlled line following robot. Good simulation results, as weel as the low energy consumption by the experimental netbook-based robot, have been also obtained considering the synthetic reduced resolution "grayscale" image constructed using the Monte Carlo method, where the relative areas of the objects in each block have been used instead of the luminance level (after normalisation).

The planned future work is related to the physical application of the proposed approach for some other types of robot control algorithms based on video analysis, e.g. SLAM. Another direction of the future research should be the development of a dedicated binary image quality assessment method, which could be used for the optimisation of the proposed algorithm.

Another interesting direction of the future work is the extension of the proposed method towards direct operations using the colour images. For this purpose a verification of the usefulness of the typical colour models is necessary by means of the chroma keying used for the thresholding in the Monte Carlo method.

7. References

Cowan, N., Shakernia, O., Vidal, R. & Sastry, S. (2003). Vision-based follow-the-leader, *Proceedings of IEEE/RSJ International Conference on Intelligent Robots and Systems*, Vol. 2, Las Vegas, Nevada, pp. 1796–1801.

Cucchiara, R., Grana, C., Piccardi, M. & Prati, A. (2003). Detecting moving objects, ghosts, and shadows in video streams, *IEEE Transactions on Pattern Analysis and Machine Intelligence* 25(10): 1337–1342.

Dupuis, J.-F. & Parizeau, M. (2006). Evolving a vision-based line-following robot controller, *Proceedings of the 3rd Canadian Conference on Computer and Robot Vision*, Quebec City, Canada, p. 75.

Geiger, C., Reimann, C., Stocklein, J. & Paelke, V. (2002). JARToolkit - a Java binding for ARToolkit, *Proceedings of the The First IEEE International Workshop on Augmented Reality Toolkit*, Darmstadt, Germany, p. 5.

Hähnel, D., Triebel, R., Burgard, W. & Thrun, S. (2003). Map building with mobile robots in dynamic environments, *International Conference on Robotics and Automation*, Vol. 2, Taipei, Taiwan, pp. 1557–1563.

Hugues, L. & Bredeche, N. (2006). Simbad: An autonomous robot simulation package for education and research, *in* S. Nolfi, G. Baldassarre, R. Calabretta, J. C. T. Hallam, D. Marocco, J.-A. Meyer, O. Miglino & D. Parisi (eds), *SAB*, Vol. 4095 of *Lecture Notes in Computer Science*, Springer, pp. 831–842.

Kalyan, B., Lee, K. & Wijesoma, W. (2010). FISST-SLAM: Finite set statistical approach to simultaneous localization and mapping, *International Journal of Robotics Research* 29(10): 1251–1262.

Mazurek, P. & Okarma, K. (2006). Car-by-light tracking approach based on log-likelihood track-before-detect algorithm, *Proceedings of the 10th International Conference "Computer Systems Aided Science Industry and Transport" TRANSCOMP*, Vol. 2, Zakopane, Poland, pp. 15–20.

Okarma, K. & Lech, P. (2008a). Monte Carlo based algorithm for fast preliminary video analysis, *in* M. Bubak, G. D. van Albada, J. Dongarra & P. M. A. Sloot (eds), *ICCS (1)*, Vol. 5101 of *Lecture Notes in Computer Science*, Springer, pp. 790–799.

Okarma, K. & Lech, P. (2008b). A statistical reduced-reference approach to digital image quality assessment, *in* L. Bolc, J. L. Kulikowski & K. W. Wojciechowski (eds), *ICCVG*, Vol. 5337 of *Lecture Notes in Computer Science*, Springer, pp. 43–54.

Okarma, K. & Lech, P. (2009). Application of Monte Carlo preliminary image analysis and classification method for automatic reservation of parking space, *Machine Graphics and Vision: International Journal* 18(4): 439–452.

Okarma, K. & Lech, P. (2010). A fast image analysis technique for the line tracking robots, *in* L. Rutkowski, R. Scherer, R. Tadeusiewicz, L. A. Zadeh & J. M. Zurada (eds), *ICAISC (2)*, Vol. 6114 of *Lecture Notes in Computer Science*, Springer, pp. 329–336.

Piccardi, M. (2004). Background subtraction techniques: a review, *Proceedings of the IEEE International Conference on Systems, Man & Cybernetics*, Vol. 4, The Hague, Netherlands, pp. 3099–3104.

Rahman, M., Rahman, M., Haque, A. & Islam, M. (2005). Architecture of the vision system of a line following mobile robot operating in static environment, *Proceedings of the 9th International Multitopic Conference, IEEE INMIC 2005*, Karachi, pp. 1–8.

Se, S., Lowe, D. & Little, J. (2001). Local and global localization for mobile robots using visual landmarks, *Proceedings of the IEEE/RSJ International Conference on Intelligent Robots and Systems*, Vol. 1, Maui, Hawaii, pp. 414–420.

Se, S., Lowe, D. & Little, J. (2002). Mobile robot localization and mapping with uncertainty using scale-invariant visual landmarks, *International Journal of Robotics Research* 21: 735–758.

Tamimi, H. & Zell, A. (2004). Vision based localization of mobile robots using kernel approaches, *Proceedings of the 2004 IEEE/RSJ International Conference on Intelligent Robots and Systems*, Sendai, Japan, pp. 1896–1901.

18

An Embedded Type-2 Fuzzy Controller for a Mobile Robot Application

Leonardo Leottau[1] and Miguel Melgarejo[2]
[1]*Advanced Mining Technology Center (AMTC), Department of Electrical Engineering, University of Chile, Santiago*
[2]*Laboratory for Automation, Microelectronics and Computational Intelligence (LAMIC), Faculty of Engineering, Universidad Distrital Francisco José de Caldas, Bogotá D.C.*
[1]*Chile*
[2]*Colombia*

1. Introduction

Fuzzy logic systems (FLS) have been used in different applications with satisfactory performance (Wang, 1997). The human perception cannot be modelled by traditional mathematical techniques, thus, the introduction of fuzzy set (FS) theory in this modelling has been suitable (John & Coupland, 2007). When real-world applications are treated, many sources of uncertainty often appear. Several natures of uncertainties would influence the performance of a system. It is independent from what kind of methodology is used to handle it (Mendel, 2001).

Type-1 Fuzzy logic systems (T1-FLS) have limited capabilities to directly handle data uncertainties (Mendel, 2007). Once a type-1 membership function (MF) has been defined, uncertainty disappears because a T1-MF is totally precise (Hagras, 2007). Type-2 fuzzy logic systems (T2-FLS) make possible to model and handle uncertainties. These are rule based systems in which linguistic variables are described by means of Type-2 fuzzy sets (T2-FSs) that include a footprint of uncertainty (FOU) (Mendel, 2001). It provides a measure of dispersion to capture more about uncertainties (Mendel, 2007). While T2-FSs have non-crisp MFs, T1-FSs have crisp membership grades (MGs) (John & Coupland, 2007).

A representation of the inference model for T2-FLS is depicted in Figure 1 (Mendel, 2007). It begins with fuzzification, which maps crisp points into T2-FSs. Next, inference engine computes the rule base by making logical combinations of antecedent T2-FS, whose results are implicated with consequent T2-FS to form an aggregated output type-2 fuzzy set. Afterwards, Type-Reduction (TR) takes all output sets and performs a centroid calculation of this combined type-2 fuzzy set, which leads to a type-1 fuzzy set called *type-reduced set*. That reduced set is finally defuzzyfied in order to obtain a crisp output (Mendel, 2001; Karnik & Mendel 2001). The computational complexity of this model is reduced if interval type-2 fuzzy sets are used (Mendel, 2001), it is convenient in the context of hardware implementation in order to make softer the computational effort and sped up the inference time.

Type-2 fuzzy hardware is a topic of special interest, since the application of T2-FLS to particular fields that demand mobile electronic solutions would be necessary. Some recent

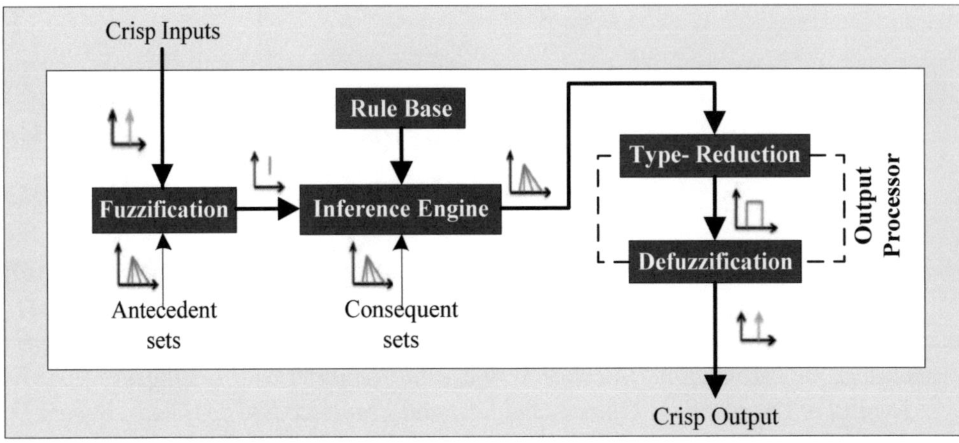

Fig. 1. Type-2 fuzzy system.

applications of T2-FLS have been developed in fields like robotics, communication and control systems among others (Castro et al., 2007; Hagras, 2007; Melgarejo & Peña, 2004, 2007; Torres & Saez, 2008). It is worth to think about the possibility of embedding T2-FLS handling these applications in order to achieve better communication speeds in smaller areas (Leottau & Melgarejo, 2010a).

Control and robotics are one of the most widely used application fields of fuzzy logic. The advantages of type-2 fuzzy logic controllers (T2-FLCs) over type-1 fuzzy logic controllers (T1-FLC) have also been demonstrated and documented in (Castro et al., 2007; Figueroa et al., 2005; Hagras 2004, 2007; Martinez et al., 2008; Torres & Saez, 2008). Although this kind of works presents improvements in the T2-FLCs performance, it is necessary to propose methodologies where these advances can be reflected in design processes as (Hagras 2008, Leottau & Melgarejo, 2010b; Melin & Castillo, 2003; Wu & Tan, 2004) and hardware implementation approaches over embedded devices as (Hagras, 2008; Leottau & Melgarejo, 2010a; Melgarejo & Peña, 2004, 2007). In this way T2-FLS would become widely used in different applicative contexts.

This work presents the implementation of an Interval Type-2 Fuzzy Logic Controller (IT2-FLC) for tracking the trajectory of a mobile robot application. The FLC is design according to the approach proposed in (Leottau & Melgarejo, 2010b), which involves some of the T1-FLC and T2-FLC properties. A hardware implementation of the designed T2-FLC is carried out over a digital signal controller (DSC) embedded platform. Different tests are performed for evaluating the performance of the IT2-FLC with respect to a T1-FLC. Simulation and emulation (i.e. with the embedded controller) results evidence that the IT2-FLC is robust to type reducer changes and exhibits better performance than its T1-FLC counterpart when noise is added to inputs and outputs. The IT2-FLC outperforms the T1-FLC in all tested cases, taking the interval time square error (ITSE) as performance index.

The chapter is organized as follows: Section 2 presents an overview of hardware implementation of IT2-FLCs. Section 3 describes the design approach followed for developing the IT2-FLC and section 4 describes its hardware implementation. Section 5 presents test, obtained results and discussion. Finally, conclusions and future work are presented in Section 6.

2. Hardware Implementation of an IT2-FLC for robot mobile applications

A FLC is a control strategy whose decisions are made by using a fuzzy inference system. Particularly, an IT2-FLC uses IT2-FSs to represent the inputs and/or outputs of the T2-FLS and these sets are completely characterized by their Footprints of Uncertainty (Mendel, 2001). So, this kind of FLC can model the uncertainty by means of the FOUs. In this way, uncertainties in sensors, actuators, operational changes, environmental changes, noise can be considered (Hagras, 2007; Mendel 2001). A FOU is described by its upper and lower MFs. Each MF is composed by as many membership grades (MGs) as discretization levels are considered in antecedents and consequents universes of discourse (Baturone et al., 2000).

This work considers two alternatives for computing the MGs of IT2-FSs: memory based approach and function computing method, which are involved in fuzzyfier and inference engine stages. These two alternatives are the most commonly used in fuzzy hardware implementations (Baturone et al., 2000). Moreover, centroids and center of sets methods are considered for TR stage. As TR algorithm, the Enhanced Karnik-Mendel algorithm is used (Wu & Mendel, 2009). For the sake of clarity, figure 1 depicts the stage structure of an IT2-FLC. These mentioned methods and alternatives are briefly introduced below:

Function Computing Approach (FCA):

This method carries out a direct computation of MF by using numeric algorithms that avoid constructing and storing look up tables (Leottau & Melgarejo, 2010a). It reduces memory usage and facilitates the implementation of MF. However, its execution could require several machine cycles depending on the complexity of MF (Baturone et al., 2000).

Memory Based Approach (MBA):

This approach stores the MGs of every input value into a memory. This strategy is executed considerably fast, because it uses the input value as the pointer to the memory and to directly retrieve the MG (Leottau & Melgarejo, 2010a).

Centroid Type-Reduction (Cent):

The centroid TR combines all the rule-output T2-FSs using union and then finds the centroid of this T2-FS (Karnik & Mendel, 2001). Therefore, this method presents high accuracy but low computational performance (Mendel, 2001). Cent TR is chosen in order to obtain the best accurate output while the hardware platform is forced to the highest computational effort.

Center-of-sets type-reduction (CoS):

In center-of-sets TR, each type-2 consequent set is replaced by its centroid. Then, the weighted average of these centroids is found, being the weight of each centroid the degree of firing of it corresponding rule (Mendel, 2001). CoS TR is chosen because its convenient trade-off between accuracy and computational cost. It is more accurate than Heights method (Mendel, 2001) and computationally less expensive than Cent TR.

Enhanced Karnik Mendel Algorithm (EKM):

The EKM algorithm is an iterative procedure to obtain the generalized centroid of an T2-FS. It uses statistically defined values as initialization points to reduce the amount of iterations that are necessary, in this way it converges monotonically and super-exponentially fast (Wu & Mendel, 2009). In addition, as it has been mentioned in (Leottau & Melgarejo, 2010a), EKM is the fastest algorithm reported up to that date for finding the generalized centroid of a inferred IT2-FS over DSC technology.

3. Proposal to design an IT2-FLC for a truck backer-upper

A proposal for designing an IT2-FLC for tracking the trajectory of a mobile robot application involving some of the T1-FLC and T2-FLC properties is introduced in (Leottau & Melgarejo, 2010b) and briefly explained in this section. T2-FSs are used for the initial modelling taking advantage of its capability to handle linguistic uncertainty because a large number of T1-FSs are embedded. T1-FSs are used in order to make easier a fine tuning, taking into account their small amount of parameters. Finally, returning to T2-FLSs is proposed for handling the uncertainty that appears in the final system performance. The truck backer-upper (Castro et al., 2007; Nguyen & Widrow 1989, Wang, 1997) is a non-linear control problem, whose model is depicted in Figure 2. Backing a trailer truck is a difficult task which requires a great deal of practice. Usually a truck driver backing, going forward, backing again, etc. if forward movements are not permitted, a more difficult problem emerges. Thus, the problem treated in this chapter is to control the steering of a trailer truck for tracking a trajectory, where only are allowed backing movements.

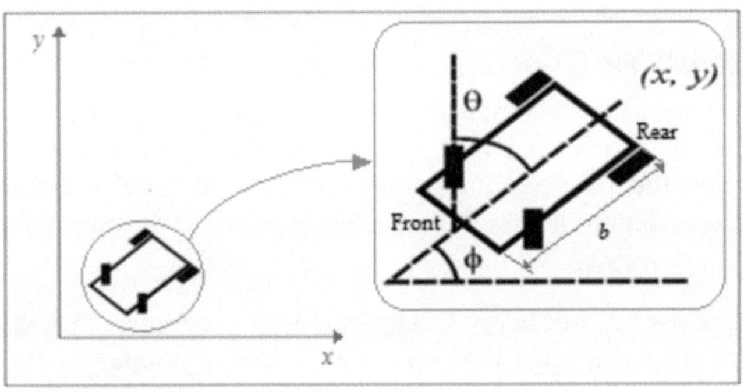

Fig. 2. The simulated truck.

3.1 Modeling the system dynamics: Truck mathematical model

In this step, the control process is studied in order to find and determine the model of the system dynamics. There are different strategies to do it (Kuo & Golnaraghi, 1996; Mendel, 2007; Wang, 1997). One of the most commonly used is by obtaining the mathematical model using physical schemes such as this case (Wang, 1997).
This problem is proposed in (Wang, 1997). The mathematical model that describes the system to control is shown below, where b is the length of the truck.

$$x(t+1) = x(t) + cos[\varphi(t) + \theta(t)] + sin[\theta(t)]sin[\varphi(t)] \tag{1}$$

$$y(t+1) = y(t) + sin(\varphi(t) + \theta(t)) - sin[\theta(t)]cos[\varphi(t)] \tag{2}$$

$$\varphi(t+1) = \varphi(t) - sin^{-1}[2sin(\theta(t))/b] \tag{3}$$

Since this work is focused for academic proposes, b=4 inch is assumed, however as it is handled in (Wang, 1997), units of that length can be considered depending on the operational environment. The truck position is determined by three state variables φ, x and y, where φ is the angle of the truck regarding the horizontal line as shows in Figure 2. The steering angle θ is the control variable to the truck which only moves backward. For the sake of simplicity, y does not have to be considered as state variable, thus position in the vertical axis is not controlled.

3.2 Initial design of an IT2-FLC for the truck

In this step, parameters as feedback variables, inputs (antecedents), outputs (consequents), universe of discourse, control action, rule base and the shape of MFs are defined. Using IT2-FS in this stage is proposed in order to obtain a preliminary design including linguistic uncertainty associated to ambiguity in the knowledge of the control process (Leottau & Melgarejo, 2010b).

Inputs of the controller are $(Ex, E\varphi \; and \; \Delta\varphi)$, where $Ex = Xref - x(t)$ is the error of x position regarding the reference trajectory *Xref*. $E\varphi = \varphi ref - \varphi(t)$ is the error of φ angle regarding the reference φref which is fixed to $\varphi ref = \pi/2$. $\Delta\varphi = \varphi(t) - \varphi(t-1)$ is the changing of φ. The output is θ, which is limited to $[-2\pi/9, 2\pi/9]$. The universe of discourse is assumed as $Ex \in [-20, 20]$, $E\varphi$ and $\Delta\varphi \in [-\pi, \pi]$ and $\theta \in [-\pi/3, \pi/3]$.

Based on intuitive reasoning, a preliminary design is obtained. Two antecedent sets for each input are defined, one to determine positive input values and their respective complement set for negative values. Eight rules are obtained by combining the antecedent sets and using four consequent sets are proposed: Positive high (+ +), positive small (+), negative small (-) and negative high (- -). The obtained rule base is shown below:

$$\begin{aligned}
&Rule\ 0: if\ Ex^- \ and\ E\varphi^- \ and\ \Delta\varphi^- \ then\ \theta^+\\
&Rule\ 1: if\ Ex^- \ and\ E\varphi^- \ and\ \Delta\varphi^+ \ then\ \theta^-\\
&Rule\ 2: if\ Ex^- \ and\ E\varphi^+ \ and\ \Delta\varphi^- \ then\ \theta^-\\
&Rule\ 3: if\ Ex^- \ and\ E\varphi^+ \ and\ \Delta\varphi^+ \ then\ \theta^{--}\\
&Rule\ 4: if\ Ex^+ \ and\ E\varphi^- \ and\ \Delta\varphi^- \ then\ \theta^{++}\\
&Rule\ 5: if\ Ex^+ \ and\ E\varphi^- \ and\ \Delta\varphi^+ \ then\ \theta^+\\
&Rule\ 6: if\ Ex^+ \ and\ E\varphi^+ \ and\ \Delta\varphi^- \ then\ \theta^+\\
&Rule\ 7: if\ Ex^+ \ and\ E\varphi^+ \ and\ \Delta\varphi^+ \ then\ \theta^-
\end{aligned} \tag{4}$$

Since negative values in antecedent sets are handled as the complement of their respective positive values, in practical terms $\mathrm{Ex}^-, \mathrm{E\varphi}^-$ and $\Delta\varphi^-$ respectively are equivalent to $\overline{\mathrm{Ex}^+}$, $\overline{\mathrm{E\varphi}^+}$ and $\overline{\Delta\varphi^+}$.

3.3 Initialization the IT2-FLC for the truck

In order to evaluate the IT2-FLC performance, it is convenient to define a figure of merit. In this way, it is possible to carry out a preliminary tuning, searching for an acceptable response. In this case, an exhaustive tuning is not carried out because to achieve stability is enough (Leottau & Melgarejo, 2010b). It could be carried out using simulations.

Determine an initial wide for FOUs in order to include the uncertainty in the system modelling is the first task of this initialization step. The main sources of uncertainty that would be present in this implementation are:

- Uncertainties in inputs to the FLC. The sensors measurements would be affected by noise levels or by variant conditions of observation (Hagras, 2007), taking into account that these kind of mobile robots applications usually are employed in outdoor environments (Hagras, 2004).

- Uncertainties in control outputs. Actuators characteristics could change with time. The mobile robots mechanics often are susceptible to wear or tear (Hagras, 2007).
- Uncertainties associated with changing operational conditions (Mendel, 2007).

The sources of uncertainty mentioned above are included in the final design. But, as it is mentioned in step 3.2 of this procedure, FOUs in this part of design, makes reference to linguistics uncertainties. This is the main advantage of use IT2-FS here, because the initial tuning would be based in some knowledge of the control process.

As it is proposed in (Leottau & Melgarejo, 2010b), it is used the middle of universe as center of antecedent sets location and the 10% of universe size as a measure for uncertainty. As universe size of $E\varphi$ and ΔE is 2π, the FOU wide of this IT2-FSs is initially defined as $2\pi \cdot 10\% = 0.2\pi$ and center of sets are located at zero. Please, see Figure 7.a for more clarity.

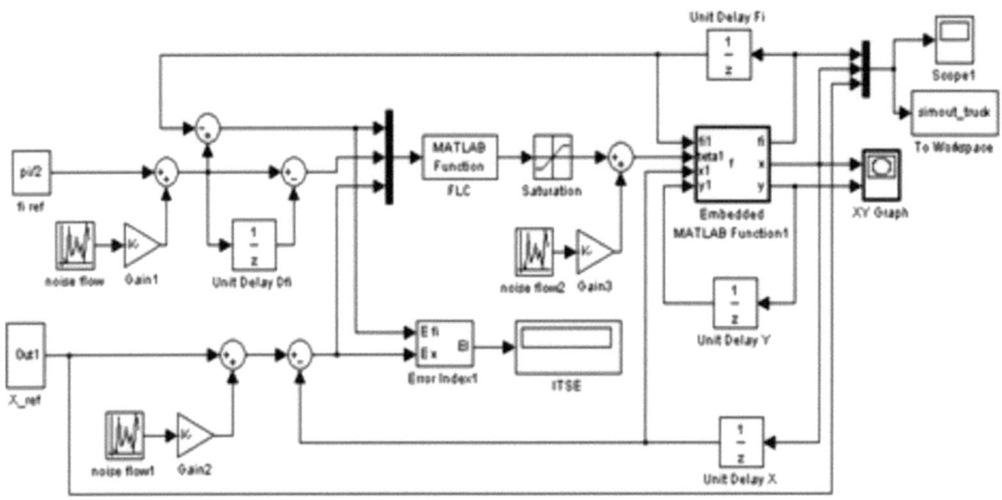

Fig. 3. Simulink model for the truck system.

The whole system is simulated using Matlab®, fuzzy Toolbox and Simulink. In addition, the IT2-FLC is designed and simulated with the IT2 Fuzzy Toolbox by the Tijuana Institute of Technology and Baja California Autonomous University (Castro et al., 2007). The Simulink model for the truck system is depicted in Figure 3.

As a figure of merit, the integral time square error (ITSE) is used (Dorf & Bishop, 1998). The objective of this example application is not to achieve an optimal performance in the controller. Thus, tuning parameters is carried out using the prove-error method, looking that the mobile robot follows a defined trajectory (see Figure 6).

3.4 The reduction of dimensionality

The reduction of dimensionality consists of converting the IT2-FLC into a T1-FLC. It is possible compressing the FOU in the IT2-FSs up to transform it in a T1-FS as is shown in Figure 4. The main objective of this procedure is making easier the tuning of fuzzy sets parameters (Leottau & Melgarejo, 2010b). The amount of parameters used for modelling a T2-FS is greater than for a T1-FS. In this way, tuning these parameters in a T1-FLC is easier and faster (Mendel, 2001).

In this application, the embedded T1-FS that are located in the middle of each FOU is used as a tuning start point. Once the IT2-FSs are converted to T1-FSs, the MF parameters' tuning is carried out using Microsoft Excel® solver tool that uses the Generalized Reduced Gradient Method as non linear optimization algorithm (Frontline Systems, 2010). A screen of used Excel® spreadsheet is shown in Figure 5.

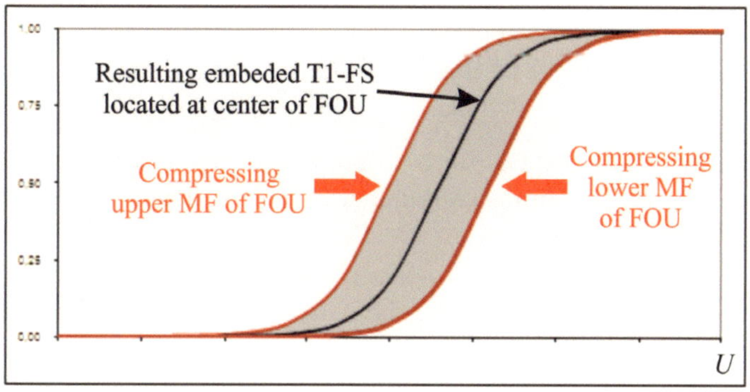

Fig. 4. Reducing the dimensionality of an IT2-FS.

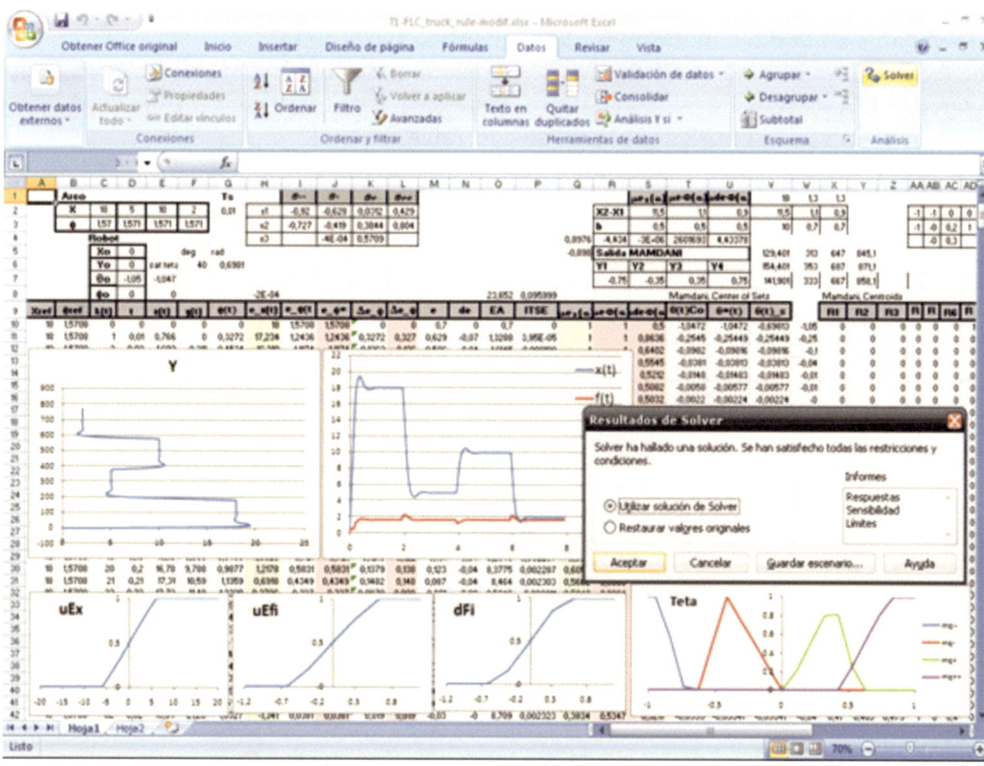

Fig. 5. A screen of used Excel® spreadsheet for MF parameters' tuning.

3.5 Returning to Type-2

Once the T1-FLC parameters are tuned, returning to T2-FLS scheme using the same uncertainties defined in step 3.3 is proposed in (Leottau & Melgarejo, 2010b), but in this case, locating the FOUs over their embedded T1-FSs obtained in previously tuning. So, By using the tuned T1-FSs obtained in step 3.4 as the central embedded T1-FS as in Figure 4, the same FOUs wide resulting by step 3.3 are included, so the FLC returns to its initial interval type-2 nature. Then a fine tuning is carried out again by using the prove-error method taking into account the system response obtained by simulation.

4. Hardware implementation of the IT2-FLC for the truck

Designed IT2-FLC showing in section 3 has the following characteristics: three inputs ($N=3$), two interval type-2 fuzzy sets (IT2-FSs) by input (M=2), six antecedent sets ($M_A=M \cdot N=6$), eight rules ($R=8$), four IT2-FSs in the consequent (M_C=4). It must be taken into account that negative values in antecedent sets are handled as the complement of their respective positive values, so in practical terms it is defined one IT2-FSs by input ($M^{\dagger} = 1$ and $M_A^{\dagger} = 3$). It is applicable for example for calculating the data memory usages. Universe of discourse' sizes have been defined in section 3.2. Its discretization levels in antecedents (D_A) and consequent (D_C) must be defined according computational and hardware resources available on the embedded platform to this implementation.

4.1 Available embedded platform for implementing the IT2-FLC

The available embedded platform for implementing the IT2-FLC is the Digital Signal Controller DSC-56F8013 evaluation board (Freescale Semiconductor, 2010). Next, some hardware resources of DSC-56F8013 are listed in Table 1.

Word length	16-bit
Core frequency	Up to 32 MIPS at 32MHz
Program Flash Memory	8KB (4KWord)
RAM	2KB (1KWord)
PWM module	One 6-channel
ADC	Two 3-channel 12-bit
Communication Interfaces	One Serial Communication Interface (SCI) with LIN slave functionality. One Serial Peripheral Interface (SPI). One Inter-Integrated Circuit (I2C) Port.
Quad Timer	One 16-bit Quad Timer.
On-Chip Emulation	JTAG/Enhanced On-Chip Emulation (OnCE™) for unobtrusive, real-time debugging.

Table 1. DSP-56F8013 basic features.

4.2 Determining computational model for implementing the IT2-FLC for the truck system

A summary of computational models for implementing the embedded IT2-FLC is shown in Table 2. The rest of this section explains why these computational models are chosen and how can be determined some of its features.

A methodological proposal for implementing interval type-2 fuzzy processors (IT2-FP) over DSC technology is reported in (Leottau & Melgarejo, 2010a). There, several computational

	IT2-FLC with centroid TR	IT2-FLC with CoS TR
Fuzzyfier	FCA	MBA
$U_A = D_A$	250	250
Inference Engine	FCA	-
U_C	1000	1000
D_C	100	4
RAM Consumption	456 Bytes (22.8%)	44 Bytes (0.28%)
DATA Mem. Consumption	4600 Bytes (57.5%)	136 Bytes (0.017%)

Table 2. Computational models features.

models have been characterized and tested over an IT2-FLS with the following characteristics: two inputs ($N=2$), three IT2-FSs by input ($M=3$), six antecedent sets ($M_A=M{\cdot}N$=6), nine rules ($R=9$), nine IT2-FSs in the consequent ($M_C=9$). Since characteristics of that IT2-FLS are similar and even more complex than the IT2-FLC designed for this mobile robot application, following considerations are based in that methodological proposal.

4.2.1 Processing time

According inference times reported in (Leottau & Melgarejo, 2010a) for a DSC-56F8013, a complete inference is carried out between 500uS and 60mS for the fastest (MBA, Dc=10) and the slowest (FCA, Dc=1000) computational model respectively. That times can be used as references for determine the IT2-FLC sample time (Ts) for the truck system. Among this time, the embedded FLC samples the inputs variables for computing a new crisp output.

Since θ output is limited to $\pm 40^{\circ}$, it is necessary to define a sampling time such that θ can rotate at least 80°. If it is used a Hitec servo motor (Hitec Servo Motors, 2010) that handling operating speed from 0.05 sec/60°, defining a sampling time of $Ts=100mS$ is enough, maintaining a convenient extra range.

4.2.2 Computational model with centroid type-reduction

Computational model chosen for the IT2-FLC with centroid TR is implemented with FCA for fuzzyfier and inference engine. Although (Leottau & Melgarejo, 2010a) evidences that processors implemented with FCA are slower than those implemented with MBA, this is not a problem taking into account that selected sampling time is enough. On the other hand, available DSC56F8013 offers limited memory resources, with that respect (Leottau & Melgarejo, 2010a) evidence that FCA is the computational model with less data memory consumption. In this way, we consider that cost-benefit trade-off of this computational model is convenient for this application.

4.2.3 Computational model with CoS type-reduction

Computational models with CoS TR offer small data memory consumption but lower accuracy regarding methods as Centroid (Mendel, 2001). On the other hand, computational models based on MBA offer the highest memory consumptions and the highest accuracy (Leottau & Melgarejo, 2010a). Thus, by combining CoS TR with a fuzzyfier based on MBA can be a convenient trade-off between accuracy and data memory usage. Thus, computational model chosen for the IT2-FLS with CoS TR is implemented with MBA for the fuzzyfier.

4.2.4 Interconnecting the emulated mobile robot plant with the embedded IT2-FLC

The truck system is emulated by using Matlab® over Simulink, thus it is necessary interconnect the DSC56F8013 with the computer in order to communicate the embedded IT2-FLC with the truck system. Since 56F8013 evaluation board include a RS-232 interface, for easy connection to a host processor the serial port is used. Since emulation is carried out over a general propose processor which serial port baud-rate is limited to 115200, it is selected a sped of 57600 bauds in order to not to force the platform.
As it is mentioned in section 3, in a context of simulation, a block of Matlab® fuzzy toolbox or it2fuzzy toolbox is included in the Simulink model for simulating the T1-FLC and the IT2-FLC respectively (Figure 3). In order to interconnect the truck system plant with the IT2-FLC embedded in the DSC56F8013, these blocks must be replaced by a Matlab function that carries out the RS-232 serial port communication.

4.2.5 Universe sizes and discretización levels regarding RS-232 communication

Universes of discourse in the process are defined as: $Ex \in [-20, 20]$ $(U_{Plant\,X} = 40)$, $E\varphi$ and $\Delta\varphi \in [-\pi, \pi](U_{(Plant\,\varphi \text{ and } \Delta\varphi)} = 2\pi)$ and $\theta \in [-\pi/3, \pi/3]$ $(U_{(Plant\,\theta)} = 2\pi/3)$.
RS-232 interface has been chosen for connecting the embedded IT2-FLC with the truck system emulated by using Matlab® over Simulink. Since RS-232 interface can send and receive up to one Byte per time (2^8=256 levels), a universe of discourse between [0, 249] is defined in order to transmit just one byte per input, so $U_A=D_A=250$. Resolution of input universes is determined as: Input_res=U_{Plant}/D_A. Thus, Ex_res=40/250=0.16, $E\phi$_res=$2\pi/250 \simeq 0.025$ and $\Delta\phi$_res=$2\pi/250 \simeq 0.025$.
Output universe is defined between $[-\pi/3, \pi/3]$. If U_C is the output universe size in the embedded IT2-FLC, calculating output resolution by using $D_A=U_C$, θ_res= $2\pi/(3 \cdot 250)$=0.00837=0.48°. Since computational models chosen particularly offer a low accuracy and output is limited to ±40°, we consider that 0.48° is a low resolution. Thus, a universe of discourse between [0, 999] is defined, so U_C=1000 and θ_res= 0.00209=0.12°.
It is necessary to transmit from the PC serial port to the IT2-FLC embedded on the DSC56F8013 three bytes, one per input of IT2-FLC. Then, it is necessary to receive one data as its output. Since the output universe of the embedded IT2-FLC has been to a thousand points. In order to receive the output of the IT2-FLC from DSC56F8013, it is necessary to divide this result. So, units, tens and cents are transmitted as three independent bytes those must be concatenated by a Matlab RS-232 function before to be injected to the truck plant.
Universes of discourse for the embedded IT2-FLC are defined in the previous paragraphs as and. So, it is necessary a scalization as:

$$V_{FLC} = V_{Plant} \cdot U_{FLC}/U_{Plant} + U_{FLC}/2 \tag{5}$$

Where V_{FLC} and U_{FLC} are respectively the scaled value and the universe size that handles the embedded FLC and U_{Plant} and V_{Plant} are respectively the universe size and the un-scaled value that handles the truck plant.

4.2.6 Discretization levels in the consequent regarding available memory resources

According (Leottau & Melgarejo, 2010a), the RAM usage expressed in Words for the inference engine with centroid type-reduction is $2 \cdot D_C + 2 \cdot R$ and for all discussed fuzzyfiers is $2 \cdot M_A$. Since DSC56F8013 offers 1KWord in RAM, the maximum discretization levels in the consequent regarding RAM resources can be calculated as:

$$D_{C(max)} = (RAM_{available} - 2 \cdot M_A - 2 \cdot R)/2 = 486 \tag{6}$$

On the other hand, the DATA memory usage for the inference engine with centroid TR and MBA is $2 \cdot D_C \cdot M_C$ and $2 \cdot M_A \cdot D_A$ for the fuzzyfier. Since DSC56F8013 offers 4KWord in Flash memory, the maximum discretization levels in the consequent regarding DATA memory resources can be calculated as:

$$D_{C(max)} = (Flash_{available} - 2 \cdot M_A^{\dagger} \cdot D_A)/(2 \cdot M_C) = 312 \tag{7}$$

Thus, with centroid TR and MBA for fuzzyfier and inference engine, *Dc* must be less than 312. On the other hand, RAM usage for inference engine with CoS TR is *2·R=16 Words.* Equation 7 evidences that *Dc* must be than three hundred twelve points. So, *Dc* is defined as one hundred taking into account the study carried out in (Leottau & Melgarejo, 2010a). This study evidence that by using *Dc=100,* results maintaining a trade-off between accuracy, memory consumption and processing time.

5. Tests, results and discussion

5.1 Obtained IT2-FLC for the truck backer-upper

The obtained IT2-FLC for tracking the trajectory of the truck Backer-Upper is presented in this sub-section. The IT2-FLC has been designed following the procedure described throughout section 3. Preliminary response of x and φ obtained in step 3.3 is presented in Figure 6. This response has been obtained with the IT2-FSs resulting in the initialization step. These IT2-FSs and resulting T1-FSs tuned in step 3.4 are shown in Figure 7.a. The final IT2-FSs obtained after step 3.5 are shown in Figure 7.b.

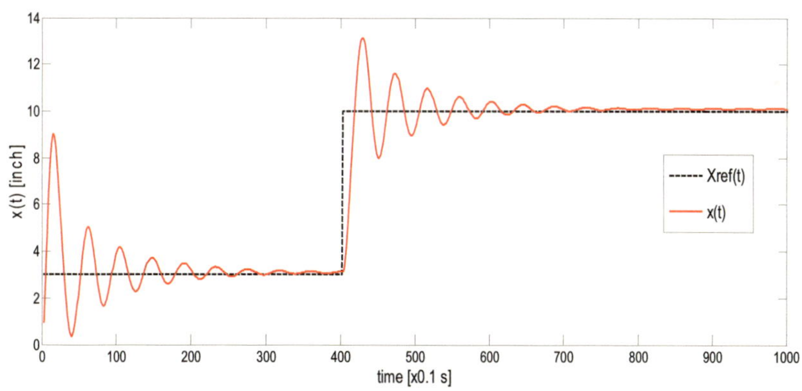

Fig. 6. IT2-FLC design procedure. Preliminary response of x and obtained in step 3.3.

5.2 Tests

Figure 8 resumes the approach outlined throughout this chapter applied for designing and implementing an IT2-FLC for the truck system. After modelling and defining parameters steps in 3.1 and 3.2, initialization and the first design in step 3.3 is carried out by simulation with Matlab® over Simulink. Subsequent to reduction of dimensionality in step 3.4, the tuning of the T1-FLC is carried out by using Excel®. Then, tuned T1-FLC is tested over simulation and emulation. Later than returning to Type-2 and the final fine tuning in step

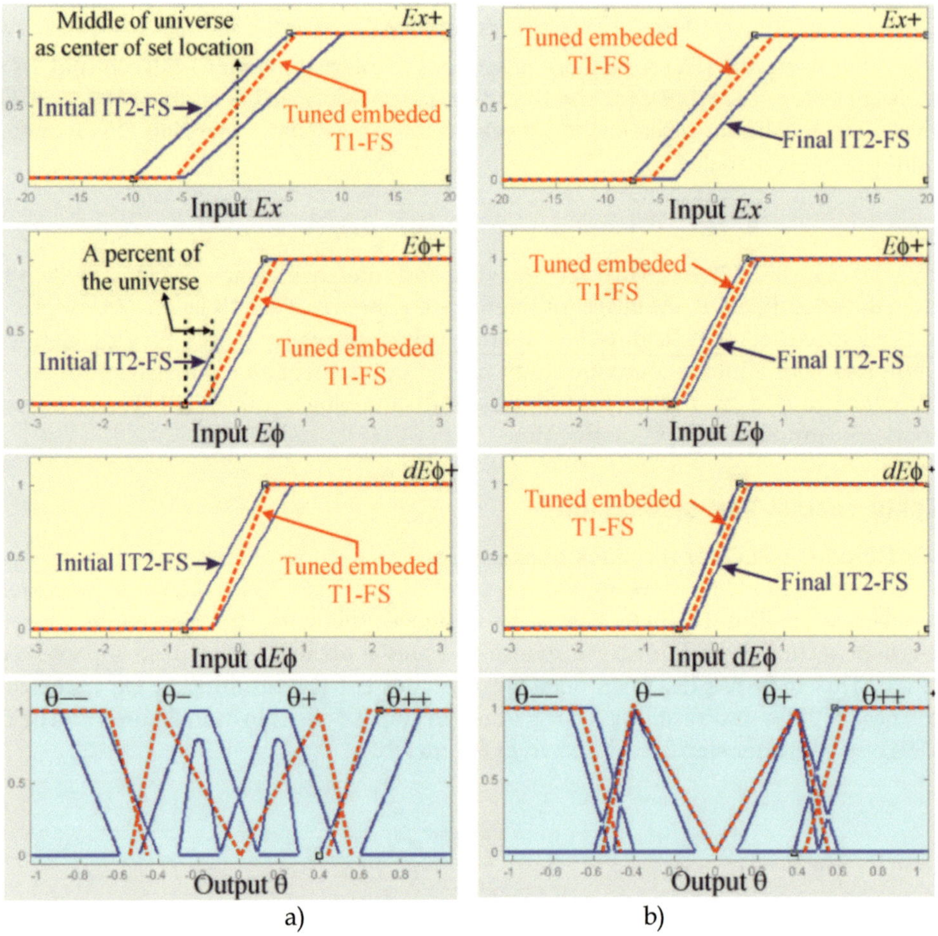

Fig. 7. IT2-FLC design procedure. a) IT2-FSs first design in step 3.3 (Blue lines) and tuned T1-FSs resulting in step 3.4. (Red dotted lines) b) IT2-FSs final implementation after step 3.5 (Blue lines) and newly tuned T1-FSs resulting in step 3.4. (Red dotted lines).

3.5, the resulting IT2-FLC is tested over simulation and emulation too. As it is mentioned before, the simulations are carried out by using fuzzy and it2-fuzzy toolboxes and emulations are carried out over the 56F8013 board whose programming is debugged and loaded by using the Codewarrior® suite for DSC56800E v8.2.3.

Different tests are achieved for evaluating the final performance of designed T1-FLC and IT2-FLC. A combined trajectory in *x* is used as reference form emulating different operating points to the FLC. In addition, random noise generators are inserted in the inputs and outputs of FLC in order to emulate some sources of uncertainty such as sensors measurements and changes of actuators characteristics. The FLCs are tested under three different noise environments: (1) without noise, (2) moderate noise with amplitude as 0.2% of maximum value that respective variable can take and (3) high noise with amplitudes as 1%. E.g. *Ex* input can take values within $|0, 20|$, then the amplitude of moderate noise is: $A_{mN}(Ex)$ =20 x 0.2% = 0.04. In order to emulate a stochastic nature of uncertainty, every noise environment is tested five times changing the initial seed of the random noise generator.

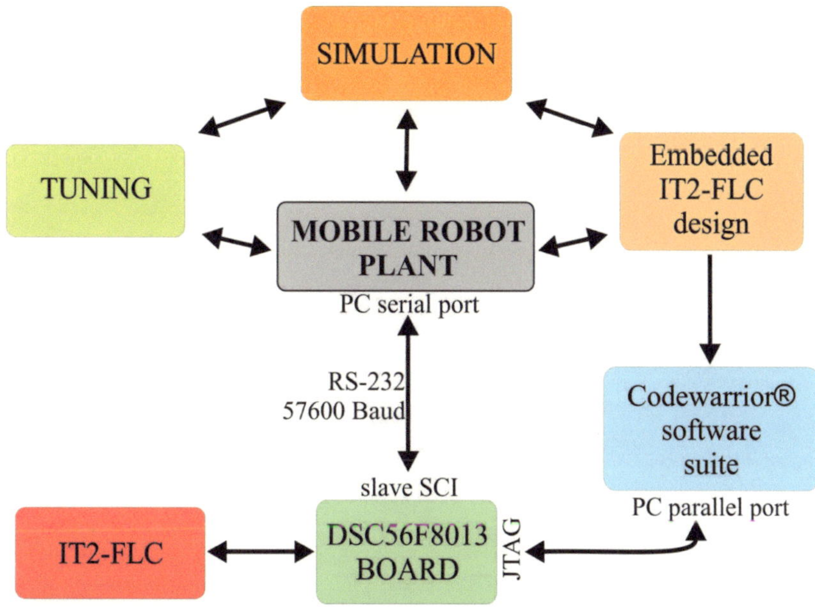

Fig. 8. Conceptual diagram of proposed approach for implementing the embedded IT2-FLC for the truck system.

As it is mentioned in section 2, there are several type reducer and defuzzyfier methods. In order to determine the impact of that in the final performance of designed T1-FLC and IT2-FLC, a test is carried out changing the TR or defuzzyfier method in each FLC. Under simulation, some of available methods in its respective Matlab® toolbox (*fuzzy* for type-1 and *it2fuzzy* for type-2) are tested: centroid and center of sets (CoS) by the T2-FLC and centroid, bisector and mean of maximum (Mom) by T1-FLC. Under emulation, considered methods in section 2 are tested: Centroid and CoS.

The tests procedure is described as follows:

1. Set a TR or defuzzification method for the FLC.
2. Set a noise environment.
3. Run one complete simulation or emulation and register ITSE.
4. Repeat the same simulation (steps 2-3) five times with a different initial noise seed in order to obtain the average and standard deviation of ITSE.
5. Repeat steeps 2-4 for the three noise environments considered.
6. Repeat steeps 1-5 for the TR and defuzzification methods considered.
7. Repeat the procedure for the T1-FLC and the IT2-FLC over simulation and over emulation.

5.3 Results

Results for x and φ are presented in Figure 9 to Figure 12. Figure 9 and 10 show the response of FLCs tested without noise, under simulation and emulation respectively. Figure 11 shows the response of simulated FLCs and Figure 12 shows the response of hardware implemented FLCs, both cases tested under the worst noise condition, emulating a high uncertainty environment. Since all FLCs are tested under three different noise environments and five different initial seeds for each noise environment, obtained results for ITSE as performance indices are presented as an average and its standard deviation. It is shown in tables 3 and 4.

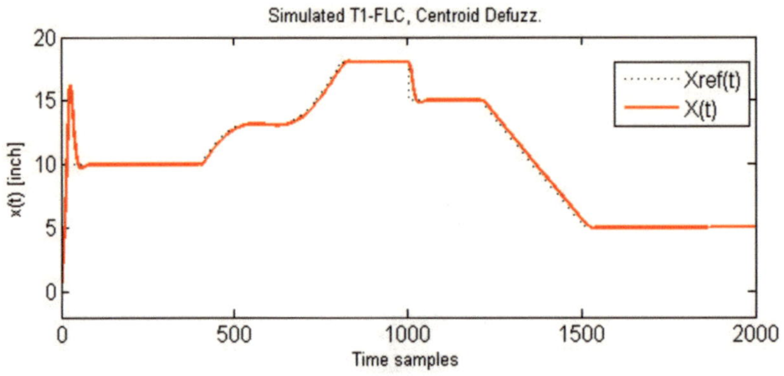

a)

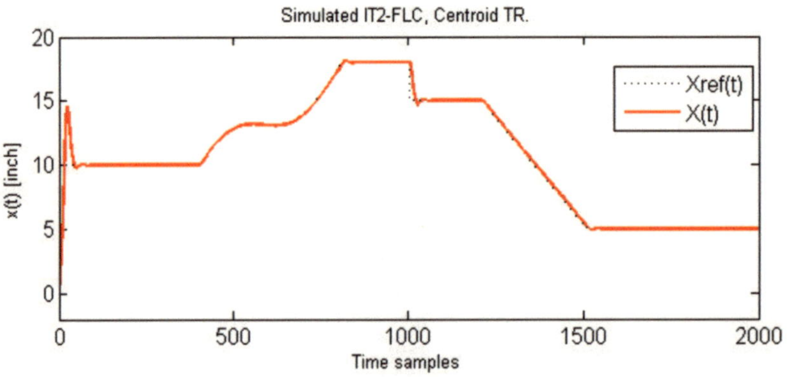

b)

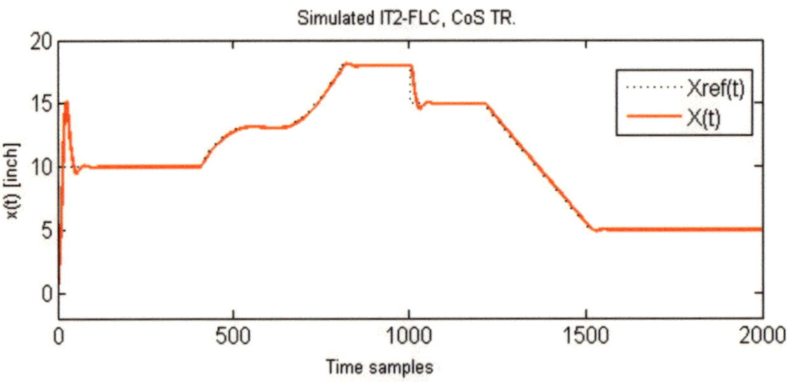

c)

Fig. 9. Without noise and under simulation, trajectory for *x* of: *a)* T1-FLC with Centroid defuzzification. *b)* IT2-FLC with Centroid type-reduction. *c)* IT2-FLC with CoS type-reduction. (*time=Time_samples/Ts*, where*Ts=0.1s*)

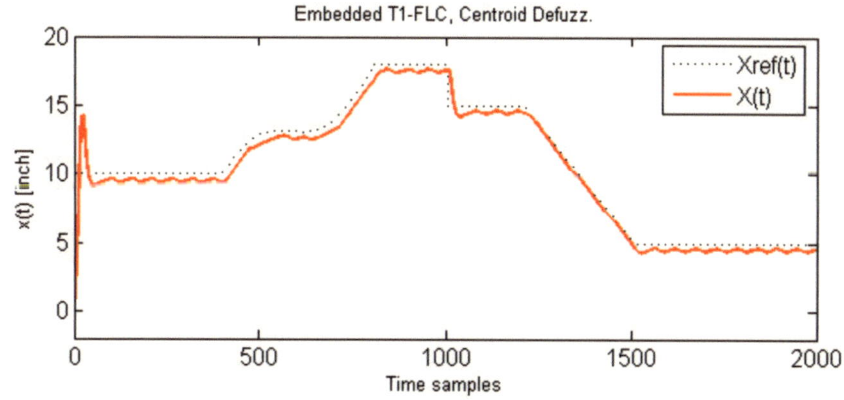

a)

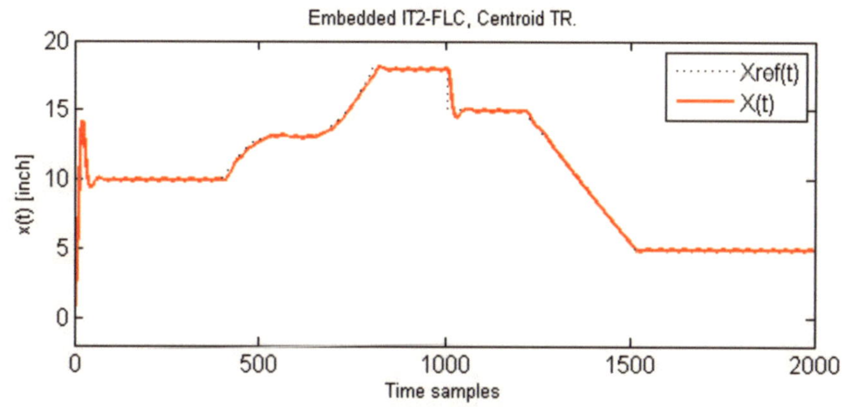

b)

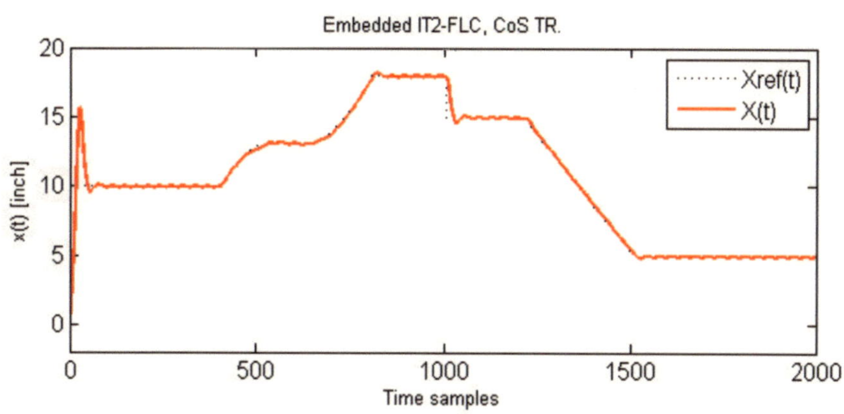

c)

Fig. 10. Without noise, trajectory for x of hardware implemented: a*)* T1-FLC with centroid defuzzification. b*)* IT2-FLC with centroid type-reduction. c*)* IT2-FLC with CoS type-reduction. (*time*=*Time_samples/Ts*, where*Ts*=*0.1s*)

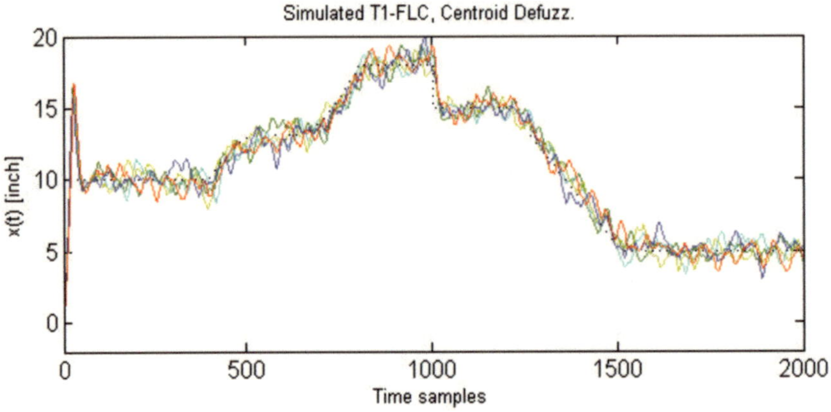

a)

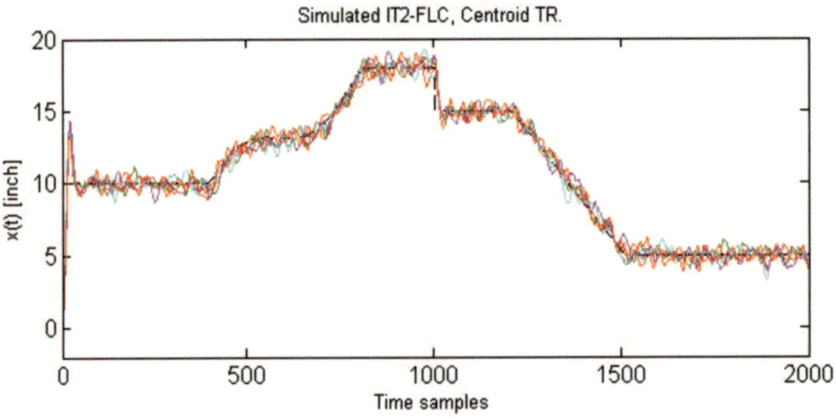

b)

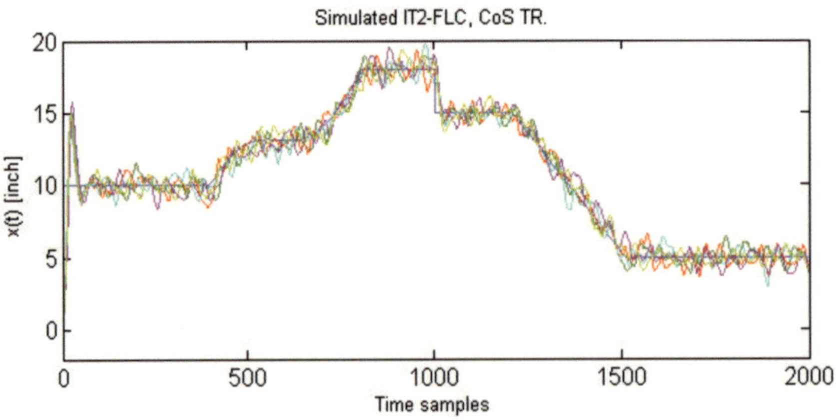

c)

Fig. 11. Under the worst noise condition and simulation, trajectory for *x* of: *a)* T1-FLC with centroid defuzzification. *b)* IT2-FLC with centroid type-reduction. *c)* IT2-FLC with CoS type-reduction. (*time=Time_samples/Ts*, where*Ts=0.1s*)

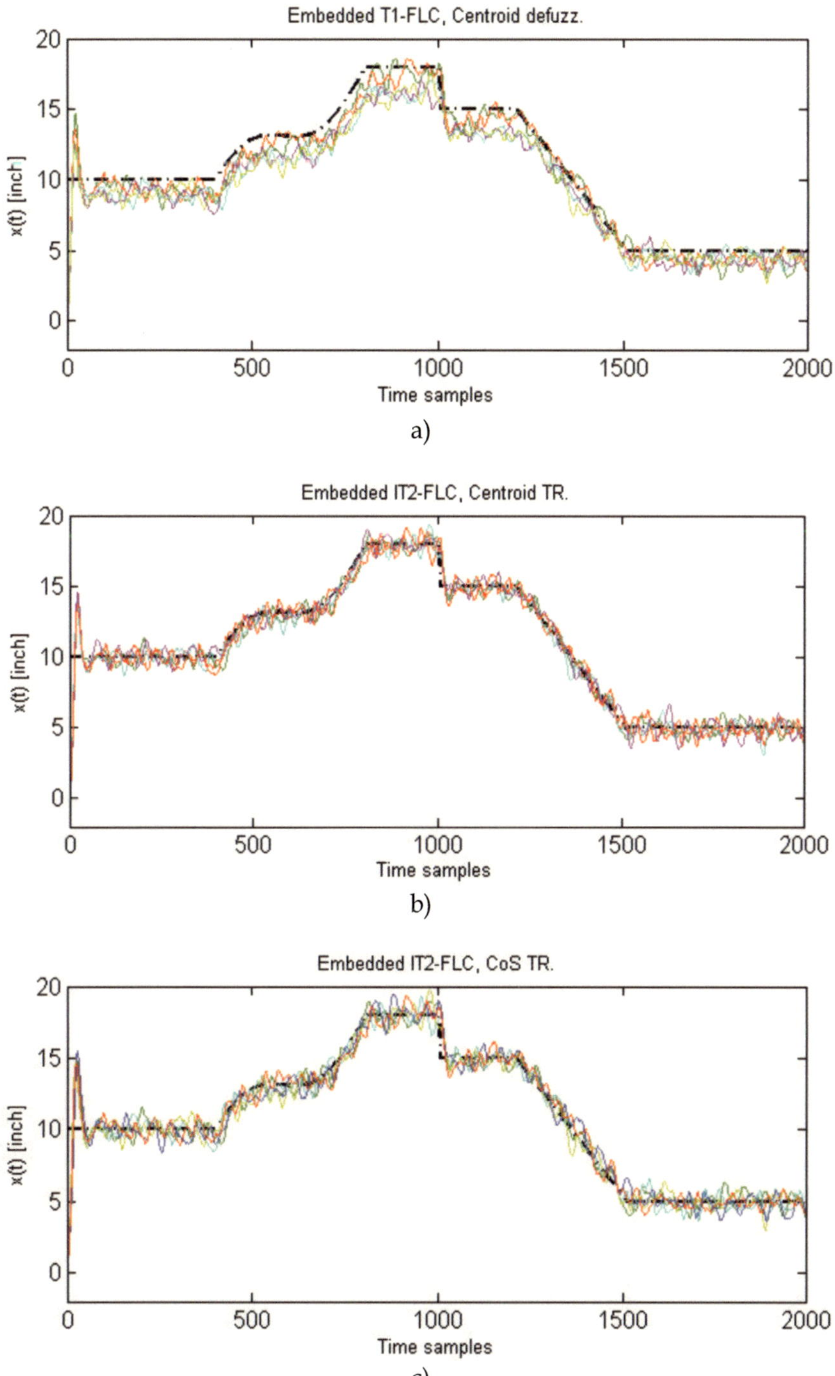

Fig. 12. Under the worst noise condition, trajectory for x of Hardware implemented: a*)* T1-FLC with Centroid defuzzification. b*)* IT2-FLC with centroid type-reduction. c*)* IT2-FLC with CoS type-reduction. (*time=Time_samples/Ts*, where*Ts=0.1s*)

Noise level	Cent.		Bisector		MoM		Cent. Hardware		CoS Hardware	
	ITSE	*Desv.*	*ITSE*	*Desv.*	*ITSE*	*Desv.*	*ITSE*	*Desv.*	*ITSE*	*Desv.*
Zero	18.48	-	†	-	††	-	68.025	-	†††	-
Mod.	22.27	2.10	†	-	††	-	84.84	5.625	†††	-
High	98.93	14.71	†	-	††	-	137.38	13.94	†††	-

† Does not reach the reference.
†† Response turns unstable.
††† Does not reach the reference and presents oscillations.
Since tests have evidenced that the T1-FLCs with MoM and Bisector defuzzyfier and the embedded T1-FLC with CoS defuzzyfier do not present a successfully performance, response for these FLCs have not shown in Figs 9 to 12.

Table 3. Results of ITSE for the T1-FLC.

Noise level	Cent.		CoS		Cent. Hardware		CoS Hardware	
	ITSE	*Desv.*	*ITSE*	*Desv.*	*ITSE*	*Desv.*	*ITSE*	*Desv.*
Zero	11.23	-	13.73	-	11.7	-	13.77	-
Mod.	13.78	0.93	18.02	1.70	13.84	0.85	18.37	0.87
High	44.57	4.81	77.41	10.49	62.122	6.45	82.92	8.76

Table 4. Results of ITSE for the IT2-FLC.

5.4 Discussion

Without noise and using centroid TR, Simulated IT2-FLC ITSE is about 39.2% better than simulated T1-FLC ITSE and embedded IT2-FLC ITSE is about 82.8% better than embedded T1-FLC ITSE. Under moderate noise and using centroid TR it can be said that: Simulated T2-FLC ITSE is about 39.4% better than simulated T1-FLC ones and embedded T2-FLC ITSE is about 83.69% better than embedded T1-FLC. Under high noise and using centroid TR Simulated T2-FLC ITSE is about 54.9% better than simulated T1-FLC ones and embeddedT2-FLC ITSE is about 54.78% better than embedded T1-FLC ones.

ITSE of simulated IT2-FLC with centroid type reduction is between 18% (without noise) and 42% (with high noise) better than the IT2-FLC with CoS. Besides, ITSE of hardware implemented IT2-FLC with centroid type reduction is between 15% (without noise) and 25% (with high noise) better than the IT2-FLC with CoS.

The ITSE of the simulated IT2-FLC with Cent type reducer is between 0.4% (with moderate noise) and 28% (with high noise) better than the embedded IT2-FLC whit the same type reducer. Besides, the ITSE of the simulated IT2-FLC with CoS type reducer is between 0.3% (without noise) and 6.7% (with high noise) better than the embedded IT2-FLC whit the same type reducer. This difference is caused by the integer numeric format that handles the hardware platform which causes lost in the accuracy of embedded FLCs regarding their models tested over simulation.

Taking into account the comparison presented above, it is possible to consider that the IT2-FLC outperforms the T1-FLC in all cases. In addition, by looking Figure 11 and 12, it can be observed that the T2-FLC presents fewer oscillations around the reference, being the IT2-FLC implemented with centroid type reduction, those with better performance.

T2-FLC shows a good immunity to type reducer changes, which is a great advantage taking into account the reduction of computational complexity and inference time when strategies as CoS or heights are implemented.

6. Conclusions

An IT2-FLS for controlling the trajectory of a truck backer-upper mobile robot application and its embedded implementation has been presented in this chapter. The IT2-FLC has been designed based on the approach proposed in (Leottau & Melgarejo, 2010b) and implemented following methodological considerations reported in (Leottau & Melgarejo, 2010a) to the hardware implementation of IT2-FLS over DSC technology.

Two computational models have been selected taking into account the available and demanded hardware and computational resources. Several tests have been carried out in order to evaluate and to compare the performance of developed T2-FLCs and a T1-FLC. Simulated and emulated results evidence that the IT2-FLC is robust to type reducer and defuzzyfier changes and exhibits better performance than a T1-FLC when noise is added to inputs and outputs emulating some sources of uncertainty.

As future work, it is considered to test the developed IT2-FLC with a real mobile robot platform in order to carry out a comparative study of simulation, emulation and real platform performance. By this way, it is possible to extend the design and implementation methodology to other applications, involving in a more formal way a modelling of uncertainty.

7. References

Baturone, I.; Barriga, A.; Sanchez, S.; Jimenez, C.J. & Lopez, D.R. (2000). *Microelectronics Design of Fuzzy Logic-Based Systems*, CRC Press LLC, London.

Castro, J.; Castillo, O. & Melin, P. (2007). An Interval Type-2 Fuzzy Logic Toolbox for Control Applications, *Proceedings of the 2007 IEEE International Conference on Fuzzy Systems, FUZZ-IEEE 2007*, London, UK.

Dorf, R. & Bishop, R. (1998). *Modern Control Systems*, Addison-Wesley.

Figueroa, J.; Posada, J.; Soriano, J.; Melgarejo, M. & Rojas, S. (2005). A Type-2 Fuzzy Logic Controller For Tracking Mobile Objects In The Context Of Robotic Soccer Game, *Proceedings of the 2005 IEEE International Conference on Fuzzy Systems, FUZZ-IEEE 2005*.

Freescale Semiconductor. (January 2010). *Digital Signal Controller 56800/E Reference*. 29.01.2010, Available from: Http://www.Freescale.Com/Dsc

Frontline Systems. (March 2010). *Smooth Nonlinear Optimization*. 17.03.2010, Available from: http://www.solver.com/technology4.htm

Hagras, H. (2004). A Type-2 Fuzzy Logic Controller for Autonomous Mobile Robots, *Proceedings of the 2004 IEEE International Conference on Fuzzy Systems, FUZZ-IEEE 2004*, Budapest, Hungary.

Hagras, H. (2007). Type-2 FLC's: a New Generation of Fuzzy Controllers, *IEEE Computational Intelligence Magazine*, February 2007, pp. 30-43.

Hagras, H. (2008). *Developing a type-2 FLC through embedded type-1 FLCs*, Proceedings of the 2008 IEEE International Conference on Fuzzy Systems, FUZZ-IEEE 2008, Hong Kong.

Hitec Servo-Motors, *Servo-motors catalogue*. 05.08.2010, Available from: http://www.hitecrcd.com/products/servos/index.html

John, R. & Coupland, S. (2007). Type-2 Fuzzy Logic A Historical View, *IEEE Computational Intelligence Magazine*, 2,1, pp. 57-62.

Karnik, N. & Mendel, J. (2001). Centroid of a Type-2 Fuzzy Set, *Information Sciences*, 132, pp. 195-220.

Kuo, B. & Golnaraghi, F. (1996). *Automatic Control Systems*, Prentice Hall.

Leottau, L. & Melgarejo, M. (2010a). A Methodological Proposal for Implementing Interval Type-2 Fuzzy Processors over Digital Signal Controllers, *Journal of Applied Computer Science Methods*, v.2-1, June 2010, pp.61-81.

Leottau, L. & Melgarejo, M. (2010b). A Simple Approach for Designing a Type-2 Fuzzy Controller for a Mobile Robot Application, *Proceedings of the North American Fuzzy Information Processing Society's NAFIPS 2010*, Toronto, Canada, July 2010.

Nguyen, D. & Widrow, B. (1989). The Truck Backer-Upper: An Example of self-learning in Neural Networks. *Proceedings of the International Joint Conference in Neural Networks*, pp. II-357-363, June 1989.

Martinez, R.; Castillo, O. & Aguilar, L. (2008). Optimization with Genetic Algorithms of Interval Type-2 Fuzzy Logic controllers for an autonomous wheeled mobile robot: A comparison under different kinds of perturbations, *Proceedings of the 2008 IEEE International Conference on Fuzzy Systems, FUZZ-IEEE 2008.*

Melgarejo, M.; Garcia, A. & Pena-Reyes, C. (2004). Computational Model and architectural proposal for a hardware Type-2 Fuzzy System, *Proceedings of the 2nd IASTED conference on Neural Networks and Computational Intelligence*, Grindewald, Switzerland.

Melgarejo, M. & Pena-Reyes, C. A. (2007). Implementing Interval Type-2 Fuzzy Processors, *IEEE Computational Intelligence Magazine*, 2,1, pp. 63-71.

Melin, P. & Castillo, O. (2003). A new method for adaptive model-based control of non-linear plants using type-2 fuzzy logic and neural Networks, *Proceedings of the 2003 IEEE International Conference on Fuzzy Systems, FUZZ-IEEE 2003*, St. Louis, MO.

Mendel, J. (2001). Uncertain Rule-Based Fuzzy Logic Systems: Introduction and New Directions, Prentice Hall, New Jersey.

Mendel, J. (2007). Advances In Type-2 Fuzzy Sets and Systems, *Information Sciences*, 177,1, pp. 84-110.

Torres, P. & Saez, D. (2008). Type-2 fuzzy logic identification applied to the modelling of a robot hand, *Proceedings of the 2008 IEEE International Conference on Fuzzy Systems, FUZZ-IEEE 2008.*

Wang, L. X. (1997). *A Course in Fuzzy Systems and Control*, Prentice Hall, New Jersey.

Wu, D. & Mendel, J. (2009). Enhanced Karnik-Mendel Algorithms, *IEEE Transactions on Fuzzy Systems*, 17, pp. 923-934.

Wu, D. & Tan, W. (2004). A type-2 fuzzy logic controller for the liquid-level process, Proceedings of the 2004 IEEE International Conference on Fuzzy Systems, FUZZ-IEEE 2004.

19

LQR Control Methods for Trajectory Execution in Omnidirectional Mobile Robots

Luis F. Lupián and Josué R. Rabadán-Martin
Mobile Robotics & Automated Systems Lab, Universidad La Salle
Mexico

1. Introduction

Omnidirectional mobile robots present the advantage of being able to move in any direction without having to rotate around the vertical axis first. While simple straight-line paths are relatively easy to achieve on this kind of robots, in many highly dynamic applications straight-line paths might just not be a feasible solution. This may be the case because of two main reasons: (1) there may be static and moving obstacles between the initial and desired final position of the robot, and (2) the dynamic effects of the inertia of the robot may force it to execute a curved path. This chapter will address these two situations and present a segment-wise optimal solution for the path execution problem which is based on a Linear-Quadratic Regulator.

It must be emphasized that, rather than attempting to perform an exact path tracking, the approach presented here deals with the problem of visiting a sequence of target circular regions without specifying the path that will connect them. The freedom given to the connecting paths brings the opportunity for optimization. In fact, the path that the robot will take from one circular region to the next will emerge as the solution of an optimal control problem, hence the term *segment-wise optimal* solution.

Omnidirectional wheels have the property of sliding laterally with almost zero force while providing full traction in the rolling direction. This effect is achieved by adding a set of smaller wheels around the periphery of the main wheel, as depicted in Fig. 1.

Fig. 1. Omnidirectional wheel

By using several omnidirectional wheels distributed around the periphery of a cylindrical robot one can achieve the effect of driving the robot in a direction dependent on a vector sum of forces. This idea is illustrated in Fig. 2. The forces applied at the wheels by the motors give translational and rotational movement to the mobile robot.

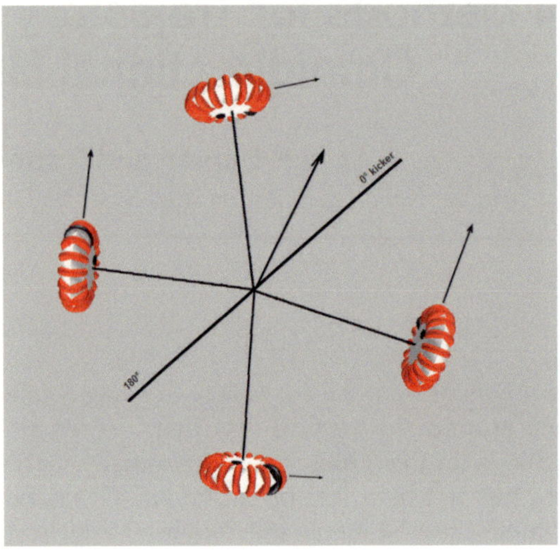

Fig. 2. Omnidirectional drive

Omnidirectional movement on a mobile robot has many applications, and thus has received attention from the scientific community over several years. Borenstein and Evans developed a control strategy that allows trajectory control on a mobile robot that uses conventional non-omnidirectional wheels and thus requires the drive wheels to rotate with respect to the vertical axis (Borenstein & Evans, 1997). With regard to omnidirectional mobile robots based on omnidirectional wheels there are several works that deserve attention. Ashmore and Barnes presented a detailed analysis of the kinematics of this kind of mobile robots and show that under certain circumstances a curved path may be faster than the straight line between two points (Ashmore & Barnes, 2002). Balkcom et al. characterized time-optimal trajectories for omnidirectional mobile robots (Balkcom et al., 2006a;b). Kalmár-Nagy et al. developed a control method to generate minimum-time trajectories for omnidirectional mobile robots (Kalmár-Nagy et al., 2004; 2002), then Purwin and D'Andrea presented the results of applying this method to a RoboCup F180 omnidirectional mobile robot (Purwin & D'Andrea, 2005).

Section 2 deals with the first necessary step towards applying LQR control for trajectory execution, which is formulating a state-space model for the dynamics of the omnidirectional mobile robot (Lupián & Rabadán-Martin, 2009).

This state-space model is non linear with respect to the control due to the fact that the robot rotates around the vertical axis and the pan angle is one of the state variables. In Sec. 3 of this chapter we show how to overcome this problem to successfully apply a Linear Quadratic Regulator for the case of three, four and five-wheeled omnidirectional mobile robots. In Sec. 4 we present a method to generate a segment-wise optimal path by solving an LQR control problem for each segment between the initial state through a sequence of target regions that ends at the desired final state of the omnidirectional mobile robot. Finally, Sec. 5 presents the results of several simulation experiments that apply the methods described in this chapter.

2. State-space dynamic model

The analysis presented pertains to a specific class of omnidirectional mobile robots that are of cylindrical shape with n omnidirectional wheels distributed around the periphery of the body of the robot, with the axes of the wheels intersecting at the geometrical vertical axis of the robot. Figure 3 shows an instance of this class of omnidirectional mobile robots for the case of $n = 5$ omnidirectional wheels. It is not necessary to have a uniform distribution of the wheels around the periphery of the robot. That is, the angle that separates the axis of one wheel to the next does not need to be the same. However, there is an obvious restriction that must be met in order to maintain stability, which is that the projection of the center of mass of the robot onto the ground must be contained within the convex hull of the set of contact points of the n wheels with the ground. For simplicity of the analysis, the mass of the robot is assumed to be distributed uniformly, so the center of mass is contained within the geometrical vertical axis of the robot.

Fig. 3. Omnidirectional robot

In the literature, there are several proposals for the dynamic model of an omnidirectional mobile robot. One of the main problems with these models is that they do not provide a complete state-space representation, so it is not possible to perform state-space control by using one of them. Most of these proposals are based on the force coupling matrix (Gloye & Rojas, 2006), which provides a direct relationship between the torques applied by the driving motors and the accelerations in the x, y and angular directions.
In compact form, these equations may be written as follows

$$
\begin{aligned}
a &= \frac{1}{M}(F_1 + F_2 + \cdots + F_n) \\
\dot{\omega} &= \frac{R}{I}(f_1 + f_2 + \cdots + f_n)
\end{aligned}
\tag{1}
$$

where a is the acceleration of the robot with respect to the inertial reference, $\dot{\omega}$ is the angular acceleration, n is the number of omnidirectional wheels around the periphery of the robot, F_i is the vector force applied by motor i, f_i is the signed scalar value of F_i (positive for counter clock-wise rotation), M is the mass of the robot, R is its radius and I is its moment of inertia.

Taking into account the driving motor distribution shown in Fig. 4, Eq. 1 may be re-written as

$$\begin{aligned} Ma_x &= -f_1 \sin\theta_1 - f_2 \sin\theta_2 - \cdots - f_n \sin\theta_n \\ Ma_y &= f_1 \cos\theta_1 + f_2 \cos\theta_2 + \cdots + f_n \cos\theta_n \\ MR\dot{\omega} &= \frac{1}{\alpha}(f_1 + f_2 + \cdots + f_n) \end{aligned} \tag{2}$$

where x and y are the axes of the inertial frame of reference, a_x is the x component of the acceleration, a_y is the y component of the acceleration, α is such that $I = \alpha MR^2$, and θ_i is the angular position of driving motor i with respect to the robot's referential frame as shown in Fig. 4. Equation 2 is expressed in matrix form as

$$\begin{bmatrix} a_x \\ a_y \\ R\dot{\omega} \end{bmatrix} = \frac{1}{M} \begin{bmatrix} -\sin\theta_1 & -\sin\theta_2 & \cdots & -\sin\theta_n \\ \cos\theta_1 & \cos\theta_2 & \cdots & \cos\theta_n \\ \frac{MR}{I} & \frac{MR}{I} & \cdots & \frac{MR}{I} \end{bmatrix} \begin{bmatrix} f_1 \\ f_2 \\ \vdots \\ f_n \end{bmatrix} \tag{3}$$

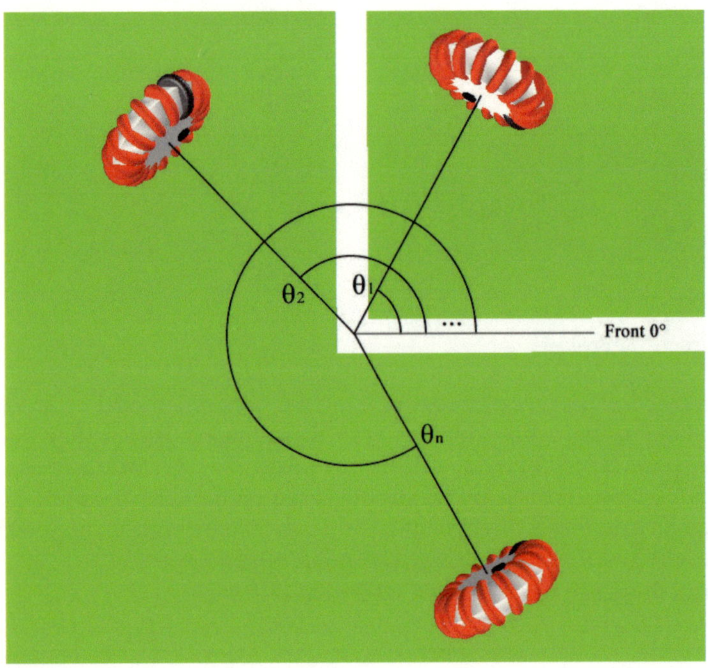

Fig. 4. Driving motor distribution

Equation 3 gives a general idea about the dynamic behavior of the mobile robot. However, it can not be considered a complete state-space model since it does not deal with the inertial effects of the mass of the robot, so we define a state vector that is complete enough to describe the dynamic behavior of the mobile robot (Lupián & Rabadán-Martin, 2009). We define the state vector z as

$$z = \begin{bmatrix} x & y & \beta & \dot{x} & \dot{y} & \dot{\beta} & \mu_1 & \cdots & \mu_n \end{bmatrix}^{\mathrm{T}} \tag{4}$$

where (x, y) is the position of the robot with respect to the inertial reference, $(\dot{x}, \dot{y})$ is the vector velocity of the robot with respect to the field plane, β is the angular position and $\dot{\beta}$ is

the angular velocity. Those are the six state variables we are interested in controlling, so it would suffice to use them to describe the state of the robot.
However, for simulation purposes we introduce variables μ_i, which correspond to the angular positions of each of the omnidirectional wheels of the robot. The decision to include these state variables was initially motivated by the fact that the dynamic response of the robot would be simulated on a virtual environment developed in OpenGL, and the motion of the wheels would make this simulation more realistic. On the physical robot, however, these state-variables will have a more important role because, unlike the other six state variables, the angular positions of the wheels can be measured directly through motor encoders, so one can implement a state observer to estimate the remaining variables.
In order to express the state equations in a compact and clear form, the state vector will be partitioned as follows:

$$\begin{aligned} z_1 &= \begin{bmatrix} x & y & \beta & \dot{x} & \dot{y} & \dot{\beta} \end{bmatrix}^{\mathrm{T}} \\ z_{11} &= \begin{bmatrix} x & y & \beta \end{bmatrix}^{\mathrm{T}} \\ z_{12} &= \begin{bmatrix} \dot{x} & \dot{y} & \dot{\beta} \end{bmatrix}^{\mathrm{T}} \\ z_2 &= \begin{bmatrix} \mu_1 & \dots & \mu_n \end{bmatrix}^{\mathrm{T}} \end{aligned} \tag{5}$$

In terms of z_1 y z_2 the state vector z can be expressed as

$$z = \begin{bmatrix} z_1 \\ z_2 \end{bmatrix} \tag{6}$$

and the state-space model becomes

$$\begin{bmatrix} \dot{z}_1 \\ \dot{z}_2 \end{bmatrix} = \begin{bmatrix} A_{11} & A_{12} \\ A_{21} & A_{22} \end{bmatrix} \begin{bmatrix} z_1 \\ z_2 \end{bmatrix} + \begin{bmatrix} B_1 \\ B_2 \end{bmatrix} u \tag{7}$$

where u is the control vector composed of the scalar forces f_i divided by the mass of the robot

$$u = \frac{1}{M} \begin{bmatrix} f_1 \\ f_2 \\ \vdots \\ f_n \end{bmatrix}, \tag{8}$$

and the matrices are defined as follows. Matrix A_{11} simply expresses the relationships among the first six state variables, and is given by

$$A_{11} = \begin{bmatrix} 0_{3\times3} & I_{3\times3} \\ 0_{3\times3} & 0_{3\times3} \end{bmatrix} \tag{9}$$

Matrix B_1 is obtained from Eq. 3 and becomes

$$B_1 = \begin{bmatrix} 0 & 0 & \cdots & 0 \\ 0 & 0 & \cdots & 0 \\ 0 & 0 & \cdots & 0 \\ -\sin\theta_1 & -\sin\theta_2 & \cdots & -\sin\theta_n \\ \cos\theta_1 & \cos\theta_2 & \cdots & \cos\theta_n \\ \frac{MR}{I} & \frac{MR}{I} & \cdots & \frac{MR}{I} \end{bmatrix} \tag{10}$$

Matrix B_1 expresses the correct influence of each force over the respective acceleration only for the case in which the angular position β of the robot is zero. However, in order to take into account the fact that the robot will rotate as time goes by this matrix should also depend on β as follows

$$B_1(\beta) = \begin{bmatrix} 0 & 0 & \cdots & 0 \\ 0 & 0 & \cdots & 0 \\ 0 & 0 & \cdots & 0 \\ -\sin(\theta_1+\beta) & -\sin(\theta_2+\beta) & \cdots & -\sin(\theta_n+\beta) \\ \cos(\theta_1+\beta) & \cos(\theta_2+\beta) & \cdots & \cos(\theta_n+\beta) \\ \frac{MR}{I} & \frac{MR}{I} & \cdots & \frac{MR}{I} \end{bmatrix} \tag{11}$$

$$B_1(\beta) = \begin{bmatrix} 0_{3\times n} \\ B_{12}(\beta) \end{bmatrix}$$

A_{12} and A_{22} are zero matrices of size $6 \times n$ and $n \times n$ respectively. Matrix A_{21} expresses the angular motion of the wheels with respect to the motion of the mobile robot and, like B_1, it is dependent on the angular position β of the robot:

$$A_{21}(\beta) = \frac{1}{r}\begin{bmatrix} 0_{n\times 3} & A_{212}(\beta) \end{bmatrix}$$
$$A_{212}(\beta) = \frac{1}{r}\begin{bmatrix} \sin(\theta_1+\beta) & -\cos(\theta_1+\beta) & -R \\ \sin(\theta_2+\beta) & -\cos(\theta_2+\beta) & -R \\ \vdots & \vdots & \vdots \\ \sin(\theta_n+\beta) & -\cos(\theta_n+\beta) & -R \end{bmatrix} \tag{12}$$

where r is the radius of the omnidirectional wheels.
Taking into account that both A_{21} and B_1 depend on β the state model in Eq. 7 may be reformulated as

$$\begin{bmatrix} \dot{z}_1 \\ \dot{z}_2 \end{bmatrix} = \begin{bmatrix} A_{11} & 0_{6\times n} \\ A_{21}(\beta) & 0_{n\times n} \end{bmatrix} \begin{bmatrix} z_1 \\ z_2 \end{bmatrix} + \begin{bmatrix} B_1(\beta) \\ 0_{n\times n} \end{bmatrix} u \tag{13}$$

3. Global linearization of the state-space dynamic model

Since the angular position β is one of the state variables this implies that the model in Eq. 13 is non-linear, and that represents an important difficulty for the purpose of controlling the robot. This is why it became necessary to find a way to linearize the model.
The solution to this problem required a shift in perspective in relation to the model. The only reason why the angle β has a non-linear effect on the dynamics of state variables in z_1 is because as the robot rotates also the driving forces, which are the control variables, rotate.
Let $F = F_1 + F_2 + \ldots + F_n$ be the resulting vector force applied by the omnidirectional wheels to the robot when it is at the angular position β and the control vector is u. Let $\breve{u}$ be the control vector that will produce the same resulting vector force F when the robot is at the angular position $\beta = 0$. This idea is explained by Fig. 5.
The control vectors u and $\breve{u}$ are then related by

$$\begin{aligned} \begin{bmatrix} \sin(\theta_1) & \cdots & \sin(\theta_n) \end{bmatrix} \breve{u} &= \begin{bmatrix} \sin(\theta_1+\beta) & \cdots & \sin(\theta_n+\beta) \end{bmatrix} u \\ \begin{bmatrix} \cos(\theta_1) & \cdots & \cos(\theta_n) \end{bmatrix} \breve{u} &= \begin{bmatrix} \cos(\theta_1+\beta) & \cdots & \cos(\theta_n+\beta) \end{bmatrix} u \end{aligned} \tag{14}$$

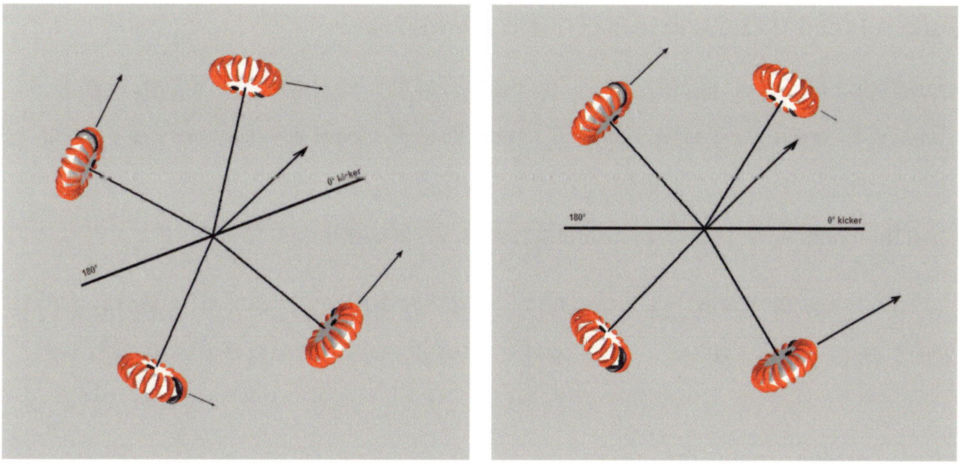

Fig. 5. Change of variable in control vector

Moreover, in order to ensure that the angular acceleration will remain unchanged it is necessary that the resulting scalar force $f = f_1 + f_2 + \ldots + f_n$ is the same in both cases, which translates to

$$\sum_{i=1}^{n} \breve{u}_i = \sum_{i=1}^{n} u_i \tag{15}$$

What we need from Eqs. 14 and 15 is a one-to-one mapping that can take us from u to $\breve{u}$ back and forth. Since u is n-dimensional and we have three equations the system is under-determined whenever $n > 3$. Although the pseudo-inverse would provide a linear transformation from one domain to the other, this transformation would be rank-deficient and thus would not be one-to-one. Our solution requires then to add $n - 3$ *complementary equations*. In order to avoid unnecessary numerical problems, these equations should be chosen so that the linear transformation is well-conditioned for any value of the angle β.

3.1 Transformation matrix for $n = 3$

The simplest case comes when the number of omnidirectional wheels of the robot is $n = 3$, see Fig. 6. For this particular case the number of equations provided by Eqs. 14 and 15 is equal to the number of control variables, so there is no need for additional equations to complete the transformation.

Fig. 6. Omnidirectional drive for $n = 3$

Equations 14 and 15 can be written in matrix form as follows

$$\begin{bmatrix} \sin(\theta_1) & \sin(\theta_2) & \sin(\theta_3) \\ \cos(\theta_1) & \cos(\theta_2) & \cos(\theta_3) \\ 1 & 1 & 1 \end{bmatrix} \check{u} = \begin{bmatrix} \sin(\theta_1+\beta) & \sin(\theta_2+\beta) & \sin(\theta_3+\beta) \\ \cos(\theta_1+\beta) & \cos(\theta_2+\beta) & \cos(\theta_3+\beta) \\ 1 & 1 & 1 \end{bmatrix} u \quad (16)$$

We can then define the transformation matrix $\Omega_3(\beta)$ according to

$$\Omega_3(\beta) = \begin{bmatrix} \sin(\theta_1) & \sin(\theta_2) & \sin(\theta_3) \\ \cos(\theta_1) & \cos(\theta_2) & \cos(\theta_3) \\ 1 & 1 & 1 \end{bmatrix}^{-1} \begin{bmatrix} \sin(\theta_1+\beta) & \sin(\theta_2+\beta) & \sin(\theta_3+\beta) \\ \cos(\theta_1+\beta) & \cos(\theta_2+\beta) & \cos(\theta_3+\beta) \\ 1 & 1 & 1 \end{bmatrix} \quad (17)$$

3.2 Transformation matrix for $n = 4$

For the particular case of $n = 4$ (Fig. 7), it is easy to see that Eq. 18 satisfies the requirement of completing a well-conditioned 4×4 transformation since it is orthogonal to two of the other equations and still sufficiently linearly independent from the third one.

$$\begin{bmatrix} 1 & -1 & 1 & -1 \end{bmatrix} \check{u} = \begin{bmatrix} 1 & -1 & 1 & -1 \end{bmatrix} u \quad (18)$$

Fig. 7. Omnidirectional drive for $n = 4$

Equations 14, 15 and 18 can be put in matrix form as follows

$$\begin{bmatrix} \sin(\theta_1) & \sin(\theta_2) & \sin(\theta_3) & \sin(\theta_4) \\ \cos(\theta_1) & \cos(\theta_2) & \cos(\theta_3) & \cos(\theta_4) \\ 1 & 1 & 1 & 1 \\ 1 & -1 & 1 & -1 \end{bmatrix} \check{u} =$$

$$\begin{bmatrix} \sin(\theta_1+\beta) & \sin(\theta_2+\beta) & \sin(\theta_3+\beta) & \sin(\theta_4+\beta) \\ \cos(\theta_1+\beta) & \cos(\theta_2+\beta) & \cos(\theta_3+\beta) & \cos(\theta_4+\beta) \\ 1 & 1 & 1 & 1 \\ 1 & -1 & 1 & -1 \end{bmatrix} u \quad (19)$$

We can then define the transformation matrix $\Omega_4(\beta)$ according to

$$\Omega_4(\beta) = \begin{bmatrix} \sin(\theta_1) & \sin(\theta_2) & \sin(\theta_3) & \sin(\theta_4) \\ \cos(\theta_1) & \cos(\theta_2) & \cos(\theta_3) & \cos(\theta_4) \\ 1 & 1 & 1 & 1 \\ 1 & -1 & 1 & -1 \end{bmatrix}^{-1} \begin{bmatrix} \sin(\theta_1+\beta) & \sin(\theta_2+\beta) & \sin(\theta_3+\beta) & \sin(\theta_4+\beta) \\ \cos(\theta_1+\beta) & \cos(\theta_2+\beta) & \cos(\theta_3+\beta) & \cos(\theta_4+\beta) \\ 1 & 1 & 1 & 1 \\ 1 & -1 & 1 & -1 \end{bmatrix} \quad (20)$$

3.3 Transformation matrix for $n = 5$

For the particular case of $n = 5$, shown in Fig. 8, an adequate choice for the complementary equations is shown in Eqs. 21.

$$\begin{aligned} \begin{bmatrix} 1 & -1 & 0 & 1 & -1 \end{bmatrix} \check{u} &= \begin{bmatrix} 1 & -1 & 0 & 1 & -1 \end{bmatrix} u \\ \begin{bmatrix} 1 & -1 & 0 & -1 & 1 \end{bmatrix} \check{u} &= \begin{bmatrix} 1 & -1 & 0 & -1 & 1 \end{bmatrix} u \end{aligned} \quad (21)$$

Fig. 8. Omnidirectional drive for $n = 5$

Equations 14, 15 and 21 can be put in matrix form as follows

$$\begin{bmatrix} \sin(\theta_1) & \sin(\theta_2) & \sin(\theta_3) & \sin(\theta_4) & \sin(\theta_5) \\ \cos(\theta_1) & \cos(\theta_2) & \cos(\theta_3) & \cos(\theta_4) & \cos(\theta_5) \\ 1 & 1 & 1 & 1 & 1 \\ 1 & -1 & 0 & 1 & -1 \\ 1 & -1 & 0 & -1 & 1 \end{bmatrix} \check{u} = \begin{bmatrix} \sin(\theta_1+\beta) & \sin(\theta_2+\beta) & \sin(\theta_3+\beta) & \sin(\theta_4+\beta) & \sin(\theta_5+\beta) \\ \cos(\theta_1+\beta) & \cos(\theta_2+\beta) & \cos(\theta_3+\beta) & \cos(\theta_4+\beta) & \cos(\theta_5+\beta) \\ 1 & 1 & 1 & 1 & 1 \\ 1 & -1 & 0 & 1 & -1 \\ 1 & -1 & 0 & -1 & 1 \end{bmatrix} u \quad (22)$$

We can then define the transformation matrix $\Omega_5(\beta)$ according to

$$\Omega_5(\beta) = \begin{bmatrix} \sin(\theta_1) & \sin(\theta_2) & \sin(\theta_3) & \sin(\theta_4) & \sin(\theta_5) \\ \cos(\theta_1) & \cos(\theta_2) & \cos(\theta_3) & \cos(\theta_4) & \cos(\theta_5) \\ 1 & 1 & 1 & 1 & 1 \\ 1 & -1 & 0 & 1 & -1 \\ 1 & -1 & 0 & -1 & 1 \end{bmatrix}^{-1} \begin{bmatrix} \sin(\theta_1+\beta) & \sin(\theta_2+\beta) & \sin(\theta_3+\beta) & \sin(\theta_4+\beta) & \sin(\theta_5+\beta) \\ \cos(\theta_1+\beta) & \cos(\theta_2+\beta) & \cos(\theta_3+\beta) & \cos(\theta_4+\beta) & \cos(\theta_5+\beta) \\ 1 & 1 & 1 & 1 & 1 \\ 1 & -1 & 0 & 1 & -1 \\ 1 & -1 & 0 & -1 & 1 \end{bmatrix} \quad (23)$$

Further generalization is possible by taking into account that the set of $n-3$ complementary equations must be chosen such that the transformation matrix is full rank and of minimum condition number.

3.4 Global linearization using the transformation matrix $\Omega_n(\beta)$

The change of variable from u to $\check{u}$ for any number of wheels n can now be expressed as

$$\begin{aligned} \check{u} &= \Omega_n(\beta)u \\ u &= \Omega_n(\beta)^{-1}\check{u} \end{aligned} \quad (24)$$

Matrix $\Omega_n(\beta)$ has the property of canceling the non-linear effect of β on matrix $B_1(\beta)$ since $B_1(0) = B_1(\beta)\Omega_n(\beta)^{-1}$ hence the model can be linearized at least with respect to the state variables in z_1, which are in fact the only variables we need to control. It is important to note that this is not a local linearization but a global one, so it is equally accurate in all of the control space.

By applying this change of variable to the state-space model of Eq. 13 we obtain the following model

$$\begin{aligned} \begin{bmatrix} \dot{z}_1 \\ \dot{z}_2 \end{bmatrix} &= \begin{bmatrix} A_{11} & 0_{6\times n} \\ A_{21}(\beta) & 0_{n\times n} \end{bmatrix} \begin{bmatrix} z_1 \\ z_2 \end{bmatrix} + \begin{bmatrix} B_1(\beta) \\ 0_{n\times n} \end{bmatrix} \Omega_n(\beta)^{-1}\check{u} \\ \begin{bmatrix} \dot{z}_1 \\ \dot{z}_2 \end{bmatrix} &= \begin{bmatrix} A_{11} & 0_{6\times n} \\ A_{21}(\beta) & 0_{n\times n} \end{bmatrix} \begin{bmatrix} z_1 \\ z_2 \end{bmatrix} + \begin{bmatrix} B_1(\beta)\Omega_n(\beta)^{-1} \\ 0_{n\times n}\Omega_n(\beta)^{-1} \end{bmatrix} \check{u} \\ \begin{bmatrix} \dot{z}_1 \\ \dot{z}_2 \end{bmatrix} &= \begin{bmatrix} A_{11} & 0_{6\times n} \\ A_{21}(\beta) & 0_{n\times n} \end{bmatrix} \begin{bmatrix} z_1 \\ z_2 \end{bmatrix} + \begin{bmatrix} B_1(0) \\ 0_{n\times n} \end{bmatrix} \check{u} \end{aligned} \quad (25)$$

If we separate this model according to the state vector partition proposed in Eq. 5 it is easy to see that the transformed model is linear with respect to state variables in z_1 and control variables $\check{u}$

$$\begin{aligned} \dot{z}_1 &= A_{11}z_1 + B_1(0)\check{u} \\ \dot{z}_2 &= A_{21}(\beta)z_1 \end{aligned} \quad (26)$$

From the state-space model in Eq. 26 it is possible to formulate a wide variety of linear state-space controllers for state variables in z_1.

4. Segment-wise optimal trajectory control

Our solution to the trajectory execution control problem requires the specification of the desired trajectory as a sequence of target regions as shown in Fig. 9. Each of the target regions is depicted as a red circle. This figure shows the robot in its initial state near the top left corner. The robot will have to visit each of the target regions in sequence until it reaches the final target which contains the red ball near the bottom right corner. As soon as the center of the robot enters the current target region it will start moving towards the next one. In this way, the trajectory is segmented by the target regions. The desired precision of the trajectory is determined by the size of the next target region. A small target requires higher precision than a larger one. Each target region makes additional specifications for the desired final state of the robot as it reaches the target. These specifications include the desired final scalar speed and the desired final heading direction. This last specification is useful for highly dynamic applications, such as playing soccer, since it allows the robot to rotate as it moves along the trajectory in order to align its manipulating device with the location of the target (a ball in the case of soccer).

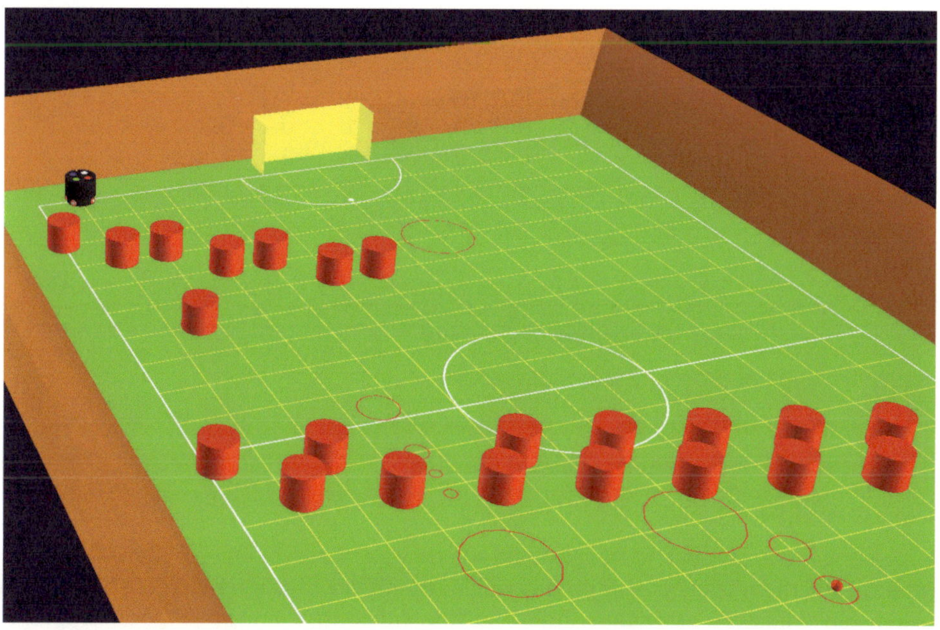

Fig. 9. Sequence of target regions

Each of the segments of the trajectory can be treated as a different control problem in which the final state of the previous segment is the initial state of the current one. Each segment has its own desired final state and in this way the mobile robot is forced to loosely follow the specified trajectory.

Once the trajectory has been segmented by this sequence of target regions the problem becomes how to force the robot to move from one target to the next. Our proposal is to solve an infinite-horizon LQR control problem for each of the segments. Although this approach may provide a sub-optimal solution rather than the optimal solution, that would be obtained from the corresponding finite-horizon LQR problem formulation, it has the advantage of requiring the solution of an algebraic Riccati equation rather than the more computationally demanding

differential Riccati equation. Computational efficiency is of course an important concern in this application since the solution will have to be obtained in real time in a high speed mobile robot environment.
The infinite-horizon LQR control problem consists on finding the state-feedback matrix K such that $\breve{u} = -Kz_1$ minimizes the performance index J given by

$$J = \int_0^{\infty} (z_1{}^{\mathrm{T}} Q z_1 + \breve{u}^{\mathrm{T}} R \breve{u}) dt \tag{27}$$

taking into account that z_1 and $\breve{u}$ are restricted by the dynamics of the mobile robot given by Eq. 26.
The performance index J specifies the total cost of the control strategy, which depends on an integral quadratic measure of the state z_1 and control $\breve{u}$. Q and R represent positive definite matrices that give a weighted measure of the cost of each state variable and control variable, respectively.
For simplicity, in our solution Q and R are defined to be diagonal matrices

$$Q = \begin{bmatrix} w_{xy} & 0 & 0 & 0 & 0 & 0 \\ 0 & w_{xy} & 0 & 0 & 0 & 0 \\ 0 & 0 & w_{\beta} & 0 & 0 & 0 \\ 0 & 0 & 0 & w_v & 0 & 0 \\ 0 & 0 & 0 & 0 & w_v & 0 \\ 0 & 0 & 0 & 0 & 0 & w_{\dot{\beta}} \end{bmatrix} \tag{28}$$
$$R = w_m I_{n \times n}$$

where

- w_{xy} : cost weight of the XY position of the robot
- w_v : cost weight of the XY speed of the robot
- w_{β} : cost weight of the angular position of the robot
- $w_{\dot{\beta}}$: cost weight of the angular speed of the robot
- w_m : cost weight of the torque of the driving motors

This set of cost weights has to be specified for each of the segments of the trajectory. The weights specify the relative cost of each variable, and by an appropriate choice of their values one can easily adjust the optimality index for different control strategies. For example, if w_m is very large in comparison to the other weights then our strategy will be to save energy, if w_{xy} is large in comparison to the rest then the strategy dictates that the robot should reach the target region as soon as possible without regard to a specific target final speed, angle β or energy consumption. It would make sense to keep w_{β} very low during most of the trajectory except for those segments in which the robot is approaching the final target region or, in the case of soccer applications, orienting itself to receive a pass from a teammate robot.
The LQR approach is much more natural from the point of view of the designer as compared to other linear state-feedback controller design techniques such as *pole placement*. Although pole placement may help in specifying a desired time-domain transient behavior, the resulting feedback control may turn out to be higher than what the actuators can actually achieve. In LQR, on the other hand, the solution strikes a balance between the transient behavior of the state variables and the energy consumed by the actuators. In the case of LQR the resulting

poles are implicitly determined by the choice of matrices Q and R, rather than being explicitly specified.
Our segment-wise optimal solution to the trajectory execution control problem allows then to formulate control strategies that can easily adapt to the circumstances of the application. So, for example, the robot may be forced to move quickly in segments of the trajectory that do not require precision, and it may be left free to rotate in those segments in which its orientation is not of concern. In this way, its limited energy is consumed only for those objectives that are important at the moment.

5. Experimental results

A benchmark sequence of target regions was used for the purpose of testing our proposed solution. This sequence is shown in Fig. 9, and has a total of nine segments. Several obstacles were placed on the testing field to give more realism to the 3D OpenGL simulation of the resulting trajectory. A four wheel omnidirectional robot was used for this simulation with wheels distributed around its periphery according to Table 1. Wheels 1 and 4 are the frontal wheels. They are intentionally separated wider apart than the rest in order to allow for a manipulating device to be installed between them.

Table 1. Omnidirectional wheels distribution

θ_1	θ_2	θ_3	θ_4
$60°$	$135°$	$-135°$	$-60°$

The specific optimality-index parameters used for each of the segments of the trajectory are shown in Table 2. Throughout the whole trajectory, parameter w_m is kept at a moderate value in order to conserve energy without degrading the time-domain performance too much. Parameter w_β is kept at a low value for all segments except for the final two. This saves the energy that would otherwise be required to maintain the orientation of the robot towards a specific angle in the first seven segments where that would not be of interest, but then forces the robot to align itself with the location of the final target region in order to capture the object of interest located at the center of such region. Parameter $w_{\dot{\beta}}$ is kept at a low value throughout the whole trajectory. Parameter w_{xy} is set to a high value in segments 2, 3, 4, 5 and 9. The first four of these segments correspond to the part of the trajectory that takes the robot through the narrow opening in the obstacle barrier, where high precision is required in order to avoid a collision. High precision is also desirable in the last segment to ensure a successful capture of the object of interest. Finally, parameter w_v is given a moderate value throughout the first seven segments of the trajectory and then is lowered in the last two segments in order to allow for the energy to be used for the more important objective of orienting the robot towards the final target region. In this analysis what matters is the relative value of the weights. However, in order to give a more precise idea, in this analysis *moderate* means 1, *low* means 0.1 and *high* means 5.
The resulting trajectory is shown in Fig. 10. The continuous black line shows the path of the robot, while the smaller black straight lines show the orientation of the robot at each point. This gives an idea of how the robot rotates as it moves along the trajectory. One surprising result of this simulation experiment is that the robot started the second segment almost in reverse direction and slowly rotated to approach a forward heading motion. The final section of the trajectory shows how the object of interest gets captured by the robot. Figure 11 shows

Par.	1	2	3	4	5	6	7	8	9
w_{xy}	1	5	5	5	5	1	1	1	5
w_v	1	1	1	1	1	1	1	0.1	0.1
w_β	0.1	0.1	0.1	0.1	0.1	0.1	1	5	5
$w_{\dot{\beta}}$	1	0.1	0.1	0.1	0.1	1	0.1	0.1	0.1
w_m	1	1	1	1	1	1	1	1	1

Table 2. Optimality index parameters for each segment

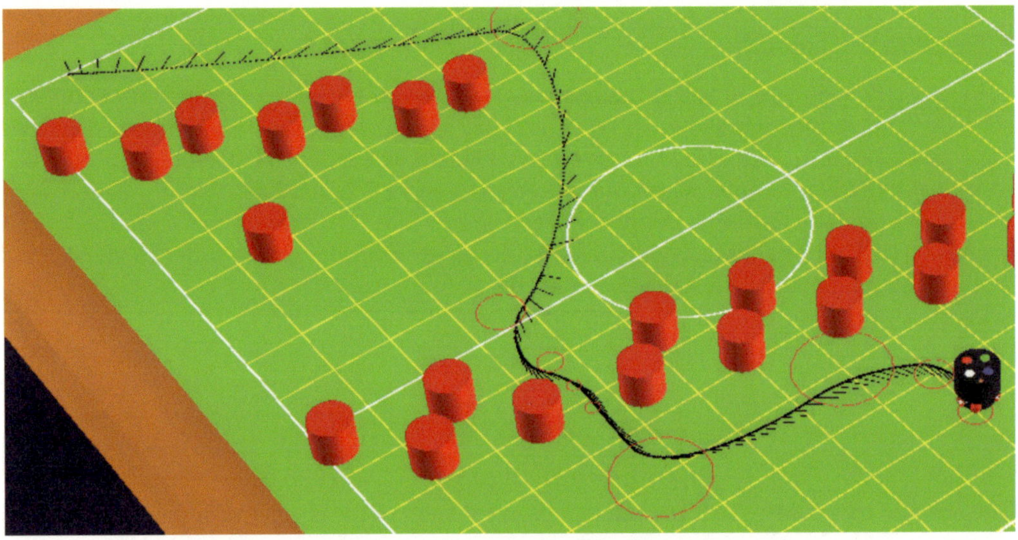

Fig. 10. Simulation of the segment-wise optimal trajectory execution

how the speeds and torques of each driving motor evolve against time. From these graphs we can interpret that there are only two required high energy transients. The first one comes at the initial time, when the robot must accelerate from a motionless state. The second high energy transient comes at the transition from the first to the second segment, where the robot must abruptly change direction while moving at a relatively high speed.

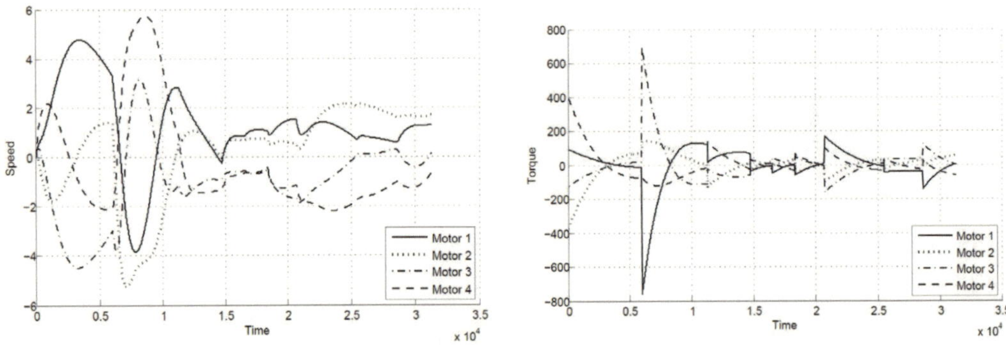

Fig. 11. Speed and torque vs. time for each wheel

A second experiment that we performed to test the segment-wise optimal solution to the trajectory execution problem is that of performing an 8-shaped trajectory, which was successfully executed as shown in Fig. 12. Although the resulting 8 is not perfectly smooth the advantage here is that the path is achieved without the need for a computationally intensive path tracking algorithm but rather only loosely specified by a small set of target regions.

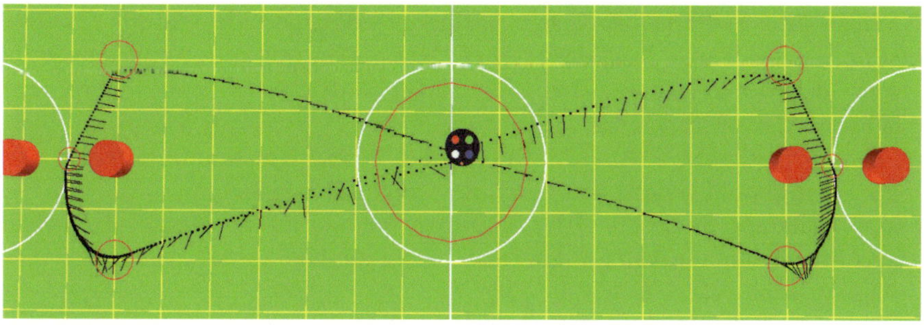

Fig. 12. Simulation of the segment-wise optimal 8-shaped trajectory

6. Conclusion

This work presents the formulation of a complete state-space model for the omnidirectional mobile robot. Although the model is nonlinear with respect to the control vector, it is shown how it can be globally linearized by a change of variables on the control vector. It has been shown how to apply this linearization for the case of $n = 3$, $n = 4$ and $n = 5$ omnidirectional wheels. The distinction among number of wheels is far from trivial since there is the need of introducing *complementary equations* that will make the proposed change of variable invertible.

The model used for analysis in this paper assumes an idealized omnidirectional robot with uniform density. Nonetheless, the same fundamental ideas that were tested under simulation in this work would still apply for a real robot.

There are still two major considerations to take into account before our proposed control method can be applied in the real world. First, our solution assumes availability of all state variables in z_1. However, in reality these state variables would have to be estimated from the available measurements using an approach similar to that presented in (Lupián & Avila, 2008). These measurements would come from the encoders of the driving motors and from the global vision system. The global vision system provides a measurement of variables x, y and β, but this measurement comes with a significant delay due to the computational overhead, so in actuality these variables would have to be predicted (Gloye et al., 2003). The second major consideration is that this solution will be implemented on a digital processor so, rather than a continuous-time model, the problem should be formulated in discrete-time.

The benchmark sequence of target regions used in the "Simulation results" section assumes the environment is static. In a real highly dynamic application the sequence of target regions will have to be updated as the trajectory execution progresses. This dynamic path-planning problem would be the task of a higher-level artificial intelligence algorithm which is left open for future research.

The first of the two simulation experiments can be better appreciated on the 3D OpenGL animation developed for this purpose. The reader can access a video of this animation online at (Lupián, 2009).

7. References

Ashmore, M. & Barnes, N. (2002). Omni-drive robot motion on curved paths: The fastest path between two points is not a straight-line, *AI '02: Proceedings of the 15th Australian Joint Conference on Artificial Intelligence*, Springer-Verlag, London, UK, pp. 225–236.

Balkcom, D. J., Kavathekar, P. A. & Mason, M. T. (2006a). The minimum-time trajectories for an omni-directional vehicle, *Seventh International Workshop on the Algorithmic Foundations of Robotics*, New York City.

Balkcom, D. J., Kavathekar, P. A. & Mason, M. T. (2006b). Time-optimal trajectories for an omni-directional vehicle, *Int. J. Rob. Res.* 25(10): 985–999.

Borenstein, J. & Evans, J. (1997). The omnimate mobile robot: Design, implementation, and experimental results, *International Conference on Robotics and Automation, IEEE*, Albuquerque, NM, pp. 3505–3510.

Gloye, A. & Rojas, R. (2006). *Künstliche Intelligenz*, Springer, chapter Holonomic Control of a Robot with an Omnidirectional Drive.

Gloye, A., Simon, M., Egorova, A., Wiesel, F., Tenchio, O., Schreiber, M., Behnke, S. & Rojas, R. (2003). Predicting away robot control latency, *Technical report B-08-03*, FU-Berlin.

Kalmár-Nagy, T., D'Andrea, R. & Ganguly, P. (2004). Near-optimal dynamic trajectory generation and control of an omnidirectional vehicle, *Robotics and Autonomous Systems* 46: 47–64.

Kalmár-Nagy, T., Ganguly, P. & D'Andrea, R. (2002). Real-time trajectory generation for omnidirectional vehicles, *American Control Conference*, Anchorage, AK, pp. 286–291.

Lupián, L. F. (2009). Simulación de robot F180 en AIR-GL, World Wide Web electronic publication.
URL: *http://www.youtube.com/watch?v=-rBP3VwDPAs*

Lupián, L. F. & Avila, R. (2008). Stabilization of a wheeled inverted pendulum by a continuous-time infinite-horizon LQG optimal controller, *Robotic Symposium, IEEE Latin American*, Bahia, Brazil, pp. 65–70.

Lupián, L. F. & Rabadán-Martin, J. R. (2009). Segment-wise optimal trajectory execution control for four-wheeled omnidirectional mobile robots, *Robotics Symposium (LARS), 2009 6th Latin American*, Valparaíso, Chile, pp. 1 – 6.

Purwin, O. & D'Andrea, R. (2005). Trajectory generation for four wheeled omnidirectional vehicles, *American Control Conference* pp. 4979–4984.

20

Feedback Equivalence and Control of Mobile Robots Through a Scalable FPGA Architecture

G.P. Moustris[1], K.M. Deliparaschos[2] and S.G. Tzafestas[1]
[1]*National Technical University of Athens*
[2]*Cyprus University of Technology*
[1]*Greece*
[2]*Cyprus*

1. Introduction

The control of mobile robots is an intense research field, having produced a substantial volume of research literature in the past three decades. Mobile robots present a challenge in both theoretical control design as well as hardware implementation. From a theoretical point of view, mobile robots exert highly non-linear kinematics and dynamics, non-holonomy and complex operational environment. On the other hand, hardware designs strive for scalability, downsizing, computational power and low cost.

The main control problems in robot motion control can be classified in three general cases; stabilization to a point, trajectory tracking and path following. Point stabilization refers to the stabilization of the robot to a specific configuration in its state space (pose). For example, if the robot at $t = t_0$ is at $p_0 = (x_0, y_0, \theta_0)$, then find a suitable control law that steers it to a goal point p_g=(x_g, y_g, θ_g). Apparently p_0 must be an equilibrium point of the closed loop system, exerting asymptotic stability (although practical stability can also be sought for). In the path following (or path tracking) problem, the robot is presented with a reference path, either feasible or infeasible, and has to follow it by issuing the appropriate control commands. A path is defined as a geometric curve in the robot's application space. The trajectory tracking problem is similar, although there is a notable difference; the trajectory is a path with an associated timing law i.e. the robot has to be on specific points at specific time instances.

These three problems present challenges and difficulties exacerbated by the fact that the robot models are highly non-linear and non-holonomic (although robots that lift the non-holonomic rolling constraint do exist and are called *omni-directional robots*. However the most interesting mobile robots present this constraint). The non-holonomy means that there are constraints in the robot velocities e.g. the non-slipping condition, which forces the robot to move tangentially to its path or equivalently, the robot's heading is always collinear to its velocity vector (this can readily be attested by every driver who expects his car to move at the direction it is heading and not sideways i.e. slip). For a more motivating example of holonomy, consider a prototypical and pedagogical kimenatic model for motion analysis and control; the unicycle robot, described by the equations,

$$\Sigma : \begin{bmatrix} \dot{x} \\ \dot{y} \\ \dot{\theta} \end{bmatrix} = \begin{bmatrix} \cos\theta \\ \sin\theta \\ 0 \end{bmatrix} v + \begin{bmatrix} 0 \\ 0 \\ 1 \end{bmatrix} \omega \qquad (1)$$

This model is linear on the inputs and describes a linear combination of vector fields which evolve on a 3D configuration space $M = \mathbb{R}^3 \times S^1$ (a three dimensional manifold). The robot is controlled by two inputs v and ω, expressing the linear and angular velocities respectively. The generalized velocities live on the manifold's tangent space T_qM at each point q, thus the system's equations express the available directions of movement. The non-holonomic no-slipping constraint is expressed by the Pfaffian equation,

$$0 = \dot{x}\sin(\theta) - \dot{y}\cos(\theta) \tag{2}$$

which can be put into the general form,

$$G(q)\dot{q} = 0 \tag{3}$$

Here $q \in M$ is the state vector and $G(q) \in \mathbb{R}^{1\times3}$ is the constraint matrix (although in this case is just a vector). Each row vector (covector) of G lives in the manifold's cotangent space T_q^*M at q, which is the dual space of T_qM. Equation 3 describes restrictions on the available directions of movement. Apparently, since the velocities must satisfy (3), it is evident that they live in the null space of G. One can move from Eq.(3) to Eq.(1) by solving (3) with respect to the velocities $\dot{q}$. Since the system is underdetermined (note that G is a one by three "matrix"), two generalized velocities can vary freely, which are precisely the two inputs of (1). The non-holonomy of the system derives from the fact that Eq.(2) is not integrable i.e. does not express the total derivative of some function. By the Frobenius theorem, if Δ is the distribution spanned by the two vector fields of Σ, the system is holonomic if Δ is involutive under Lie bracketing, a condition that is not satisfied by (1).
Due to these challenges, the path following problem has been attacked by several researchers from many angles, ranging from classical control approaches (Altafini, 1999; Kamga & Rachid, 1997; Kanayama & Fahroo, 1997), to nonlinear control methodologies (Altafini, 2002; Egerstedt et al., 1998; Koh & Cho, 1994; Samson, 1995; Wit et al., 2004) to intelligent control strategies (Abdessemed et al., 2004; Antonelli et al., 2007; Baltes & Otte, 1999; Cao & Hall, 1998; Deliparaschos et al., 2007; El Hajjaji & Bentalba, 2003; Lee et al., 2003; Liu & Lewis, 1994; Maalouf et al., 2006; Moustris & Tzafestas, 2005; Rodriguez-Castano et al., 2000; Sanchez et al., 1997; Yang et al., 1998). Of course, boundaries often blend since various approaches are used simultaneously. Fuzzy logic path trackers have been used by several researchers (Abdessemed et al., 2004; Antonelli et al., 2007; Baltes & Otte, 1999; Cao & Hall, 1998; Deliparaschos et al., 2007; El Hajjaji & Bentalba, 2003; Jiangzhou et al., 1999; Lee et al., 2003; Liu & Lewis, 1994; Moustris & Tzafestas, 2011; 2005; Ollero et al., 1997; Raimondi & Ciancimino, 2008; Rodriguez-Castano et al., 2000; Sanchez et al., 1997) since fuzzy logic provides a more intuitive way for analysing and formulating the control actions, which bypasses most of the mathematical load needed to tackle such a highly nonlinear control problem. Furthermore, the fuzzy controller, which can be less complex in its implementation, is inherently robust to noise and parameter uncertainties.
The implementation of Fuzzy Logic Controllers (FLC) in software suffers from speed limitations due to the sequential program execution and the fact that standard processors do not directly support many fuzzy operations (i.e., minimum or maximum). In an effort to reduce the lack of fuzzy operations several modified architectures of standard processors supporting fuzzy computation exist (Costa et al., 1997; Fortuna et al., 2003; Salapura, 2000). Software solutions running on these devices speed up fuzzy computations by at least one order of magnitude over standard processors, but are still not fast enough for some real-time applications. Thus, a dedicated hardware implementation must be used (Hung, 1995).

Due to the increased number of calculations necessary for the path tracking control, a high performance processing system to efficiently handle the task is required. By using a System-on-a-Chip (SoC) realised on an FPGA device, we utilize the hardware/software re-configurability of the FPGA to satisfy the needs of fuzzy logic path tracking for autonomous robots for high-performance onboard processing and flexible hardware for different tasks.
FPGAs provide several advantages over single processor hardware, on the one hand, and Application Specific Integrated Circuits (ASIC) on the other. FPGA chips are field-upgradable and do not require the time and expense involved with ASIC redesign. Being reconfigurable, FPGA chips are able to keep up with future modifications that might be necessary. They offer a simpler design cycle, re-programmability, and have a faster time-to-market since no fabrication (layout, masks, or other manufacturing steps) time is required, when compared to ASICs.
The use of FPGAs in robotic applications is noted in (Kongmunvattana & Chongstivatana, 1998; Leong & Tsoi, 2005; Li et al., 2003; Reynolds et al., 2001). A review of the application of FPGA's in robotic systems is provided be Leong and Tsoi in (Leong & Tsoi, 2005). A notable case study is the use of FPGA's in the Mars Pathfinder, Mars Surveyor '98, and Mars Surveyor '01 Lander crafts, analysed in (Reynolds et al., 2001).

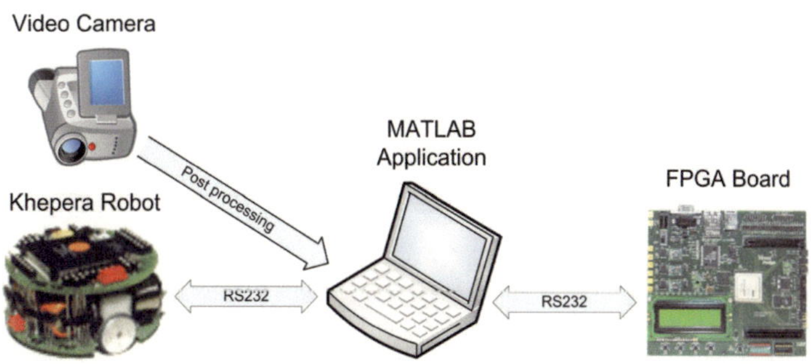

Fig. 1. Overview of the general system.

In this chapter we analyse a SoC implementation for the non-holonomic robot path tracking task using a fuzzy logic controller, along with a non-linear feedback-equivalence transformation which reduces path tracking to straight line tracking. The major components of the SoC are a parametrized Digital Fuzzy Logic Controller (DFLC) soft IP core Deliparaschos et al. (2006) Deliparaschos & Tzafestas (2006), implementing the fuzzy tracking algorithm, and Xilinx's Microblaze soft processor core as the top level flow controller. The system was tied to a differential drive robot and experiments were performed in order to asses the efficacy and performance of the overall control scheme. This was facilitated using an image analysis algorithm, presented in the following sections, which calculated the robot's position from a video stream captured using an overhead camera. The analysis was made off-line. The overall system setup can be seen in Fig 1.

2. Kinematics & odometry of the Khepera II robot

The mobile robot used in this work, is the Khepera II differential drive robot, described by the equations,

$$\begin{bmatrix} \dot{x} \\ \dot{y} \\ \dot{\theta} \end{bmatrix} = \begin{bmatrix} \frac{r}{2}\cos\theta \\ \frac{r}{2}\sin\theta \\ \frac{r}{L} \end{bmatrix} u_r + \begin{bmatrix} \frac{r}{2}\cos\theta \\ \frac{r}{2}\sin\theta \\ -\frac{r}{L} \end{bmatrix} u_l \tag{4}$$

Here, x, y are the coordinates of the middle point of the axis, L is the axis length (the distance between the two wheels), r is the wheel radius and u_l, u_r the angular wheel velocities. A diagram of the model is seen in Fig.(2). Equations 4 can be transformed to a model more akin to the unicycle by first noting that the linear velocity of a point on the wheel's circumference is $v = r\omega$ ($\omega_i \triangleq u_i$). It can be easily shown that the linear velocity of the wheel's center equals the velocity of its circumference. Thus, denoting the centres' velocities as v_r, v_l, then,

$$\begin{aligned} v_r &= r u_r \\ v_l &= r u_l \end{aligned} \tag{5}$$

and substituting them into Eq.4, the system takes the form,

$$\begin{bmatrix} \dot{x} \\ \dot{y} \\ \dot{\theta} \end{bmatrix} = \begin{bmatrix} \cos\theta/2 \\ \sin\theta/2 \\ 1/L \end{bmatrix} v_l + \begin{bmatrix} \cos\theta/2 \\ \sin\theta/2 \\ -1/L \end{bmatrix} v_r \tag{6}$$

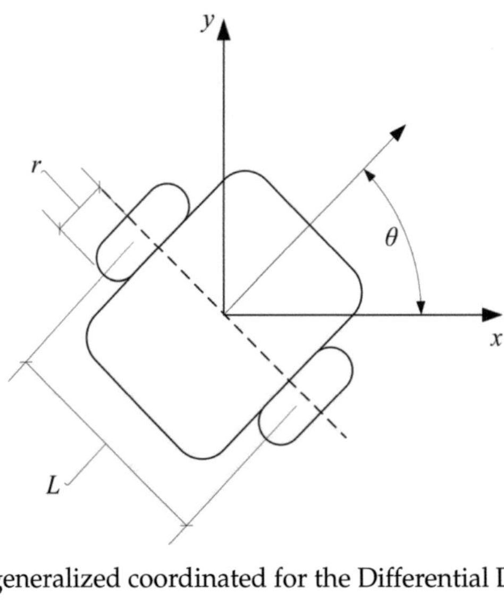

Fig. 2. Depiction of the generalized coordinated for the Differential Drive model

If we further apply a new input transformation,

$$\begin{aligned} u_s &= \frac{v_r + v_l}{2} \\ u_\theta &= \frac{v_r - v_l}{2} \end{aligned} \tag{7}$$

we get the familiar unicycle model, i.e.

$$\begin{bmatrix} \dot{x} \\ \dot{y} \\ \dot{\theta} \end{bmatrix} = \begin{bmatrix} \cos\theta \\ \sin\theta \\ 0 \end{bmatrix} u_s + \begin{bmatrix} 0 \\ 0 \\ 1 \end{bmatrix} u_\theta \tag{8}$$

For the physical interpretation of the inputs u_s, u_θ, consider that the robot performs a turn of radius R with respect to the axis' middle point (point *K*), centred at the point O (Fig 3). The point O is called the *Instantaneous Centre of Rotation* or ICR. If ω is the robot's angular velocity (actually the angular velocity of *K*), then its linear velocity is $v_s = \omega R$. It also holds that,

$$\begin{aligned} v_r &= \omega(R + L/2) \\ v_l &= \omega(R - L/2) \end{aligned} \tag{9}$$

By adding (9) together and solving for the linear velocity v_s, we get,

$$v_s = \frac{v_r + v_l}{2} \tag{10}$$

Subtracting (9) we come up with the robot's angular velocity ω,

$$v_\theta = \frac{v_r - v_l}{L} = \omega \tag{11}$$

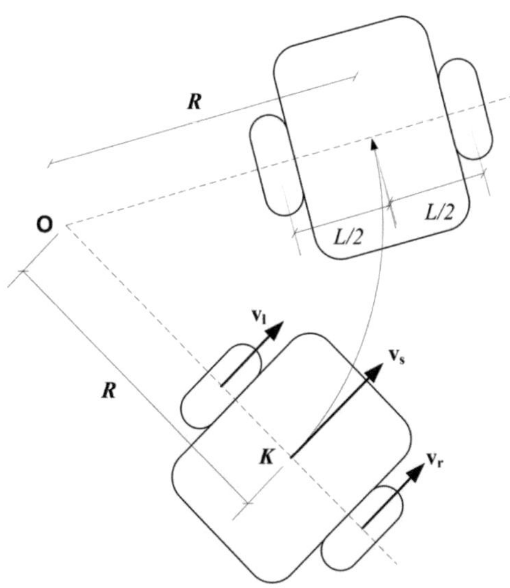

Fig. 3. Depiction of the velocities of the Differential Drive model

Observe that v_s, v_θ are actually u_s, u_θ, thus the new input variables in Eq.7 are actually the robot's linear and angular velocities. What we have proved so far is that the unicycle (or the Dubins Car, which is the unicycle with a constant speed) is related to the Differential Drive by an input transformation, hence they are equivalent. This means that the Differential Drive can *emulate* these models. Consequently, we can develop a controller for either system and apply it to the others by using this transformation (this is feasible if the input transformation is actually *bijective*. If it is not, then the controller can by ported to one direction i.e. from model A to model B, and not the other way around). As mentioned earlier, the Dubins Car is a simplified model where the velocity is considered constant. Furthermore, the path following problem involves the tracking of a purely geometric curve, where indeed the velocity is irrelevant. Hence, the model (8) can be transformed to a more suitable control form by using the curvature $\kappa \triangleq u_\kappa$. The curvature is related to the angular and linear velocities by the well known formula $\omega = \kappa v$, or, using our nomenclature,

$$u_\theta = u_\kappa u_s \tag{12}$$

Since the linear velocity $u_s = v$ is constant, by applying this transformation to the system (8), we get,

$$\begin{bmatrix} \dot{x} \\ \dot{y} \\ \dot{\theta} \end{bmatrix} = \begin{bmatrix} v\cos\theta \\ v\sin\theta \\ 0 \end{bmatrix} + \begin{bmatrix} 0 \\ 0 \\ v \end{bmatrix} u_\kappa \tag{13}$$

This model is of control-affine form, with a non-vanishing drift term, where the only input is the curvature u_κ. By controlling the curvature in this model, we expect to control the actual system i.e. the Khepera robot, which is a differential drive and has two inputs. Thus, starting with the curvature, in order to calculate the true input vector to the robot, we need a second equation. This is of course the velocity equation $u_s = v$, which is considered known. By combining (10), (11), (12), the wheel velocities are calculated as,

$$\begin{aligned} v_r &= v(1 + u_\kappa L/2) \\ v_l &= v(1 - u_\kappa L/2) \end{aligned} \tag{14}$$

Equation 14 produces the linear wheel velocities of the robot, given its linear velocity and curvature. Since the linear velocity is constant, the only input is the curvature which is output by the fuzzy controller implemented on the FPGA.

In order to calculate its position, MATLAB queries the robot about its wheel encoder readings every 12.5 msec. The robot returns the 32bit encoder position, expressed in pulse units, with each unit corresponding to 0.08 mm. Consequently, by multiplying the units with 0.08 we get the *total length* each wheel has travelled since the beginning of the experiment. Now let $S_R(t), S_L(t)$ be the travel length of the right and left wheels at time t, and $S_R(t-1), S_L(t-1)$ be the length at $t-1$. We assume that in the interval Δt the robot moves with a constant curvature, and thus traces an arc. This translates to constant wheel velocities (Eq. 14). Then, using (11) we have,

$$\omega = \frac{\Delta\theta}{\Delta t} = \frac{v_r - v_l}{L} = \frac{\Delta S_R - \Delta S_L}{\Delta t L} \Leftrightarrow \Delta\theta = \frac{\Delta S_R - \Delta S_L}{L} \tag{15}$$

If the robot's initial heading θ_0, with respect to the world frame, is known, then at time t it holds that,

$$\theta(t) = \theta_0 + \sum_{\tau=0}^{t} \Delta\theta(t) = \theta_0 + \sum_{\tau=0}^{t} \frac{\Delta S_R(\tau) - \Delta S_L(\tau)}{L} \tag{16}$$

Using 10, the length travelled by the point K in Δt is found to be,

$$v_s = \frac{\Delta S}{\Delta t} = \frac{v_r + v_l}{2} \Leftrightarrow \Delta S = \frac{\Delta S_R + \Delta S_L}{2} \tag{17}$$

To calculate the robot's x, y position, we must solve the kinematics for the interval Δt, setting the input constant. The solution can be easily shown to be,

$$\begin{aligned}\Delta x(t) &= 2\frac{\Delta S}{\Delta\theta}\sin(\frac{\Delta\theta}{2})\cos(\theta_{t-1}+\frac{\Delta\theta}{2})\\ \Delta y(t) &= 2\frac{\Delta S}{\Delta\theta}\sin(\frac{\Delta\theta}{2})\sin(\theta_{t-1}+\frac{\Delta\theta}{2})\end{aligned} \tag{18}$$

In the case that the robot moves in a straight line, hence $\Delta\theta = 0$, taking the limit of 18 gives the equations,

$$\begin{aligned}\Delta x(t) &= \Delta S\cos(\theta_{t-1}+\frac{\Delta\theta}{2})\\ \Delta y(t) &= \Delta S\sin(\theta_{t-1}+\frac{\Delta\theta}{2})\end{aligned} \tag{19}$$

To get the absolute coordinates of the robot at t, Eq.19 must be integrated, leading to the odometric equations,

$$\begin{aligned}x(t) &= x_0 + \sum_{\tau=0}^{t}\Delta x(\tau)\\ y(t) &= y_0 + \sum_{\tau=0}^{t}\Delta y(\tau)\end{aligned} \tag{20}$$

Using the previous formulas (Eq.16 and Eq.20) we have succeeded in reconstructing the robot's state vector, i.e. the states (x, y, θ). Note that *time* is absent from the odometry equations. This has been chosen deliberately since it reduces the estimation error significantly. To consider this further, suppose that we get the wheel velocities from the robot and use the odometry equations involving the time Δt. The use of the velocities in the formulas inserts two types of errors; the first is the estimation of the velocities themselves from the robot. In the time between two queries to the robot, which is 12.5 msec, the velocity cannot be computed with adequate precision; the second error derives from the calculation of the interval Δt, which is inserted into the equations. This interval is not constant since there is always a small computational overhead in the software in order to setup and issue the command, communication delays etc. Furthermore, the queries to the robot are implemented in MATLAB using a *timer object*. The timer period however, is not guaranteed and is affected by the processes running on the computer at each instant. Thus, Δt can vary from its nominal value, something which was also seen in actual experiments and must be minimized.

3. Strip-Wise Affine Map

The Strip-Wise Affine Map (SWAM) is the first step towards constructing a feedback equivalence relation which transforms the robot's model to a suitable form, under specific requirements. The equivalence relation however, exerts the useful property of *form invariance* on the mobile robot equations. The SWAM is defined for a robot model *and* a reference path, being applied to tracking control. To begin with, consider a reference polygonal path in the original physical domain, i.e. the actual domain where the robot dwells. Denote this physical domain as D_p (w-plane) and the transformed canonical domain as D_c (z-plane). Then, the strip-wise affine map is a homeomorphism $\Phi : D_c \rightarrow D_p$ that sends the real line of the canonical domain to the reference polygonal path in the physical domain. The SWAM is a piecewise linear homeomorphism between the two spaces (Groff, 2003; Gupta & Wenger,

1997) . It acts by inducing a strip decomposition on the planes and then applying an affine map between them. Note that the map acts on the entire domains, not just on a bounded region.

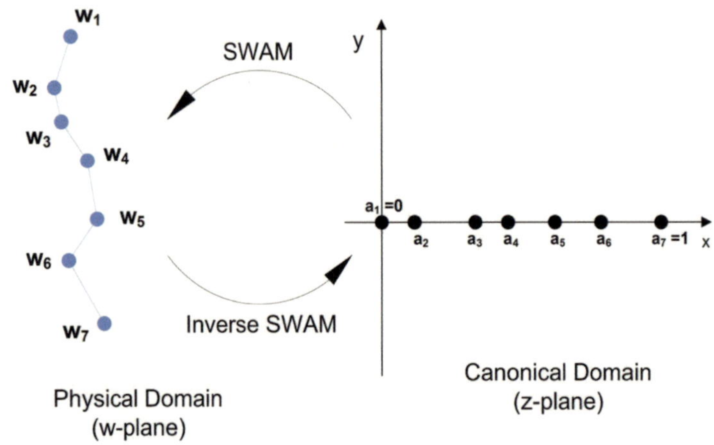

Fig. 4. Illustration of the Strip-Wise Affine Map

In more rigorous terms, let $A = \{w_1, w_2, \ldots, w_n\}, w_i \in \mathbb{C}$ be a polygonal chain on the *complex* physical domain. This represents the original reference path. Each vertex w_i of the chain is projected to a point a_i on the real axis in the canonical domain according to its normalized length,

$$a_i = \sum_{k=1}^{i} \frac{S_k}{S}, \text{i} = 1,2,3,..., \text{n} \tag{21}$$

where $S_k = |w_k - w_{k-1}|, S = \sum_{k=1}^{n} S_k$ and $S_1 = 0$. The polygon edge from w_{k-1} to w_k is linearly projected onto $[a_{k-1}, a_k]$. The transformation of $[a_{k-1}, a_k]$ onto its respective polygon edge is done using the function

$$f_{k-1}(x) = w_{k-1} + S \cdot (x - a_{k-1}) \cdot e^{j \cdot \arg(w_k - w_{k-1})} \tag{22}$$

Each interval $[a_{k-1}, a_k]$ is transformed by a respective transformation $f_1, f_2, , f_{n-1}$. Now, consider the following rectangle pulse function on the canonical real axis,

$$\psi_k = \begin{cases} 1, x \in [a_k, a_{k+1}) \\ 0, \text{elsewhere} \end{cases} \tag{23}$$

The pulse is a complex function of $z = x + jy$ in the canonical domain. Each function f_{k-1} is multiplied by the corresponding pulse and the products are summed to account for the general transformation that projects the interval [0,1) onto the polygonal chain,

$$f(x) = \sum_{k=1}^{n-1} f_k(x)\psi_k \tag{24}$$

Extension to the intervals (-∞,0) and [1,+ ∞) can be performed by appending an edge that begins from infinity ending at w_1, for the first case, and another edge starting from w_n escaping to infinity, on the second case. The edges have a direction of $\theta_{-\infty}$ and $\theta_{+\infty}$ respectively. The formulas that account for these branches are given by:

$$f_{-\infty}(x) = (w_1 + S \cdot (x - a_1) \cdot e^{j \cdot \theta_{-\infty}})\psi_0, \\ f_{+\infty}(x) = (w_n + S \cdot (x - a_n) \cdot e^{j \cdot \theta_{+\infty}})\psi_n \tag{25}$$

Here ψ_0 is an open-left pulse with a single falling edge at x=a_1, and ψ_n is the open-right pulse with a single rising edge at x=a_n. Combining the above and using the conventions $\theta_k = arg(w_{k+1} - w_k)$, $a_0 = -\infty$, $a_{n+1} = +\infty$, $w_0 \triangleq$ the point at infinity corresponding to a_0, $w_{n+1} \triangleq$ the point at infinity corresponding to a_{n+1}, $\theta_0 = arg(w_1 - w_0) = \theta_{-\infty}$, $\theta_n = arg(w_{n+1} - w_n) = \theta_{+\infty}$, the extended transformation takes the following form,

$$f(x) = \sum_{k=0}^{n} (w_k + S \cdot (x - a_k) \cdot e^{j \cdot \theta_k})\psi_k \tag{26}$$

where the functions $f_{-\infty}, f_{+\infty}$, correspond to k=0 and k=n respectively. In order to extend this transformation to the entire complex plane, let $z = x + jy$ be a complex variable in the canonical domain and consider the mapping,

$$\Phi(z) = y \cdot S \cdot e^{j \cdot \theta_s} + f(x) \tag{27}$$

where θ_s is the *shifting angle* in [$-\pi/2, \pi/2$].The complex variable $w = u + jv$ in the physical domain, is identified with the transformation $\Phi(z)$, i.e. $w = u + jv = \Phi(z)$. This transformation is the *direct strip-wise affine map* and produces a linear displacement of the polygon along the direction θ_s. Each edge of the polygon produces an affine transformation that applies only in the "strip" that the edge sweeps as it is being translated. Thus, the transformation $\Phi(z)$ can be described as a "strip-wise affine" transformation. The invertibility of the map depends firstly on the geometry of the chain and secondly on the shifting angle. It can be shown (Moustris & Tzafestas, 2008) that necessary and sufficient conditions for the mapping to be invertible are that the chain must be a strictly monotone polygonal chain (Preparata & Supowit, 1981) and the shifting angle must not coincide with the angle of any of the chain's edges i.e. the chain must not be shifted along one of its edges. The inverse strip-wise affine map can be expressed in matrix form by treating $\Phi(z)$ as an $\mathbb{R}^2$ to $\mathbb{R}^2$ mapping since it cannot be solved analytically with respect to z. The inverse mapping equations can be calculated as,

$$\begin{bmatrix} x \\ y \end{bmatrix} = \mathrm{A}^{-1} \begin{bmatrix} u \\ v \end{bmatrix} - \begin{bmatrix} C/J - a_k \\ -D/J \end{bmatrix} \tag{28}$$

where $C = S \sum_{k=0}^{n} (w_k^R \sin\theta_s - w_k^I \cos\theta_s)\psi_k$, $D = S \sum_{k=0}^{n} (w_k^R \sin\theta_k - w_k^I \cos\theta_k)\psi_k$ and J is the map Jacobian given by $J = S^2 \sum_{k=0}^{n} \sin(\theta_s - \theta_k)\psi_k$. A^{-1} is the inverse system matrix, i.e.

$$\mathrm{A}^{-1} = \begin{bmatrix} \sin\theta_s & -\cos\theta_s \\ -\sum\limits_{k=0}^{n} \psi_k \sin\theta_k & \sum\limits_{k=0}^{n} \psi_k \cos\theta_k \end{bmatrix} / S \sum_{k=0}^{n} \sin(\theta_s - \theta_k)\psi_k \tag{29}$$

Besides Eq.28, one also needs to know the activated pulse since the rectangle pulses are functions of the variable x, and thus (28) does not provide a complete solution to the inversion problem. If this information is provided, the sums in Eq.28 degenerate and the equation provides the inverse system. The activated pulse can be calculated algorithmically by doing

a point-in-strip test. Consider an axis orthogonal to the direction θ_s such that the projection of w_1 corresponds to 0. Furthermore let the projections of each w_i onto that axis be denoted as b_i, and the projection of the current mapping point denoted as b_c. The projections of w_i apparently partition the axis into consecutive line segments $[b_i b_{i+1}]$ which are into one-to-one correspondence with the edges of the polygonal chain. Then, in order to find the current pulse one needs to find the segment into which the point b_c resides. This can be performed optimally by a binary search algorithm in $O(logn)$.
Since the SWAM transforms the robot's application space, it also affects its model's equations. Denoting all the state variables in the physical and canonical domains with a subscript of p and c respectively, then the u-v plane (physical domain) is mapped to the x-y plane (canonical domain), i.e. the state vector $q_p = [x_p, y_p, \theta_p]^T$ is transformed to $q'_p = [x_c, y_c, \theta_p]^T$. The homeomorphism Φ defines an equivalence relation between the two states. Notice that the state θ_p remains unaffected. By introducing a new extended homeomorphism Ψ that also maps the heading angle θ_p, one can send the *canonical state-space* to the *physical state-space*, i.e. $q_p = \Psi(q_c)$. This transformation acts on all state variables and the new system state is $q_c = [x_c, y_c, \theta_c]^T$. The map is then defined by,

$$\begin{bmatrix} x_p \\ y_p \\ \theta_p \end{bmatrix} = \begin{bmatrix} y_c S \cos\theta_s + \mathrm{Re}(f(x_c)) \\ y_c S \sin\theta_s + \mathrm{Im}(f(x_c)) \\ \tan^{-1}\left(\frac{\sum_{\kappa=0}^{n} \sin\theta_\kappa \psi_\kappa + \sin\theta_s \tan\theta_c}{\sum_{\kappa=0}^{n} \cos\theta_\kappa \psi_\kappa + \cos\theta_s \tan\theta_c}\right) \end{bmatrix} = \Psi(q_c) \tag{30}$$

and the new system is,

$$\tilde{\Sigma} : \begin{bmatrix} \dot{x}_c \\ \dot{y}_c \\ \dot{\theta}_c \end{bmatrix} = \begin{bmatrix} v_c \cos\theta_c \\ v_c \sin\theta_c \\ 0 \end{bmatrix} + \begin{bmatrix} 0 \\ 0 \\ S^3 \gamma^3 J^{-1} v_c \end{bmatrix} \kappa_p \tag{31}$$

J is the Jacobian of Φ and $\gamma = \sqrt{1 + \sin 2\theta_c \sum_{\kappa=0}^{n} \cos(\theta_s - \theta_\kappa)\psi_\kappa}$. The input κ_p of the system remains unaffected. However, since it expresses the curvature of the physical system, it can also be transformed under Ψ. Thus by including the transformation of the input and extending the homeomorphism Ψ to $\hat{\Psi} = (\Psi, \Omega)$, where

$$\kappa_p = \Omega(\kappa_c, q_c) = S^{-3} J \gamma^{-3} \kappa_c$$

is the input map that sends the controls from the canonical input space to the controls in the physical input space, one gets the new extended system,

$$\Sigma_c : \begin{bmatrix} \dot{x}_c \\ \dot{y}_c \\ \dot{\theta}_c \end{bmatrix} = \begin{bmatrix} v_c \cos\theta_c \\ v_c \sin\theta_c \\ 0 \end{bmatrix} + \begin{bmatrix} 0 \\ 0 \\ v_c \end{bmatrix} \kappa_c \tag{32}$$

The systems Σ_p and Σ_c are feedback-equivalent (Gardner & Shadwick, 1987; 1990) since they are related by a state and input transformation. The input transformation is actually a feedback transformation of Σ_c that feeds the states q_c back to the input. A more careful look at Eq.(32) shows that it expresses the Dubins Car in the canonical domain, thus $\hat{\Psi}$ presents a kind of *form invariance* on the model.

Now, let p_{ref} be a reference path in the physical domain and let $u_c(q_c, I_c, t)$ be a straight line tracking controller for Σ_c, where I_c denotes the line segment $y_c = \{0/x_c \in [0,1]\}$ of the C-plane, i.e. the reference path in the canonical domain. This controller is transformed to a path tracking controller for Σ_p under the equation,

$$u_p(q_p, p_{ref}, t) = \Omega(u_c(q_c, I_c, t), q_c) = \Omega(u_c(\Psi^{-1}(q_p), \Phi^{-1}(p_{ref}), t), \Psi^{-1}(q_p)) \quad (33)$$

However, since Σ_c is the Dubins Car in the canonical domain, the straight line tracking controller $u_c(q_c, I_c, t)$ for Σ_c is actually a straight line tracker for the Dubins Car, and in order to build a path tracker for the Dubins Car, one has but to build a straight line tracker and use Eq.(33) to promote it to a path tracker for strictly monotone polygonal chains. Furthermore, one could also use existing path trackers for the Dubins Car to track the straight line in the canonical domain. In this case, these controllers can be simplified since in general, straight line tracking is simpler than path tracking.

4. Fuzzy Logic Controller

The controller used in this work for the path following task, is based on a Fuzzy Logic Controller (FLC) developed by the authors, which has been deployed in previous experiments. The original Fuzzy Logic tracker is described in (Moustris & Tzafestas, 2005), and further modified in (Deliparaschos et al., 2007) in order to be implemented on a FPGA. Specifically the tracker is a zero-order Takagi-Sugeno FLC with the two inputs partitioned in nine triangular membership functions each, while the output is partitioned in five singletons (Fig. 5).

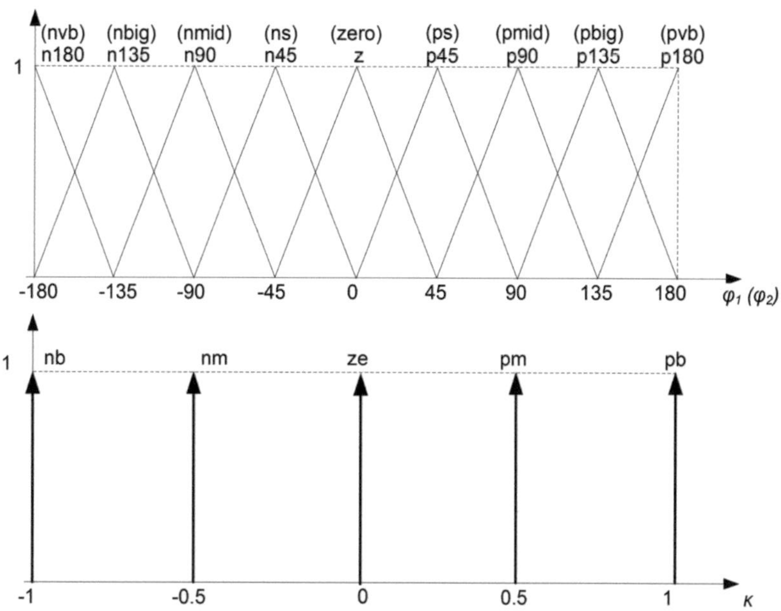

Fig. 5. Input and output membership functions of the fuzzy controller

The FL tracker uses two angles as inputs, and outputs the control input u_κ. It is assumed that the path is provided as a sequence of points on $\mathbb{R}^2$. The fuzzy rule base consists of 81 rules, which are presented in Table 1. The implication operator is the min operator.

	pvb	pbig	pmid	ps	zero	ns	nmid	nbig	nvb
p180	pb	pb	pb	pm	ze	nm	nb	nb	pb
p135	pb	pb	pb	pb	pm	nm	nb	Pb	pb
p90	pb	pb	pb	pm	pm	pm	pb	pb	Pb
p45	pb	pb	pb	pm	pm	ze	nb	pb	pb
z	pb	pb	Pb	pm	ze	nm	nb	nb	nb
n45	nb	nb	pb	ze	nm	nb	nb	nb	nb
n90	nb	nb	nb	nm	nm	nb	nb	nb	nb
n135	nb	nb	pb	pm	nm	nb	nb	nb	nb
n180	pb	pb	pb	pm	zero	nm	nb	nb	pb

Table 1. FLC Rule Base

In each control loop the closest path point is picked up and the two input angles are calculated. These angles are the angle ϕ_1 of the closest point with respect to the current robot heading and the direction ϕ_2 of the tangent of the path at that point, as depicted in Fig. 6a. Using the SWAM, we can move the tracking task to the canonical domain where the path is a straight line. In this case, the oriented straight line splits the plane into two half-planes, which present two general cases for the placement of the robot. Due to the symmetry of the cases only one will be analyzed. Consider that the robot resides in the positive half-plane (Fig. 6b) and that the distance from the closest path point P is D.

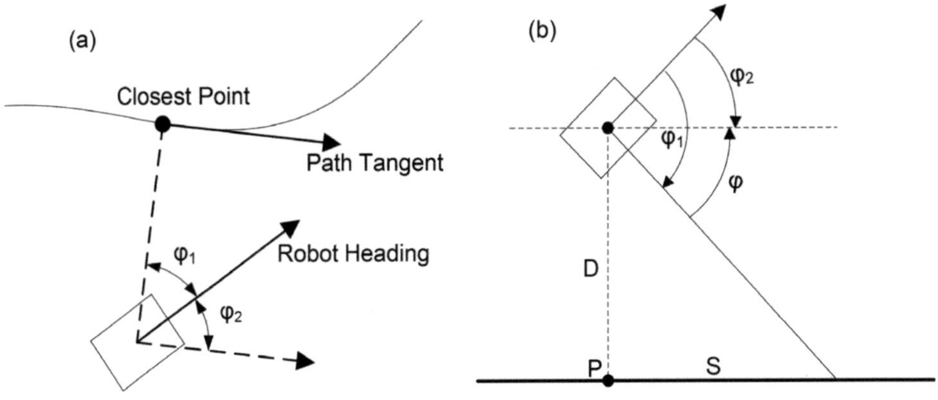

Fig. 6. Illustration of the controller inputs for path tracking in the general case (a) and in the straight line case (b).

Furthermore, one can consider that the robot tracks not the closest point P but the one found distance S ahead. The "sightline" of the robot to that point forms an angle ϕ with the path. In this situation, by varying the angle ϕ_2 one can discern four cases for the relation between the angles ϕ, ϕ_1 and ϕ_2 with the three of them being the same, namely,

$$\begin{aligned} \varphi_1 - \varphi_2 &= -\varphi \,, \varphi_2 \in [-\pi + \varphi, 0] \cup [0, \varphi] \cup [\varphi, \pi] \\ \varphi_1 - \varphi_2 &= 2\pi - \varphi \,, \varphi_2 \in [-\pi, -\pi + \varphi] \end{aligned} \tag{34}$$

When the robot resides in the positive half-plane, the angle ϕ is also positive. On the contrary, when it is in the negative half-plane, the angle changes sign although Eqs.(34) remain the same. With respect to the point being tracked, we discern two cases; either fixing the sightline, i.e. fixing the angle ϕ, or fixing the look-ahead distance S, i.e. tracking the point that is distance

S ahead of the one nearest to the robot (point P). Of course, the nearest point P can be easily found since its coordinates are $(x_c, 0)$, where x_c is the x-axis coordinate of the robot in the canonical space. In the case of a constant S, the angle ϕ varies from $[-\pi/2, \pi/2]$ and the tuple (ϕ_1, ϕ_2) is constrained in a strip. This can also lead to a rule reduction in the FLC rule base since some rules are never activated ((Moustris & Tzafestas, 2011)). For the control of the robot, the FLC and the SWAM were calculated on-line using a dedicated FPGA SoC, as described in the next section.

5. DFLC & SoC architecture

This section discusses the System on Chip (SoC) implemented on an FPGA chip for the robot path tracking task using fuzzy logic. The SoC design was implemented on the Spartan-3 MB development kit (DS-KIT-3SMB1500) by Digilent Inc. The Spartan-3 MB system board utilizes the 1.5 million-gate Xilinx Spartan-3 device (XC3S1500-4FG676) in the 676-pin fine-grid array package. A high level and a detailed architecture view of the SoC is shown in Fig.7 and 8 respectively.

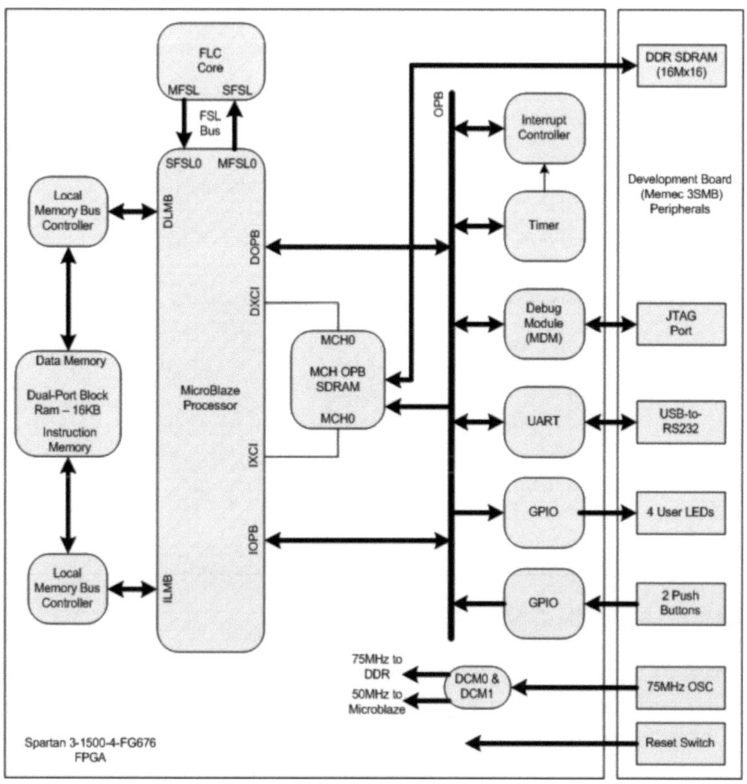

Fig. 7. Overview of the SoC's hardware architecture

A design on an FPGA could be thought as a "hard" implementation of program execution. The Processor based systems often involve several layers of abstraction to help schedule tasks and share resources among multiple processes. The driver layer controls hardware resources and the operating system manages memory and processor bandwidth. Any given processor core can execute only one instruction at a time, and processor based systems are

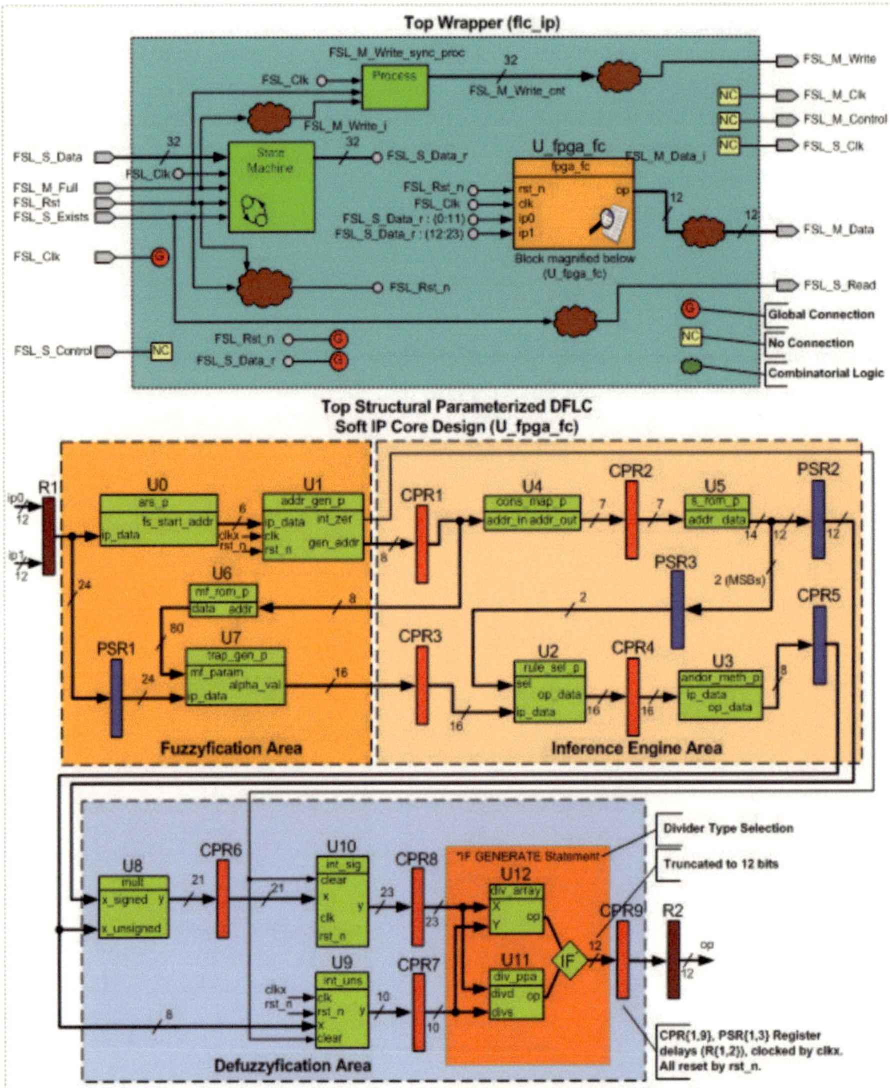

Fig. 8. Architecture of the DFLC IP core

continually at risk of time critical tasks preempting one another. FPGAs on the other hand do not use operating systems and minimize reliability concerns with true parallel execution and deterministic hardware dedicated to every task (see Fig.9).

Today's FPGAs contain hundreds of powerful DSP slices with up to 4.7 Tera-MACS throughput; 2 million logic cells with clock speeds of up to 600MHz, and up to 2.4 Tera-bps high-speed on-chip bandwidth capable to outperform DSP and RISC processors by a factor of 100 to 1,000. Taking advantage of hardware parallelism, FPGAs exceed the computing power of digital signal processors (DSPs) by breaking the paradigm of sequential execution and accomplishing more per clock cycle.

The main unit of the SoC is a parametrized Digital Fuzzy Logic Controller (DFLC) soft IP core Deliparaschos et al. (2006)Deliparaschos & Tzafestas (2006) that implements the fuzzy

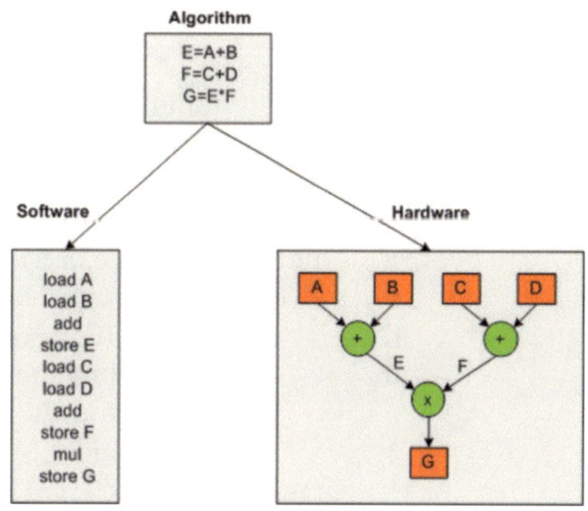

Fig. 9. Illustration of HW vs SW computation process

tracking algorithm, and a Xilinx's Microblaze soft processor core which acts as the top level flow controller. The FPGA board hosting the SoC controls the Kephera robot, used in the experiments of the tracking scheme.

In our application the DFLC facilitates scaling and can be configured for different number of inputs and outputs, number of triangular or trapezoidal fuzzy sets per input, number of singletons per output, antecedent method (t-norm, s-norm), divider type, and number of pipeline registers for the various components in the model. This parametrization enabled the creation of a generic DFLC soft IP core that was used to produce a fuzzy controller of different specifications without the need of redesigning the IP from the beginning. The fuzzy logic controller architecture assumes overlap of two fuzzy sets among adjoining fuzzy sets, and requires 2^n (n is the number of inputs) clock cycles at the core frequency speed in order to sample the input data (input sample rate of 78.2960ns), since it processes one active rule per clock cycle. In its present form the SoC design achieves a core frequency speed of 51.1 MHz. To achieve this timing result, the latency of the chip architecture involves 9 pipeline stages each one requiring 19.574ns. The featured DFLC IP is based on a simple algorithm similar to the zero-order Takagi-Sugeno inference scheme and the weighted average defuzzification method. By using the chosen parameters of Table 2, it employs two 12-bit inputs and one 12-bit output, 9 triangular membership functions (MFs) per input and 5 singleton MFs at the output with 8-bit and 12-bit degree of truth resolution respectively.

The FPGA SoC implements the autonomous control logic of the Kephera II robot. It receives odometry information from the robot and issues steering commands output by the FL tracker. The encoding and decoding of the information packets (i.e., encoding of steering control into data frames) is handled by the MATLAB application. Therefore the MATLAB application implements the actual framer/deframer for the I/O communication with the Kephera robot and downloads the tracking path to the SoC. The top-level program that supervises these tasks, treats synchronization and timing requirements, is written in C and executes in the Microblaze soft processor core. The SWAM algorithm is also implemented on the FPGA, in order to reduce the computation time.

The MATLAB application displays information about the robot's pose and speed, as well as some other data used for the path tracking control. It also calculates the robot's position relative to the world and the local coordinate frames. Another important function of the MATLAB application is to provide a path for the robot to track. The current work deals only

Parameters (VHDL generics)	Value	Generic Description
ip_no	2	Number of inputs
ip_sz	12	Input bus resolution (bits)
op_no	1	Number of outputs
op_sz	12	Output bus resolution (bits)
MF_ip_no	9	Number of input MFs (same for all inputs)
dy_ip	8	Input MFs degree of truth resolution (bits)
MF_op_no	5	Number of output MFs (singletons)
MF_op_sz	12	Output MFs resolution (bits)
sel_op	0	Antecedent method connection: 0: min, 1: prod, 2: max, 3: probor
div_type	1	Divider Model: 0: restoring array, 1: LUT reciprocal approx.
PSR		**Signal Path Route**
psr1_no	1	ip_set→psr1_no→trap_gen_p
psr2_no	4	s_rom→psr2_no→mult
psr3_no	1	s_rom→psr3_no→rul_sel_p
psr4_no	1	cpr5→psr→int_uns
CPR		**Component (Entity) Name**
cpr1_no	1	addr_gen_p
cpr2_no	1	cons_map_p
cpr3_no	3	trap_gen_p
cpr4_no	0	rule_sel_p
cpr5_no	2	minmax_p
cpr6_no	1	mult
cpr7_no	0	int_uns
cpr8_no	0	int_sig
cpr9_no	2	div_array

Table 2. DFLC soft IP core chosen parameters

with the path tracking task and not path planning. To compensate for this, the path is drawn in MATLAB, encoded properly and downloaded to the SoC. Then, the SoC begins the tracking control.

The Microblaze soft processor core is licensed as part of the Xilinx Embedded Development Kit (EDK) and is a soft core, meaning that it is implemented using general logic primitives rather than a hard dedicated block in the FPGA. The Microblaze is based on a RISC architecture which is very similar to the DLX architecture described in (Patterson & Hennessy, 1997)(Sailer et al., 1996). It features a 3-stage pipeline with most instruction completing in a single cycle. Both the instruction and data words are 32 bits. The core alone can obtain a speed of up to 100MHz on the Spartan 3 FPGA family. The Microblaze processor can connect to the OPB bus for access to a wide range of different modules, it can communicate via the LMB bus for a fast access to local memory, normally block RAM (BRAM) inside the FPGA.

Moreover, the Fast Simplex Link (FSL) offers the ability to connect user soft core IP's acting as co-processors to accelerate time critical algorithms. The FSL channels are dedicated unidirectional point-to-point data streaming interfaces. Each FSL channel provides a low latency interface to the processor pipeline allowing extending the processor's execution unit with custom soft core co-processors. In this work the DFLP IP core is playing the role of such

a co-processor and is connected to the Microblaze via the FSL bus. The architecture of the present SoC consists mainly of the DFLP that communicates with the Microblaze Processor through the Fast Simplex Bus (FSL), the utilized block RAMs (BRAM) through the LMB bus, and other peripherals such as the general purpose input/output ports (GPIO), and UART modules via the OPB bus. Here, the DFLP incorporates the fuzzy tracking algorithm, whereas the Microblaze processor mainly executes the C code for the flow control.
The parametrized zero-order TSK type Fuzzy Logic Controller (FLC) core exchanges data with the MicroBlaze processor via the FSL bus. The scope of the FLC core is to serve as high-speed fuzzy inference co-processor to the Microblaze. The DFLC core was implemented with the following parameters (see Table3).

Property	Value
Inputs	2
Input resolution	12
Outputs	1
Output resolution	12 bit
Antecedent Membership Functions (MF's)	9 Triangular MF's
Degree of Truth resolution	8 bit
Consequent MF's	5 Singletons
MF resolution	8 bit
Number of fuzzy inference rules	81
Rule activation method	MIN
Aggregation method	SUM
Implication method	PROD
MF overlapping degree	2
Defuzzification method	Weighted average

Table 3. DFLC core characteristics

Besides these two main cores and buses, the design consists of 16 KB local memory, 32 MB DDR, timer, interrupt controller, a UART, a debug peripheral (MDM) and a couple of General Purpose Inputs/Outputs (GPIOs). A Multi-CHannel (MCH) On-chip Peripheral Bus (OPB) Double Data Rate (DDR) Synchronous DRAM (SDRAM) memory controller (MCH OPB DDR with support for asynchronous DDR clock) is used in this design. This allows the MicroBlaze system to run at a lower speed of 51 MHz, which is more reasonable for Spartan-3, while the DDR is running at 75 MHz, which is the minimum required frequency for the Micron DDR chip. The on-chip Digital Clock Manager (DCM) is used to create the various clock frequencies and phases required to make this system work, all based on the 75 MHz oscillator on the 3SMB board. The FLC core runs at the same speed as the OPB and MicroBlaze, which is 51 MHz. Based on the place and route report, the design occupies 4174 out of 13312 slices of the Xilinx Spartan 3 FPGA (XC3S1500-4FG676).

6. Hardware/software co-design

On the beginning of the co-design process one starts with an architecture independent description of the system functionality. Since the description of the system functionality is independent of the HW and SW, several system modelling representations may be used, such as finite state machines (FSMs) for example. The modelled system can then be described by means of a high level language, which is next compiled into an internal representation

such as a data control flow description. This description which serves as a unified system representation allows to perform HW/SW functional partitioning. After the completion of the partitioning, the HW and SW blocks are synthesized and evaluation is then performed. If the evaluation does not meet the required objectives, another HW/SW partition is generated and evaluated (Rozenblit & Buchenrieder, 1996)(Kumar, 1995).
A general HW/SW co-design schema followed in this SoC implementation is illustrated in Fig.10).

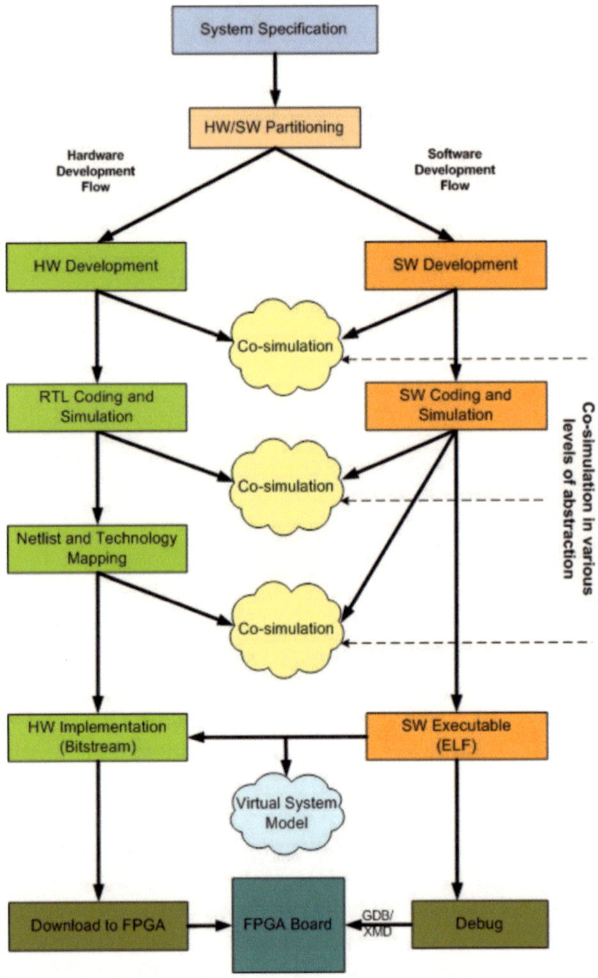

Fig. 10. HW/SW Co-design Flow

The DFLC core implementation follows a sequential design manner (see Fig.11) (Navabi, 1998). The starting point of the design process was the functional modelling of the fuzzy controller in a high level description (i.e., MATLAB/Simulink). This serves a two purpose role, first to evaluate the model and second to generate a set of test vectors for RTL and timing verification. The model was coded in register transfer level (RTL) with the use of hardware description language VHDL. Extensive use of VHDL generic and generate statements was used through out the coding of the different blocks, in order to achieve a parameterized DFLC core. The DFLC core is scalable in terms of the number of inputs/bus resolution, number of input/output fuzzy sets per input and membership resolution. More specifically A VHDL package stores the above generic parameters together with the number of necessary pipeline stages for each block. An RTL simulation was performed to ensure the correct functionality of the fuzzy controller. The DFLC core was independently synthesized with Synopsys Synplify logic synthesis tool (as it produced better synthesis results and meet timing constraints), whereas the rest of the SoC cores were synthesised with Xilinx synthesis tool XST. The Xilinx EDK studio was used for the integration flow of different SoC cores (i.e., DFLC, Microblaze, etc) and Xilinx ISE tool for the placement and routing of the SoC on the FPGA. More analytically, the place and route tool accepts the input netlist file (.edf), previously created by the synthesis tool and goes through the following steps. First, the translation program translates the input netlist together with the design constraints to a database file. After the translation program has run successfully, the logical design is mapped to the Xilinx FPGA device. Lastly, the the mapped design is placed and routed onto the chosen FPGA family and a device configuration file (bitstream) is created. Xilinx's SDK used for C programming and debugging the SoC's Microblaze soft processor. RTL and timing simulation to verify the correct functionality was handled with the use of Mentor's Modelsim simulator.

7. FPGA design and performance evaluation

The Component Pipeline Registers (CPR) blocks in Fig.8 indicate the number of pipeline stages for each component; the Path Synchronization Registers (PSR) blocks point to registers used for synchronizing the data paths, while the "U" blocks represent the different components of the DFLC Deliparaschos & Tzafestas (2006).
The U_fpga_fc component is embedded in the flc_ip top structural entity wrapper which is compliant with the FSL standard and provides all the necessary peripheral logic to the DFLC soft IP core in order to send/receive data to/from the FSL bus. The flc_ip wrapper architecture is shown in Fig.8 while the chosen (generic) parameters (VHDL package definition file) for the parameterized DFLC IP (U_fpga_fc) and its characteristics are summarized in Table 2 and Table 3 respectively.
The U_fpga_fc alone was synthesized using Synplify Pro synthesizer tool, while the rest of the design components were synthesized with Xilinx Synthesis Tool (XST) through the EDK Platform Studio. The produced .edf file for the U_fpga_fc is been seeing by the flc_ip wrapper as a blackbox during the XST flow. The placement and routing of the SoC design into the FPGA was done through the EDK by calling the Xilinx ISE tool.
According to the device utilization report from the place and route tool (see Table 4), the SoC design (including the DFLC) occupies 4,494 (16%) LUTs, 15 Block Multipliers (MULT18X18s), and 18 Block RAMs. The implemented design uses two Digital Clock Manager (DCM) Modules (DCM_0 for the system clock and DCM_1 for clocking the external DDR RAM) that produce the different clocks in the FPGA. The DFLC core itself occupies 1303 or 4% LUTs, 8 Block Multipliers, 12 64x1 ROMs (ROM64X1) and 54 256x1 ROMs (ROM256X1). The SoC achieves a minimum clock operating period of 19.574ns or a maximum frequency of $\sim$51.1 MHz respectively (the DFLC with the chosen parameters reports a frequency of 85MHz when implemented alone).

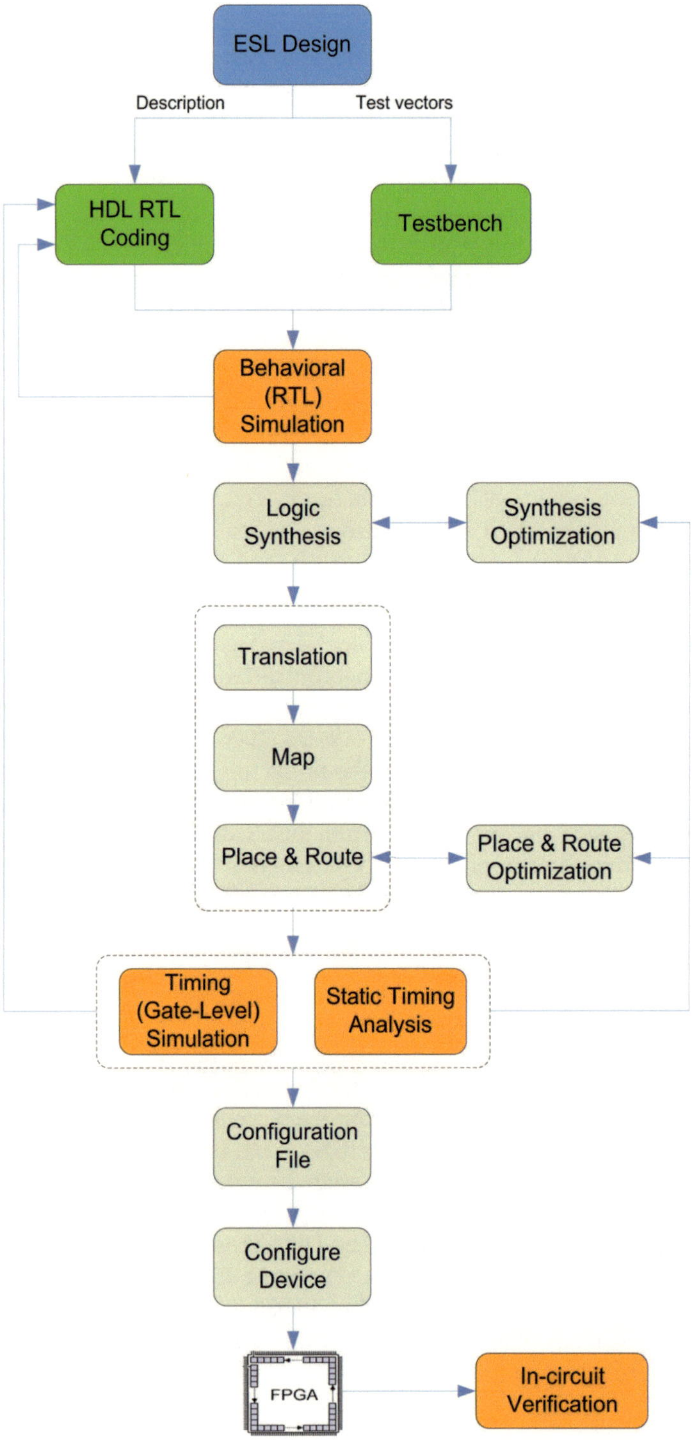

Fig. 11. HW/SW Hardware design flow

Logic Utilization		
Number of Slice Flip Flops	3,288 out of 26,624	12%
Number of 4 input LUTs	4,494 out of 26,624	16%
Logic Distribution		
Number of occupied Slices	4,174 out of 13,312	31%
Number of Slices containing only related logic	4,174 out of 4,174	100%
Number of Slices containing unrelated logic	0 out of 4,174	0%
Total Number of 4 input LUTs	5,893 out of 26,624	22%
Number used as logic	4,494	
Number used as a route-thru	165	
Number used for Dual Port RAMs	432	
(Two LUTs used per Dual Port RAM)		
Number used as 16x1 ROMs	432	
Number used as Shift registers	370	
Number of bonded IOBs	62 out of 487	12%
IOB Flip Flops	94	
IOB Dual-Data Rate Flops	23	
Number of Block RAMs	18 out of 32	56%
Number of MULT18X18s	15 out of 32	46%
Number of GCLKs	6 out of 8	75%
Number of DCMs	2 out of 4	50%
Number of BSCANs	1 out of 1	100%
Total equivalent gate count for design	1,394,323	
Additional JTAG gate count for IOBs	2,976	

Table 4. SoC design summary

8. Experimental results

The experiments consist of tracking predefined paths and analysing the displacement error. The paths are drawn by hand in the MATLAB application and downloaded to the FPGA. Then the control algorithm running on the board, calculates the steering command (curvature), relays it to MATLAB, which in turn passes it to the robot. Conversely, the MATLAB application receives odometric data from the robot which are then relayed to the FPGA. Essentially, the MATLAB application acts as an intermediary between the board and the robot, transforming commands and data to a suitable form for each party. Note also that the actual odometry is being performed by MATLAB (estimation of the robot's pose (x, y, θ) using the data from the robot's encoders). A key detail in the above process is that odometry provides an estimation of the actual pose. Thus in order to analyse the efficacy of the tracking scheme, we need to know the actual pose of the robot. Position detection of the robot is achieved using a camera hanging above the robot's activity terrain and utilizing a video tracking algorithm which extracts the robot's trajectory in post-processing. This algorithm tracks a red LED placed at the center of the robot.

The video tracking algorithm uses the high contrast of the LED with its surrounding space. Specifically, each video frame is transformed from the RGB color space to the generalized rgb

space. This space expresses the percentage of each color at each pixel, i.e.,

$$
\begin{aligned}
r &= \frac{R+G+B}{R} \\
g &= \frac{R+G+B}{G} \\
b &= \frac{R+G+B}{B}
\end{aligned}
\tag{35}
$$

It is the hyper-space of the so-called rg chromaticity space, which consists of the first two equations. Following the transformation, in order to enhance the contrast, each image pixel is raised to the 3rd power,

$$(r', g', b')_{(u,v)} = (r^3, g^3, b^3)_{(u,v)} \tag{36}$$

The new image is then re-transformed according to the rgb transform, essentially computing the color percentage of the percentage. Then, we apply a thresholding on the r channel, producing a binary image. The threshold was empirically set to "0.6". This procedure produces the "patch" of pixels corresponding to the red LED. The next step is, of course, to calculate a single pixel value from this patch, and thus get the robot's position. To this end, we calculate the median (row, col) value of the patch's pixels. This algorithm is applied to the first video frame, and is rather slow since the image dimensions are large (1280×720 pixels). In order to speed up the process, the algorithm processes an image region of interest in each consecutive frame. This ROI is a 17×17 pixel square, centred at the point extracted from the previous frame. The square dimensions are appropriate since the robot is not expected to have moved far between frames. The precision of the algorithm is about ±2 pixels, translating to 2.4 mm.

Previous to all runs, the camera was calibrated using the Camera Calibration Toolbox by J.Y. Bouguet, extracting the camera's intrinsic and extrinsic parameters. Two runs were performed; one tracking a straight line and one tracking a curved path (snapshots of the two videos are seen in Fig. 12). For the first run, the reference, odometry and camera paths are presented in Fig. 13(UP). The minimum distance versus the path length of the odometry and the camera paths are shown in Fig.13(DOWN). Likewise, for the second experiment the results are presented in Fig14.

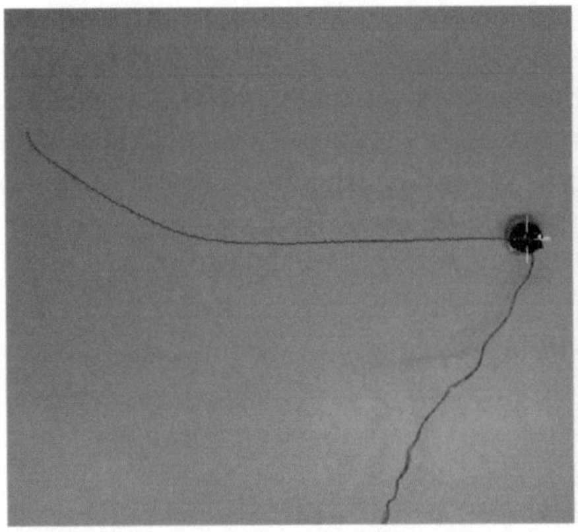

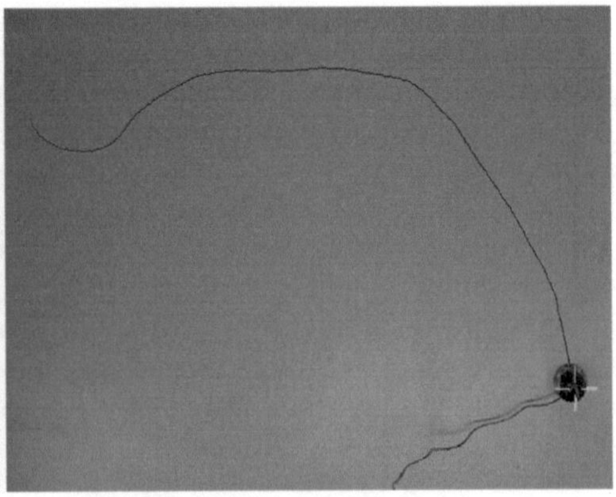

Fig. 12. Snapshots of the first (LEFT) and the second (RIGHT) experiments. The red line is the robot's path calculated off-line from the video camera.

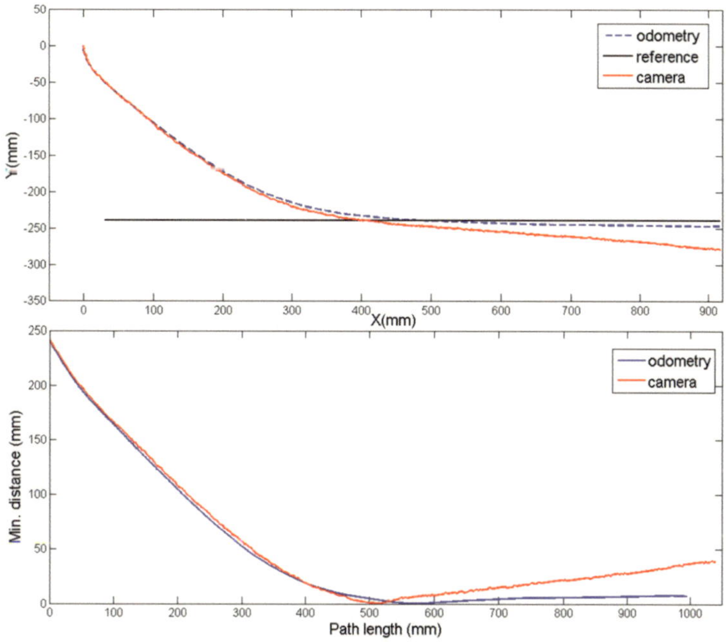

Fig. 13. (UP) The odometry (blue), camera (red) and reference (dashed black) paths for the first experiment. (DOWN) Minimum distance of the odometry and camera paths to the reference path versus path length experiment.

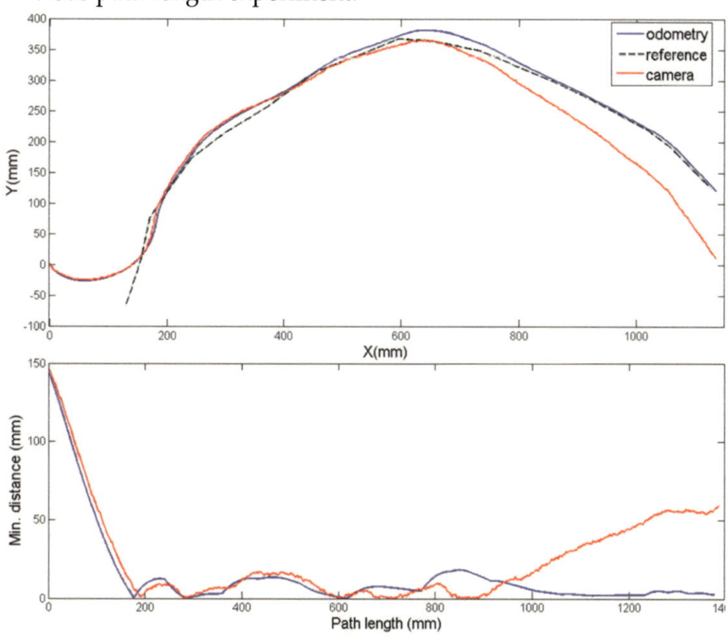

Fig. 14. (UP) The odometry (blue), camera (red) and reference (dashed black) paths for the second experiment. (DOWN) Minimum distance of the odometry and camera paths to the reference path versus path length.

As one can see, the performance of the tracking scheme is satisfactory maintaining the minimum distance to the reference path at about 50mm in the worst case. However, by taking a closer look at Figures 13 and 14, it is clear that the performance degradation is attributed not to the algorithm *per se* but to the odometry. The error accumulation of odometric data forces the robot to diverge from the actual path. But the actual odometry solution is very close to the reference path, meaning that based solely on odometry (as is the case in these experiments), the tracker maintains the robot very close to the reference path (the minimum distance is below 10mm in both cases). This implies that if a better localization technique is used, our tracking scheme would perform with more accuracy.

9. Conclusions

In this chapter we have analysed and demonstrated the applicability of the strip-wise affine transform in the path tracking task for mobile robots. The transformation was translated to hardware and implemented into an FPGA chip with the use of VHDL and advanced EDA software. The scalability of the fuzzy controller core allowed easy parameter adaptation of the theoretic fuzzy tracker model. The experiments show that the tracking scheme performs satisfactory but is degraded by the accumulation of errors of the odometry used in estimating the robots position.

10. References

Abdessemed, F., Benmahammed, K. & Monacelli, E. (2004). A fuzzy-based reactive controller for a non-holonomic mobile robot, *Robotics and Autonomous Systems* 47(1): 31–46.

Altafini, C. (1999). A Path-Tracking criterion for an LHD articulated vehicle, *The International Journal of Robotics Research* 18(5): 435–441.

Altafini, C. (2002). Following a path of varying curvature as an output regulation problem, *Automatic Control, IEEE Transactions on* 47(9): 1551–1556.

Antonelli, G., Chiaverini, S. & Fusco, G. (2007). A Fuzzy-Logic-Based approach for mobile robot path tracking, *Fuzzy Systems, IEEE Transactions on* 15(2): 211–221.

Baltes, J. & Otte, R. (1999). A fuzzy logic controller for car-like mobile robots, *Computational Intelligence in Robotics and Automation, 1999. CIRA '99. Proceedings. 1999 IEEE International Symposium on*, Monterey, CA , USA, pp. 89–94.

Cao, M. & Hall, E. L. (1998). Fuzzy logic control for an automated guided vehicle, *Intelligent Robots and Computer Vision XVII: Algorithms, Techniques, and Active Vision* 3522(1): 303–312.

Costa, A., Gloria, A. D., Giudici, F. & Olivieri, M. (1997). Fuzzy logic microcontroller, *IEEE Micro* 17(1): 66–74.

Deliparaschos, K. M., Nenedakis, F. I. & Tzafestas, S. G. (2006). Design and implementation of a fast digital fuzzy logic controller using FPGA technology, *Journal of Intelligent and Robotics Systems* 45(1): 77–96.

Deliparaschos, K. M. & Tzafestas, S. G. (2006). A parameterized T-S digital fuzzy logic processor: soft core VLSI design and FPGA implementation, *International Journal of Factory Automation, Robotics and Soft Computing* 3: 7–15.

Deliparaschos, K., Moustris, G. & Tzafestas, S. (2007). Autonomous SoC for fuzzy robot path tracking, *Proceedings of the European Control Conference 2007*, Kos, Greece.

Egerstedt, M., Hu, X. & Stotsky, A. (1998). Control of a car-like robot using a dynamic model, *Robotics and Automation, 1998. Proceedings. 1998 IEEE International Conference on*, Vol. 4, pp. 3273–3278 vol.4.

El Hajjaji, A. & Bentalba, S. (2003). Fuzzy path tracking control for automatic steering of vehicles, *Robotics and Autonomous Systems* 43(4): 203–213.

Fortuna, L., Presti, M. L., Vinci, C. & Cucuccio, A. (2003). Recent trends in fuzzy control of electrical drives: an industry point of view, *Proceedings of the 2003 International Symposium on Circuits and Systems*, Vol. 3, pp. 459–461.

Gardner, R. B. & Shadwick, W. F. (1987). Feedback equivalence of control systems, *Systems & Control Letters* 8(5): 463–465.

Gardner, R. B. & Shadwick, W. F. (1990). Feedback equivalence for general control systems, *Systems & Control Letters* 15(1): 15–23.

Groff, R. E. (2003). *Piecewise Linear Homeomorphisms for Approximation of Invertible Maps*, PhD thesis, The University of Michigan.

Gupta, H. & Wenger, R. (1997). Constructing piecewise linear homeomorphisms of simple polygons, *J. Algorithms* 22(1): 142–157.

Hung, D. L. (1995). Dedicated digital fuzzy hardware, *IEEE Micro* 15(4): 31–39.

Jiangzhou, L., Sekhavat, S. & Laugier, C. (1999). Fuzzy variable-structure control for nonholonomic vehicle path tracking, *Intelligent Transportation Systems, 1999. Proceedings. 1999 IEEE/IEEJ/JSAI International Conference on*, pp. 465–470.

Kamga, A. & Rachid, A. (1997). A simple path tracking controller for car-like mobile robots, *ECC97 Proc.*

Kanayama, Y. & Fahroo, F. (1997). A new line tracking method for nonholonomic vehicles, *Robotics and Automation, 1997. Proceedings., 1997 IEEE International Conference on*, Vol. 4, pp. 2908–2913 vol.4.

Koh, K. & Cho, H. (1994). A path tracking control system for autonomous mobile robots: an experimental investigation, *Mechatronics* 4(8): 799–820.

Kongmunvattana, A. & Chongstivatana, P. (1998). A FPGA-based behavioral control system for a mobile robot, *Circuits and Systems, 1998. IEEE APCCAS 1998. The 1998 IEEE Asia-Pacific Conference on*, pp. 759–762.

Kumar, S. (1995). *A unified representation for hardware/software codesign*, PhD thesis, University of Virginia. UMI Order No. GAX96-00485.

Lee, T., Lam, H., Leung, F. & Tam, P. (2003). A practical fuzzy logic controller for the path tracking of wheeled mobile robots, *Control Systems Magazine, IEEE* 23(2): 60– 65.

Leong, P. & Tsoi, K. (2005). Field programmable gate array technology for robotics applications, *Robotics and Biomimetics (ROBIO). 2005 IEEE International Conference on*, pp. 295–298.

Li, T., Chang, S. & Chen, Y. (2003). Implementation of human-like driving skills by autonomous fuzzy behavior control on an FPGA-based car-like mobile robot, *Industrial Electronics, IEEE Transactions on* 50(5): 867– 880.

Liu, K. & Lewis, F. (1994). Fuzzy logic-based navigation controller for an autonomous mobile robot, *Systems, Man, and Cybernetics, 1994. 'Humans, Information and Technology'., 1994 IEEE International Conference on*, Vol. 2, pp. 1782–1789 vol.2.

Maalouf, E., Saad, M. & Saliah, H. (2006). A higher level path tracking controller for a four-wheel differentially steered mobile robot, *Robotics and Autonomous Systems* 54(1): 23–33.

Moustris, G. P. & Tzafestas, S. G. (2011). Switching fuzzy tracking control for mobile robots under curvature constraints, *Control Engineering Practice* 19(1): 45–53.

Moustris, G. & Tzafestas, S. (2005). A robust fuzzy logic path tracker for non-holonomic mobile robots., *International Journal on Artificial Intelligence Tools* 14(6): 935–966.

Moustris, G. & Tzafestas, S. (2008). Reducing a class of polygonal path tracking to straight line tracking via nonlinear strip-wise affine transformation, *Mathematics and Computers in Simulation* 79(2): 133–148.

Navabi, Z. (1998). *VHDL: analysis and modeling of digital systems*, McGraw-Hill Professional.

Ollero, A., Garcia-Cerezo, A., Martinez, J. L. & Mandow, A. (1997). Fuzzy tracking methods for mobile robots, *in* M. Jamshidi, A. Titli, L. Zadeh & S. Boverie (eds), *Applications of fuzzy logic: Towards high machine intelligence quotient systems*, Prentice-Hall, New Jersey.

Patterson, D. A. & Hennessy, J. L. (1997). *Computer Organization and Design: The Hardware/Software Interface*, 2 edn, Morgan Kaufmann.

Preparata, F. P. & Supowit, K. J. (1981). Testing a simple polygon for monotonicity, *Info. Proc. Lett.* 12(4): 161–164.

Raimondi, F. & Ciancimino, L. (2008). Intelligent neuro-fuzzy dynamic path following for car-like vehicle, *Advanced Motion Control, 2008. AMC '08. 10th IEEE International Workshop on*, pp. 744–750.

Reynolds, R., Smith, P., Bell, L. & Keller, H. (2001). The design of mars lander cameras for mars pathfinder, mars surveyor '98 and mars surveyor '01, *Instrumentation and Measurement, IEEE Transactions on* 50(1): 63–71.

Rodriguez-Castano, A., Heredia, G. & Ollero, A. (2000). Fuzzy path tracking and position estimation of autonomous vehicles using differential GPS, *Mathware Soft Comput* 7(3): 257–264.

Rozenblit, J. & Buchenrieder, K. (1996). *Codesign: Computer-aided Software/Hardware Engineering*, I.E.E.E.Press.

Sailer, P. M., Sailer, P. M. & Kaeli, D. R. (1996). *The DLX Instruction Set Architecture Handbook*, 1st edn, Morgan Kaufmann Publishers Inc.

Salapura, V. (2000). A fuzzy RISC processor, *IEEE Transactions on Fuzzy Systems* 8(6): 781–790.

Samson, C. (1995). Control of chained systems application to path following and time-varying point-stabilization of mobile robots, *Automatic Control, IEEE Transactions on* 40(1): 64–77.

Sanchez, O., Ollero, A. & Heredia, G. (1997). Adaptive fuzzy control for automatic path tracking of outdoor mobile robots. application to romeo 3R, *Fuzzy Systems, 1997., Proceedings of the Sixth IEEE International Conference on*, Vol. 1, pp. 593–599 vol.1.

Wit, J., Crane, C. D. & Armstrong, D. (2004). Autonomous ground vehicle path tracking, *J. Robot. Syst.* 21(8): 439–449.

Yang, X., He, K., Guo, M. & Zhang, B. (1998). An intelligent predictive control approach to path tracking problem of autonomous mobile robot, *Systems, Man, and Cybernetics, 1998. 1998 IEEE International Conference on*, Vol. 4, pp. 3301–3306 vol.4.

21

Control Architecture Design and Localization for a Gas Cutting Robot

KiSung You, HwangRyol Ryu and Chintae Choi
Research Institute of Industrial Science and Technology
Korea

1. Introduction

Conventional control architecture which has been employed in mobile robot control software is divided into two categories such as knowledge-based and behavior-based control architecture. Early implementation of control architecture was mainly focused on building for sensing the environment, modeling it, planning based on this perceived model and executing the planned action to achieve a certain task. This design approach is called sense-model-plan-act (SMPA) or knowledge-based. A mobile robot making use of a knowledge-based controller tries to achieve its goal by following closely the sense-model-plan-act procedure. SMPA controller also needs initial knowledge, required to model its task environment prior for the robots to executing the planned task. Hence, if the initial knowledge is suited to its working environment, the resulting tasks guarantee success. Although the resulting overall behavior is predictable, the controller often suffers from being slow and becomes complex as it deals with a dynamic environment because most of the controller processing time is consumed in building a model, doing general perception and planning. Therefore, it is suitable controller for robots to require high-level intelligence and work in static and predictable environment.
Brooks proposed a radically different approach in the design of mobile robot control architecture to address the drawback of knowledge-based control architecture. This control architecture functions a horizontal computation scheme so that each behavior is a fixed action pattern with respect to the sensory information. When the mobile robot is confronted by a dynamic environment, a behavior-based robot can react fast because of the direct coupling between its behaviors and the sensed states. The robot controller can be built incrementally, thus making it highly modular and easy to construct. However, this reactive approach still suffers from planning the productive and efficient actions in an unstructured environment because they are only confined to reactions to sensors and the changing states of other modules.
In this chapter, we address the functional safety problems potentially embedded in the control system of the developed mobile robot and introduce a concept of Autonomous Poor Strip Cutting Robot (APSCR) control architecture with a focus on safety in order to design the walkthrough procedure for each behavior. In section 2, we explain the working environment where the robot will be manually or autonomously operated. Section 3 explains the control architecture of APSCR. In section 4, we explain the localization system of APSCR. Finally, section 5 shows some experimental result and in section 6 conclusions will be lastly addressed.

2. Working environment explanation

This section makes an explanation of the robot's working environment in Hot Rolling facility, Gwangyang steel work and describes the challenges and difficulties which the robot shall confront based on the results of CCR (customer critical requirements) and TCR (technical critical requirements) are.

2.1 Robot's working environment

Hot Rolling, used mainly to produce sheet metal or simple cross sections from billets is the most efficient process of primary forming used for the mass production of steel. The first operation of hot rolling process is to heat the stock to the proper deformation temperature. During heating, a scale forms on the stock surface which must be systematically removed and thus descaling is performed mechanically by crushing during a hot forming operation or by spraying with water under high pressure.

Fig. 1. Picture of poor strip produced during hot rolling.

During rolling, deformation of material occurs in between in the forms of rotation, driven rolls. The transporting force during rolling is the friction between rolls and a processed material. Due to this tremendous force applied to the slabs into wrong direction, the poor strips are produced as shown in Figure 1. Due to these poor strips, the whole facility has to be stopped until those poor strips are to be taken away from the roller conveyor by cutting them into the right size and weight so that the heavy-weight crane can hold and carry them out. In order for the workers to cut the poor strips (generally the poor strip is 1.5m x 60m), they have to put on the fire resistant work wear and do the oxygen cutting on the surface of the poor strip whose temperature is over 200 °C.

During inspecting the working environment, we gathered the strong demands and voices about the development of new type of automation which can provide the safety and the cutting operation. We carefully determined the requirements in the respects to the workers or technical barrier. We concluded that the APSCR must be equipped with all the safety requirements and be designed to respond to the severe faults that might be done by operators or malfunctioned by the mechanical faults.

2.2 Technical requirement

We must take into consideration robot's functionalities and safety with respect to all cutting operations because the working environment is extremely bad to the robot for itself. For the purpose of fulfilling all the tasks, we analyzed the survey results in technically critical requirements in the engineer's point of view and took into consideration task achievements in the customer's point of view as well. Below is a list of important information required for the robot design in the customer's point of view.

- Robot must cut a poor strip at single cutting operation, which means not to move back to the points where cutting is done.
- When a flame is not detected due to the malfunction of flame detection sensor or clogged by the heavy dust, robot must stop all the operations.
- Poor strips should be cut within 10 ~15 minutes regardless of the thickness of the poor strips in the width direction.
- Poor strip (approximately 1.5 m x 60 m) cutting operation should be done within 1 hour.
- Robot should be equipped with a backfire protection device.
- We take into consideration repulsiveness of torching at the moment of emitting the gas and oxygen.
- Maximum thickness of the poor strips is 150 mm.
- Electrical spark must be sealed off because gas tank and oxygen tank are loaded on the robot.

We also take the design of the robot into consideration in the engineer's point of views as follows:

- Torch tip should be designed to rotate automatically.
- It should be an automatic ignition system.
- Robot should move along the guide fence installed at the edge of the conveyor stand.
- Backfire protection device should be mounted in the pipe of gas.
- Workers should control a flow of gas and oxygen manually.
- Robot body should sustain a fully stretching arm (full reach: 3.8 m) so that links of arm segment should not hang downwards with no firmness.
- Cutting operation is divided by two steps: pre-heating and blowing. Each step must be accurately controlled by a system or worker.
- For the safety protection against gas explosion, a flame detection sensor should be interrupted by a hardware and software.
- Object detection mechanism, implemented by high sensitive ultrasonic sensor must be considered in order to avoid collision on moving next cutting point.
- Weight balance mechanism should be taken into consideration since robot's total weight is approximately 700 kg.
- Intensity of flame shouldn't be strong enough to cut the bottom plate under a poor strip.

We had collected many opinions and suggestions regarding to robot's functionalities and behaviors. Above lists are a part of surveys, which workers thought of the fundamental tasks and operations which the robot developer should consider.

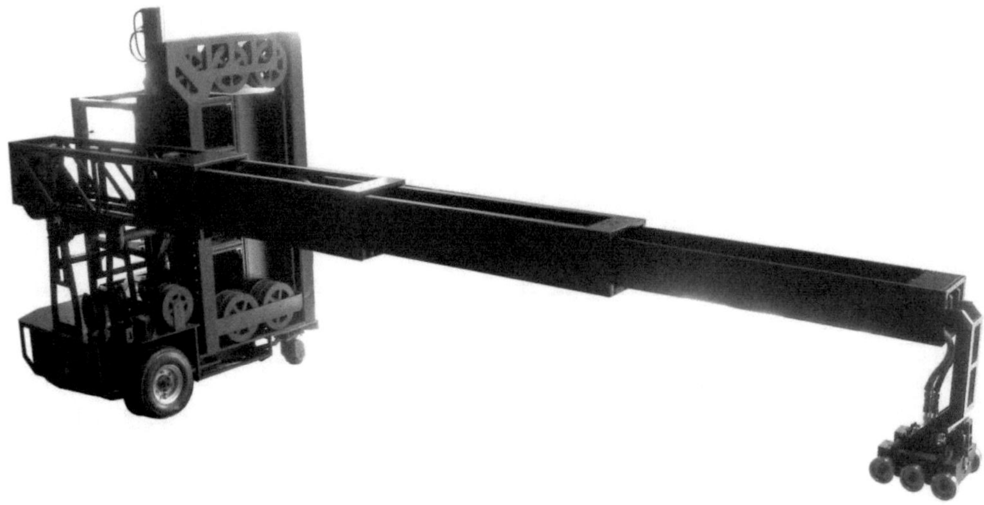

Fig. 2. APSCR stretches its arm up to 3.8 m.

3. System configuration for APSCR

When we develop industrial robots under an unpredictable environment such as steelworks, there are many critical requirements in the sense that hardware parts would comprise of the supplementary functionalities with a help of software parts. On the contrary, software and hardware are separately developed and combined without considering control architecture. The proposed control architecture mainly makes use of the advantages of conventional reactive control architecture. In this section we focus on the software architecture assuming that the hardware architecture is well designed and describe how the safety walkthrough procedure is sufficiently cooperated within the reactive controller by utilizing hardware and software interrupts as well. Figures 3 through 4 are pictures illustrating the proposed gas cutting robot as mentioned in section 2.

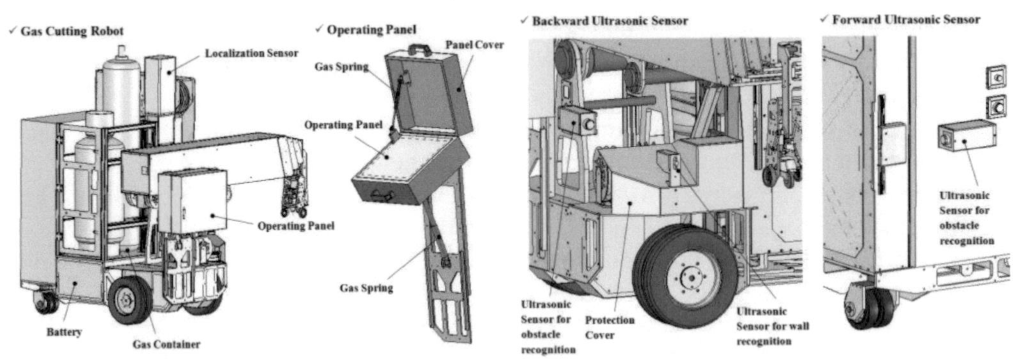

Fig. 3. Layout of a gas cutting robot.

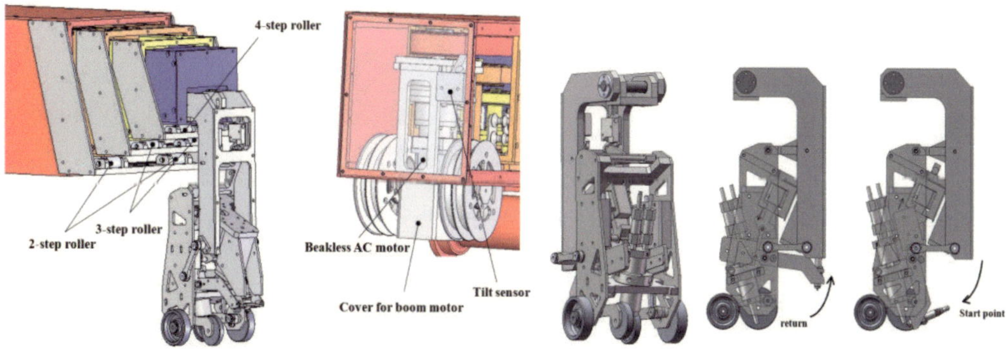

Fig. 4. Structure of boom(robot arm) and touch holder.

3.1 Design of safety interaction for software interrupts

In the context of developing safety-oriented software, safety protection functionalities in control architecture become a key to making a robot function as was designed with a focus on the safety in this project. The safety protection functionalities are embodied within the task-achieving behaviors as follows.

- *BEHAVIOR_SelfWalkThrough*
- *BEHAVIOR_CheckDistance10cm*
- *BEHAVIOR_InitialAlignmentWithLane*
- *BEHAVIOR_MeasureDepthOfPoorStrip*
- *BEHAVIOR_SetStartingPosition4Cutting*
- *BEHAVIOR_StartFiring*
- *BEHAVIOR_SetFinishingPosition4Cutting*
- *BEHAVIOR_MovingNextCuttingPosition*
- *BEHAVIOR_CheckSensorStatus4Navigation*
- *BEHAVIOR_CheckSensorStatus4Cutting*
- *BEHAVIOR_Preheating4Cutting*
- *BEHAVIOR_SettingCuttingParameter*

The task-achieving behaviors, which are described above provide the functional mechanism for APSCR to provide the poor strip cutting procedure autonomously. The following Figure 5 shows how behaviors modules depend on each other and are grouped according to their functions.

To be able to operate the task-achieving behaviors sufficiently as defined in Figure 6, the APSCR starts with checking its system by *BEHAVIOR_SelfWalkThrough* and do the alignment job in order to previously position cutting procedure by calling the following functions *CheckDistance10cm* and *InitialAlignemtnwithStand* in Before Cutting. After initial setup is done, the robot starts to measure the thickness of the poor strip (*MeasureDepthOfPoorStrip*), finds the starting position in which the edge of the poor strip is detected by the roller mounted on the torch holder (*SetStartingPostion4Cutting*), cuts the poor strip with controlling the speed of Boom and Derrick and maintaining the gap between the torch tip and surface of the strip (*StartFiring, Preheating4Cutting, SettingCuttingParameter*), and finally detects the finishing edge of the strip (*SetfinishigPosition4cutting*). After finishing the cutting procedure, while monitoring the alignment between the robot and the poor strip stand and making obstacle avoidance modules running (*CheckSensorStatus4Navigation*), the robot moves to the next cutting position

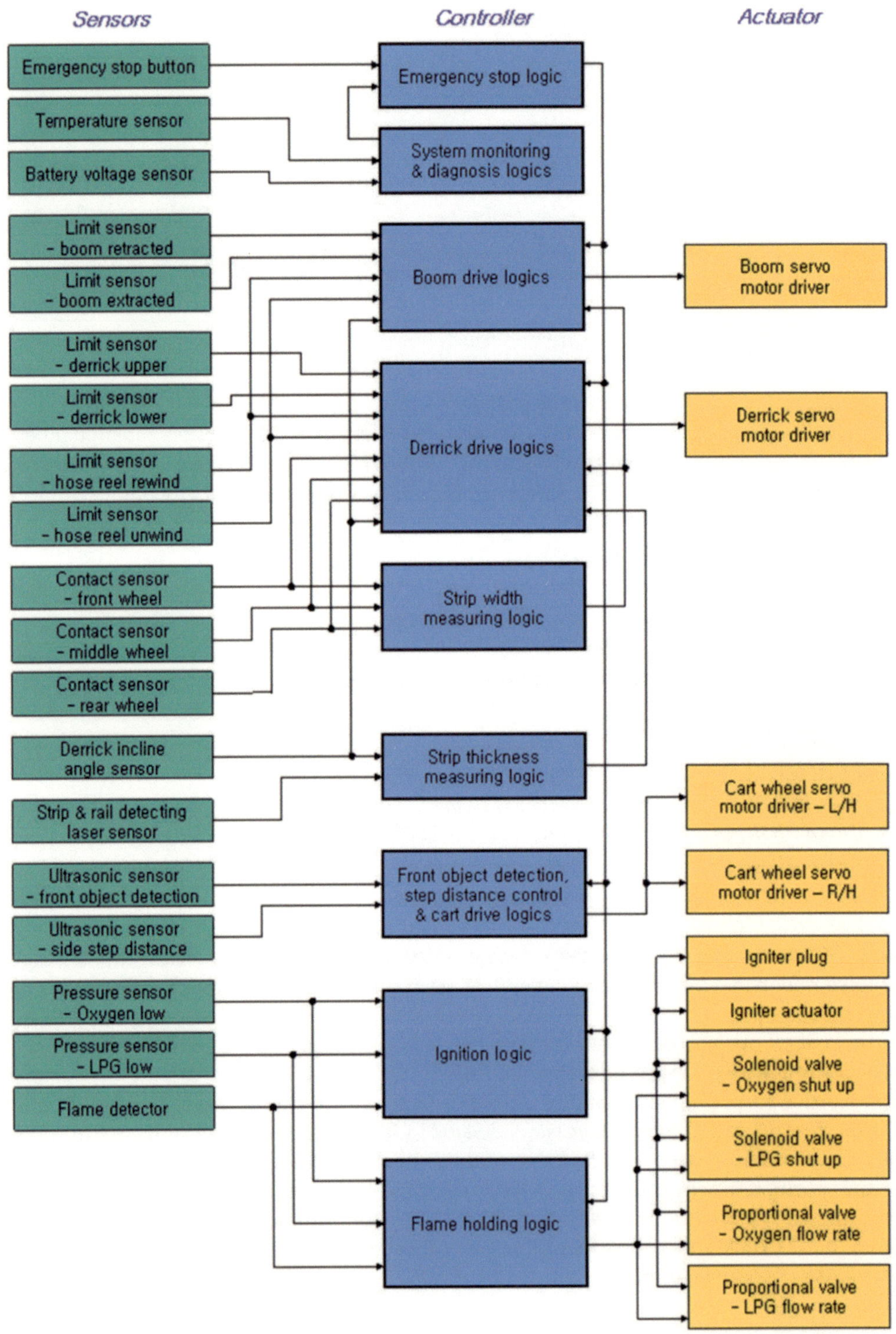

Fig. 5. Interaction between sensors and actuators.

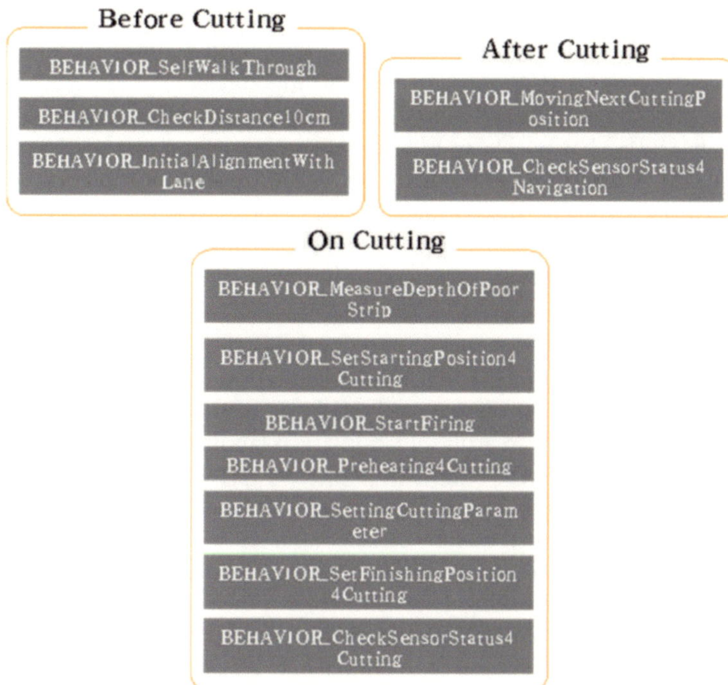

Fig. 6. Autonomous poor strip cutting procedure.

(*MovingNextCuttingPosition*). Each of these task-achieving behaviors comprises of the primitive action modules to support the functional mechanism of the robot and sort out the interactions between the behaviors. Every primitive action module has its own timer or interrupts because each behavior or primitive action should be considerably taken for the safety.

3.2 Design of motion controller

As shown in Figure 6, every primitive action modules are implemented in the manner that each sensor considers a different property and I/O interface. Controller integrates all modules discussed previously using the safety-oriented reactive control architecture. One of the significant advantages for encapsulating the behavior modules is that when programming, we can hide a system level architecture such as sensor calibration, hardware or software interrupts, and so on. More importantly, we do not have to program the detailed safety modules of relevance. The motion controller for a gas cutting robot is composed of a microprocessor and peripheral circuits with a digital input and output (DIO), an analog to digital converter (ADC), a digital to analog converter (DAC), a serial communication circuit, and timers. The microprocessor of the controller is the Atmega128 made in Atmel Co., Ltd.

4. Localization for a gas cutting robot

Localization system for a gas cutting robot is a unique sensor system for indoor localization of industrial mobile robots. It analyzes infrared ray image which is reflected from a passive landmark with an independent ID. The output of position and heading angle of a robot is given with very precise resolution and high speed. It is seldom affected by surroundings such as an infrared ray, a fluorescent light and sunshine. It is composed of an IR Projector part and an image processing unit. It can calculate high resolution and high speed localization of position and heading angle. Also, landmark is used by being attached on ceiling. This localization system doesn't require any device for any synchronization or communication between a robot and a landmark. The area that localization system covers is extended by only adding landmarks to ceiling. Then each section can be distinguished easily each other by using landmarks with different IDs. This system automatically measure and calibrate distance between landmarks and ceiling height. The greatest advantage of this system is that it is nearly not affected in environment such as lamp and sunlight and works excellent localization function at night as well as in the daytime.

4.1 Background art

In general, to control an indoor mobile robot, it is required to recognize a position of the robot. There are two self-localization calculation methods performed by a robot itself using a camera. First, there is a method of using an artificial landmark. a landmark having a certain meaning is installed on a ceiling or a wall, the landmark is photographed by a CMOS image sensor, the landmark is extracted from an image, and coordinates on a screen are allowed to be identical with coordinates of a mobile robot, thereby calculating a self-localization of the mobile robot by itself On the other hand, the landmark is installed on the top of the mobile robot and the CMOS image sensor is installed on the ceiling. Second, there is a method of using a natural landmark. A ceiling is photographed by a camera; information of structures such as lightings installed on the ceiling and straight lines and edges of interfaces between the ceiling and walls is extracted, thereby calculating a self-localization of a mobile robot by itself using the information. However, when using the artificial landmark, the artificial landmark may be affected by lightings and color information of the landmark may be distorted by sunlight. Also, when using the natural landmark, since the natural landmark is much affected by brightness of an ambient light and there is required odometer information or another robot position reader when recording a position of a feature of the landmark, a large memory is required and an additional device is essential. Particularly, when there is no illumination, it is very difficult to use the natural landmark. Accordingly, it is required a new self-localization recognition method of a mobile robot that is capable of be disaffected by lightings and reduces a calculation time of the image processing. Also, when using the two conventional methods described above, since coordinates of a camera and coordinates of a landmark attached to a ceiling calculate position information of a mobile robot while assuming that there is no rotation of in directions excluding the direction of gravity, the robot position information calculated using an image obtained by the camera may have many errors when the mobile robot goes over a small mound or is inclined by an external force or an inertial force of rapid acceleration or deceleration. On the other hand, though there may be an initial correction for an inclination occurring when attaching one of a CMOS or CCD sensors used for a camera device to a robot, the initial correction is merely for an error occurring when it is initially installing in the robot but not for an error caused by an inclination occurring while the robot actually is driving.

An aspect of the present localization system provides a landmark for recognizing a position of a mobile robot, the landmark capable of allowing one of position and area information of the mobile robot to be detected from a landmark image photographed by a camera and being recognized regardless of indoor lightings. Also it provides an apparatus and method for recognizing a position of a mobile robot, in which an image of the landmark is obtained by an infrared camera and the position and area information of a gas cutting robot can be obtained without particular preprocessing the obtained image. This proposed method provides an apparatus and method for recognizing a position of a mobile robot, the apparatus having an inclination correction function to precisely recognizing the position of the mobile robot by correcting position information of an image of a landmark photographed by an infrared camera by measuring inclination information of the mobile robot, such as a roll angle and a pitch angle, by using two axis inclinometer.

4.2 Technical solution

According to aspect of the present localization system, the landmark is used as recognizing coordinates and azimuth information of a mobile robot. The landmark including a position recognition part is formed of a mark in any position and at least two marks on an X axis and Y axis centered on the landmark. The landmark may further include an area recognition part formed of a combination of a plurality of marks to distinguish an individual landmark from others. According to another aspect of the present localization system, there is an apparatus provided for recognizing a position of a mobile robot. The apparatus including: an infrared lighting unit irradiating an infrared ray to a landmark formed of a plurality of marks reflecting the infrared; an infrared camera photographing the landmark and obtaining a binary image; a mark detector labeling a partial image included in the binary image and detecting the mark by using a number and/or dispersion of labeled pixels for each the partial image; and a position detector detecting coordinates and an azimuth of the mobile robot by using centric coordinates of the detected mark. The landmark may include a position recognition part formed of a mark in any position and at least two marks located on an X axis and Y axis centered on the mark. Also, it is provided a method of recognizing a position of a mobile robot having an inclination correction function, the method including: (a) obtaining a binary image by irradiating an infrared ray to a landmark including a position recognition part formed of a mark in any position and at least two marks located on an X axis and Y axis centered on the mark to reflect the infrared ray and photographing the landmark; (b) detecting two-axis inclination information of the mobile robot to the ground and obtaining a binary image again when the detected two-axis inclination information is more than a predetermined threshold; (c) labeling a partial image included in the binary image and detecting the mark by using a number and/or dispersion of labeled pixels for each the partial image; and (d) detecting coordinates and an azimuth of the mobile robot by using centric coordinates of the detected mark. Coordinates and an azimuth of the mobile robot can be detected by correcting the centric coordinates of the detected mark by using a coordinate transformation matrix according to the two axis inclination information. Artificial landmark to enhance reflective gain is shown in Figure 7.

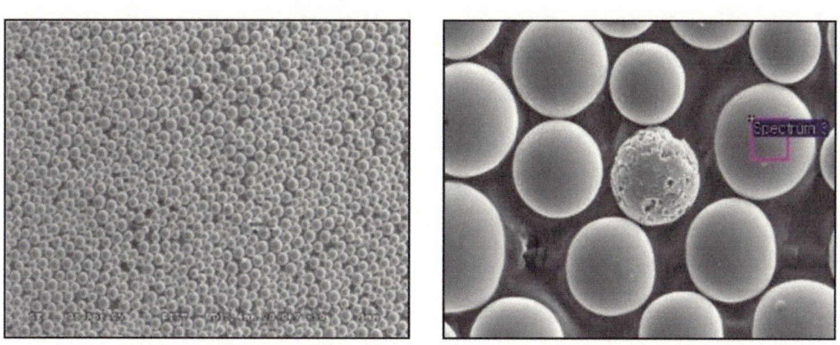

(a) Landmark (b) enlarged landmark

Fig. 7. Retro high gain reflective landmarks

4.3 Camera calibration

A pinhole model generally uses cameral model when describing an image process. Figure 8 is a configuration diagram illustrating the pinhole model. Referring to Figure 8, a point m_r of a point M_c on a three dimensional space corresponds to a point at which a straight line connecting the point M_c to a point C meets with a plane r. In this case, the point C is designated as an optical center, and the plane r is designated as a retinal plane. A straight line passing the point C and vertical to the plane r may exist, which is designated as an optical axis. Generally, the point C is allowed to be an origin point of camera coordinates, and the optical axis is allowed to be identical with Z axis of an orthogonal coordinate system. After the camera model is determined, a structure of the camera may be expressed with various parameters. The parameters may be divided into two kinds of parameters used for describing a camera, intrinsic parameters and extrinsic parameters.

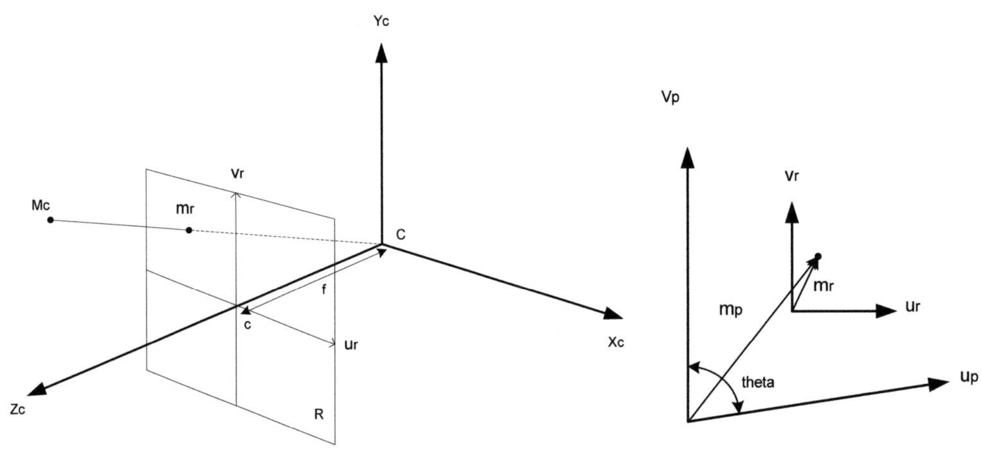

Fig. 8. Camera Model & In case: out of square in CCD array.

The intrinsic parameters describe corresponding relationships between points on the camera coordinates, which is expressed within three dimensional coordinates and the retinal plane with two dimensional coordinates, where the points are projected. The extrinsic parameters describe a transform relationship between the camera coordinates and world coordinates. Hereinafter, the intrinsic parameters will be described.

Referring to Figure 8, it may be known that a relationship between points $M_C = [X_C\ Y_C\ Z_C]^T$ on the camera coordinates and a point $m_r = [u_r\ v_r]^T$ on the corresponding retinal plane is provided as follows.

$$\begin{cases} u_r = f\dfrac{X_C}{Z_C} \\ v_r = f\dfrac{Y_C}{Z_C} \end{cases} \tag{1}$$

wherein f indicates a focal length that is a distance between the optical center C and a point c at which the optical axis meets the retinal plane. The point c indicates a principal point.

A phase formed on the retinal plane is sampled by the CCD or CMOS array, converted into a video signal, outputted from the camera, and stored in a frame buffer. Accordingly, a finally obtained coordinate value is not a coordinate value of the retinal plane but a pixel coordinate value. When pixel coordinates corresponding to m_r is $m_p = [u_p\ v_p]^T$, a transform relationship between the two coordinates is provided as follows.

$$\begin{cases} u_p = k_u u_r + u_0 \\ v_p = k_v v_r + v_0 \end{cases} \tag{2}$$

wherein α and β are values indicating a scale transformation between the two coordinates and u_0 and v_0 are pixel coordinates values of the principal point c.
The relationship given by Eq. (2) is effective when an array of the CCD array is formed by a perfect right angle. However, since it is actually difficult to form a perfect right angle, it is required to obtain a relationship equation considering the difficulty. As shown in Figure 8, when an angle formed by both axes forming the pixel coordinates is designated as θ, there is a relationship between coordinates on the retinal plane and the pixel coordinates, as follows.

$$\begin{cases} u_p = k_u u_r - k_u \cot\theta v_r + u_0 \\ v_p = k_v \csc\theta v_r + v_0 \end{cases} \tag{3}$$

When applying Eq. (1) to Eq. (3), a relationship equation between the three dimensional coordinates on the camera coordinates and the pixel coordinates is finally obtained as follows.

$$\begin{cases} u_p = \alpha_u \dfrac{X_C}{Z_C} - \alpha_u \cot\theta \dfrac{Y_C}{Z_C} + u_0 = \alpha \dfrac{X_C}{Z_C} + \gamma \dfrac{Y_C}{Z_C} + u_0 \\ v_p = \alpha_v \csc\theta \dfrac{Y_C}{Z_C} + v_0 = \beta \dfrac{Y_C}{Z_C} + v_0 \end{cases} \tag{4}$$

As described above, the intrinsic parameters are formed of five such as α, β, γ, u_0, and v_0. Hereinafter, the extrinsic parameters will be described.
Generally, points on a three dimensional space are described in different coordinates from the camera coordinates, the coordinates generally designated as world coordinates. Accordingly, a transformation equation from the world coordinates to the camera coordinates is required, the transformation equation capable of being shown by a displacement vector indicating a relative position between origin points of respective coordinates and a rotation matrix showing a rotation amount of each coordinate axis. When a point shown in world coordinates is $M_w = [X_w\ Y_w\ Z_w]^T$ and is converted into M_c in the camera coordinates, a relationship equation between Mw and Mc is shown as follows.

$$M_C = R(M_w + t) \tag{5}$$

wherein R indicates the rotation matrix and t indicates the displacement vector. Since R includes three independent parameters and t also includes three independent parameters, the number of extrinsic parameters is six. Hereinafter, it will be described to show a camera mode using projective geometry.
A pinhole model of a camera may be linearly shown by using a concept of homogeneous coordinates. When a point on two dimensional pixel coordinates is defined as $m = [u\ v]^T$ and the coordinates on three dimensional world coordinates corresponding to the point, are

defined as $M=[X\,Y\,Z]^T$, homogeneous coordinates formed by adding 1 to the last term of the coordinates become $\tilde{m}=[u\,v\,1]^T$, and $\tilde{M}=[X\,Y\,Z\,1]^T$.
A relationship equation between the three dimensional point M and m that is formed by projecting the point M is expressed using the described pinhole model as follows.

$$s\tilde{m} = A[R\ \ t]\tilde{M} \tag{6}$$

wherein s is a scale factor and R and t are a rotation matrix and a displacement vector, respectively, which are extrinsic parameters. A is a matrix of the intrinsic parameter and is designated as a calibration matrix.

$$A = \begin{bmatrix} \alpha & \gamma & u_0 \\ 0 & \beta & v_0 \\ 0 & 0 & 1 \end{bmatrix} \tag{7}$$

wherein α and β correspond to scale values to u and v axes, γ corresponds to a skewness of two image axes, and u_0 and v_0 are principal points. We use the abbreviation A^{-T} for $\left(A^{-1}\right)^T$ or $\left(A^T\right)^{-1}$.
Homography between the model plane and its image without loss of generality, we assume the model plane is on $Z=0$ of the world coordinate system. Let's denote the ith column of the rotation matrix R by r_i. From Eq. (6), we have

$$s\begin{bmatrix} u \\ v \\ 1 \end{bmatrix} = A\begin{bmatrix} r_1 & r_2 & r_3 & t \end{bmatrix}\begin{bmatrix} X \\ Y \\ 0 \\ 1 \end{bmatrix} = A\begin{bmatrix} r_1 & r_2 & t \end{bmatrix}\begin{bmatrix} X \\ Y \\ 1 \end{bmatrix} \tag{8}$$

By abuse of notation, we still use a point on the model plane, but $M=\begin{bmatrix} X & Y \end{bmatrix}^T$ since Z is always equal to 0. In turn, $\tilde{M}=\begin{bmatrix} X & Y & 1 \end{bmatrix}^T$. Therefore, a model point M and its image m is related by a homography H:

$$s\tilde{m} = H\tilde{M} \ with \ H = \begin{bmatrix} r_1 & r_2 & t \end{bmatrix} \tag{9}$$

As is clear, the 3×3 matrix is defined up to a scale factor. Given an image of the model plane, an homography can be estimated. Let's denote it by $H=\begin{bmatrix} h_1 & h_2 & h_3 \end{bmatrix}$. From Eq. (7), we have

$$\begin{bmatrix} h_1 & h_2 & h_3 \end{bmatrix} = \lambda A \begin{bmatrix} r_1 & r_2 & t \end{bmatrix} \tag{10}$$

where λ is an arbitrary scalar. Using the knowledge that r_1 and r_2 are orthonormal, we have

$$h_1^T A^{-T} A^{-1} h_2 = 0 \tag{11}$$

$$h_1^T A^{-T} A^{-1} h_1 = h_2^T A^{-T} A^{-1} h_2 \tag{12}$$

These are the two basic constraints on the intrinsic parameters, given one homography. Because a homography has 8 degrees of freedom and there are 6 extrinsic parameters (3 for

rotation and 3 for translation), we can only obtain 2 constraints on the intrinsic parameters. Note that $A^{-T}A^{-1}$ actually describes the image of the absolute conic(Luong and Faugeras, 1997). We use the closed-form solution in order to solve camera calibration problem among nonlinear optimization methods.
Let

$$B = A^{-T}A^{-1} \equiv \begin{bmatrix} B_{11} & B_{12} & B_{13} \\ B_{21} & B_{22} & B_{23} \\ B_{31} & B_{32} & B_{33} \end{bmatrix}$$
$$= \begin{bmatrix} \frac{1}{\alpha^2} & -\frac{\gamma}{\alpha^2\beta} & \frac{v_0\gamma - u_0\beta}{\alpha^2\beta} \\ -\frac{\gamma}{\alpha^2\beta} & \frac{\gamma^2}{\alpha^2\beta^2}+\frac{1}{\beta^2} & -\frac{\gamma(v_0\gamma - u_0\beta)}{\alpha^2\beta^2}-\frac{v_0}{\beta^2} \\ \frac{v_0\gamma - u_0\beta}{\alpha^2\beta} & -\frac{\gamma(v_0\gamma - u_0\beta)}{\alpha^2\beta^2}-\frac{v_0}{\beta^2} & \frac{(v_0\gamma - u_0\beta)^2}{\alpha^2\beta^2}+\frac{v_0^2}{\beta^2}+1 \end{bmatrix} \tag{13}$$

Note that B is symmetric, defined by a 6D vector

$$b = \begin{bmatrix} B_{11} & B_{12} & B_{22} & B_{13} & B_{23} & B_{33} \end{bmatrix}^T \tag{14}$$

Let the ith column vector of H be $h_i = \begin{bmatrix} h_{i1} & h_{i2} & h_{i3} \end{bmatrix}^T$. Then, we have

$$h_i^T B h_j = v_{ij}^T b \tag{15}$$

with $v_{ij} = [h_{i1}h_{i1} \quad h_{i1}h_{i2} + h_{i2}h_{i1} \quad h_{i2}h_{i2} \quad h_{i3}h_{i1} + h_{i1}h_{i3} \quad h_{i3}h_{i2} + h_{i2}h_{i3} \quad h_{i3}h_{i3}]^T$.
Therefore, the two functional constraints (11) and (12), from a given homography, can be rewritten as 2 homogeneous equations in b:

$$\begin{bmatrix} v_{12}^T \\ (v_{11} - v_{22})^T \end{bmatrix} b = 0 \tag{16}$$

If n images of the model plane are observed, by stacking n such equations as (17) we have

$$Vb = 0 \tag{17}$$

where V is a $2n \times 6$ matrix. If $n \geq 3$, we will have in general a unique solution b defined up to a scale factor. If $n = 2$, we can impose the skewless constraint $\lambda = 0$. If b is solved, we can calculate A, $R = \begin{bmatrix} r_1 & r_2 & r_3 \end{bmatrix}$ and t as follows:

$$v_0 = \left(B_{12}B_{13} - B_{11}B_{23}\right) / \left(B_{11}B_{22} - B_{12}^2\right)$$
$$\lambda = B_{33} - \left[B_{13}^2 + v0\left(B12B13 - B11B23\right)\right] / B11$$
$$\alpha = \sqrt{\lambda / B_{11}}$$
$$\beta = \sqrt{\lambda B_{11} / \left(B_{11}B_{22} - B_{12}^2\right)}$$

$$
\begin{aligned}
&\gamma = -B_{12}\alpha^2\beta / \lambda \\
&u0 = \gamma v_0 / \beta - B_{13}\alpha^2 / \lambda \\
&r_1 = sA^{-1}h_1 \\
&r_2 = sA^{-1}h_2 \\
&r_3 = r_1 \times r_2 \\
&t = sA^{-1}h_3 \\
&s = 1 / \left\|A^{-1}h_1\right\| = 1 / \left\|A^{-1}h_2\right\|
\end{aligned}
\tag{18}
$$

The above solution is obtained through minimizing an algebraic distance which is not physically meaningful. We can refine it through maximum likelihood inference. We are given n images of a model plane and there are m points on the model plane. Assume that the image points are corrupted by independent and identically distributed noise. The maximum likelihood estimation can be obtained by minimizing the following functional:

$$
\sum_{i=1}^{n}\sum_{j=1}^{m}\left\|m_{ij} - \hat{m}(A, R_i, t_i, M_j)\right\|^2 \tag{19}
$$

where $\hat{m}(A, R_i, t_i, M_j)$ is the projection of point M_j in image i, according to Eq. (7). A rotation R is parameterized by a vector of 3 parameters, denoted by r, which is parallel to the rotation axis and whose magnitude is equal to the rotation angle. R and r are related by the Rodrigues formula (Faugeras, 1993). Minimizing Eq. (20) is a nonlinear minimization problem, which is solved with the Levenberg-Marquardt algorithm (More, 1977). It requires an initial guess of A, $\{R_i \;\; t_i | i = 1 \cdots n\}$ which can be obtained using the technique described in the previous subsection. Up to now, we have not considered lens distortion of a camera. However, a desktop camera usually exhibits significant lens distortion, especially radial distortion. Therefore, we only consider the first two terms of radial distortion.

Let $(u, \;\; v)$ be the ideal (nonobservable distortion-free) pixel image coordinates, and $(\hat{u}, \;\; \hat{v})$ the corresponding real observed image coordinates. The ideal points are the projection of the model points according to the pinhole model. Similarly, $(x, \;\; y)$ and $(\hat{x}, \;\; \hat{y})$ are the ideal(distortion-free) and real(distorted) normalized image coordinates. When we represent pixels of camera coordinates to $M_C = [XC \;\; YC \;\; ZC]^T$, normalized pixels is changed as follows:

$$
M_n = \begin{bmatrix} X_C / Z_C \\ Y_C / Z_C \end{bmatrix} = \begin{bmatrix} x_n \\ y_n \end{bmatrix}, \tag{20}
$$

where x_n, y_n is normalized coordinates vale.

Let $r = x_n^2 + y_n^2$. If M_d is assumed by distorted coordinates,

$$
M_d = \begin{bmatrix} x_d \\ y_d \end{bmatrix} = M_n\left(1 + k_1 r + k_2 r^2 + \cdots\right) + t(M_n). \tag{21}
$$

Where $M_n\left(1 + k_1 r + k_2 r^2 + \cdots\right)$ is the radial distortion and $t(M_n)$ is the tangential distortion. Finally, $t(M_n)$ can be solved as follows:

$$t(M_n) = \begin{bmatrix} 2p_1 x_n y_n + p_2(r + 2x_n^2) \\ p_1(r + 2y_n^2) + 2p_2 x_n y_n \end{bmatrix}. \tag{22}$$

Experimentally, we found the convergence of the above alternation technique is slow. A natural extension to Eq. (20) is then to estimate the complete set of parameters by minimizing the following functional:

$$\sum_{i=1}^{n}\sum_{j=1}^{m}\left\| m_{ij} - \hat{m}(A,k_1,k_2,p_1,p_2,R_i,t_i,M_j)\right\|^2 , \tag{23}$$

where $\hat{m}\left(A,k_1,k_2,R_i,t_i,M_j\right)$ is the projection of point M_j in image i according to Eq. (6), followed by distortion according to Eq. (11) and (12) (Zhang, 1997). This is a nonlinear minimization problem, which is solved with the Levenberg-Marquardt Algorithm (More, 1977). A rotation is again parameterized by a 3 vector r. An initial guess of A and $\{R_i \quad t_i | i = 1 \cdots n\}$ can be obtained using the prescribed technique. An initial guess of k_1 and k_2 can be obtained with the technique described in the last paragraph technique, or simply by setting them to 0.

A vector allowing the coordinates and the azimuth of the mobile robot to be known may be obtained by using a calibration equation, which is disclosed in detail in several theses as follows. A transform relationship of projecting a point on world coordinates to camera pixel coordinates will be described referring to Figure 9.

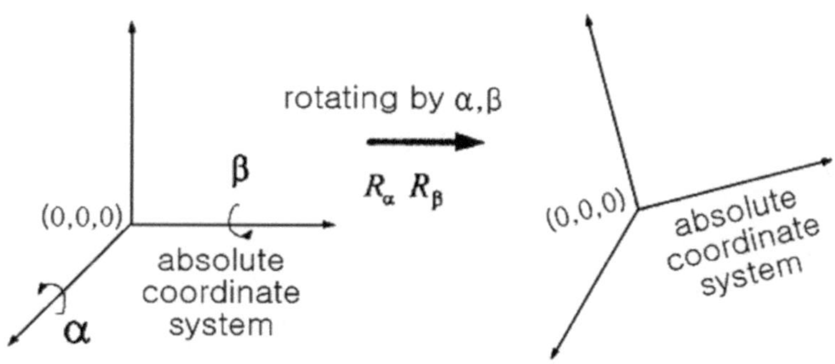

Fig. 9. Diagram illustrating a coordinate system when a camera rotates.

When a roll angle and a pitch angle corresponding to inclination of a camera are α and β, respectively, such a degree of inclination as α and β is expressed in a matrix as follows.

$$R_\alpha = \begin{bmatrix} 1 & 0 & 0 \\ 0 & \cos\alpha & \sin\alpha \\ 0 & -\sin\alpha & \cos\alpha \end{bmatrix} \tag{24}$$

$$R_\beta = \begin{bmatrix} \cos\beta & 0 & \sin\beta \\ 0 & 1 & 0 \\ -\sin\beta & 0 & \cos\beta \end{bmatrix} \tag{25}$$

wherein homogeneous pixel coordinates $\tilde{m}_{\alpha\beta}$ corresponding to a scale parameter are obtained as follows.

$$s\,\tilde{m}_{\alpha\beta} = R_\alpha R_\beta A\begin{bmatrix} R & t \end{bmatrix}\tilde{M} \tag{26}$$

When assuming the displacement vector t to be known, the point M on the world coordinates may be obtained as follows.

$$M = sR^T\left(A^{-1}R_\beta^T R_\alpha^T \tilde{m}_{\alpha\beta} - t\right) \tag{27}$$

When the point M corresponding to a reference point is known, the displacement vector t that is finally to be calculated is obtained as follows, thereby calculating a self-localization of the mobile robot.

$$t = sA^{-1}R_\beta^T R_\alpha^T \tilde{m}_{\alpha\beta} - RM \tag{28}$$

As described above, the vector amount allowing the coordinates and azimuth of the mobile robot to be simultaneously known may be obtained by the vector operation using the three detected marks of the position recognition part and the calibration equation, thereby embodying a microprocessor at a low price.

4.4 Coordinates calculation for mobile robot

The proposed localization system includes a landmark indicating position information such as coordinates and an azimuth of a mobile robot, a position recognition apparatus, and method. A landmark indicating a position of a mobile robot will be described with reference to Figure 10.

According to an embodiment of the proposed localization system, the landmark is attached to a ceiling of a space, in which the mobile robot moves, and is photographed by a camera installed on the mobile robot or attached to a top of the mobile robot and photographed by a camera installed on the ceiling to be used for recognizing the position of the mobile robot. A landmark according to an exemplary embodiment of the present invention includes a position recognition part formed of three marks to recognize essential position information such as coordinates, an azimuth of a mobile robot, and an area recognition part formed of a plurality of marks to distinguish an individual landmark from others to recognize additional area information of the mobile robot. The position recognition part is formed of one mark B in any position and two marks C and A located on an X axis and Y axis, respectively, centered on the mark B. The three marks B, A, and C provide the landmark with a reference point and reference coordinates. Though there is mark shown in Figure 10, the number of the marks is not limited to this and more than two marks may be used. Though the area recognition part is formed of 4×4 marks inside the position recognition part as shown in Figure 10, a position and the number of the marks forming the position recognition part may be varied according to purpose. By giving an ID corresponding to the number and the position of each of the marks forming the area recognition part, each individual landmark may be distinguished from others. As shown in Figure 10, when the area recognition part is formed of the 3×3 marks, IDs of 512 is given. In this case, the position of the mark forming the area recognition part may be determined according to the reference coordinates provided by the position recognition part and each of the IDs may be

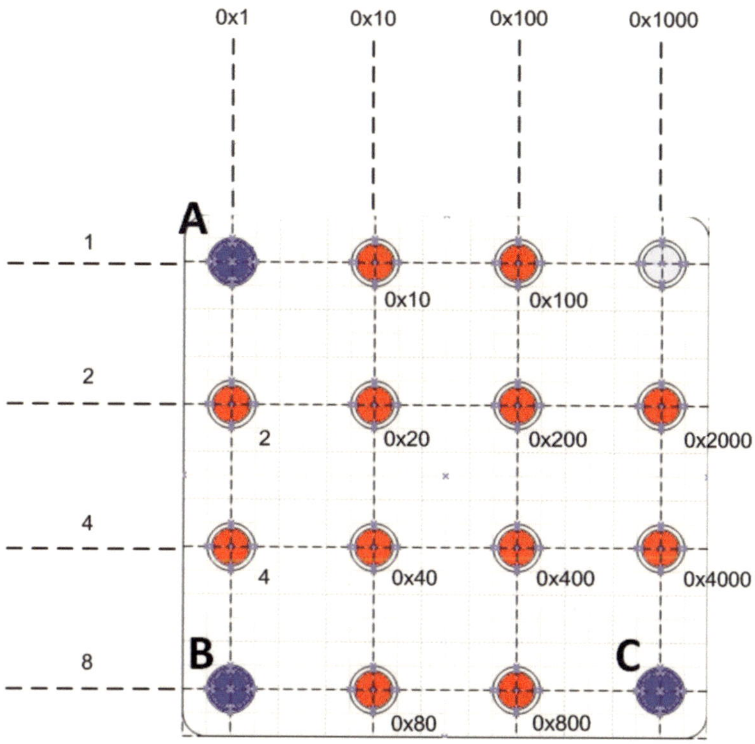

Fig. 10. Design of artificial landmark.

binary coded, thereby quickly recognizing area information of the mobile robot. On the other hand, an infrared reflection coating may be applied or a reflection sheet may be attached to the marks forming the landmark in order to diffusely reflect an infrared ray in a certain wavelength band, particularly, a wavelength band of 800 to 1200 nm. Accordingly, not only in the night but also when there exist reflected lights, only an infrared ray reflected by the mark is detected by an infrared camera, thereby quickly recognizing the position of the mobile robot without using other image processing methods. In this case, the mark may be formed in the shape of only a circle of a predetermined size or may be formed in the protruded shape of one of a circle on a plane and a hemisphere from a plane. The mark formed in the shape of one of the circle and the hemisphere may be used for easily obtaining a number, dispersion, and centric coordinates of pixels when detecting the mark. Though the marks may be formed identically with each other, the marks for the position recognition part are formed different from those for the area recognition part in size and/or color, thereby easily distinguishing the position recognition part from the area recognition part. The mark forming the landmark, described above, may be applied to a conventional mobile robot position recognition apparatus without using an infrared camera, and a use of the marks is not limited to a position recognition apparatus according to an exemplary embodiment of the proposed localization system.

Next, an apparatus and method for recognizing a position of a mobile robot, according to an exemplary embodiment of the proposed localization system, will be described in the order of operations. The embodiment may be applied to when a space in which the mobile robot moves or a space to which a landmark is attached has no bend and is flat. An infrared light emitting diode (LED) irradiates an infra ray to the landmark and an image reflected by the mark forming the position recognition part is photographed by a camera, thereby obtaining a binary image. Namely, the mark in the image obtained by the camera is set up as a bright light close to white and is converted into the binary image by selecting a predetermined threshold brightness value. Considering the camera in detail, the camera includes a plurality of infrared LEDs, an infrared light controller, a CMOS array, and a image processing controller around a wide angle lens. The camera is installed on one of the mobile robots and a ceiling of a space in which the mobile robot moves, to obtain an image of the landmark attached to one of the ceiling, a wall, and a top of the mobile robot. A partial image brightly displayed in the binary image is labeled, and the mark is detected from the number and/or dispersion of the labeled pixel. In this case, labeling indicates a procedure of recognizing an individual image, giving a reference number to the individual image, and making a label list to know a position and a size of the partial image brightly displayed in the binary image. After the labeling, centric coordinates are obtained for each label and the mark is detected from the number and/or dispersion of the labeled pixels. There may be various methods of detecting a mark from a label list. For example, one method may limit the number of pixels forming a label. Namely, since the mark is formed in the shape of a circle and has a uniform size, only a label having a certain number of pixels is selected as a mark candidate and labels having pixels more or less than the certain number are deleted from the label list.

Another method may determine a predetermined dispersion value corresponding to a dispersion index with respect to centric coordinates from the labels and delete labels in which pixels are not clustered from the label list, thereby determining a mark candidate, since the marks are clustered in the shape of a circle. The two methods of detecting a mark from labels, described above, may be used selectively or simultaneously if necessary.

On the other hand, when only the marks of the position recognition part exist in the landmark, three marks may be detected by using the above methods. However, when there are the marks of the area recognition part, whose size is identical with the mark of the position recognition part, only the marks corresponding to the position recognition part may be separately detected from the total marks by performing an additional process as follows. Namely, three labels whose distances from each other are similar and located in the shape of a right angle are detected from the determined mark candidates, thereby detecting the marks of the position recognition part. For example, an inner product of vectors connecting labels is obtained and a label whose inner product value is closest to a largest valid inner product value is detected, thereby detecting only the marks of the position recognition part from the total marks. When indexes of labels corresponding to A, B, and C of Figure 10 are designated by i, j, and k and a largest valid value of an inner product of vectors between the labels is the indexes whose difference of magnitudes is smallest among indexes whose inner product value corresponds to a range is obtained by using Eq. (6).

$$\begin{aligned} D(i,j,k) = \left\| \left\| \vec{ij} \right\| - \left\| \vec{kj} \right\| \right\|,\ \textit{where}\ \{i,j,k\} &= \arg_{i,j,k} \min D(i,j,k) \\ &\equiv \arg_{i,j,k} \left| \vec{ij} \bullet \vec{kj} \right| \prec \delta_{th} \end{aligned} \tag{29}$$

When an existence and a position of the mark has been recognized by using Eq. (29), an identifier (ID) of the mark may be easily obtained by calculating the position by using a sum of position values and detecting whether the label exists in the position. Position information such as coordinates and an azimuth and area information of the mobile robot are detected by

using the detected mark. The ID determined according to the number and position of the marks corresponding to the area recognition part from the detected marks may be quickly obtained, and the area information of the mobile robot may be obtained. In this case, the area information of the mobile robot is allocated to the ID and is an approximate position in which the mobile robot is located. Detailed position information of the mobile robot, such as the coordinates and the azimuth, may be obtained by using centric coordinates of the detected three marks A, B, and C forming the position recognition part. According to an exemplary embodiment of the present invention, coordinates of the mobile robot may be obtained by considering any point obtained from the centric coordinates of each of the three marks A, B, and C shown in Figure 10 as reference coordinates. In this case, the any point may be a center of gravity obtained by the centric coordinates of the three marks. In this case, since the center of gravity is an average of errors with respect to the centric coordinates of the three marks, an error with respect to the coordinates of the mobile robot obtained by using the center of gravity may be reduced. An azimuth of the mobile robot may be obtained based on one direction vector obtained by three centric coordinates, for example, a direction vector obtained by summation of a vector from B to A and a vector from B to C. A vector allowing both the coordinates and the azimuth of the mobile robot to be known may be obtained by using a calibration equation, which is disclosed in detail in several theses as follows (Hartley, 1994, Liebowitz and Zisserman, 1998).
The calibration equation is shown as follows.

$$\begin{aligned} s\tilde{m} &= A\begin{bmatrix} R & t \end{bmatrix}\tilde{M} \\ \Rightarrow \tilde{m} &= \frac{1}{s}A\left(RM + t\right) \end{aligned} \tag{30}$$

where $\tilde{m}_0$ is projected pixel coordinates corresponding to a reference position of the mobile robot, R_0 and t_0 are a rotation matrix and a displacement vector, respectively, s is a scale value, A is a calibration matrix, $\tilde{m}_1$ is pixel coordinates of a position to which the mobile robot rotates and moves, R_1 is a matrix corresponding to a rotation angle amount, and t_1 is a displacement vector, Eq. (31) and (32) may be obtained by using Eq. (30).

$$\tilde{m}_0 = \frac{1}{s} A\left(R_0 M + t_0\right) \tag{31}$$

$$\tilde{m}_1 = \frac{1}{s} A\left(R_0 R_1 M + R_1 t_1\right) \tag{32}$$

A coordinate value M is obtained by using Eq. (31), and the obtained value is assigned to Eq. (32). In Eq. (32), R_1 may be calculated by using a sum of vectors of the recognized marks. Since all values in Eq. (32) excluding t_1 are known, the displacement vector t_1 may be calculated. That is, the displacement vector of the mobile robot may be obtained by using Eq. (32).

$$t_1 = s\, R_1^{-1} A^{-1} \tilde{m}_1 - R_0 M \tag{33}$$

As described above, the vector allowing the coordinates and the azimuth of the mobile robot to be simultaneously known may be obtained by using the detected three marks of the position recognition part and a vector operation using the calibration equation, thereby embodying a microprocessor at a low price. Also, brief area information and detailed position information such as the coordinates and azimuth of the mobile robot may be converted into code information. The code information is transmitted to the mobile robot to perform necessary operations. An apparatus and method of recognizing a position of a mobile robot having an inclination correction function, according to another embodiment of the proposed localization system will be described referring to Figure 11.

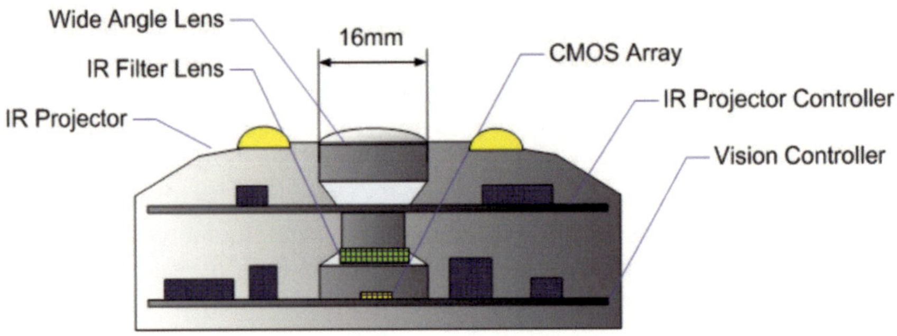

Fig. 11. The layout of proposed localization system.

The apparatus of Figure 11 further includes two axis inclinometers. An infrared light emitting diode (LED) irradiates an infra ray to the landmark and an image reflected by the mark forming the position recognition part is photographed by a camera, thereby obtaining a binary image. Namely, the mark in the image obtained by the camera is set up as a bright light close to white and is converted into the binary image by selecting a predetermined threshold brightness value. Considering the camera in detail, as shown in Figure 11, the camera includes a plurality of infrared LEDs, an infrared light controller, a CMOS array, a vision controller, and two axis inclinometers around a wide angle lens. The camera is installed on the mobile to obtain an image of the landmark attached to one of the ceiling and a wall.

Entire flowchart for localization algorithm is depicted in Figure 12. Position information such as coordinates and an azimuth of the mobile robot is detected by using the detected mark and the two-axis inclination information. The detailed position of the mobile robot, such as the coordinates and the azimuth of the mobile robot, may be obtained by using centric coordinates of the detected three marks A, B, and C forming the position recognition part.

4.5 Evaluation

A mobile robot can identify its own position relative to landmarks, the locations of which are known in advance. The main contribution of this research is that it gives various ways of making the self-localizing error smaller by referring to special landmarks which are developed as high gain reflection material and coded array associations. In order to prove the proposed localization system, we develop the embedded system using TMS320DM640 of Texas Instrument Co., Ltd. And then, the proposed localization system has been tested on mobile robot. The schematic diagram for embedded system is depicted in Figure 13. And embedded system on mobile robot is shown in Figure 14. This localization system is composed of a microprocessor and CMOS image sensor with a digital bus, a serial communication circuit, and infrared LED driver as shown in Figure 14. Calibration system for camera distortion has 3-axis motion controller in order to control gesture of reference

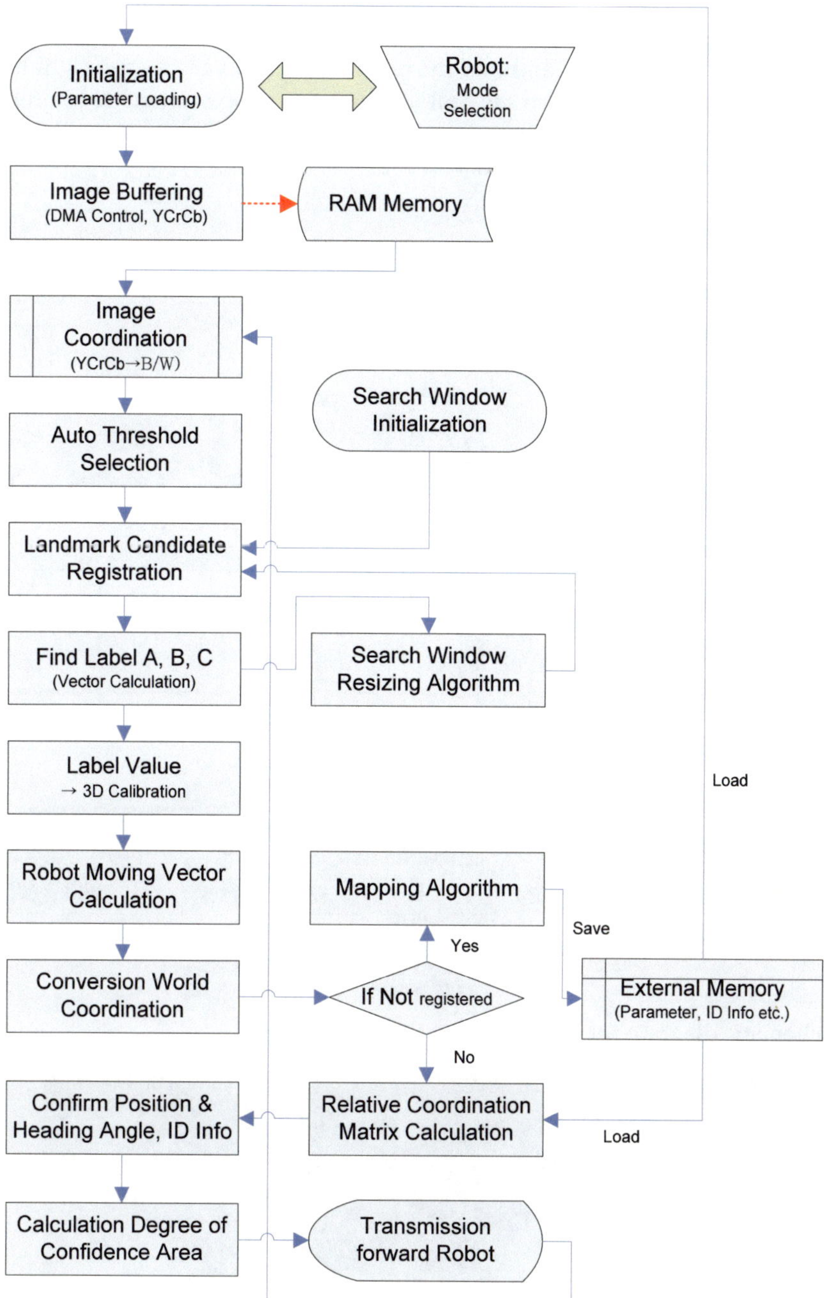

Fig. 12. Flowchart of localization algorithm.

plates. And reference images for camera calibration are show in Figure 15, which these can be acquired by calibration system as shown in Figure 26. In order to find label in reference images automatically, we use the Javis' March algorithm. This is perhaps the most simple-minded algorithm for the convex hull, and yet in some cases it can be very fast. The basic idea is as follows: Start at some extreme point, which is guaranteed to be on the hull. At each step, test each of the points, and find the one which makes the largest right-hand turn. That point has to be the next one on the hull. Because this process marches around the hull in counter-clockwise order, like a ribbon wrapping itself around the points, this algorithm also called the "gift-wrapping" algorithm.

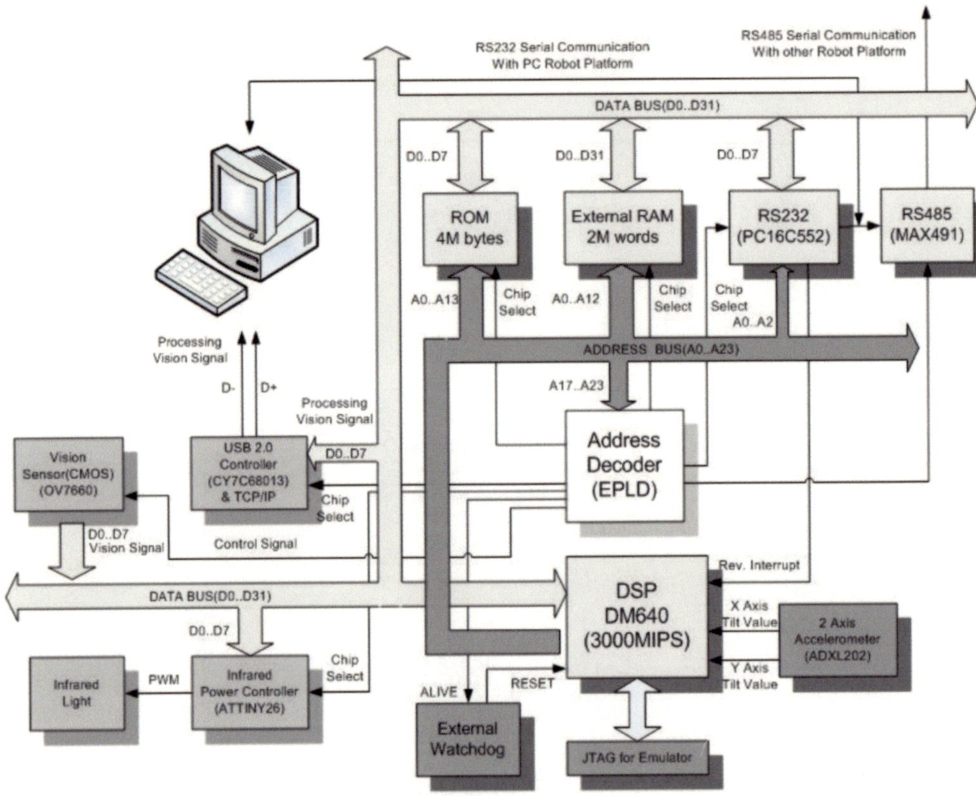

Fig. 13. Schematic diagram for embedded system.

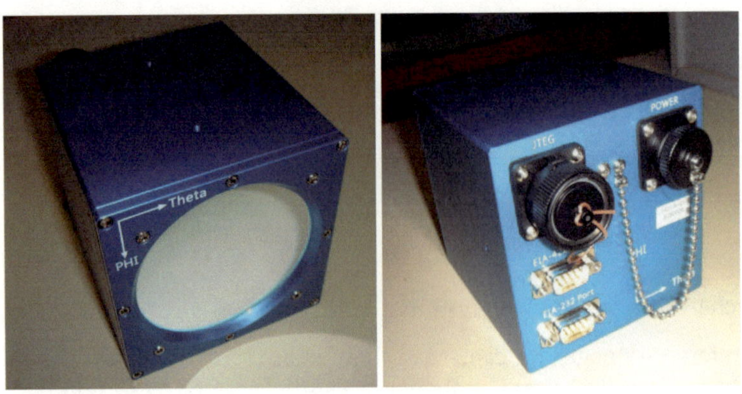

Fig. 14. Embedded system for localization.

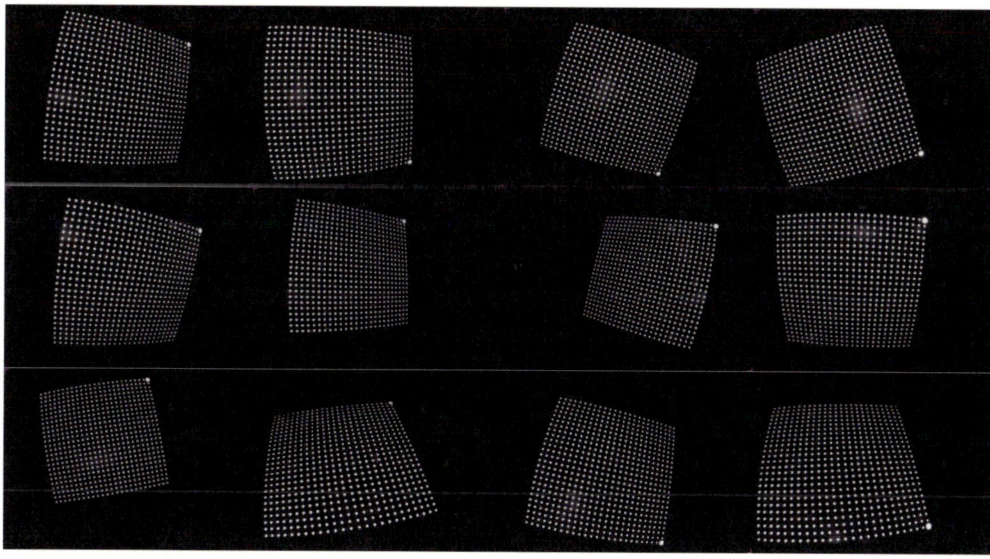

Fig. 15. Reference images for camera calibration.

The accuracy of self-localizing a mobile robot with landmarks based on the indices is evaluated as shown in Figure 17 and a rational way to minimize to reduce the computational cost of selecting the best self-localizing method. The simulation results show a high accuracy and a good performance as depicted in Figure 16. Also, the localization errors for the proposed algorithm are shown in Figures 17(a) and (b), respectively. In results of evaluation, the peak error is less than 3 cm as shown in Figure 17.

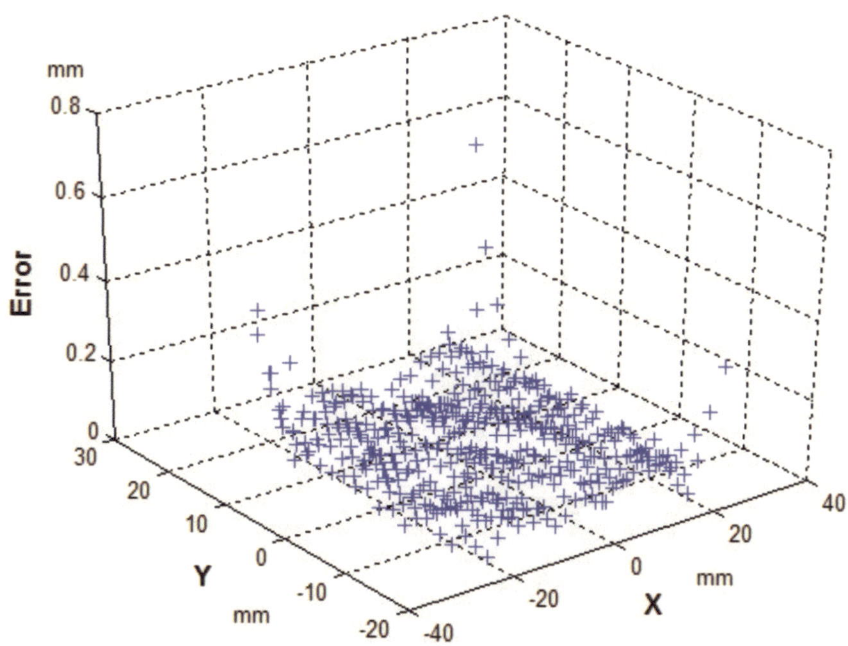

Fig. 16. Calibration errors for camera.

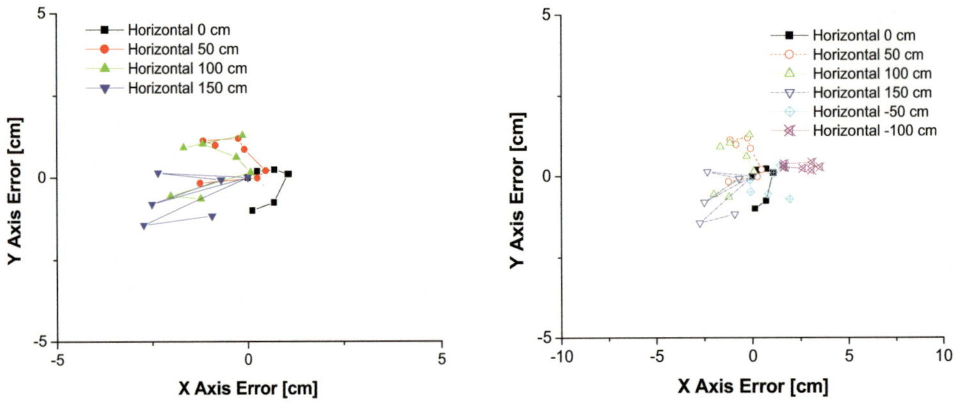

(a) Horizontal error (b) Vertical error

Fig. 17. Localization error.

5. Simulations and experiments

One of the most important cutting procedures taken into consideration is to properly use the cutting torch according the hot slab thickness ranged between 50 mm to 300 mm because choosing the right torch tip plays the major effects on the performance of cut quality and amount of usage for oxygen and gas. For automated oxygen cutting to blow the hot slab cobble into two pieces in the way of oxygen cutting equipped in the end of the robot arm, there are several things to consider as follows:

Step1: way of being lit with spark
Step2: control of flammable gas mixed with oxygen without human intervention
Step3: status of the metal hot enough to be melt down
Setp4: Status of end of blowing out of the melting

In order for the torch to be lit with spark, we installed the stepping motor with 300mm long guided arm equipped with the engine ignition plug used in the car engine. When signal is on, it is rolled out all the way to the bottom of torch tip and sparked until the gas is lit and then begin automatically adjusting the oxygen based on the oxygen insertion table [Table 1] that is previously decided.

Slab Thickness(mm)	Gas Pressure		Oxygen Flow rate
	Oxygen	LPG	
1~10	2.5	0.2~0.4	550
10~30	2.5	0.2~0.4	1,100
30~50	3.0	0.2~0.4	1,900
50~70	3.5	0.2~0.4	2,800
70~100	4.0	0.2~0.4	4,300

Table 1. Oxygen and LPG Insertion Table.

Before blowing out of the way, we must preheat starting area. In order to locate the starting area, we used limit switch mounted inside the torch holder wheel. The decision signal to determine the preheating status depends on the human decision. The most challenging task in the oxygen cutting is to hold the flame with blowing out of the focused area with the proper speed along the way to the end. To determine the proper speed according to the slab thickness, we designed the table that maps between the speed and slab thickness. Figures 18 are the pictures of gas cutting experiments. In summary, the first procedure in the developed oxygen cutting robot for the hot slab cobble is to choose the proper torch tip based on the slab thickness. Secondly, ignition procedure is operated based on the table explained in the Table 1. Thirdly, locating the starting area by using the limit switch inside the torch holder is automatically done. At last, the human inspects the right condition for preheating and send command for "Moving Forward" to the robot.

(a) Step 1: Start

(b) Step 2: Ignition

(c) Step 3: Cutting

(d) Step 4: Boom control

(e) Step 5: Flame control

(f) Step 6: Finish

Fig. 18. Experiment for cutting process.

6. Conclusion

In this chapter, we proposed the safety-oriented reactive controller implemented in an innovative and robust industrial mobile robot designed for cutting a poor strip in the steelworks. This mobile robot, called "APSCR", which provides the fully-autonomous operation and wireless controlled operation, is designed to take into consideration the functional safety because it controls the gas and oxygen without human intervention. To able to support the safety guarantee, we encapsulated the robot controller such as behavior and primitive action modules so that the behaviors do not care about how safety protection mechanism works behind the hardware part or software part. In this chapter, we propose a set of indices to evaluate the accuracy of self-localizing methods using the selective reflection landmark and infrared projector, and the indices are derived from the sensitivity enhancement using 3D distortion calibration of camera. And then, the accuracy of self-localizing a mobile robot with landmarks based on the indices is evaluated, and a rational way to reduce the computational cost of selecting the best self-localizing method is proposed. The simulation results show a high accuracy and a good performance. With the preliminary results, we have proved the robustness and reliability of the proposed control architecture. To prove that APSCR is able to use in the steelworks, we still have to perform further experiments with mechanical modification.

7. References

Faugeras, O. (1993). Three-Dimensional Computer Vision: a Geometric Viewpoint, *MIT Press*, 1993.

Faugeras, O., Luong, T., & Maybank, S. (1992). Camera self-calibration: theory and experiments, In G. Sandini, editor, *Proc 2nd ECCV, volume 588 of Lecture Notes in Computer Science*, pages 321–334, Santa Margherita Ligure, Italy, May 1992. Springer-Verlag.

Faugeras, O. , & Toscani, G. (1986). The calibration problem for stereo, *In Proceedings of the IEEE Conference on Computer Vision and Pattern Recognition*, pages 15–20, Miami Beach, FL, June 1986. IEEE.

Hartley, R. (1994). Self-calibration from multiple views with a rotating camera, In J.-O. Eklundh, editor, *Proceedings of the 3rd European Conference on Computer Vision*, volume 800-801 of Lecture Notes in Computer Science, pages 471–478, Stockholm, Sweden, May 1994. Springer-Verlag.

Hartley, R. (1995). In defence of the 8-point algorithm, *In Proceedings of the 5th International Conference on Computer Vision*, pages 1064–1070, Boston, MA, June 1995. IEEE Computer Society Press.

Hartley, R. I. (1994). An algorithm for self calibration from several views, *In Proceedings of the IEEE Conference on Computer Vision and Pattern Recognition*, pages 908–912, Seattle, WA, June 1994. IEEE.

Liebowitz, D. & Zisserman, A. (1998). Metric rectification for perspective images of planes, *In Proceedings of the IEEE Conference on Computer Vision and Pattern Recognition*, pages 482–488, Santa Barbara, California, June 1998. IEEE Computer Society.

Zhang, Z. (1997). Motion and structure from two perspective views: From essential parameters to euclidean motion via fundamental matrix, *Journal of the Optical Society of America A*, 14(11):2938–2950, 1997.